Introducing Category Theory

Introducing Category Theory

Second edition

Peter Smith

LOGIC MATTERS

Published by Logic Matters, Cambridge.

ISBN 978-1-0683467-0-5 Paperback

Send corrections, please, to peter_smith@logicmatters.net.

Visit logicmatters.net/categories for more resources related to the topic of this book.

Contents

Part III: Toposes as generous arenas

Preface

A little background A few years ago I put together some notes on elementary category theory, initially as an exercise in getting things clearer in my own mind. I later posted versions online. Rather to my surprise, these were steadily downloaded hundreds of times a month – which was both embarrassing and encouraging. Embarrassing because those earlier draft efforts were very half-baked. But encouraging enough for me to resolve to do better once other projects were off my desk. Hence this expanded and much revised version.

Who are these notes for? I originally got interested in category theory because of its connections with issues in the foundations of mathematics, broadly construed. This angle of approach no doubt influences the shape of these notes in various ways (someone approaching category theory from the direction of theoretical computer science, say, would cover a rather different selection of topics with different emphases). But despite my interest in foundations, there is little overt 'philosophical' discussion here – it is mostly mathematics, served up quite straight. And I expect that the most likely reader is going to be a student of pure mathematics who wants an elementary first introduction to some category theory, perhaps as a preliminary warm-up before taking on an industrial-strength graduate-level course.

Still, I hope that other readers, perhaps with less mathematical background, might also find something useful here. I have tried to give a reasonably accessible exposition of core categorial[1] ideas, enough to give an initial sense of what some of the fuss is about, and to provide a launchpad for further explorations, both conceptual and more technical.

One thing will be quite obvious from the outset: I *do* go at a leisurely pace. I don't apologize at all for this: if you find the pace *too* slow, there are plenty of faster-track alternative introductions available. However, it is not just my own experience which suggests that, for many, getting a secure understanding of category-theoretic ways of thinking by initially taking things pretty gently can make later adventures going beyond the basics much more manageable.

But of course, whether my angle of approach and the moderately-relaxed-but-fairly-traditional mode of exposition will satisfy *you* must in the end depend

[1] Logicians already have a quite different use for 'categorical'. So when talking about categories, I much prefer the adjectival form 'categorial', even though it is the minority usage.

entirely on your particular interests, background, and preferences in matters of mathematical style.

What do you need to bring to the party? One crucial thing that category theory does is give us a story about the ways in which different parts of modern abstract mathematics hang together. Obviously, you can't be in a good position to appreciate this if you really know almost nothing beforehand about modern mathematics! But I do try to presuppose relatively little detail. Suppose you know a few basic facts about groups (there's some revision in Chapter 2), know a little about different kinds of orderings, are acquainted with some elementary topological ideas, and know a few more bits and pieces; then you should be able to cope fairly easily with the introductory discussions here. And if some illustrative examples pass you by, don't panic. I usually try to give multiple illustrations of important concepts and constructions; so feel free simply to skip those examples that happen not to work so well for you.

How far do we aim to get? You can think of Part I (Chapters 2–25) as providing a relatively undemanding prologue, introducing some first categorial ideas. Part II (Chapters 26–43) then gets down to work on characteristic categorial themes like the behaviour of functors, natural transformations, the Yoneda Lemma, and adjunctions.

By the end of Part I, however, we already have everything we need to define the idea of an elementary topos, a category with a particular combination of nice properties. The short Part III (Chapters 44 to 49) returns to develop this idea, with the modest aim of taking it just far enough to entice you into further explorations.

Despite the length of the book, though, we don't really get beyond the very beginnings of category theory. What count as 'beginnings'? I suppose that's debatable. But I note that the famous introduction to topos theory by Mac Lane and Moerdijk (1992) starts with a fourteen page chapter of 'Categorical Preliminaries'. That isn't supposed to be a stand-alone exposition so much as a checklist of assumed basics. And their checklist turns out to correspond pretty closely to the coverage of Parts I and II of these notes, which suggests that my menu of topics there is sensible enough.

The order in which we tackle things If you have already glanced at some other introductions to category theory, you'll immediately spot that while the overall coverage might be pretty standard, I do present the various topics of Parts I and II in a somewhat unusual order.

Again, I make no apology. As I say in the Introduction, the gadgets of basic category theory interconnect in multiple ways, so there is no one best exploratory trail to follow. I do think, however, that there is a logical attraction to the route I take, and it surely makes for a rather gentler ascent to the categorial heights than some alternatives.

Theorems as exercises What follows is still best regarded as a set of notes; it isn't a textbook with the usual apparatus of collections of exercises at the ends of chapters. However, almost all the proofs of the theorems you meet here are *very* straightforward, particularly at the outset. Almost always, you just have to 'do the obvious thing' in the context. So you can think of the statement of a theorem as in fact presenting you with an exercise that you should ideally attempt to work through for yourself in order to fix ideas; the ensuing proof which I spell out is then the answer (or at least, *an* answer) to the exercise. Sometimes a few theorems are explicitly stated as a series of 'challenges', with the proofs coming a bit later.[2]

So, together with the easy-going pace, that's another reason why this is a long book – in effect, I give fully worked answers to nearly all the exercises. (And if anything, I probably often err on the side of giving over-detailed proofs. But better that, say I, than giving under-developed proof hints that can leave one or another reader unnecessarily puzzled.)

Notation I should flag a related pair of notational innovations. Upright variables such as 'G' are introduced in early chapters, and are intended to be read *plurally*, as typically denoting many things (so are to be contrasted with a variable like 'G' which is *singular* and might stand for a set or other single thing). Associated with plural variables, 'ε' is to be read as 'is one of' or 'is among', so $x \varepsilon \mathrm{G}$ says that x is one of the objects G. Correspondingly, 'for any $x \varepsilon \mathrm{G}, \ldots$' means 'for any x that is one of the objects G, ...', and so on.

'Iff' is of course short for 'if and only if'. '□' is used as an end-of-proof marker or to conclude the statement of a theorem whose proof needn't be further spelt out. I also use '△' as an end-of-definition marker.

And from now on, I mostly follow the usual mathematicians' practice of omitting quotation marks when mentioning symbolic expressions, if no confusion is likely to result. Logicians can get irritatingly fussy about this sort of thing, and let's try to avoid that.

This edition Compared with the previous edition, I have rearranged the order of some chapters. There are also some new (sub)sections and quite a few smaller changes aiming for greater clarity or reader-friendliness. The current text is certainly not set in stone, however. Indeed, you should think of what you are reading as still a 'beta version', functional though surely not bug-free. So all corrections and suggestions for further improvement will continue to be very gratefully received.

Thanks! I am always struck by the kindness of logical strangers generously providing comments and corrections on versions of these notes, including (at various stages) Andrew Bacon, Matthew Bjerknes, Sam Butchart, Ruiting Jiang, Malcolm F. Lowe, Laureano Luna, Phil Nguyen, David Ozonoff, Leonardo Pacheco, Simon Schneider, Mariusz Stopa, Jan Thiemann, Zoltán Tóth, Adrian Yee,

[2] I've borrowed the label 'challenges' from Bartosz Milewski's terrific series of category theory blogposts at bartoszmilewski.com, though (to be frank) few of my challenges are very taxing. You could cheerfully skip them: they are mostly tests of basic understanding rather than of ingenuity.

Hongyu Zhang, and particularly Rowsety Moid. Georg Meyer found an embarrassing number of typos and also some more serious mistakes still lurking in late drafts. Extended exchanges with John Zajac led to a large number of significant improvements to the first full book version. Very warm thanks to everyone.

But, as ever, the person I must thank the most is my wife Patsy, who makes it all possible.

1 Introduction

Mathematical science is in my opinion an indivisible whole, an organism whose vitality is conditioned upon the connection of its parts. For with all the variety of mathematical knowledge, we are still clearly conscious of the similarity of the logical devices, the relationship of the ideas in mathematics as a whole and the numerous analogies in its different departments.
(Hilbert 1900)

[O]ur theory provides general concepts applicable to all branches of abstract mathematics, and so contributes to the current trend towards uniform treatment of different mathematical disciplines. In particular, it provides opportunities for the comparison of constructions and of the isomorphisms occurring in different branches of mathematics; in this way it may occasionally suggest new results by analogy.
(Eilenberg and Mac Lane 1945)

1.1 The categorial imperative

(a) Modern pure mathematics explores mathematical structures. And these structures cluster in families.

Take a family of structures together with a good helping of the structure-respecting maps between them. Then we can think of this inter-related family as forming a further structure – a structure-of-structures, if you like – something else to explore mathematically.

(1) Here's a basic example. A particular *group* is a structure that comprises some objects equipped with a binary operation defined on them, where the operation obeys the well-known requirements. But we can also think of a whole family of groups, together with appropriate maps between them – i.e. homomorphisms that respect group structure – as forming a further structure-of-structures.

(2) Another example: any particular *topological space* is a structure, classically conceived as comprising some objects, 'points', that are equipped with a

topology. But again, a family of these spaces, together with appropriate maps between them – this time, the continuous functions that respect topological structure – forms another structure-of-structures.

(3) And so it goes. Perhaps what interests you are *some objects equipped with an order*: these constitute another type of mathematical structure – with different kinds of ordering giving us, of course, different kinds of structure. Perhaps it is well-orderings in particular that you are concerned with. There is a whole family of well-ordered structures together with order-respecting maps between them, and we are interested in the structure of this family (perhaps in the guise of the theory of ordinals, the theory of order-types of well-orderings). We want to know too about other kinds of families of ordered objects and the relations between them.

In each of these various cases, then, we not only investigate *individual* structures (the particular groups, particular topological spaces, particular collections of ordered objects), but we can also explore *families* of such structures (families of groups, families of topological spaces, families of ordered pluralities), with a family itself structured by the maps or morphisms between its members.

An obvious point: we see similar relationships recurring within different families. For example, some groups are products of others and some spaces are products of others ('a cylinder is the product of a line and a circle'). Likewise, we can form products of e.g. well-ordered collections to get a longer well-ordering. Again, some groups can be seen as the result of quotienting another group by a suitable equivalence relation (in effect, we identify equivalent objects); similarly we can form quotient spaces (as when we in effect identify opposite edges of a rectangle to get the surface of a torus). And we can quotient orderings by equivalence relations too. It is entirely natural, then, to want an account – one that applies across different families of structures – about what makes for products in general, what makes for quotients in general, etc. As we will see in Part I (Chapters 2–25), entry-level category theory gives us just such an account, because structured families of structures are prime examples of categories.

(b) A central categorial motif is that we learn about the objects inside a category by considering the structure-respecting maps between them. The same idea applies to categories themselves, which do not exist in glorious isolation from each other. We learn more about categories by looking at an additional level of structure-respecting maps, the so-called *functors*, this time linking whole structures-of-structures – as when, for example, we get information about a family of (pointed) topological spaces by using a suitable functorial map between the spaces and their corresponding fundamental groups.[1]

And even this is not the end of it. Going up another level of abstraction, we will find ourselves wanting to consider whole families of functors together with

[1]Relatedly, it was the exploration of the functorial connection between spaces and their cohomology groups that was one of the original prompts towards the development of category theory.

operations that map one functor to another while preserving their functorial character (in ways we will eventually need to explain).

So here we are encountering *one* central imperative of modern mathematics: to explore these levels of increasingly abstract structure.

Let's agree straight away that this project certainly doesn't appeal to all – or even most – mathematicians. A vast amount of pure mathematics is of course carried on at very much less exalted levels. Still, the eventually hyper-abstracting project can resonate with a certain systematizing cast of mind. And evidently, if we *are* going to set out on such an enquiry, we will want a framework for dealing with these upper layers of abstraction in a disciplined and illuminating way.

As Part II (Chapters 26–43) shows, category theory provides what we need as we first set out to explore this territory: its distinctive ideas and constructions provide a toolkit for systematically probing not only structures-of-structures but structures-of-structures-of-structures and more. And it is the theory in *this* role that will be our main concern in most of this beginners' guide.

1.2 From a bird's eye view

(a) But what do we gain by ascending through those levels of abstraction and by developing tools for imposing some order on what we find?

For a start, we should get a richer conceptual understanding of how various parts of mathematics relate to each other. And I suppose we might reasonably say that this will be a 'philosophical' gain, in *one* sense of that contested label. After all, many philosophers, pressed for a crisp characterization of their discipline, like to quote a famous remark by Wilfrid Sellars:

> The aim of philosophy, abstractly formulated, is to understand how things in the broadest possible sense of the term hang together in the broadest possible sense of the term. (Sellars 1963, p. 1)

Category theory indeed provides us with a suitable unifying framework for exploring in depth some of the ways in which a lot of mathematics hangs together. That's why it should be of considerable interest to philosophers of mathematics as well as to mathematicians interested in the conceptual shape of their own discipline.

(b) But category theory does much more than give us an insightful way of organizing already-familiar relations between structures. As Tom Leinster so very nicely puts it, the theory

> ... takes a bird's eye view of mathematics. From high in the sky, details become invisible, but we can spot patterns that were impossible to detect from ground level. (Leinster 2014, p. 1)

Category theory crucially enables us, then, to reveal *new* connections we hadn't made before; so-called 'adjunctions' are a prime example, as we will eventually find.

Seeing recurrent patterns in different families of structures and making new connections between them in turn enables new mathematical discoveries. And it was because of the depth and richness of the resulting discoveries in e.g. algebraic geometry that category theory first came to prominence. To keep things accessible, however, in these notes I will stick to very much more elementary concerns, with a particular emphasis on unification and conceptual clarification. We will still have more than enough to discuss.

(c) As we go along, however, we will find that category theory increasingly changes the way we look at mathematical structures, foregrounding the morphisms between structures. There is also an accompanying shift towards a perhaps more democratic conception of the mathematical universe, as we eventually come to see the universe of sets (as conventionally understood) as perhaps not uniquely 'foundational' (whatever exactly that means). In fact, any rich enough category – any 'elementary topos with a natural numbers object', as they say – forms an arena in which we can develop a good deal of mathematics. In Part III (Chapters 44 to 49), I aim to say enough about this idea to intrigue and to encourage further exploration.

1.3 A slow ascent

The gadgets of basic category theory do fit together rather beautifully in multiple ways. These interconnections mean that there certainly isn't a single best route into the theory. Different linear narratives will take topics in significantly divergent orders, all illuminating in their various ways.

I will follow the simplest plan, however, and make a slow ascent to the categorial heights. We begin just one level up from talking about particular structures. In other words, we start by talking about *categories* – for, as I said, many paradigm cases of categories are in fact structured-families-of-structures. And in Part I we go on to develop ways of describing what happens inside a category. In this setting, we revisit many familiar ideas about maps between structures, and about ways of creating new structures, by (for instance) forming products or taking quotients.

Only after exploring inside categories taken singly in Part I do we move up another level to consider *functors*, maps between categories (typically, maps between families of structures). And only after we have spent a number of chapters at the beginning of Part II thinking about how particular functors work do we move up a further level to define operations that map one functor to another – these are the so-called *natural transformations* and *natural isomorphisms*. We then explore these notions, and arrive at the famed Yoneda Lemma, before we at last start exploring the key notion of *adjunctions*.

In summary, my chosen route into the basics of category theory in Parts I and II steadily ascends through increasing levels of abstraction (before dropping down from the heights again in Part III). This route is perhaps not the most usual one, but I find that it has considerable logical appeal, and it does very nicely reveal what depends on what. True enough, this means that we don't meet some of the most important and characteristic categorial ideas for over two hundred pages, until Part II. However, this disadvantage will (I hope) be more than counterbalanced – at least for enough readers – by the real gains in understanding that come from taking our gently sloping path. I will have to do my best to make even the more elementary ideas we encounter along the way already seem pretty interesting and fruitful.

2 Groups, and categories of groups

Category theory gives us a framework for thinking systematically about structured families of mathematical structures. One paradigm case of such a family, I said, comprises some groups organized by homomorphisms between them; and at the end of this chapter, in §2.9, I will explain more carefully what it takes to form a category of groups.

We'll warm up, however, by rehearsing a few facts about groups at a decidedly elementary level (hardly a lecture's worth from Algebra 101). So you will *very* probably be able to skim-read most of this chapter at considerable pace – be my guest. The main point of the exercise leading up to the final section is to remind ourselves of a few key themes which are already there in familiar mathematics, before we see them taking on an explicitly categorial guise in later chapters.[1]

Do note however that – for good reasons! – most of the definitions I give in this chapter are in fact not *quite* the conventional ones you have previously met, so don't skip them entirely. There is a real point to their mildly deviant character, which I will explain in Chapter 4, highlighting some further themes which impact on category theory. Bear with me until then.

2.1 Groups defined

(a) Take 'G' to stand for one or more objects; and (as indicated in the Preface) read 'for any $x \varepsilon$ G' as meaning 'for any x which is one of the objects G', and so on. Then here is my preferred version of the usual definition:

Definition 1. The objects G equipped with a binary operation $*$ form a *group* iff

(i) G are closed under $*$, i.e. for any $x, y \varepsilon$ G, $x * y \varepsilon$ G;
(ii) $*$ is associative, i.e. for any $x, y, z \varepsilon$ G, $(x * y) * z = x * (y * z)$;
(iii) there is a distinguished object $e \varepsilon$ G which acts as a group identity, i.e. for any $x \varepsilon$ G, $x * e = x = e * x$;
(iv) every group object has an inverse, i.e. for any $x \varepsilon$ G, there is at least one object $y \varepsilon$ G such that $x * y = e = y * x$.

---

[1] If you are impatient, though, you *could* now just read the final section, §2.9, followed by §4.3, before tackling Chapter 5 where the categorial action really starts!

A group is *abelian* iff its binary operation is commutative, i.e. for all $x, y \varepsilon$ G, $x * y = y * x$. △

Don't read too much into 'equipped'. It is a standard turn of phrase ('endowed' is a common alternative); it means no more than that we are dealing with some objects G together with an operation defined over them.

It is, of course, immediate that if e and e' are both identities for the group formed by the objects G equipped with $*$, then $e = e * e' = e'$; so group identities are unique.

Likewise, if y and y' are both inverses of x, then $y = y * e = y * (x * y') = (y * x) * y' = e * y' = y'$; so group inverses are unique too.

(b) Note the huge variety of objects and operations that can form a group. For a start, *any* item e, whatever you like, together with the only possible binary operation $*$ such that $e * e = e$, forms a one-object group. Similarly, any two items e, j, whatever you like, form a group when they are equipped with the binary operation $*$ for which e is the identity and $j * j = e$.

Less trivially there are, for example, groups of integers (equipped, say, with addition mod n, with 0 as the identity), and groups of complex numbers (e.g. the non-zero ones equipped with multiplication, with 1 as the identity). There are groups of functions, such as the group of permutations of the first n naturals, with functional composition as the group operation and the do-nothing permutation as the group identity (if $n > 2$, that group is non-abelian). Or consider groups of geometrical transformations – for instance, there is the non-abelian group of symmetries of a regular polygon (i.e. the group of rotation and reflection operations which map the given polygon to itself).

Then there are, for example, various groups of real invertible matrices. More intriguingly, perhaps, there are groups of closed directed paths through a base point in a topological space (with concatenation of paths as the group operation). And so on and on it goes. But of course you knew all that!

(c) Entirely familiar though the point is, it is still worth having explicitly noted this variety in the sorts of things which can form groups.

For compare: there is a view, introduced into modern logic by Frege, according to which there are *absolute* type-theoretic distinctions to be made between objects (individual things) and first-level functions (sending objects to objects) and second-level functions (sending first-level functions to first-level functions), etc. Whatever the virtues of that view, I want to emphasize that when we talk about the objects of a group, the notion of object in play is a *relative* one.

A group involves a group operation (a binary function of some kind or other, whose inputs and outputs must be at the same type-level); and then this group's 'objects' – or 'elements' to use a more customary term – are the items (of whatever shared type level) which are the inputs and outputs for that operation. These items can be objects-as-individuals (like numbers); but the items can equally well be first-level functions (like permutations of some numbers, i.e. bijections between those numbers); or they can be of other types too.

I stress this point in the familiar context of groups because, looking ahead, we will meet the same relative use of the notion of object when we get round to defining the general notion of a category.

2.2 Fixing notation

I will use '(G, $*$, e)' to abbreviate '(the objects) G equipped with the operation $*$ and with the distinguished object e'. Similarly, of course, for e.g. '(H, $\star$, d)' and so on. Note then that the parentheses here are just helpful punctuation. They are *not* being used to form a term for a new entity such as a set-theoretic triple.

If (G, $*$, e) satisfy the conditions for forming a group, then I'll briskly write e.g. 'the group (G, $*$, e)' rather than 'the group formed by (G, $*$, e)'. More briefly still, when I want to refer to a group without going into details about its structure, I'll simply use a letter like 'G'.[2]

As we've seen, the group operation can be very different from case to case – all that's required is that the operation satisfies Defn. 1. But we will default to using multiplicative notation and talk generically of group 'products'; we will correspondingly default to denoting the unique inverse of a group object x by x^{-1}. (By tradition, however, additive notation is commonly used when dealing with abelian groups.)

2.3 New groups from old

(a) Given one or more groups, we can form further groups from them in various natural ways. For a start, there are subgroups, in the entirely predictable sense:

Definition 2. The group S is a subgroup of the group G iff (i) all S's objects belong to G too, and (ii) S's group operation is the restriction of G's operation to S's objects. △

Simple example: the even integers (still with addition as the group operation, and with zero as the group identity) form a subgroup of the additive group of integers. For another example: the complex numbers on the unit circle form a subgroup of the multiplicative group of non-zero complex numbers.

(b) Next, products of groups. And, as a preliminary, we first need the general idea of a *pairing scheme*:

Definition 3. Given the objects X and the objects Y, a scheme for pairing up X with Y comprises

(i) some pair-objects O (which can be any suitable objects, and which may or may not be disjoint from X and/or Y);

[2] Looking forward, a standard notation – the one I adopt – uses italic capital letters for objects in categories, objects which might be structures like groups. Hence for consistency I'll use 'G' (syntactically a singular term, taking a singular verb) for a group, contrasting with 'G' (a plural term) for the objects which form the structure. I'll say a little more about what this distinction might or might not come to in the next chapter.

(ii) a binary pairing function which we can notate '$\langle\ ,\ \rangle$' which sends $x \in \mathrm{X}$ and $y \in \mathrm{Y}$ to a pair-object $\langle x, y\rangle \in \mathrm{O}$ (where every $o \in \mathrm{O}$ is indeed some such $\langle x, y\rangle$);

(iii) a couple of unpairing functions which send any pair-object $\langle x, y\rangle$ to x and y respectively. △

Note, it is immediate from this definition that the pairing function sends distinct ordered pairs x, y and x', y' to distinct pair-objects $\langle x, y\rangle$ and $\langle x', y'\rangle$.

Don't jump to over-interpreting the notation here. The angle-brackets might remind you of some standard set-theoretic construction of ordered pairs. But all we need for a pairing scheme are *some* objects to 'code' for pairs together with interlocking pairing and unpairing functions. For example, if both X and Y are the natural numbers, then we could perfectly well take suitable pair-objects $\langle m, n\rangle$ to be the numbers $2^m 3^n$, with the obvious pairing and unpairing functions. So pair-objects need not be sets: what matters is not their intrinsic nature but the role they play (a categorially-flavoured point I press again in Chapter 10).

With Defn. 3 to hand, we can now define the notion of a product group:

Definition 4. Suppose G and H are respectively the groups $(\mathrm{G}, *, e)$ and $(\mathrm{H}, \star, d)$. And suppose we have some scheme for pairing the objects G and H using the pair-objects K – so $x \in \mathrm{G}$ and $y \in \mathrm{H}$ are mapped to a pair-object $\langle x, y\rangle \in \mathrm{K}$.

Put $k = \langle e, d\rangle$, and define multiplication of pairs componentwise, so $\langle x, y\rangle \diamond \langle x', y'\rangle = \langle x * x', y \star y'\rangle$. Then $(\mathrm{K}, \diamond, k)$ form a group K, a *product* of the groups G and H, which we can notate $G \times H$. △

It is routine to check that $(\mathrm{K}, \diamond, k)$ really do form a group.

For a very simple example, suppose the group J comprises just the two objects e, j. If a group K_1 is to be a product of J with itself, it will need to comprise four distinct objects $\langle e, e\rangle$, $\langle e, j\rangle$, $\langle j, e\rangle$, $\langle j, j\rangle$, with the first of these objects being the group identity. For brevity's sake, call these four pair-objects $1, a, b, c$ respectively. K_1's group operation $\diamond$ is then defined by the following table (read the table entry as giving the value of row-object $\diamond$ column-object):

$\diamond$	1	a	b	c
1	1	a	b	c
a	a	1	c	b
b	b	c	1	a
c	c	b	a	1

The symmetry of the table reflects the fact that K_1 is abelian.

Note that we speak here of 'a' product of the group J with itself, not 'the' product. Why? Because there are unlimitedly many alternative schemes for coding pairs of objects, and different schemes will give rise to different product groups. In this present example, *any* four distinct objects we like can play the role of the required pair-objects, as long as we have pairing and unpairing functions to match. However, the resulting different groups *will* be equivalent-as-groups: any way of forming a product group from a two-object group and itself gives us a group describable by reinterpreting the same table.

The point of course generalizes. Products produced by using different pairing schemes will always be equivalent, in a familiar sense we'll clarify shortly.

(c) Now for a third, rather more interesting, way of forming new groups. We start with another general idea, and define a *quotient scheme*:

Definition 5. Given the objects G and an equivalence relation $\sim$ defined over them, a scheme for quotienting G by $\sim$ comprises

(i) some quotient-objects Q (which can be any suitable objects, which may or may not be disjoint from G),

(ii) a unary function which we can notate '[ ]' which sends $x \varepsilon \mathrm{G}$ to a quotient-object $[x] \varepsilon \mathrm{Q}$ (with every $q \varepsilon \mathrm{Q}$ being some such $[x]$), where

(iii) for all $x, y \varepsilon \mathrm{G}$, $[x] = [y]$ iff $x \sim y$. △

So $[x]$ behaves in the crucial respect like a $\sim$-equivalence class containing x. But note, we do *not* require $[x]$ to be an equivalence class or other set. For example, take the integers and consider the equivalence relation $\equiv_8$, i.e. congruence mod 8. Then in this case we can simply put $[x]$ to be the remainder when x is divided by 8, since (thus defined) $[x] = [y]$ iff $x \equiv_8 y$.

Again, what matters about quotient-objects is not their 'internal' nature but their 'external' liaisons, the role they serve in a quotient scheme. And again, as with the parallel point about pairs, this point about quotients illustrates what will turn out to be a quite central motif of category theory, namely the crucial importance of 'external' relations in pinning down what we care about in various constructions.

Next, we want the idea of an equivalence relation on the objects of a group which respects the structure of the group in the following sense:

Definition 6. Given a group $(\mathrm{G}, *, e)$, then $\sim$ is a *congruence* relation for the group iff (i) $\sim$ is an equivalence relation on G, and (ii) for any objects $x, y, z \varepsilon \mathrm{G}$, given $x \sim y$, then $x * z \sim y * z$ and $z * x \sim z * y$ (that is to say, 'multiplying' equivalent objects by the same object yields equivalent results). △

And now we can use a quotient scheme to, as it were, collapse congruent objects together to form a new group:

Definition 7. Suppose that we have a group $(\mathrm{G}, *, e)$, and $\sim$ is a congruence relation for the group. And suppose we also have a quotient scheme for $\sim$, which sends a group object x to $[x]$ (so the function notated '[ ]' in effect ignores the distinction between congruent objects). Let G/$\sim$ be all the objects $[x]$ for $x \varepsilon \mathrm{G}$, and put $[x] \star [y] = [x * y]$.

Then G/$\sim$ equipped with the operation $\star$ and with $[e]$ as the operation's identity also form a group, which we'll denote *G*/$\sim$, a *quotient* of the original group *G* with respect to $\sim$. △

For this definition to work, $\star$ has to be a genuine function. So we need to show that the result of $\star$-multiplication does not depend on how we pick out the multiplicands. In other words – *without* yet assuming $\star$ is a function so we can substitute identicals! – we need to show that if $[x] = [x']$ then (1) $[x] \star [y] = [x'] \star [y]$, and (2) $[y] \star [x] = [y] \star [x']$. But for (1), simply note that if $[x] = [x']$, then by definition $x \sim x'$, hence (since $\sim$ is a congruence respecting group structure) $x * y \sim x' * y$, hence $[x * y] = [x' * y]$, hence by definition $[x] \star [y] = [x'] \star [y]$. We derive (2) similarly. It remains, then, to check that $(\mathrm{G}/\!\!\sim, \star, [e])$ do form a group – but that's straightforward.

A quick example. Let Z be the group $(\mathbb{Z}, +, 0)$ formed by the integers $\mathbb{Z}$ under addition:[3] and consider again the equivalence relation of congruence mod 8. This equivalence relation respects the additive structure of the integers; for if $x \equiv_8 y$ then $x + z \equiv_8 y + z$ and $z + x \equiv_8 z + y$. As suggested before, we can take our quotient scheme for this equivalence relation simply to send x to the remainder on dividing x by 8; this gives us as quotient-objects the eight numbers from 0 to 7, which we will together denote $\overline{8}$. Then $(\overline{8}, +_8, 0)$ – where $+_8$ is addition mod 8 – form a group we can call $Z/\!\equiv_8$, which is a quotient of Z by $\equiv_8$.

Note, we again talk of 'a' quotient of a group by a given equivalence relation rather than of 'the' quotient group. There will be many ways of finding quotient schemes for a congruence $\sim$ defined over objects G, hence there can be many alternative candidates $\mathrm{G}/\!\!\sim$ from which to build a quotient group (though, as with product groups, quotient groups constructed using different quotient schemes will all 'look the same').

(d) A word more about quotient groups. Let's recall another important notion (very familiar if you have done even a little group theory):

Definition 8. The group $(\mathrm{N}, *, e)$ is a *normal subgroup* of $(\mathrm{G}, *, e)$ iff it is a subgroup and, for any $n \mathrel{\varepsilon} \mathrm{N}$ and any $x \mathrel{\varepsilon} \mathrm{G}$, then $x * n * x^{-1} \mathrel{\varepsilon} \mathrm{N}$.[4] △

And here's a nice two-part theorem relating normal subgroups to congruences and hence quotients. Every congruence induces a normal subgroup, and conversely every normal subgroup induces a congruence:

Theorem 1. *(i) If $\sim$ is a congruence for the group $(\mathrm{G}, *, e)$, and $\mathrm{N}_\sim$ are the objects among G such that $n \mathrel{\varepsilon} \mathrm{N}_\sim$ iff $n \sim e$, then $(\mathrm{N}_\sim, *, e)$ is a normal subgroup of $(\mathrm{G}, *, e)$.*

(ii) If $(\mathrm{N}, *, e)$ is a normal subgroup of $(\mathrm{G}, *, e)$, and for any $x, y \mathrel{\varepsilon} \mathrm{G}$ we put $x \sim_N y$ iff there is an $n \mathrel{\varepsilon} N$ such that $x = n * y$, then $\sim_N$ is a congruence for the group $(\mathrm{G}, *, e)$.

Proof of (i). It is easily seen that the objects $\mathrm{N}_\sim$ are closed under the group operation. Note too that if $n \mathrel{\varepsilon} \mathrm{N}_\sim$ (so $e \sim n$) and n^{-1} is its $*$-inverse, then $n^{-1} = n^{-1} * e \sim n^{-1} * n = e$, so $n^{-1} \mathrel{\varepsilon} \mathrm{N}_\sim$.

[3] I'll recycle familiar set-theoretic notation like '$\mathbb{Z}$' for plural use when convenient.

[4] Are we really going to fuss about notationally distinguishing the group operation defined over the objects G from its restriction to the objects N?

Hence $(\mathrm{N}_\sim, *, e)$ do form a group. And as for normality, we note that for any $x \,\varepsilon\, \mathrm{G}$ and $n \,\varepsilon\, \mathrm{N}_\sim$, $x * n * x^{-1} \sim x * e * x^{-1} = e$, therefore $x * n * x^{-1} \,\varepsilon\, \mathrm{N}_\sim$. □

Proof of (ii). It is more or less immediate that $\sim_N$ is an equivalance relation. So we only need to show that for any $x, y, z \,\varepsilon\, \mathrm{G}$, then given $x \sim y$ it follows that $x * z \sim y * z$ and $z * x \sim z * y$.

The first of those is trivial. For the second, we note that if $x = n * y$, then $z * x = z * n * y$. But by the normality assumption, $z * n * z^{-1} = n'$ for some $n' \,\varepsilon\, \mathrm{N}$, i.e. $z * n = n' * z$. Whence $z * x = n' * z * y$, so $z * x \sim z * y$ as required. □

In short, a normal subgroup N of G gives rise in a natural way to a quotient group of G respect to the congruence $\sim_N$ (i.e. the quotient group conventionally denoted G/N).

2.4 Group homomorphisms

(a) Let's move on and equally briskly recall some basic facts about structure-respecting maps between groups.

Definition 9. A *group homomorphism* from the group $(\mathrm{G}, *, e)$ as source to the group $(\mathrm{H}, \star, d)$ as target is a function f defined over G with values among H such that for every $x, y \,\varepsilon\, \mathrm{G}$, $f(x * y) = fx \star fy$. △

In sum, such a homomorphism sends products of objects in the source group to corresponding products in the target group.

When we want to make explicit the structure of the groups G and H which a homomorphism connects, then we can explicitly write $f\colon (\mathrm{G}, *, e) \to (\mathrm{H}, \star, d)$. But when the structural details are not germane, we will simply write $f\colon G \to H$.

It is of course immediate from the definition that a homomorphism sends a group identity to another group identity, and sends inverses to inverses. In other words, suppose f is a group homomorphism from $(\mathrm{G}, *, e)$ to $(\mathrm{H}, \star, d)$; then (i) $fe = d$, and (ii) for any $x \,\varepsilon\, \mathrm{G}$, $f(x^{-1}) = (fx)^{-1}$.

For (i), we have $fe = fe \star d = fe \star (fe \star (fe)^{-1}) = (fe \star fe) \star (fe)^{-1} = f(e * e) \star (fe)^{-1} = fe \star (fe)^{-1} = d$.

And for (ii), we note $f(x) \star f(x^{-1}) = f(x * x^{-1}) = f(e) = d$, and similarly $f(x^{-1}) \star f(x) = d$. So $f(x^{-1})$ is the (unique) inverse of $f(x)$.

(b) Some simple initial examples:

(1) Let $(\mathrm{G}, *, e)$ form a group. Then there is a unique homomorphism f from that group to any given one-object group, which sends every object from G to the sole object of the target group.

(2) Likewise, there is a unique homomorphism g in the opposite direction, from a given one-object group to $(\mathrm{G}, *, e)$. It's the function which sends the sole object of the first group to e, the group identity of the second.

(3) Relatedly, there is always a 'collapse' homomorphism h from a group $(\mathrm{G}, *, e)$ to itself which sends every object from G to the group identity e.

These cases remind us that, although homomorphisms are often described as *preserving* group structure, this does not mean replicating *all* structure. A homomorphism from G to H can compress many or most aspects of the group structure on G simply by mapping distinct G-objects to one and the same H-object. It is better, then, to talk of homomorphisms as *respecting* group structure.

Three more interesting but still elementary examples:

(4) There is a homomorphism from Z, the additive group of integers $(\mathbb{Z}, +, 0)$, to any two object group J which sends even numbers to J's identity, and sends odd numbers to J's other object. Thought of just as a function from Z to J, the homomorphism here is surjective but not injective.

(5) There is a homomorphism from Z to Q, the additive group of rationals $(\mathbb{Q}, +, 0)$, which sends an integer n to the corresponding rational $n/1$. As a function from integers to rationals, this is injective but not surjective.

(6) The reals $\mathbb{R}$ form a group under addition, and the non-zero complex numbers $\mathbb{C}^*$ form a group under multiplication. Define the homomorphism $j\colon (\mathbb{R}, +, 0) \to (\mathbb{C}^*, \times, 1)$ by putting $j(x) = \cos x + i \sin x$. Then the function from $\mathbb{R}$ to $\mathbb{C}^*$ is neither injective nor surjective.

(c) Let's pause to see what can be said about group homomorphisms in general, very various though they have already proved to be. We have:

Theorem 2. *(1) Any two homomorphisms $f\colon G \to H$, $g\colon H \to J$, with the target of the first being the source of the second, will compose to give a homomorphism $g \circ f\colon G \to J$.*

(2) Composition of homomorphisms is associative. In other words, if f, g, h are group homomorphisms which can compose so that one of $h \circ (g \circ f)$ and $(h \circ g) \circ f$ is defined, then the other composite is defined, and the two composites are equal.

(3) For any group G, there is an identity homomorphism $1_G\colon G \to G$ which sends each object to itself. Then for any $f\colon H \to J$ we have $f \circ 1_H = f = 1_J \circ f$.

Proof sketch. For (1) we simply take $g \circ f$ ('g following f') applied to an object x from the group G to be $g(f(x))$ and then check that $g \circ f$ so defined does satisfy the condition for being a homomorphism given that g and f do.

For (2), associativity of homomorphisms is inherited from the associativity of ordinary functional composition for the underlying functions simply thought of as mapping objects to objects.

(3) is also immediate. □

(d) That was so very easy! But do note the important fact that this – only our second numbered theorem – is not (repeat, *not*) a mere logical consequence of our definitions of groups and group homomorphisms. Our proof sketch here plainly depends on invoking background assumptions about functions, such as the assumption that functional composition is associative. These assumptions may be utterly uncontentious, but that doesn't mean that they aren't needed.

And so it goes. Contrary to what is sometimes far-too-casually said, almost *nothing* in group theory follows *merely* from the definitions alone.

2.5 Isomorphisms, automorphisms

(a) Now we highlight the special case where a homomorphism is both injective and surjective, so gives rise to a nice structure-respecting bijection between two groups (or between a group and itself).

Definition 10. A *group isomorphism* $f\colon G \xrightarrow{\sim} H$ is a homomorphism where the underlying function is a one-to-one correspondence, a bijection, between the respective objects of the two groups.[5]

We say that the groups G and H are *isomorphic* as groups iff there is a group isomorphism $f\colon G \xrightarrow{\sim} H$, and then write $G \cong H$.

A *group automorphism* is a group isomorphism $f\colon G \xrightarrow{\sim} G$ whose source and target are the same. △

(b) Let's have some elementary examples:

(1) Any two two-object groups are isomorphic. Take e, j equipped with the only possible group operation $*$, and e', j' equipped with $*'$. Then the map which sends the group identity e to the group identity e' and sends j to j' is obviously a group isomorphism.

(2) There are two automorphisms from the additive group $(\mathbb{Z}, +, 0)$ to itself. One is the identity homomorphism; the other is the function which sends an integer j to $-j$.

(3) There are infinitely many automorphisms from the group $(\mathbb{Q}, +, 0)$ to itself. For any non-zero rational q the map $x \mapsto qx$ 'stretches/compresses' the rationals, maybe reversing their order but respecting additive structure.

(4) Let K_2 be the group consisting in $1, 3, 5, 7$ equipped with multiplication mod 8 as the group operation. And let K_3 be the group of symmetries of a non-equilateral rectangle whose four 'objects' are the operations of leaving the rectangle in place, vertical reflection, horizontal reflection and rotation through $180°$, with the group operation being simply composition of geometric operations. Then $K_2 \cong K_3$.

The easiest way to see this is by constructing an abstract 'multiplication table'. First, take $1, a, b, c$ to be respectively the numbers $1, 3, 5, 7$, and take the group operation $\diamond$ to be multiplication mod 8. Second, take $1, a, b, c$ to be the geometric operations on a rectangle in the order just listed and take $\diamond$ to be composition. Both times we get the same table: in fact we get the same table again as for K_1 that we met in §2.3. Matching up the two new

[5] I find myself still rather liking that vivid old-school talk of one-to-one correspondences!

interpretations of $1, a, b, c$ and the two corresponding interpretations of $\diamond$ gives us the claimed isomorphism $f\colon K_2 \xrightarrow{\sim} K_3$. By the same reasoning, both groups are isomorphic to K_1.

This illustrates an obvious general point. Groups that can interpret the same 'multiplication table' are isomorphic; conversely, isomorphic groups can be described by the same (possibly infinite) table.

(c) In defining a product of two groups, we were allowed to invoke any scheme for coding pairs of objects from the two groups. But whichever scheme we choose, the resulting product (I said) will 'look the same', and have the same multiplication table. We can now put it like this: suppose J_1 and J_2 are both products of G with H; then $J_1 \cong J_2$.

For take the map $j\colon J_1 \to J_2$ which sends the pair-object $\langle x, y\rangle_1$ to $\langle x, y\rangle_2$ – where $\langle x, y\rangle_1$ pairs x from G and y from H according to the pairing scheme used in constructing J_1, and $\langle x, y\rangle_2$ pairs the same objects according to the pairing scheme used in constructing J_2. This bijection j is evidently a group isomorphism, so $J_1 \cong J_2$.

Likewise, suppose J_1' and J_2' are now different quotients of a group G with respect to a congruence relation $\sim$, different because they rely on different quotient schemes for, in effect, representing $\sim$-equivalent classes of objects from G. Take the bijection j' that sends the quotient-object $[x]_1$ according to the first quotient scheme to the corresponding object $[x]_2$ according to the second scheme. Then this is a group isomorphism, and so we again have $J_1' \cong J_2'$.

(d) A group is isomorphic to itself (by the identity homomorphism). The composition of two group isomorphisms is again an isomorphism. And it is easy to check that the inverse of an isomorphism is an isomorphism (this also immediately follows from Theorem 4 below). Therefore, exactly as we want,

Theorem 3. *Being isomorphic is an equivalence relation between groups.* □

2.6 Redefining isomorphisms

(a) Here's another very easy result, which gives us an alternative characterization of isomorphisms:

Theorem 4. *A group homomorphism $f\colon G \to H$ is an isomorphism iff it has a two-sided inverse, i.e. there is a homomorphism $g\colon H \to G$ such that $g \circ f = 1_G$ and $f \circ g = 1_H$.*

Proof. Suppose $f\colon (\mathrm{G}, *, e) \to (\mathrm{H}, \star, d)$ is a group isomorphism. Then the underlying function $f\colon \mathrm{G} \to \mathrm{H}$ is a bijection and therefore has a two-sided inverse $g\colon \mathrm{H} \to \mathrm{G}$. So we only need to confirm that this inverse function g gives rise to a homomorphism $g\colon (\mathrm{H}, \star, d) \to (\mathrm{G}, *, e)$.

But since f is a homomorphism, $(fgx \star fgy) = f(gx * gy)$; and so, since g is a two-sided inverse for f, we have $g(x \star y) = g(fgx \star fgy) = gf(gx * gy) = gx * gy$. Therefore g is a homomorphism.

Conversely, suppose f is a homomorphism with a two-sided inverse. Then as a function between objects it must have a two-sided inverse; but it is a routine result that a function with a two-sided inverse is a bijection. □

(b) This last theorem already illustrates what will turn out to be a key categorial theme: a kind of morphism which is initially characterized by how it acts on *objects* (matching them up one-to-one) gets re-characterized in terms of how it interacts with other *morphisms*.

Let's have another illustration of the same theme. We noted that some group homomorphisms are injective (another idea defined in the first place by how the morphisms act on objects). Can we characterize this kind of morphism too in terms of how it interacts with other morphisms?

We might guess that, since a bijective homomorphism can be redefined in terms of having a two-sided inverse, an injective one can be redefined in terms of having a one-sided inverse. And it is trivial that if the group homomorphism f has a left inverse g, then f is injective – for then if $fx = fx'$, $gfx = gfx'$ and hence $x = x'$.

But the converse is false: f can be injective without having a left inverse. For a toy example, consider the homomorphism $f\colon (\mathbb{Z}, +, 0) \to (\mathbb{R}, +, 0)$ which sends an integer n to the corresponding real number n. That's injective, but doesn't have a left-inverse.[6] So back to the drawing board! Is there another way of characterizing injective group homomorphisms in terms of the way they interact with other homomorphisms?

There is indeed. Let's say:

Definition 11. A group homomorphism is a *monomorphism* if and only if it is *left-cancellable*. In other words, $f\colon G \to H$ is a monomorphism iff for any group homomorphisms $g, h\colon J \to G$, if $f \circ g = f \circ h$, then $g = h$.

Then it is quite easy to show:

Theorem 5. *A group homomorphism is injective as a function if and only if it is a monomorphism.*[7]

For convenience, however, I'll delay the proof until §8.1.

As you would expect, there's a dual notion of being an epimorphism, i.e. being right-cancellable, and the group epimorphisms are the surjective homomorphisms. But more about all this in due course.

---

[6] Suppose otherwise, so there's a $g\colon (\mathbb{R}, +, 0) \to (\mathbb{Z}, +, 0)$ such that $gf(1) = 1$. Then $gf(1) = g1 = g(1/2 +_{\mathbb{R}} 1/2) = g(1/2) +_{\mathbb{Z}} g(1/2) = 1$. But that's impossible, since g takes no non-integral values.

[7] There is a terminological tangle here. Some older books define monomorphisms to be injections – see e.g. Hungerford (1974, p. 30) – and then of course the theorem we need is that monomorphisms thus defined are left-cancellable. More recent books tend to define monomorphisms our way, which turns out to be the categorial way – see e.g. Aluffi (2009, p. 14).

2.7 Homomorphisms and constructions

In §2.3 we considered some basic ways of forming new groups from old, yielding subgroups, product groups and quotient groups. In §2.4 we introduced structure-respecting maps between groups. We now bring the two themes together, foreshadowing what will be another absolutely key motif of category theory.

(a) For the simplest case, start by noting how homomorphisms give rise to subgroups and vice versa.

Theorem 6. *Suppose* $f\colon (\mathrm{G}, *, e) \to (\mathrm{H}, \star, d)$ *is a group homomorphism, and let* $f[\mathrm{G}]$ *be all the objects which are an* f*-image of some object from* G*. Then those objects equipped with (the restriction of) the operation* $\star$*, form a group – the* f*-image of* $(\mathrm{G}, *, e)$ *– which is a subgroup of the group* $(\mathrm{H}, \star, d)$*.*

Proof. (i) Suppose $y_1, y_2 \mathrel{\varepsilon} f[\mathrm{G}]$. By assumption, these objects are f-images of some objects $x_1, x_2 \mathrel{\varepsilon} \mathrm{G}$. So we have $y_1 \star y_2 = fx_1 \star fx_2 = f(x_1 * x_2)$, and hence $y_1 \star y_2 \mathrel{\varepsilon} f[\mathrm{G}]$ as required.

(ii) $d = f(e) \mathrel{\varepsilon} f[\mathrm{G}]$.

(iii) Since $\star$ is associative and d an identity for that operation, it only remains to show that if $y \mathrel{\varepsilon} f[\mathrm{G}]$ then its inverse belongs to $f[\mathrm{G}]$ too. But y is by assumption $f(x)$ for some object $x \mathrel{\varepsilon} \mathrm{G}$, and homomorphisms send inverses to inverses. So y^{-1}, i.e. $(fx)^{-1}$, is $f(x^{-1})$ and hence $y^{-1} \mathrel{\varepsilon} f[\mathrm{G}]$.

Those three points establish that $(f[\mathrm{G}], \star, d)$ form a group, and it is trivially a subgroup of the group $(\mathrm{H}, \star, d)$. □

The reverse theorem is even easier:

Theorem 7. *For any subgroup S of a given group H, there is a homomorphism $f\colon G \to H$ such that S is the f-image of G.* □

Simply put $G = S$ and take the inclusion map which sends a G-object to itself as an H-object.

Combining those results, we can characterize all the subgroups of a given group using homomorphisms with that group as their target. Putting it roughly, then, we can trade in claims about what goes on *inside* various groups when forming subgroups for claims about corresponding homomorphisms *between* groups.

(b) Let me mention two more theorems, the first linking homomorphisms with subgroups in another way, the second making a link with quotient groups.

Theorem 8. *Suppose* $f\colon (\mathrm{G}, *, e) \to (\mathrm{H}, \star, d)$ *is a group homomorphism, and let* K *be the objects among* G *which* f *maps to the identity element* d*. Then* $(\mathrm{K}, *, e)$ *form a normal subgroup of* $(\mathrm{G}, *, e)$*, the* kernel *of* f*.*

Proof. We need to establish the closure of K under the operation $*$. But suppose $k_1, k_2 \mathrel{\varepsilon} \mathrm{K}$. Then $f(k_1 * k_2) = fk_1 \star fk_2 = d \star d = d$. Hence $(k_1 * k_2) \mathrel{\varepsilon} \mathrm{K}$.

By the definition of a homomorphism, $f(e) = d$, so $e \mathrel{\varepsilon} \mathrm{K}$. Then recall that homomorphisms send inverses to inverses. Therefore if $k \mathrel{\varepsilon} \mathrm{K}$ so $f(k) = d$, then

$f(k^{-1}) = d^{-1} = d$ ensuring $k^{-1} \varepsilon$ K. Hence (K, $*, e$) form a subgroup of the group (G, $*, e$).

For normality, we simply note that for any $k \varepsilon$ K and $g \varepsilon$ G, $f(g * k * g^{-1}) = f(g) \star f(k) \star f(g^{-1}) = f(g) \star d \star f(g)^{-1} = d$. Therefore $g * k * g^{-1} \varepsilon$ K. □

There is a converse theorem too, that every normal subgroup for a group G is the kernel of some homomorphism with the source G.

Theorem 9. *Given a group homomorphism $f\colon G \to H$, and x, y from among G's objects, put $x \sim y$ iff $fx = fy$. Then $\sim$ is a congruence on G and the f-image of the group G is a quotient group $G/{\sim}$. Conversely, given a quotient group of G with respect to a congruence relation $\sim$, we can find a homomorphism f with the source G such that $G/{\sim}$ is the f-image of G.*

Proof. The relation $\sim$ of being equalized-by-f is an equivalence relation. But we need to check that $\sim$ respects G's group operation $*$ so that $G/{\sim}$ exists. In other words, we need to show that for any group objects x, y, z, given $x \sim y$, then (i) $x * z \sim y * z$ and (ii) $z * x \sim z * y$.

For (i), if $x \sim y$, then $fx = fy$, hence $f(x * z) = fx \star fz = fy \star fz = f(y * z)$, so $x * z \sim y * z$ (here, $\star$ is H's group operation). Case (ii) is exactly similar.

By the definition of $\sim$, the f-images of G's objects act like quotient-objects with respect to $\sim$; hence G's image under f is a quotient group $G/{\sim}$.

For the converse result, suppose $G/{\sim}$ is a quotient of G with respect to some equivalence relation $\sim$, with $f_\sim\colon x \mapsto [x]$ giving us the relevant quotient scheme. Then $f_\sim\colon G \to G/{\sim}$ is easily checked to be our required homomorphism. □

In sum: given a homomorphism $f\colon G \to H$, the f-image of G is a quotient of G and a subgroup of H.

Our last two theorems again show that we can trade in certain claims about the internal structure of groups for corresponding claims about homomorphisms between groups. The ways in which we can probe structures by looking at the maps between them will turn out to be a pivotal theme of category theory.

2.8 'Identical up to isomorphism'

(a) Let's pause over another important point.

We have met the groups K_1, K_2, K_3 which are isomorphic to each other. They are also isomorphic to any other group whose four objects can be labelled $1, a, b, c$ in such a way that the same 'multiplication table' in §2.3 applies again. Call such groups *Klein four-groups*. And note, the way in which the various Klein four-groups might differ from each other, namely in the internal constitution of their various *objects*, is not relevant to their core behaviour as groups, for that depends only on the *functional relations between the objects* induced by the group operation. In other words, despite the differences between their objects, the groups are the same at least as far as their structural properties – i.e. the properties as determined by their shared 'multiplication table' – are concerned.

A bit of care is needed in describing the situation, however. Consider, for example, the following from a rightly well-regarded algebra text:

> ... the groups G and H are isomorphic if there is a bijection between them which preserves the group operations. Intuitively, G and H are the same group except that the elements and the operations may be written differently in G and H. (Dummit and Foote 2004, p. 37)

But that surely isn't a happy way to put things. We have just reminded ourselves that K_1, K_2 and K_3 are isomorphic groups. But K_2, for example, comprises four *numbers* as its elements, and K_3 comprises four *operations* on a non-equilateral rectangle; and there is no reasonable sense in which numbers and geometric operations are the same things 'written differently'. If anything, then, it is exactly the other way around: we have here distinct groups with different elements and different group operations which, however, can be 'written the same', being represented by the same table under different interpretations.

A seemingly rather happier, and widely used, way of putting things is this: our Klein groups K_1, K_2 and K_3 are *identical up to isomorphism*. And indeed, for many purposes, group theory will simply not care about the differences between isomorphic groups.[8] Looking ahead, we'll find that ignoring differences between isomorphic widgets becomes a lead theme of category theory too.

(b) When not caring about the differences between isomorphic groups, we may fall into talking about *the* Klein group, as in 'The Klein group is abelian'. This can be blandly interpreted as generalizing talk about such groups (compare, for example, talk about 'the barn owl' – as in 'The barn owl hunts by night' – which can similarly be interpreted as generalizing talk about such birds). But we do also meet the more radical idea (originating perhaps with Dedekind) that, as well as 'concrete' Klein groups whose elements have an independent nature (which could be numbers, pairs of numbers, rotations and reflections, whatever), there is also a more purely 'abstract', purely 'structural', Klein group. This has the right multiplication table, but is supposedly built up from objects with no properties at all over and above being distinct from each other and being interrelated by the group operation according to the given table. And it is this group comprising abstract de-natured elements – which are, as it were, 'point-like', just nodes or positions in the structure – which is then said to be, properly speaking, *the* abstract Klein group.

Now, it is one thing to say that when we talk about 'the' Klein group we are generalizing, abstracting away from the specific non-group-theoretic properties of the elements of particular Klein groups: it is another thing to say that we are referring to a special group whose elements actually lack any non-group-theoretic properties. Does the latter suggestion really make sense? The idea here has been defended by some philosophers of mathematics, and has been vigorously criticized by others (that's philosophers for you).

[8]Though sometimes, for other purposes as in group representation theory, we will care a lot about finding isomorphic copies of groups when the copies come with particularly nice-to-handle presentations (e.g. as groups of matrices).

However, this is certainly not the time for us to get distracted by such debates! I'm mentioning them now because the Dedekindian idea of purely abstract structures can resurface in discussions about category theory.[9]

2.9 Categories of groups

(a) That will have to do by way of an initial review of some *very* basic facts about groups and the homomorphisms between them.

Now, we could pause here to take in a similar lightning-speed review of some equally basic facts e.g. about topological spaces and the continuous maps between *them*; and the parallels would be instructive. But perhaps setting the scene in that case would take a bit too long, and you are by now wanting to hear about categories! So instead let's straight away ask: given some groups and some homomorphisms between them, what does it take for these to form a structured family which counts as a category of groups?

(b) We in fact impose just two very natural conditions. First, the homomorphisms in the category should be closed under composition. In other words, if the homomorphism f takes us from G_1 to G_2, and g takes us from G_2 to G_3, then we want the homomorphism $g \circ f$ to be available to take us from G_1 to G_3. Second, for each of the groups in the category, its homomorphism needs to be included.

And this is *all* it takes: it really is that simple! Still, let's say the same thing again, but this time in more laborious detail, for clarity's sake:

Definition 12. A *category of groups* comprises

(1) some groups, Grp, and
(2) some homomorphisms, Hom,

where Grp and Hom are governed by the following conditions:

Sources and targets If $f\colon G \to H$ is one of the homomorphisms Hom – is among Hom, for short – then both its source G and its target H are among Grp.

Composition If $f\colon G \to H$, $g\colon H \to J$ are both among Hom, where the target of f is the source of g, then $g \circ f\colon G \to J$ is also among Hom.

Identity homomorphisms If G is among Grp, the corresponding identity homomorphism $1_G : G \to G$ is among Hom.

Further, we have:

Associativity of composition. For any $f\colon G \to H$, $g\colon H \to J$, $h\colon J \to K$ among Hom, $h \circ (g \circ f) = (h \circ g) \circ f$.

Identity homomorphisms do behave as identities. For any $f\colon G \to H$ among Hom, $f \circ 1_G = f = 1_H \circ f$. △

---

[9] For an extended defence of the idea of abstract structures whose objects have no properties other than occupying a certain place in the structure, see in particular Shapiro (1997). A very useful entry-point to some of the discussions round and about this and related sorts of structuralism is provided by Hellman (2005).

Of course, we know the last two conditions will automatically be satisfied because of Theorem 2. But I'm redundantly mentioning those conditions again here so that our definition of categories of groups matches up nicely with our general definition of categories in Chapter 5.

(c) As we've reminded ourselves, groups are many and various; so too are categories of groups.

For example, a single group G together with its identity homomorphism $1_G : G \to G$ trivially counts as a category of groups. So too does any uncommunicative bunch of groups equipped only with their identity homomorphisms.

But those are *very* boring cases! Things can get more interesting when the groups in a category start to communicate (so to speak).

Consider next, then, the category that collects all the finite groups whose objects are some natural numbers, together with all the isomorphisms between those groups. Now there is a *bit* of structure to this category, with the isomorphic groups at least connected together by the maps between them. But this is still relatively unexciting: we have different islands of isomorphic groups, and a group inhabiting one island knows nothing about groups inhabiting other islands.

So let's move on to consider the category comprising those same finite groups but this time combined with *all* the homomorphisms between them (whether isomorphisms or not). And *now* non-isomorphic groups can 'see' each other. So we have enough homomorphisms in play to be able to distinguish, for example, the one-object groups in the category by saying that these are the groups which have one and only one homomorphism to and from every other group, as noted in §2.4. We can also use these homomorphisms to tell a story about e.g. subgroups and quotient groups living in the category, as indicated in our preliminary sketch in §2.7.

(d) We can multiply examples of categories of groups without limit. Some of these are again relatively restricted, small-scale, families of structures. For example, there is the category of symmetry groups of finite regular polygons and the homomorphisms between them, there is the category of infinite abelian groups of natural numbers and their homomorphisms, there is the category of groups of invertible real matrices and *their* homomorphisms. Then we can consider larger and larger categories of groups too. And going for broke, we can in particular ask: is there also an inclusive mega-category of *all* groups and *all* the possible homomorphisms between them?

That's a very good question! We'll work up to an answer by the end of Chapter 4. But first, we need quickly to agree on some terminology and then to comment on one notable feature of our definition of a category of groups.

3 Sets, classes, plurals

Before proceeding, then, some remarks about the notions of set, class, and collection, and some related remarks about the use of plural idioms.

3.1 Sets and (virtual) classes

Following Cantor, I'll understand a set – strictly so called – to be a single object, a thing in itself over and above its members (so the 'set of' operator takes zero, one, or many things, and outputs a distinct new thing).

However, if this is the guiding conception, then the very first thing to say is that a great deal of elementary informal talk of sets or classes is no more than a façon de parler. Yes, it is a useful and now very familiar idiom for talking about many things at once. But in a whole range of elementary contexts informal talk of a set or class doesn't actually carry any serious commitment to there being any *additional* object over and above those many things. Singular talk of *the set/class of widgets* can very often be traded in without significant loss for plural talk of *the widgets*.[1]

Here is Paul Finsler writing a century ago, emphasizing the key distinction we need (and adding a bit of linguistic stipulation):

> It would ... be inconvenient if one always had to speak of many things in the plural; it is much more convenient to use the singular and speak of them as a class. ... A class of things is understood as being the things themselves, while the set which contains them as its elements is a single thing, in general distinct from the things comprising it. ... Thus a set is a genuine, individual entity. By contrast, a class is singular only by virtue of linguistic usage; in actuality, it almost always signifies a plurality. (Finsler 1926, p. 106, quoted in Incurvati 2020, p. 3.)

Finsler writes 'almost always', I take it, because a class term may actually denote just one thing, or even – perhaps by misadventure – none.

[1] In other words, the replacement of informal plural locutions (e.g. 'the premisses $P_1, P2$, and P_3 entail C iff ...') by talk of sets (as in '$\{P_1, P2, P_3\} \vDash C$ iff ...') is often doing no real work, and involves no essential commitment to the introduced sets as distinct entities over and above their members.

Nothing at all hangs, of course, on the stipulative choice of the particular words 'set' vs 'class' to mark the distinction.[2] What matters is the contrast between uneliminable talk of sets in Cantor's sense of entities in their own right and, on the other hand, non-committal talk, eliminable in favour of plural locutions.

And here is Quine making the key point in a later and much more famous passage:

> Much ... of what is commonly said of classes with the help of '∈' can be accounted for as a mere manner of speaking, involving no real reference to classes nor any irreducible use of '∈'. ... [T]his part of class theory ... I call the virtual theory of classes. (Quine 1963, p. 16)

This same usage plays an important role in set theory itself in some treatments of so-called 'proper classes' as distinguished from sets. For example, in his standard book *Set Theory*, Kenneth Kunen writes

> Formally, proper classes do not exist, and expressions involving them must be thought of as abbreviations for expressions not involving them. (Kunen 1980, p. 24)

However, to complicate matters, other developments of set theory do allow for proper classes (classes which are 'too big to be sets') to count as entities in their own right. For example, this is how things go in NBG set theory where every set is a class (so classes have the same existence status as sets), but not every class is a set.[3] Hence we can't reliably use 'class' and expect to be understood in Finsler's or Kunen's way.

So let me go in for a minor bit of linguistic stipulation of my own. When I talk of a 'set' I will, by default, understand this in Cantor's sense to be referring to a single object, something over and above its members. I'll avoid talk of 'classes' as in practice dangerously ambiguous (unless repeatedly accompanied by explanatory glosses). And I'll use 'collection' – a term which carries minimal theoretical baggage – when I want a non-committal singular term for talking about many things at once.

3.2 Don't try to eliminate plurals!

(a) Finsler perhaps rather exaggerates the supposed inconvenience of plural talk.

After all, there is nothing at all unusual or forced (or inconvenient!) about the use of plural terms in everyday mathematics. Consider, for example, terms such as 'the complex fifth roots of 1', 'the real numbers between 0 and 1', 'the points where line L intersects curve C', 'the finite groups of order 8', 'the premisses'

[2] Bertrand Russell had earlier contrasted a 'class as one' with a 'class as many' to mark a version of the same distinction (Russell 1903, §70).

[3] A classic textbook presentation of von Neumann-Bernays-Gödel set theory is provided by Mendelson (1964, and later editions).

(of a certain argument), 'Hilbert's axioms for geometry', 'the symmetries of a rectangle', 'the ordinals', etc., etc. Mathematicians habitually use such terms which, taken at face value, refer plurally to many things; and they use them all the time without the slightest sense of strain or impropriety.

And now compare

(i) 13, 17, 21, 25, and 29 are odd numbers,
(ii) 13, 17, 21, 25, and 29 form an arithmetical progression,

You might propose that (i), with its apparently plurally denoting list '13, 17, 21, 25, and 29', is *really* an abbreviation for a conjunction of the separate singular sentences '13 is an odd number' and '17 is an odd number' etc. But that's a rather special case. Such a 'disguised conjunction' account is not available for (ii) with its collective predicate 'form an arithmetical progression'. Similarly, no version of the 'disguised conjunction' story is available to eliminate the plural terms in such perfectly ordinary mathematical claims as

(iii) The roots of equation E form an arithmetical progression,
(iv) The points of intersection of C_1 and C_2 are colinear,
(v) The cardinal numbers are linearly ordered.

(b) But some might still be tempted by the 'singularist' thought that in those last cases too we can reconstrue informal plural talk about *the widgets* as disguised singular talk, albeit in a different way. You might propose that this time, while such claims don't unpack into a conjunction of singular claims about individual widgets, we still only have singular reference – for the apparently plurally denoting term actually refers to just one thing, e.g. to *the set of widgets.*

But in fact, you already know that we can't *always* construe plural terms as denoting some corresponding set. Take the familiar truth taught us by Russell's paradox:

> The sets which are not members of themselves do not form a set.

Plainly, the plural term 'the sets which are not members of themselves' can't here be singularly referring to the (non-existent) set of sets which are not members of themselves! And mutatis mutandis, the same line of argument will defeat the idea that the use of plural terms *always* involves disguised singular reference to (non-virtual) classes or other candidates.

But set aside the extreme cases which are vulnerable to such a Russellian argument. By my lights, the more fundamental point is that – when we actually get down to details – it is impossible to smoothly and systematically eliminate plural talk across the board in favour of singular reference to something like sets, while even more piecemeal attempts still require the most procrustean and implausible contortions.

Sadly, it would be *far* too distracting to set off to pursue the twists and turns of the case for that last claim here. But the upshot of a lot of relatively recent logical work is that we do much better to suppose that *plural talk is in perfectly good logical order as it is* and doesn't need to be parsed away. And – importantly – this is still true, even if your measure of being in good logical order is formalizability. What look on the surface to be plural terms referring to many things at once can and should be taken at face value.[4]

3.3 Why fuss?

But why am I fussing about this point? *Because it can matter when we turn to talking about categories.*

We've just reminded ourselves that there are, for example, too many sets to form a set. And we'll find very soon that we'll want to consider interesting categories – relatively *large* categories – which also comprise too many structures to form a set. So we can't straightforwardly define a category of widgets in general as comprising a *set* of widgets suitably equipped with maps between them.

There are various ways of handling this point (as we will be noting again later in Chapter 30). But at least initially, the simplest and best option is to use frankly plural idioms in our definition. This is in fact a common enough tack, even if it isn't always loudly announced as such. This is exactly the line I took at the end of the last chapter in introducing the idea of a category of groups: I said some groups (one or more) and some homomorphisms (one or more) form a category if they together satisfy certain conditions – no talk yet of sets or classes (so we can allow really large categories of groups, which include too many groups to form a set).

And the point I want to urge again is that the use of the plural idiom here in defining a category is in good logical standing: it is not, so speak, a temporary expedient, demanding to be cashed out in a singularist spirit. We can take it at face value.

[4]For a lot more on why we shouldn't try to eliminate plurals in a singularist spirit, see Oliver and Smiley (2016, Ch. 1 and then particularly Ch. 3 'Changing the subject') and McKay (2006, particularly Ch. 2 'Against singularism'). For an extended formal treatment of how to argue with plural terms and plural quantifiers, taking them at face value, see the later chapters of Oliver and Smiley's tour de force. For a more accessible introduction to some key themes of plural logic together with pointers to many more debates and issues, also see for example Linnebo (2022).

4 Where do categories of groups live?

We left a question hanging at the very end of Chapter 2: can we sensibly suppose that there is an inclusive mega-category – call it Grp – comprising *all* groups and *all* the possible homomorphisms between them?

It will help first to think more generally about groups and the families of interconnected groups that form categories of groups: where do they live?

4.1 One 'generous arena' in which to pursue group theory

(a) In my account of some (very) elementary group theory in Chapter 2, the word 'set' didn't occur – except in saying that the objects in pairing and quotient schemes need *not* be thought of as sets. I instead proceeded using a plural idiom. This was, needless to say, the mildly deviant aspect of the mode of presentation.

By contrast, it is of course usual to say things like 'A group is a set $\underline{G}$, together with a binary operation $*$ on $\underline{G}$ which has the following properties ...' (Beardon 2005, p. 2) or 'A group is an ordered pair $(\underline{G}, *)$ where $\underline{G}$ is a set and $*$ is a binary operation on $\underline{G}$ satisfying the following axioms ...' (Dummit and Foote 2004, p.16).[1] And it might be added that binary operations on sets are themselves, strictly speaking, more sets too. But we should pause to ask: what real work is the notion of *set* doing there?

I agree with Paulo Aluffi, who explicitly acknowledges at the beginning of his fine book *Algebra, Chapter 0* that the informal set idiom which he adopts in the standard way is actually "little more than a system of notation and terminology" (Aluffi 2009, p. 1). That surely seems right. At least at the outset, the story of group theory will unfold in basically the same way in either system of notation and terminology, whether we habitually use singular talk of sets, or alternatively adopt a plural idiom as I did (taking that idiom at face value, as recommended in the last chapter). Yes, we can specify a group G as being a collection or set of objects $\underline{G}$ equipped with a suitable binary operation and distinguished object. But, at least initially, we don't need to construe talk of a 'collection' or 'set' here in more than a very attenuated sense: it just serves as a useful-but-eliminable way of talking in the singular about many things at once. So, I claim, we can equally well use a plural idiom and talk of G as being some suitably equipped

[1] I've slightly altered the notation used in those books, by writing the underlined '$\underline{G}$' for the underlying set of objects (the 'carrier set') of a group.

objects G. Part of the point of the presentation in Chapter 2 was to make this thought begin to seem plausible.

(b) Still, some might very well object that – however things might appear at the outset – a theory of sets-as-entities-in-their-own-right is all the same required to be there, even if hovering off-stage at the start, in any serious development of group theory. But why so?

Perhaps we can pick out a couple of questions here which might seem to invite genuinely set-theoretic answers.

Reflect that even as soon as we reach our trite Theorem 2, we are going beyond the purely logical consequences of our definitions of groups and group homomorphisms. So what do we in fact need to bring to the table to get group theory going? Answer:

(i) the usual mathematical stock-in-trade of a body of assumptions about *functions*, together with

(ii) a repertoire of available *constructions*.

For example, we assume that functions always do compose when they can (i.e. when the target of the first is the source of the second), and that composition is associative. We assume that a function with a two-sided inverse is a bijection. We assume that it makes determinate sense e.g. to talk about *all* the permutations of some given objects, or *all* the automorphisms on a given group. And so on, and so forth. These, of course, look pretty unproblematic assumptions – but as I emphasized before, they are needed all the same.

Again, we typically assume that we can construct what will serve as ordered pairs ad libitum; and we assume that, whenever an equivalence relation partitions some objects, we can somehow represent these partitions. More carefully, using our earlier terminology, we assume pairing schemes and quotient schemes are available whenever we want them. And going forward, we will assume that we can not only construct pairs, triples, and finite tuples more generally, but we can form infinite sequences too. We also need to assume that we can freely construct multiple 'copies' of whatever structures we already have. And so on, through more sophisticated constructions.

I'm leaving the details vague, but quite intentionally so. I am merely gesturing at the way that standard textbook developments of group theory simply help themselves to a bunch of unproblematic background assumptions as needed as we go along. Which is fair enough. But this does raise an obvious first question. *What if we want to start getting more explicit and methodical about these background assumptions? Suppose we want to regiment these assumptions and organize them into a neat package – what package would suffice?*

(c) Shelve that issue for a moment, and consider a different issue arising from what we earlier said about groups.

Here's a silly-sounding question. Suppose I cut out a cardboard non-equilateral rectangle: have I hereby brought into a being a new Klein four-group, the group of *this* new rectangle's own (approximate!) rotation/reflection symmetries?

Well, we were previously entirely permissive about where we can find our groups: on our Defn. 1, we just need some new 'objects' (in a very broad sense) and a suitable operation on them, and then we get a new group. But on the other hand, a new physically realized Klein group is surely neither here nor there as far as the mathematics of groups is concerned. Group theory will for most purposes abstract from the differences between groups which are identical up to isomorphism.

OK: suppose that there is a capacious enough fixed mathematical universe in which we can find copies of all the different kinds of groups we will ever want to study. Then we can focus on what happens in that universe and won't care about any additional copies of its groups which are (as it were) roaming outside in the wild, or popping into existence when I cut up a new bit of cardboard.[2]

But this last thought prompts another – and this time, more sensible – question. *Where can we find a suitably rich mathematical universe in which there are at least copies of all the groups we want?*

(d) We have two related questions on the table: what package of assumptions about available functions and about structure-building constructions will suffice for group theory? where can we find (at least copies of) all the groups we want, neatly corralled together?

And now suppose we had started out in Chapter 2 not by talking informally about groups (and families of groups with structure-respecting maps between them) but by talking instead about e.g. topological spaces (and families of spaces with structure-respecting maps between *them*). Then we could have similarly reached the point of asking: what package of assumptions about available functions and about structure-building constructions will suffice for topology? and where can we find (at least copies of) all the spaces we want?

Importantly, we'd also want these packages for supporting group theory and supporting topology – and for supporting other areas of theorizing about mathematical structures – to be at least mutually consistent and preferably nicely unified. Because, as is very familiar, results from one branch of mathematics can get applied in solving problems from another, and we need to be able transfer results without getting entangled in contradiction.[3]

So where can we find (copies of) all the groups, topological spaces and other mathematical structures that we want? Where can we find a suitably unified package of functions and construction kits?

(e) There is of course a *very* familiar answer! A suitable universe of sets provides exactly the sort of generous arena in which we can find a plenitude of groups, topological spaces, etc., together with all the functions and construc-

[2]When thinking about applied mathematics, we will want to say something about how abstract groups can be physically exemplified. But we won't care about physically realized groups for the purposes of pure group theory.

[3]Burgess (2015, pp. 56–63) presses this point eloquently in a section on 'The Unity of Mathematics'.

tions involving them that we want for ordinary mathematical purposes.[4]

If we are to put no limit on the complexity of available constructions, we'll need a rich enough universe of sets. So take a universe with an unbounded hierarchy of levels. And once we have enough levels in our hierarchy we get set-theoretic surrogates for natural numbers, real numbers, etc., for free. That's why it is usually supposed that we can keep our set-theory-for-foundations 'pure', meaning that its sets only have other sets as members all the way down. So we are looking at the kind of set universe described, for example, by ZFC or some close relation.[5]

(f) In summary, then. Yes, perhaps it is the case that at the outset a rather non-committal naive set-theoretic idiom can be replaced by a plural idiom at a relatively informal level. It might reasonably be claimed, however, that – when we reflect on what is needed to systematically develop group theory – it is in fact entirely appropriate to cleave to the conventional line and fall into talking about sets from the beginning. Because once the wraps are off, once we aim to regiment the background assumptions which we really need to get our theory up and running, we will find that our theory is exposed as at bottom set-theoretic through and through.

4.2 Alternative implementations?

Or so the story might go. A little reflection, however, suggests that the argument does proceed much too fast at the end. Yes, a universe of sets as described by ZFC or some extension (for example) may provide *one* generous arena in which we can implement all the gadgets we need in developing group theory or topology. *But why suppose that it is the only option?*

(a) We are so used to being told that various mathematical widgets and what-nots are officially to be defined as sets of one kind or another that it can take a bit of work to loosen the grip of that doctrine. But for reasons which will emerge, in the context of this book it is worth putting in the effort.

Consider, for example, the standard set-theoretic story about the nature of one-place functions. So take some way of implementing ordered pairs as sets, for instance as Kuratowski pairs $\langle x, y\rangle_K = \{\{x\}, \{x, y\}\}$. Then here is a familiar definition:

Definition 13. Given a function f with domain X and codomain Y, its *graph* is the set $\hat{f}$ of ordered pairs $\langle x, y\rangle_K$ where $x \varepsilon \mathrm{X}$ and $y \varepsilon \mathrm{Y}$, and $fx = y$. △

[4]The phrase 'generous arena' is borrowed from the very helpful discussion of the idea of set-theoretic foundations in Maddy (2017).

[5]Compare, then, the sort of relatively naive set-theory-for-ordinary-mathematical-applications which is outlined in the introductory chapters of a hundred textbooks. This allows 'urelements' (a.k.a. 'individuals' or 'atoms'), i.e. set-members which aren't themselves sets, and is also silent about the extent of the universe of sets. For a classic sample presentation of sets-for-application, see the beautifully lucid first chapter of Munkres's classic topology text (2000). Munkres at various points allows natural numbers, real numbers, and even the pieces of pasteboard of a pack of cards as urelements.

Then an equally familiar orthodoxy, at least in its baldest and most unqualified form, *identifies* a function f with its graph $\hat{f}$.

(b) An initial point. The graph of a function fixes its domain but doesn't fix a codomain. So if we want a determination of its codomain to be part of the data of a function – as in fact category theorists will do – then we'll need to set-theoretically identify a function with e.g. a *triple* whose members are the function's graph, its domain-as-a-set, and its codomain-as-a-set.

We'll return to this point later, but we can put it to one side for the present. Because our real worries about the simple set-theoretic orthodoxy (functions are graphs) will carry over, mutatis mutandis, to the fancier story (functions are set-theoretic triples involving graphs).

(c) Here's a first argument why we should resist always identifying a function with its graph (or with a triple involving its graph).

To play on the set-theorists' own turf, consider the function which maps an object to its singleton. Then – by the set-theorists' own lights – *that function doesn't have a graph*: the totality of pairs $\langle x, \{x\}\rangle_K$, pairing-up every set x with its singleton, is the size of the universe of sets and so is 'too big' to form a set. Likewise the function which maps every ordinal to its successor lacks a graph.

Therefore not all functions can be identified with their graphs.

(d) One counterexample is enough to defeat a universal claim. It might be suggested, though, that the cases where a function relates too many things to be a set are in some sense exceptional cases needing special treatment. So, in a concessive spirit, let's put aside those cases and see where that gets us.

Well, next note that treating a function as a set of ordered pairs involves arbitrary choices of implementation scheme.

(i) It is arbitrary to fix on Kuratowski's implementation of pairs. Other set-theoretic pairing schemes will work equally well.
(ii) Even relative to a choice of pairing scheme, we could equally well model a function by the set of pairs $\langle y, x\rangle$ where $f(x) = y$, rather than by the set of pairs $\langle x, y\rangle$ – some textbooks do this. Again other choices are possible.

However, if various permutations of choices at stages (i) and (ii) are pretty much as workable as each other, then we surely can't suppose that – when we choose to equate a function with its graph as we just defined it – we have made the uniquely *right* choice, i.e. the choice that correctly identifies which set that function 'really' is. And if there is no fact of the matter about which set a given function is, then we can't flat-out identify the function with its graph.

(e) But we can dig still deeper: the key observation is that *a function and its graph belong to different logical types* – and that is fundamentally why they *can't* be identical. Alonzo Church makes the essential point when he writes that

> it lies in the nature of any given [one-place] function to be applicable to certain things and, when applied to one of them as argument, to yield a certain value. (Church 1956, p. 15)

For example, a function such as the factorial defined over the natural numbers is, of its nature, the type of gadget which yields a numerical value for a given number as argument: by contrast a set doesn't, of its nature, take an argument or yield a value.[6] And what holds of sets in general holds of sets of ordered pairs (graphs of functions) in particular: such a set can't *by itself* do the work of a function f, taking arguments and yielding output values. As the mathematician Terence Tao, who has no philosophical axe to grind, briskly puts it in his introductory book on analysis,

> Strictly speaking, functions are not sets, and sets are not functions; it does not make sense to ask whether an object x is an element of a function f, and it does not make sense to apply a set A to an input x to create an output $A(x)$. (Tao 2016, p. 51)

Which *of course* isn't to deny that we can make use of the graph of a function (a glorified input-output look-up table) in mapping an input object to an output value. But to do this, we need to deploy *another* function, namely a two-place evaluation function which takes an object x and a graph, and outputs y if and only if the pair $\langle x, y\rangle_K$ is in the graph. And unless we are planning to set off on an infinite regress, we had better not seek to again trade in this evaluation function for another set.

(f) So a function, strictly speaking, isn't in itself a set. But what we can do in a set-theoretic environment is *implement* functions as graphs, or treat graphs as *proxies* for functions.[7] And we can then transmute a claim about a function into a corresponding set-theoretic claim about some set of ordered pairs. For example, suppose we are considering, say, a one-place function of natural numbers. Then yes, we can implement this as a set of ordered pairs in a suitable universe of pure sets. Though these won't be ordered pairs of numbers – since strictly speaking numbers aren't themselves sets either.[8] Rather we will be dealing with pairs of whatever-sets-we-choose-for-implementing-numbers (another locus for arbitrary choices). So if $\hat{m}$ is our preferred set-implementation for the natural number m, then a numerical claim $f(m) = n$ will be mirrored by a set-theoretic claim of the form $\langle \hat{m}, \hat{n}\rangle_K \in \hat{f}$, with $\hat{f}$ a suitable set of pairs.

(g) We can make similar points about the conventional story that identifies a relation with its extension (where the extension of R is the set of ordered pairs $\langle x, y\rangle_K$ such that xRy).

[6] Church is here following Frege, whose metaphor of 'unsaturatedness' might be helpful. The picture is that functions of their nature are 'unsaturated', have a certain number of empty slots waiting to be filled appropriately when the function is applied to the right number of arguments. By contrast, an object like a set is already 'saturated', with no empty slots waiting to be filled.

[7] No word seems ideal – whichever we choose, talk of 'implementations', 'proxies', 'surrogates' or 'representations' can be potentially misleading. I'll mostly prefer 'implementations', as perhaps the most colourless term.

[8] The locus classicus for this point is Paul Benacerraf's – very readable! – 'What numbers could not be' (1965).

Size considerations block some relations from having extensions. For example, ironically, the membership relation built into the universe of sets can't be identified with its extension because it doesn't have one (the pairs $\langle x, y\rangle_K$ such that $x \in y$ are as plentiful as all the sets, so too many to form a set). More generally, it is arbitrary to identify relations with their extensions as conventionally defined as opposed to other candidate sets (and again, if there is no fact of the matter about which set a given relation is, then we can't flat-out identify a relation with its extension). And most fundamentally, there is a type difference between relations and their extensions: it is of the nature of e.g. a binary relation to hold true or false of pairs of things, while sets don't hold true or false of things.

So relations aren't extensions. Though, as with functions and their graphs, we can of course treat extensions as implementations or proxies for relations (other than for the relation of set membership itself), as when we e.g. mirror a numerical relational claim mRn by a set-theoretic relational claim $\langle \hat{m}, \hat{n}\rangle_K \in \hat{R}$, with $\hat{R}$ a suitable set of pairs.

(h) What is the point of insisting that the story about functions-as-graphs and relations-as-extensions doesn't tell us what functions and relations 'really' are, but rather reports one way of implementing them in the universe of sets? Am I *very* boringly splitting hairs?

I do hope not! To repeat: the point of trying to loosen the grip of the idea that functions must be identified with their graphs-as-sets, and relations must identified with their extensions-as-sets, is to make room for the thought that there might be *other* attractive ways of theorizing about the functions and relations of ordinary mathematics in different foundational frameworks.

Return to thinking about groups, for example. Take e.g. the additive group of integers. Strictly, the objects of this group – the integers – aren't sets: and the binary addition function isn't a set either. The same goes for other groups of ordinary group-theory: typically, their objects won't intrinsically be sets, and their binary operation can't be. So what we can find in the universe of pure sets should properly be regarded as implementations of, or proxies for, groups.

Which is all fine! I am not for a moment wanting to deny that these proxies can serve certain theoretical purposes brilliantly well – I am certainly *not* in the business of scorning the business of modelling or implementing mathematical structures in a set-theoretic framework (or in some rival foundational framework). I am just emphatically highlighting that we *are* here in the implementation business. And once we look at things that way, we can now rather more easily see that we shouldn't too hastily assume that a set-theoretic framework provides the *only* possible kind of generous arena, the only possible foundational framework in which we can find a plenitude of surrogates or proxies for implementing the mathematical structures and constructions which we want to regiment and study.

Indeed, we can't rule it out that an alternative choice of general framework *might* even do the job rather better in some respects (maybe with different costs and benefits accruing to the different choices). For example, treating all mathematical widgets and whatnots as sets seemingly gives rise to such daft questions as 'is the square root function for complex numbers a member of π?', 'Does any simple group appear as a zero of the Riemann Zeta function?'. We can block such foolish questions by using a more type-disciplined framework which strongly distinguishes types of entities in our mathematical universe in the sort of way that practicing mathematicians habitually do. So a modern type theory *could* be the way to go. And note, incidentally, that types are often said to be 'collections' in some sense, with a type theory thus offering us a different theory of collections to standard set theory.

Or perhaps the general foundational framework we want will not just be broadly type-theoretic but will be essentially category-theoretic in flavour. Here's the logician Dana Scott:

> What we are probably seeking is a 'purer' view of functions: a theory of functions in themselves, not a theory of functions derived from sets. What, then, is a pure theory of functions? Answer: category theory. (Scott 1980, p. 406)

Scott quickly goes on to remark that the general notion of a category won't give us enough. But arguably a suitable *topos* (that's a particular sort of category which we'll eventually meet) can provide another sort of 'generous arena', another universe in which we can better regiment much of our ordinary mathematics. Such a suggestion – which of course we won't be in a position to understand for quite a while! – would be very puzzling if we have already jumped too quickly to fall in with the conventional trope that ordinary abstract mathematics is already quite fixedly set-theoretic through and through.[9]

4.3 'The' category of groups?

(a) With those reflections in place, let's now return to the question left hanging at the end of §2.9 and raised again at the outset of this chapter. What can we make of talk of a category Grp of *all* groups and *all* the homomorphisms between them?[10]

[9] Having said all this, I'm not ruling out the possibility that (i) what are presented as sufficiently rich alternative framework arenas will in the end turn out to be intertranslatable and equivalent in some sense strong enough to warrant the claim that there is in fact just one mathematical universe here which can be conceptualized in various ways, including set-theoretically. And – a further step – it might also be arguable that (ii) there is no mathematical reason to prefer another of the re-conceptualizations to the familiar set-theoretic story. In that case, the conventional treatment of the denizens of our foundational framework as sets would then turn out to be as defensible as any alternative – and it will have the advantage of established familiarity.

But if things do pan out so as to warrant (i) and (ii) and thus the claim that set-theoretic foundations are at least an equal 'best buy', then we surely ought to arrive at that conclusion the hard way, by arguing through the alternatives.

[10] Grp is commonly included in introductory lists of examples of categories, and see e.g. Roman (2017, p. 3) and Agore (2023, p. 4) for their explicit emphasis that this category comprises *all* groups and *all* the homomorphisms between them; see also McLarty (1992,

Suppose we cast the net widely, treating suitably equipped objects of any old kind (maybe physically realized) as forming a group. Then it seems that talk of 'all' groups will range much more widely than we need to care about (as we keep finding new concrete instances of the same kinds of groups popping up in odd corners of the universe). Relatedly, thought of in this sweepingly inclusive inclusive way, groups will not constitute a fixed totality.

But suppose instead that we concentrate our attention on *some* preferred 'generous arena' for regimenting ordinary mathematics. Then we can hope to find there copies of any of the groups and group-theoretic gadgets we want for mathematical purposes (along with all the apparatus of functions and constructions we need). And, relative to our chosen arena, it can now make sense to assume that there is a determinate category, a fixed structure-of-structures, which brings together all of the various implementations of groups living in *that* chosen universe and all the homomorphisms between *them*.

(b) Unsurprisingly, then, we find that one approach in introductions to category theory is to suppose that the constituents of inclusive categories like category Grp – the category of 'all' groups and their homomorphisms – are to be found in a particular ambient universe, with the predictable default choice being a universe of sets (in fact, a universe of pure sets).[11]

For example, in his hugely influential classic *Categories for the Working Mathematician*, originally published in 1971, Saunders Mac Lane presents the axioms for category theory as in §5.1 in our next chapter, before he bluntly defines a category proper as "any interpretation of the category axioms within set theory". Then in an early section titled 'Foundations' he makes a first pass at outlining the particular kind of set theory he initially has in mind, namely a certain extension of ZF.

Much more recently, Emily Riehl in her *Category Theory in Context* can write: "common practice among category theorists is to work in an extension of the usual Zermelo–Fraenkel axioms of set theory ...". And as Horst Schubert emphasizes in his excellent *Categories*, "One has to be aware that the set theory used here [an extension of ZF] has no 'primitive (ur-)elements'; elements of sets, or classes ... are always themselves sets".[12]

On this approach, then, the mega-category Grp, to stick with that example, will be taken to comprise all groups-implemented-as-pure-sets together with the group-homomorphisms-implemented-as-pure-sets. True, the contents of Grp will vary with the extent of our favoured universe of pure sets. But this relativity is an acceptable feature not a bug:

> For applications ... one doesn't need a category of literally all groups ... It is always enough to have a category of 'enough' groups, though

p. 4), Leinster (2014, p. 11), Riehl (2017, p. 4), and many others.

[11]Careful! We are not saying that Grp, for example, is itself a set: it can comprise too many groups for form a set.

[12]See Mac Lane (1997, p.10, pp. 22–23), Riehl (2017, p. 6), Schubert (1972, §3.1).

how many is enough may vary from application to application.[13] (Burgess 2015, p. 174)

(c) An alternative view, trailed in the quotation from Dana Scott in the last section, is that category theory itself can provide – perhaps in the shape of Lawvere's Elementary Theory of the Category of Sets (ETCS) – an autonomous and significantly different realm which provides a better arena in which to reconstruct classical mathematics.[14] Now, there are very marked differences between the objects of ETCS and sets as ordinarily conceived. Indeed, some would argue that the objects of this topos are really not sufficiently set-like to be properly described as sets at all; but let that pass for now. It is certainly the case that implementing Grp in the universe described by ETCS involves a significantly different story from the standard set-theoretic one. But this is for much later, though by the end of the book we will be able to say more about the virtues or otherwise of ETCS.

(d) The views just canvassed give Grp (and comparable categories that we will meet in the next chapter) ingredients which all have a 'home address', so to speak, in a wider universe of ZFC-style sets or of ETCS-style sets-and-functions. Some would argue that we go off-track in thinking of inclusive categories like Grp in this way, and instead should accord them some other status. But again it is far too soon to try to tangle with such ideas.[15]

How should we proceed? We could try – at least from the outset – to remain as neutral as possible about the nature of 'the category of all groups' and the like. But with some regrets, I think that the least distracting policy is actually to start off conservatively marching in step with most other introductions to category theory (this way you can then smoothly move from one presentation to another). So when we talk about a category like Grp, you can take it pro tempore that we too are by default talking about the category of all the relevant structures implemented in some suitably capacious, if perhaps not-yet-fully-specified, universe of pure sets of a familiar kind.

But I'm flagging up here that in the end we might well want to loosen this initial anchorage in a specifically set-theoretic setting as standardly understood. And eventually we'll be in a position to think a little about at least one possible alternative. However, there is a *great* deal of ground we need to cover first. We'll have to dive in and see how things work out.

---

[13]Perhaps we should note that there are questions of group theory which will get different answers in different set universes; for some examples see tinyurl.com/groupqns. But we really, *really*, don't want to get delayed by such issues at this early point!

[14]And then we can find the suggestion that "Generally in category theory one doesn't worry too much about its set-theoretic foundations because one thinks that categories (namely toposes) provide a better foundation for mathematics", Streicher (2004, p. 20).

[15]Hellman (2003) poses what he calls the 'problem of the home address', and Awodey (2004) responds that Hellman's view of category theory is misguided. However, now isn't the time to follow up these references!

5 Categories in general

We have met categories of only one kind so far, namely categories comprising some groups and enough homomorphisms between them. Here, 'enough' stands in for some pretty minimal requirements – essentially that compositions of homomorphisms in the category are also in the category, and the identity homomorphism for each group in the category is also present.

We now make our real start on category theory by generalizing to ...

5.1 The very idea of a category

(a) We said that many paradigm examples of categories are – as in our first illustrative case of categories of groups – families of structures with structure-respecting maps between them. But what can we say about nicely behaved such families, abstracting across cases?

One sufficiently general thought is this: if, within a family of structures including A, B, and C, we have a structure-respecting map f from A to B and another structure-respecting map g from B to C, then we want to be able to compose these maps. That is to say, the first map f followed by the second map g should also count as a structure-respecting map $g \circ f$ from A to C.

What principles will govern such composition of maps? Associativity, surely. Using a natural diagrammatic notation, if we are given maps

$$A \xrightarrow{\ f\ } B \xrightarrow{\ g\ } C \xrightarrow{\ h\ } D$$

it ought not matter how we carve up the journey from A to D. It ought not matter whether we apply the map f followed by the composite g-followed-by-h, or alternatively first apply the composite map f-followed-by-g and then afterwards apply h.

What else can we say at the same level of stratospheric generality about nice families of structures and structure-respecting maps? Very little! Except that (as with our groups) there presumably will always in principle at least be the limiting case of a 'do nothing' identity map, which applied to any structure A leaves it untouched.

That apparently doesn't give us a great deal to work with. But it is already enough to shape our following definition of categories. However, it is helpful to abstract even further from the idea of structures with structure-respecting maps between them, and – using less specific terminology – we'll now speak very generally of *objects* and of *arrows* between them. Then we say:

Definition 14. A category C comprises two kinds of things:

(1) C-objects (which we will typically notate by $A, B, C, \ldots$),
(2) C-arrows (which we typically notate by $f, g, h, \ldots$).

These C-objects and C-arrows are governed by the following axioms:

Sources and targets For each arrow f, there are unique associated objects $src(f)$ and $tar(f)$, respectively the *source* and *target* of f, not necessarily distinct. We write $f\colon A \to B$ or $A \xrightarrow{f} B$ to notate that f is an arrow with $src(f) = A$ and $tar(f) = B$.

Composition For any two arrows $f\colon A \to B$, $g\colon B \to C$, where $src(g) = tar(f)$, there exists an arrow $g \circ f\colon A \to C$, 'g following f', which we call the *composite* of f with g.

Identity arrows For any object A, there is an arrow $1_A\colon A \to A$ called the *identity arrow* on A.

We also require the arrows to satisfy the following further axioms:

Associativity of composition. For any $f\colon A \to B$, $g\colon B \to C$, $h\colon C \to D$, we have $h \circ (g \circ f) = (h \circ g) \circ f$.

Identity arrows behave as identities. For any $f\colon A \to B$ we have $f \circ 1_A = f = 1_B \circ f$. △

Evidently, any category of groups as defined in §2.9 will be a category in this sense. And given what we have already said, the objects which are mathematical structures of a particular kind taken together with enough arrows which are structure-respecting maps between them should also satisfy those axioms, and hence should count as forming a category too: I'll give examples in a moment.

(b) Note that – as before, in defining categories of groups – we haven't defined a category as necessarily involving a *set* of objects and a *set* of arrows: we want to allow categories which have too many objects and arrows to form sets (at least on a standard set-theoretic story). Hence my use of a non-committal plural idiom.[1]

Some want to accommodate categories with too many objects/arrows to form sets by instead defining a category as involving a *class* of objects and *class* of arrows. But 'class' in exactly what sense? – for note again the ambiguity pointed out in §3.1. We don't need, at least at this stage in the proceedings, to be tangling with a theory of classes-that-needn't-be-sets but which are still entities in their own right. While if it is only virtual classes which are in play, so talk of classes

[1] I am *not* going out on a limb here! Respected texts which similarly define categories in plural terms – without making a song and dance about it – include those by Awodey (2010, p. 4), Lawvere and Schanuel (2009, p. 21) and McLarty (1992, p. 13). Note, though, that I do revisit the question of how to define categories in §30.1.

here is just a way of talking in the singular about many things at once, then it is clearer to use a frankly plural idiom from the start. So that's what we will do.

Though it won't matter for now, we will however stretch a point and read the plural definition as covering the zero case, allowing for an empty category with no objects and no arrows. We have already countenanced the one-object, one-arrow case, when we allowed the boring category consisting in a single group and its identity homomorphism.

(c) Here are eight(!) more quick remarks on terminology and notation:

(i) The objects and arrows of a category are often called the category's *data*. That's a helpfully non-committal term if you don't read too much into it, and I will occasionally adopt this common way of speaking.

(ii) But note, fixing its objects and arrows is not enough to pin down a category. We need to know too the sources and targets of the arrows. For sources and targets are not intrinsic to arrows – in other words, the same items can appear as arrows in different categories with different sources and targets: for an example see §7.3(b).

We also need to know how arrows compose. Given arrows $f\colon A \to B$ and $g\colon B \to C$, which of the arrows $A \to C$ in the category (if more than one) counts as the composite $g \circ f$?

So in a broader sense of the term, facts about sources, targets and composition are also part of the basic data of a category.

(iii) The label 'objects' for a category's first kind of data is quite standard. But note that, as with the 'objects' of groups, the 'objects' in categories needn't be objects-as-individuals in a type-theoretic sense which contrasts objects with entities like relations or functions. There are perfectly good categories whose objects are actually relations, and other categories where they are functions. And in a category of groups, an object is of course a structure, a group.

(iv) Borrowing the notation $f\colon A \to B$ for 'arrows' in categories is entirely natural given that in many categories arrows *are* (structure-respecting) functions: in fact, that is the motivating case. But again, as we'll soon see, not all arrows in categories are functions or morphisms in the usual sense of that term. We'll find examples where arrows are relations, or matrices, or elements of a monoid, etc. Which is why I generally prefer the colourless 'arrow' to the equally common term 'morphism' for the second sort of data in a category. (Not that that will stop me talking of morphisms or maps when context makes it natural!)

(v) In keeping with the functional paradigm, the source and target of an arrow are frequently called, respectively, the 'domain' and 'codomain' of the arrow – for usually, when arrows are functions, that's what the source and target are. But that usage has the potential to mislead when arrows aren't functions (or aren't functions 'in the right direction', cf. §7.2), which is why I prefer our common alternative terminology.

Note, arrows always have determinate targets/codomains. But compare, for example, the common set-theoretic understanding where functions are identified with their graphs; so understood, functions don't have determinate targets/codomains. Hence: when arrows are thought of as functions they need, rather more carefully, to be thought of as functions-assigned-specific-codomains. I'll return to this point shortly.

(vi) I have adopted the more usual convention for notating a composite arrow.[2] It is again suggested by the functional paradigm. Suppose the two arrows $A \xrightarrow{f} B \xrightarrow{g} C$ are both functions in the ordinary sense, then $(g \circ f)(x) = g(f(x))$. Occasionally, to reduce clutter, we may write simply 'gf' rather than '$g \circ f$'. Note the inversion here: g follows f in our mini-diagram, but is written before f in the notation '$g \circ f$'.

(vii) Initially, we will explicitly indicate which object an identity arrow has as both source and target, as in '1_A'. Again to reduce clutter, we will later allow ourselves simply to write '1' when context makes it clear which identity arrow is in question.

(viii) Finally for now, a very general point about naming categories. As Emily Riehl nicely puts it:

> It is traditional to name a category after its objects; typically, the preferred choice of accompanying structure-preserving morphisms [arrows] is clear. However, this practice is somewhat contrary to the basic philosophy of category theory: that mathematical objects should always be considered in tandem with the morphisms between them. (Riehl 2017, p. 3).

We have in fact already seen that there can be different categories whose objects are the same groups but whose arrows are different selections of the structure-respecting morphisms between them.

5.2 Identity arrows

The definition of a category implies our first mini-result:

Theorem 10. *Identity arrows on a given object are unique; and the identity arrows on distinct objects are distinct.*[3]

Proof. For the first part, suppose A has identity arrows 1_A and $1'_A$. Then applying the identity axioms for each, we immediately have $1_A = 1_A \circ 1'_A = 1'_A$.

For the second part, we simply note that $A \neq B$ entails $src(1_A) \neq src(1_B)$, which entails $1_A \neq 1_B$. □

[2]Some computer scientists writing about categories do things the other way about.

[3]As in this case, the most trivial of lemmas, as well as run-of-the-mill propositions, interesting corollaries, and the weightiest results, will all continue to be labelled 'theorems' without distinction. I did initially try to mark a distinction between, as-it-were, capital-'T' theorems and unexciting lemmas and the rest, but that didn't work out well.

So there's a one-to-one (bijective) correspondence between objects in a category and their identity arrows; and we can pick out the identity arrows by the special way they interact with all the other arrows. Hence we could in principle give a variant definition of categories framed entirely in terms of arrows.[4] But I am not unusual in finding this bit of trickery quite unhelpful. A central theme of category theory is indeed the idea that we should probe the objects in a category by considering the arrows between them; but that's no reason to write the objects out of the story altogether.

5.3 Monoids and preordered collections

Let's continue by looking at two simple but instructive types of categories, one algebraic, one order-theoretic.

(a) We have already met the example of various categories of groups. But it is worth thinking now about cases where the algebraic structure is cut nearer to the bone. Consider, say, the finite strings of symbols from some given alphabet, including the limiting case of the empty string, together with the operation of concatenation. This operation is evidently associative, $s_1 ^\frown (s_2 ^\frown s_3) = (s_1 ^\frown s_2) ^\frown s_3$. And concatenating with the empty string leaves us where we were, so the empty string acts like an identity element for concatenation. This structure gives us a paradigm example of a *monoid* – which is, so to speak, a group minus the requirement for inverses. And a monoid homomorphism is then a function which respects monoid structure.

More carefully, we have:[5]

Definition 15. The objects M with a distinguished object e, equipped with a binary operation $*$, form a *monoid* $(\mathrm{M}, *, e)$ – M for short – iff

(i) the binary operation $*$ maps monoid-objects to monoid-objects, i.e. for any $x, y \mathrel{\varepsilon} \mathrm{M}$, $x * y \mathrel{\varepsilon} \mathrm{M}$;
(ii) $*$ is associative, i.e. for any $x, y, z \mathrel{\varepsilon} \mathrm{M}$, $(x * y) * z = x * (y * z)$;
(iii) $e \mathrel{\varepsilon} \mathrm{M}$, and e acts as a monoid unit or identity, i.e. for any $x \mathrel{\varepsilon} \mathrm{M}$, $x * e = x = e * x$.[6]

Further, a *monoid homomorphism* from M, i.e. $(\mathrm{M}, *, e)$, to N, i.e. $(\mathrm{N}, \star, d)$, is a function f defined over M with values among N such that:

(i) for every $x, y \mathrel{\varepsilon} \mathrm{M}$, $f(x * y) = fx \star fy$,
(ii) $f(e) = d$. △

---

[4] For an account of how to do this, see Adámek et al. (2009, pp. 41–43).

[5] On notation, in case you have skipped forward: upright letters like 'M' are plural variables, and '$x \mathrel{\varepsilon} \mathrm{M}$' can be read *x is among (the objects)* M.

[6] A factoid: any non-empty collection of objects can trivially be equipped with a binary operation to give a monoid. The one-object case is obvious. So suppose we have two or more objects, call them $0, e, a, b, c, \ldots$. Then define the operation $*$ so that $e * o = o * e = o$ for any object o (so e is our identity) while $o * o' = 0$ for any objects o, o' other than e.

By contrast, it isn't at all trivial that every collection of objects can be equipped with a binary operation making it a group. That is actually an equivalent of the Axiom of Choice.

(b) It is evident that monoid homomorphisms $f\colon M \to N$ and $g\colon N \to O$ compose to give a homomorphism $g \circ f\colon M \to O$. Composition of homomorphisms is associative. And the identity function on M is a homomorphism $f\colon M \to M$ which acts as an identity with respect to composition.

Hence, as with groups, some monoids together with enough homomorphisms will form a category – where by 'enough' we mean as before that (i) compositions of homomorphisms in the category are also in the category, and (ii) the identity homomorphism for each monoid in the category is also present.

So far, that makes no assumptions about where categories of monoids are to be found. But assume now that we are working in some sufficiently capacious universe – by default let's suppose a suitable universe of sets – which contains (implementations for) all the monoids we want together with (implementations for) their homomorphisms. We can then sensibly say:

(C1) Mon is the category whose objects are all the monoids and whose arrows are all the monoid homomorphisms (living in our chosen universe).

Fine print: yes, we should emphasize again that what we have in Mon are strictly speaking implementations of monoids and their homomorphisms (cf. §4.24.2). But still, these proxies for monoids and their homomorphisms can count perfectly well as objects and arrows in a category. In particular, note that arrows in a category don't have to be kosher functions. So, in a slogan: Mon will be a genuine category of proxies for monoids, and not a proxy category!

(c) Next, an example involving ordered objects; and again we'll cut structure to the bone by considering the simplest case, preorderings.

For convenience, let's use 'collection' to give us a way of speaking in the singular about perhaps many things – see the end of §3.1. Then we can say

Definition 16. The objects P equipped with a relation $\preccurlyeq$ form a *preordered collection* $(\mathrm{P}, \preccurlyeq)$ iff, for all $a, b, c \in \mathrm{P}$,

(i) if $a \preccurlyeq b$ and $b \preccurlyeq c$, then $a \preccurlyeq c$,
(ii) $a \preccurlyeq a$.

A monotone map $f\colon (\mathrm{P}, \preccurlyeq) \to (\mathrm{Q}, \sqsubseteq)$ between such preordered collections is then defined to be a function f from the objects P into Q which respects order, i.e. such that for any $a, b \in \mathrm{P}$, if $a \preccurlyeq b$, then $fa \sqsubseteq fb$. △

Monotone maps between preordered collections will compose to give monotone maps; and the identity map on some preordered objects gives rise to an identity monotone map. Evidently, then, we can have categories with the following kind of data:

(1) objects: various preordered collections $(\mathrm{P}, \preccurlyeq)$,
(2) arrows: enough monotone maps between these various objects,

where (and we won't keep repeating this) 'enough' means the maps are closed under composition and each object gets its own identity map.

OK, now assume again that we are working in some capacious universe where collections are treated as sets. So there we can have a *set* $\underline{P}$ whose members are (perhaps suitable proxies for) the objects P, and a corresponding set-implementation $\leqslant$ for the preorder relation $\preccurlyeq$. Then we can sensibly say

(C2) Preord is the category whose objects are all the preordered sets $(\underline{P}, \leqslant)$ and whose arrows are all the monotone set-functions between them (living in our chosen set universe).

5.4 Some rather sparse categories

(a) So far, so very unsurprising.

However, let's remark straight away that monoids can get into the story in a second way. As we've seen, monoids as objects taken together with enough monoid homomorphisms as arrows can form a category. However, any single monoid by itself can also be thought of giving rise to a category. Here's how:

(C3) Take any monoid $(\mathrm{M}, *, e)$. Then define a corresponding category M whose data is as follows:

(1) M's sole object is some arbitrary entity – choose whatever you like, it *doesn't* have to be among the monoid's objects, and dub it '$\bullet$';

(2) Then any object $a \in \mathrm{M}$ counts as an M-arrow $a\colon \bullet \to \bullet$ (in other words, we put $src(a) = tar(a) = \bullet$). Composition of arrows $a \circ b$ is defined to be the monoid product $a * b$, and the identity arrow $1_\bullet$ is defined to be the monoid identity e.

It is then immediate that the axioms for a category are satisfied (check this!).

Note in this case, since the 'object' in the category M can be anything you like, it needn't be an object in any ordinary sense (let alone be a structure). And unless the objects of the original monoid happen to be functions, the arrows of the associated category M will also not be functions or morphisms or maps in any ordinary sense. So this sort of single-monoid-as-a-category won't usually be *anything* like a 'structure of structures'.

Note too that there is a sort of converse to (C3). Any one-object category M gives rise to an associated monoid built from M's arrows, with multiplication in the associated monoid being composition of arrows. Hence we can think of many-object categories as, in a sense, generalizing from the case of the one-object categories which are tantamount to monoids.

(b) Similarly, while we can put preordered collections and the monotone maps which interrelate them together to form a category, we can also think of a single collection of objects equipped with a preorder as itself giving us a category. Here's how:

(C4) Take any preordered collection $(\mathrm{P}, \preccurlyeq)$. Then define a corresponding category P whose data is as follows:

(1) P's objects are the objects P again;

(2) there is a (single) P-arrow from A to B if and only if $A \preccurlyeq B$ – this arrow might as well be identified as a pair $\langle A, B\rangle$ (according to *some* pairing scheme), which is assigned the 'source' A and 'target' B. We define composition by putting $\langle B, C\rangle \circ \langle A, B\rangle = \langle A, C\rangle$. Take the identity arrow 1_A to be $\langle A, A\rangle$; there is always such an arrow since $\preccurlyeq$ is reflexive.

It is immediate that, so defined, the arrows for P satisfy the associativity and identity axioms, so we do have another category here (check this!). And again, this isn't a category comprising structures and structure-respecting maps.

Conversely, any category with objects Obj and where there is at most one arrow between two objects $A, B \in$ Obj can be regarded as a preordered collection (Obj, $\preccurlyeq$), where $A \preccurlyeq B$ just in case there is an arrow from A to B in the category. It is therefore natural to say

Definition 17. A *preorder category* is a category such that, for any objects A and B, there is at most one arrow from A to B. △

Hence we can think of many-arrow categories as, in a sense, a generalization from the case of preorder categories.

(c) Monoids-as-categories and preordered-collections-as-categories can give us very small categories with few objects and/or arrows. And here are some more sparse categories.

(C5) For any objects we take, we get the *discrete category* on those by adding as few arrows as possible, i.e. just an identity arrow for each of the objects we started with.

As noted before, we can for convenience allow the empty category, with zero objects and zero arrows. Otherwise, the smallest discrete category is **1** which has exactly one object and one arrow (the identity arrow on that object). Let's picture it in all its glory!

(C6) And having mentioned the one-object category **1**, here's another very small category, this time with two objects, the necessary identity arrows, and one further arrow between them. We can picture it like this:

Call this category **2**.

We could think of this category as arising from the von Neumann ordinal 2, i.e. the set $\{\emptyset, \{\emptyset\}\}$; take the ordinal's members as objects of the category, and let there be an arrow between objects when the source is a subset of the target. Other von Neumann ordinals, finite and infinite, similarly give rise to other categories.

But hold on! Should we in fact talk about *the* category 1 (or *the* category 2, etc.)? Won't different choices of object make for different one-object categories, etc.? Well, yes and no! Suppose we are working in our chosen set universe. There can be different choices of a single object (a set X) equipped with an identity arrow (set-function from X to X) to play the role of 1 – *but they will be indiscernible from within category theory.* So as far as category theory is concerned, they are all 'essentially the same' – in exactly the same spirit as e.g. different Klein four-groups are 'essentially the same' in group theory. We will want to return to this point.

5.5 More categories

Note, I'm already falling into a notational habit which I will try to stick to pretty systematically. I'll use a sans serif font (as in 'Grp', 'Mon', etc.) for names of categories, and will also use the same font for informal variables for categories (as in 'C', 'J' etc.).

Let's continue, then, our list of varieties of category, first generalizing from our basic examples in §5.3, and then adding some geometric and other categories. And now, both for brevity's sake and also to stay in step with presentations you'll find elsewhere, in most cases we will jump straight to the maximal version living in our favourite mathematical universe, which we'll take by default to be a universe of sets (so this maximal version will be the category which stands to other instances of the same general sort as e.g. Grp does to other categories of groups).

Categories of monoids and categories of groups are only the first of a whole family of cases, where the object-data are algebraic structures and the arrows are the homomorphisms respecting the relevant amount of structure. For example, we also have

(C7) Ab is the category whose objects are all the (implementations of) abelian groups, and whose arrows are all the (implementations of) group homomorphisms living in our default universe.

(C8) Ring is the category of rings, whose objects are predictably enough all (implementations of) rings and whose arrows are all (implementations of) ring homomorphisms living in our default universe.

(C9) And Bool is the category of Boolean algebras and homomorphisms between them which preserve binary meets and joins, inverses and top and bottom elements. Likewise CABool is the category of complete atomic Boolean algebras and homomorphisms between *them* which preserve arbitrary meets and joins, etc. (It would be very boring to keep on repeating the reference to implementations of structures and morphisms between them in our favourite universe – so now take this as read, here and below!).

We similarly have further categories of ordered objects. Enrich the notion of a preorder, take as structures objects-equipped-with-the-richer-order, take enough order-respecting functions as arrows, and we get another kind of category. For one example,

(C10) Pos is the category whose objects are all the collections-equipped-with-some-partial-order (that's a preorder $\preccurlyeq$ such that if $A \preccurlyeq B$ and $B \preccurlyeq A$ then $A = B$), and the arrows are all order-respecting maps again. As implemented in our favoured universe of sets, this will be the category of posets.

Now for another paradigm type of case, namely geometric categories (even more central to the original development of category theory than the cases of algebraic categories or order categories).

(C11) Top is the category with

(1) objects: all the topological spaces;
(2) arrows: the continuous maps between spaces.

(C12) Met is also a category. This has

(1) objects: metric spaces, which we can take to be a set of points S equipped with a real metric d;
(2) arrows: the non-expansive maps, where – in an obvious notation – $f\colon (S, d) \to (T, e)$ is non-expansive iff $d(x, y) \geq e(f(x), f(y))$.

(C13) Vect_k is a category with

(1) objects: vector spaces over the field k (each such space comprising vectors equipped with vector addition and multiplication by scalars in the field k);
(2) arrows: linear maps between the spaces.

And finally in this section, let's have a logical example.

(C14) Suppose L is a first-order formal language (the details don't matter). Then there is a category of propositions Prop_L with

(1) objects: propositions, closed sentences $X, Y, \ldots$ of the formal language;
(2) arrows: there is a (unique) arrow from X to Y iff $X \vDash Y$, i.e. X semantically entails Y.

The reflexivity and transitivity of semantic entailment means we get the identity and composition laws which ensure that this is a category.

5.6 The category of sets

(a) Assuming that we are thinking about our inclusive mega-categories as living in some universe of sets, the monoids which are objects of Mon comprise sets equipped with not-very-much structure. Likewise for the preordered sets in Preord.

Going then in one direction, we get various categories whose objects are sets equipped with some richer structure and whose arrows are functions (ok, better, functions with assigned codomains) constrained to respect this richer structure. Going in the other direction, we get categories of naked sets – i.e. categories whose objects are simply sets (equipped with no additional structure at all) and whose arrows are functions between these sets (any old functions so long as they are closed under composition, and we include the relevant identity functions: there is no requirement that functions respect structure because there is no structure to respect).

And here's the limiting case, the most inclusive category of sets:

(C15) Set is the category with

(1) objects: all sets in our favoured universe (characteristically thought of as a universe of pure sets which is a model of ZFC or some extension of it),

(2) arrows: for any sets X, Y, in our universe, every (total) function f with source X and target Y is an arrow.

There's an identity function on any set. And functions $f\colon A \to B$, $g\colon B \to C$ (where the source of g is the target of f) always compose. And so the axioms for being a category are satisfied.

NB: We haven't determinately fixed our default background universe of (pure) sets, and hence haven't determinately fixed which category Set is. But, for the moment at any rate, we can leave that up for grabs.

(b) Some elementary comments:

(i) Note that the arrows in Set, like any arrows, must come with determinate targets/codomains. But we reminded ourselves in §4.2 that the standard way of treating functions set-theoretically is simply to define a function f as its *graph* $\hat{f}$, i.e. the set of pairs $\langle x, y\rangle$ such that $f(x) = y$. And this definition is lop-sided in that it fixes the function's source/domain, the set of first elements in the pairs, but it doesn't determine the function's target. So we can't simply identify an arrow in Set with a function's graph.

Perhaps set theorists themselves ought to define a set-function $f\colon A \to B$ as a triple $\langle A, \hat{f}, B\rangle$. But anyway, that's how category theorists should officially regard arrows $f\colon A \to B$ in Set, and in other categories too where arrows are functions.

(ii) We should perhaps also remind ourselves why there *is* an identity arrow for $\emptyset$ in Set. Vacuously, for *any* target set Y, there is exactly one set-function

$f\colon \emptyset \to Y$, i.e. the one whose graph is the empty set. Hence in particular there is a function $1_\emptyset\colon \emptyset \to \emptyset$.

(iii) Note that in Set, the empty set is the one and only set such that there is exactly one arrow *from* it to any other set. This gives us a particularly simple example of how we can characterize a significant object in a category not by its internal constitution, so to speak, but by what arrows it has to and from other objects.

For another example, note that we can define singletons in Set by relying on the observation that there is exactly one arrow from any set *to* a singleton, and conversely if there is exactly one arrow from any set to S, then S must be a singleton (why?).

(iv) So now choose a singleton $\{\bullet\}$, it won't matter which one (treat the bullet symbol here as a wildcard). Call this chosen singleton '1'. And consider the possible arrows (i.e. set-functions) from 1 to A.[7]

We can represent the arrow from 1 to A which sends the element of the singleton 1 to $x \in A$ as $\vec{x}\colon 1 \to A$ (the over-arrow here is simply a helpful reminder that we are notating an arrow). Then there is evidently a one-to-one correspondence between these arrows $\vec{x}$ and the elements $x \in A$. So talk of such arrows $\vec{x}$ is available as a category-speak surrogate for talking about elements x of A.

More on this sort of thing in §9.3: but we have another glimpse ahead of how we might trade in talk of sets-and-their-elements for categorial talk of sets-and-arrows-between-them.

(c) A quick note on some terminology which I think I should at least mention here in passing. You will find that categories like Grp, Preord and Set whose objects are sets, perhaps equipped with some structure, and whose arrows are structure-respecting set-functions, are often called *concrete* categories. As we have seen, lots of categories are not concrete in *this* informal sense – for example, neither a monoid-as-category nor a preordered-collection-as-category will usually count.[8]

(d) But what if you eventually want to work in some generous enough mathematical universe which is not conventionally set-theoretic? You'll still want a story about unstructured pluralities and maps between them, and will still want to consider the inclusive mega-category of *those*. By mild abuse of notation, you could continue to call this category 'Set' so that your category theory can unfold in a standard-looking way. But let's not fuss about this sort of thing right now. As I said, we'll keep things marching in step with other presentations by assuming by default that our mega-categories are indeed living in some universe of (pure) sets as conventionally understood.

[7] We are overloading notation – here '1' is a special object, while in other contexts '1' is a special arrow, an identity arrow. You'll need to get used to this sort of thing, where we rely on context to disambiguate shared notations for objects and arrows.

[8] Later, in §27.5, we'll note a sharper technical notion of concrete category which, however, doesn't align well with the informal idea.

5.7 Yet more examples

(a) Here are another four inclusive mega-categories, variants on the category of sets:

(C16) There is a category FinSet whose objects are sets with finitely many members, and whose arrows are the set-functions between such objects.

(C17) FinOrd is the category whose objects are the finite ordinals – you can take these to be the von Neumann ordinals – and whose arrows are the set-functions between them.

(C18) Pfn is the category of sets and *partial* functions. Here, the objects are all the sets again, but an arrow $f\colon A \to B$ is a function which is not necessarily everywhere defined on A (one way to think of such an arrow is as a total function $f\colon A' \to B$ where $A' \subseteq A$). Given arrows-qua-partial-functions $f\colon A \to B$, $g\colon B \to C$, their composition $g \circ f\colon A \to C$ is defined in the obvious way, though you need to check that this succeeds in making composition associative.

(C19) $\mathsf{Set}_\star$ is the category of 'pointed sets' with

(1) objects: all the non-empty sets, with each set A having a distinguished member $\star_A$;

(2) arrows: all the total functions $f\colon A \to B$ which map $\star_A$ to $\star_B$, for any non-empty sets A, B.

(b) Let's now mention another kind of category, which arises when we give sets not a distinguished member but some distinguished endofunction (i.e. function whose domain and codomain are the same). We can think of a set X equipped with an endofunction $f\colon X \to X$ as a discrete dynamical system, where at each tick of the clock an element $x \in X$ gets sent on to $fx \in X$. Then we say:

(C20) $\mathsf{Set}^{\circlearrowright}$ is the category of discrete dynamical systems:

(i) objects: any (X, f) where X is a set equipped with an endofunction $f\colon X \to X$;

(ii) arrows: an arrow from $(X, f\colon X \to X)$ to $(Y, g\colon Y \to Y)$ is any function $j\colon X \to Y$ such that $j \circ f = g \circ j$.

The idea is that successive applications of f in X get mapped by j to successive applications of g in Y – so starting from some point in X, it doesn't matter whether we update according to f, and then use j to map across to Y, or first jump across to Y using j and then update according to g. You need to check that the composition of arrows in $\mathsf{Set}^{\circlearrowright}$, defined in the natural way, works as it should.

Generalizing, there are also categories whose objects are sets equipped with multiple endofunctions. In particular, we can choose a monoid $M = (\mathrm{M}, *, e)$ and form the category of M-sets. Here, an M-set is a set X equipped with a family of endofunctions $f_m\colon X \to X$, one for each object $m \varepsilon \mathrm{M}$, such that the

f_m combine like the elements of the monoid M: so $f_e = 1_X$ and $f_m \circ f_n = f_{m*n}$. Call such an M-like family f_M. Then

(C21) For a given monoid $M = (\mathrm{M}, *, e)$, the corresponding category M-Set is the category with

(1) objects: any (X, f_M) where X is a set and f_M is an M-like family of functions $f_m \colon X \to X$;
(2) arrows: an arrow $j \colon (X, f_M) \to (Y, g_M)$ is any 'equivariant' function $j \colon X \to Y$ which respects the action of the corresponding operations on X and Y. In other words, for each $m \varepsilon \mathrm{M}$, if the operation f_m on X sends x to x', then the corresponding operation g_m on Y will send $j(x)$ to $j(x')$ – i.e. $j \circ f_m = g_m \circ j$.

As you would expect, composition of arrows is ordinary functional composition. The identity arrow on (X, f_M) is the function 1_X.

Here's a special case which will crop up in a number of later examples.[9] Consider the two-object monoid M_2 whose objects are the numbers $0, 1$, and whose monoid operation is multiplication (so 1 is the monoid unit). Then an object of the corresponding category of M_2-sets will be a set X equipped with an endomorphism f_0 (where $f_0 = f_0 \circ f_0$) and an endomorphism f_1 which will be the identity 1_X. Since the identity arrow is automatically supplied, we can more briskly define the category like this:

(C22) The category M_2 is the category with

(1) objects: any (X, f) where X is a set and the function $f \colon X \to X$ is such that $f \circ f = f$;
(2) arrows: the arrows $j \colon (X, f) \to (Y, g)$ are the equivariant functions $j \colon X \to Y$, i.e. the functions such that $j \circ f = g \circ j$.

(c) Now let's have a couple more examples where the arrows in a category are *not* functions:

(C23) The category Rel again has naked sets as objects, but this time an arrow $A \to B$ in Rel is (not a function but) any relation R between A and B. We can take this officially to be a triple $(A, \hat{R}, B)$, where $\hat{R} \subseteq A \times B$ is R's extension, the set of pairs $\langle a, b \rangle$ such that aRb.[10]

The identity arrow on A is then the diagonal relation whose extension is $\{\langle a, a \rangle \mid a \in A\}$.

And $S \circ R \colon A \to C$, the composition of arrows $R \colon A \to B$ and $S \colon B \to C$, is defined by requiring $a\, S \circ R\, c$ if and only if $\exists b(aRb \wedge bSc)$. It is easily checked that composition is associative.

[9] Inspired by Goldblatt (1984).

[10] Note, we need the arrow to be the triple, as the extension $\hat{R}$ by itself wouldn't determine either a source or a target. Note too that we must allow the case where $\hat{R}$ is empty and the arrow is a triple of the form $(A, \emptyset, B)$.

(C24) The category Mat has natural numbers as its objects, and an arrow $m \to n$ is an $m \times n$ matrix of real numbers, the composition of arrows defined as ordinary matrix multiplication.

Finally, let's recall that there are various kinds of graphs[11], depending on whether we allow more than one edge between nodes, whether edges are directed, and whether we allow edges looping round from a node back to itself. But here let's take a graph to comprise zero or more nodes together with zero or more directed edges between them, loops allowed. So the objects/arrows of any category can be thought of as constituting the nodes/edges of a special sort of graph in this sense, one where every node has its own identity loop, and where edges compose (i.e. if there is an edge from A to B and an edge from B to C, then there is also an edge from A to C).[12]

Now, in Part II of these notes, we'll have to ask whether it makes sense to talk about a category of all categories. But it makes sense to talk about a maximal category of graphs living in our default universe of sets. Thus

(C25) The category Graph has the following data:

(1) objects: all the graphs G living in our favoured universe (so nodes and edges are identified as sets).
(2) arrows: all the graph homomorphisms – these are functions f between graphs G_1 and G_2, which respect graph structure.

In more detail, a graph homomorphism f from G_1 to G_2 will have two components, f_N acting on nodes, and f_E acting on edges, where these components fit together in the obvious way. So if f_N maps the nodes a and b to $f_N(a)$ and $f_N(b)$ respectively, then f_E maps an edge from a to b to an edge from $f_N(a)$ to $f_N(b)$.

(d) And that will surely do for the moment as an introductory list. There most certainly is no shortage of categories of various kinds, then!

In fact, by this stage, you might very reasonably be wondering whether it isn't far *too* easy to be a category. If such very different sorts of structures as (i) a particular very small monoid, and (ii) a whole universe of topological spaces and their continuous maps, equally count as categories, then how much mileage can there possibly be theorizing in general about categories and their interrelations?

Well, that's exactly what we hope to find out over the coming chapters.

[11]'Graphs' in the sense of graph theory *of course*, not in the sense of extensions of functions.

[12]'So is category theory just a department of graph theory?' Well, a glance at e.g. Béla Bollobás's classic 1998 text shows how very different the concerns of graph theory and category theory are. One key reason is that, for a graph in general, where edges don't always compose, we get all kinds of highly non-trivial extremal problems about the length of paths between nodes. And as Bollobás remarks, "Extremal problems are at the very heart of graph theory."

6 Diagrams, informally

We can diagrammatically represent objects related by arrows in a very natural way – we've already seen some mini-examples. And in particular, we can represent facts about the equality of arrows using so-called commutative diagrams. We'll soon be using such diagrams a great deal: so I'd better make some headline points about them straight away. These points are important enough to deserve a brief chapter to themselves.

6.1 Diagrams, in two senses

Talk of diagrams is commonly used in three related ways. Later, in §28.1, we will give a sharp characterization of a more technical notion of a diagram. But for the moment, we can be informal and work with two looser but more immediately intuitive notions. Firstly:

Definition 18. A *representational diagram* is a directed graph with nodes representing objects from a given category C, and directed edges between nodes (drawn as arrows!) which represent arrows of C. Nodes and edges will normally be appropriately labelled, to make it clear what is being represented.

Two different nodes in a diagram can be joined by zero, one, or more directed edges. There can also be edges looping round from a node to itself, representing the identity arrow on an object or representing some other arrow whose source and target is the same.

A directed edge (drawn arrow) labelled 'f' going from the node labelled 'A' to the node 'B' of course represents the arrow $f\colon A \to B$ of C. △

And then, relatedly:

Definition 19. A *diagram in a category* C is what is represented by a representational diagram – in other words, it will be some C-objects and some C-arrows between them. △

Note, diagrams (in either sense) needn't be *full*. That is to say, a diagram-as-a-picture need only represent *some* of the objects and arrows in a category; and a diagram-as-what-is-pictured need only be a portion of the whole category in question.

6.2 Commutative diagrams

(a) Within a representational diagram, we may be able to follow a directed path through more than two nodes, walking along the connecting directed edges. So a path in a representational diagram from a node labelled 'X' to a node labelled 'Y' might look like this:

$$X \xrightarrow{f} Z_1 \xrightarrow{g} Z_2 \xrightarrow{h} Z_3 \xrightarrow{j} Y$$

This path-as-picture represents a connected chain of arrows (with the target of one arrow being the source of the next). The axiom about composition tells us that, in the represented category, there will also be an arrow $j \circ (h \circ (g \circ f))$ from the object X to the object Y: we will say that this arrow is obtained by composing along the path.

Two points. First, because of the associativity of composition we needn't actually worry about bracketing here, and can simply describe that composite as $j \circ h \circ g \circ f$ (or even plain $jhgf$). From now on, then, we freely insert or omit brackets in writing composites, doing whatever promotes local clarity.

Second, in §5.1(c) we explained the rationale for our notational choice for the order in which we write the composite of two arrows. But note again that this does mean that the components in the notation '$j \circ h \circ g \circ f$' occur in the opposite order to the path diagram.

(b) We now say – as our initial shot –

Definition 20. A representational diagram *commutes* iff, for any nodes X and Y and any two directed paths from X to Y, the arrow you obtain by composing along the first path is equal to the arrow you obtain composing along the second path.

Relatedly, a diagram in a category commutes iff it can be represented by a commutative diagram – i.e., iff composites taken along two chains of arrows between a source and final target are always equal.[1] △

Hence, for example, the associativity axiom $h \circ (g \circ f) = (h \circ g) \circ f$ can be presented by saying that diagrams like the following always commute:

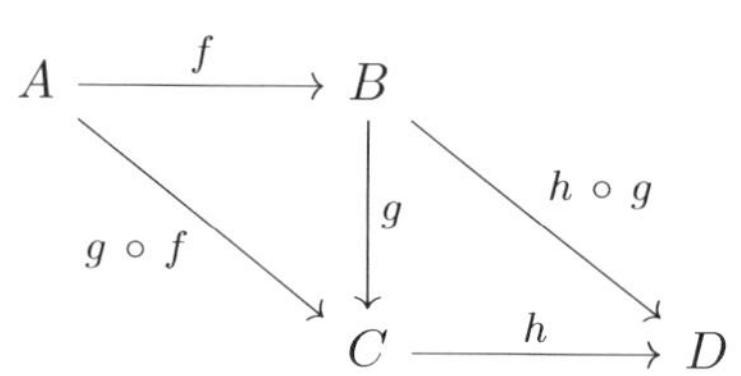

Each of the two triangles here commutes by the definition of composition. And then by associativity we can paste the triangles together to get a larger commutative diagram.

[1]Arbib and Manes (1975, p. 2) put it nicely: "*commutare* is the Latin for *exchange*, and we say that a diagram commutes if we can exchange paths, between two given points, with impunity."

Here's another example. If the two squares on the left commute, then by associativity we can paste them together along the common arrow to get the larger commutative diagram on the right:

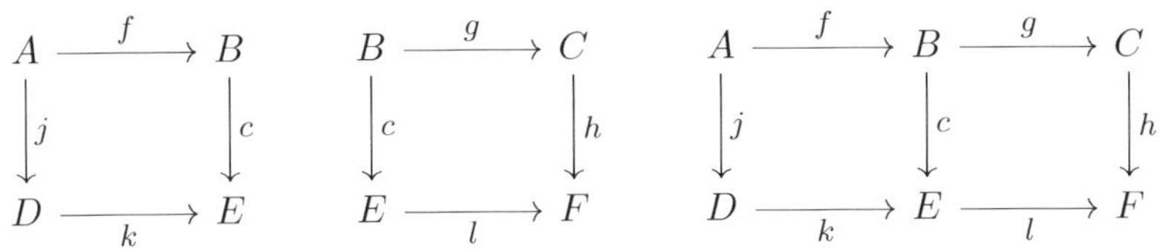

To check this, note that

$$h(gf) = (hg)f = (lc)f = l(cf) = l(kj)$$

with the equations holding alternately by associativity and by the assumed commutativity of the squares as we chase around the right-hand diagram.

These two cases illustrate a general claim: because of associativity, a diagram commutes if each minimal polygon in the diagram commutes. That can be shown by induction on the number of polygons. But since we won't need to invoke the claim in full generality, we need not pause to prove it. (Particular examples later where we claim to build up a commuting diagram from commuting triangles or squares can be checked piecemeal.)

A third example diagram to illustrate a further point:

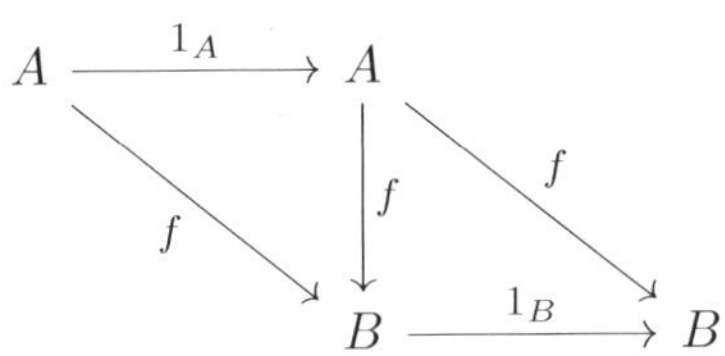

The claim that this diagram commutes neatly encapsulates (an instance of) the basic categorial axiom that identity arrows behave as identities. Note carefully, though, that the fact that there are two occurrences of the label 'A' (or 'B') in the diagram does *not* mean that we are dealing with two separate duplicates of the designated object A (or B). Likewise the three occurrences of 'f' all designate one and the same arrow. The diagram is about only two objects and three arrows. (Motto: duplicate representations do not mean represented duplicates.)

(c) We will meet many more examples of commutative diagrams in the coming chapters, so I won't give more illustrations yet. But here are two immediate points to note.

First, I've just been a bit fussy in explicitly distinguishing the two ideas, a diagram-as-representation, and a diagram-as-what-is-represented. But having made the distinction, we will rarely need to bother about it, and can let context determine a sensible reading of informal claims about diagrams.

Second, do note – this is obvious but important! – that merely drawing a diagram with different routes from X to Y in the relevant category emphatically *doesn't* always mean that we have a commutative diagram: the identity of the composites along the paths in each case needs to be argued for (e.g. by showing that each component subdiagram commutes).

6.3 A revised definition

OK, I have given a basic definition of a commutative diagram. But it turns out to be useful to tweak it very slightly. Why so?

Well, later we will quite often be encountering 'fork' diagrams like this:

$$E \xrightarrow{\ e\ } A \underset{g}{\overset{f}{\rightrightarrows}} B$$

And it is quite convenient to count such a diagram as commuting so long as $f \circ e = g \circ e$, *without* also requiring the 'parallel' arrows from A to B to be equal. So, on our tweaked definition, we won't require arrows along *every* path between any two nodes to be equal: instead, we'll now say this:

Definition 20*. A representational diagram *commutes* iff, for any nodes X and Y and any two directed paths from X to Y (so long as at least one of these paths contains more than one edge), the arrow you obtain by composing along the first path is equal to the arrow you obtain composing along the second path.

Relatedly, a diagram in a category commutes iff it can be represented by a commutative diagram. △

Of course nothing important hangs on explicitly tweaking the definition this way: some authors do, most don't.[2]

[2]For some that do, see Arbib and Manes (1975, p. 4) and Roman (2017, p. 17).

7 Categories beget categories

We already know that categories are very plentiful. And in this chapter we are going to introduce yet more, by describing a number of general constructions which give us new categories from old. We'll meet additional constructions later, but these first examples will suffice to be going on with.

7.1 Subcategories, products, quotients

Three familiar ways of getting new widgets from old involve extracting sub-widgets, forming products of widgets, and quotienting by a suitable equivalence relation. We met these sorts of constructions on groups in §2.3. And we can do the same constructions with categories.

(a) We get a new subgroup from an old group by slimming down the group while retaining enough group structure. Similarly, the simplest way of getting a new category is by slimming down an old one while retaining enough categorial structure:

Definition 21. Given a category C, if S consists of the data

(1) objects: some or all of the C-objects,
(2) arrows: some or all of the C-arrows,

subject to the conditions

(i) any S-arrow has the same source and target it has as a C-arrow,
(ii) for each S-object A, the C-arrow 1_A is also an S-arrow,
(iii) for any S-arrows $f\colon A \to B$, $g\colon B \to C$, the C-arrow $g \circ f\colon A \to C$ is also an S-arrow,

then, with composition of arrows in S defined as in the original category C, S is a *subcategory* of C. △

Plainly, the conditions in the definition – in particular, still containing the identity arrows for the remaining objects and being closed under composition – are there to ensure that the slimmed-down S is still a category.

Many cases where we prune an existing category will leave us with constructions of no particular concern. Other cases can be more interesting:

(1) All categories of groups-implemented-as-sets will be subcategories of Grp.

Note that, as we've set things up, not *every* category of groups will be a subcategory of Grp – for we don't rule out that suitably structured families of groups and their homomorphisms can live outside the particular universe of sets in which all the objects of Grp are to be found. Though by design Grp should contain *copies* of all the groups we want.

By contrast, we can say outright

(2) Ab is a subcategory of Grp.

For having fixed on a background universe as a home for both Ab and Grp, all the abelian groups implemented in that universe will of course be among the groups living in that universe. Similarly to that case,

(3) FinOrd is a subcategory of FinSet;

(4) FinSet is a subcategory of Set;

(5) Set is a subcategory of Pfn.

And trivially,

(6) The discrete category on the objects of C is a subcategory of C for any category.

So, we can shed objects and/or arrows in moving down from a category to a subcategory. In examples (5) and (6) we keep all the objects but shed some or all of the non-identity arrows. While in cases (2), (3) and (4) we drop some objects while keeping all the existing arrows between the remaining objects, and there is a standard label for such cases:

Definition 22. If S is a subcategory of C where, for all S-objects A and B, the S-arrows from A to B are *all* the C-arrows from A to B, then S is said to be a *full subcategory* of C. △

(b) It might help to fix ideas if I briefly pause to emphasize that – even if we are thinking of our categories as living in a universe of sets – a category like Grp, for example, is *not* a subcategory of Set.

Why not? Suppose we implement a group G set-theoretically as a triple $\langle \underline{G}, *, e\rangle$ where $\underline{G}$ is the underlying set of (proxies for) G's objects, while $*$ corresponds to the group operation and e is an element of $\underline{G}$. Then, an arrow between $\langle \underline{G}, *, e\rangle$ and $\langle \underline{G}', *', e'\rangle$ in Grp is a suitable function $f\colon \underline{G} \to \underline{G}'$: but that's not an arrow between $\langle \underline{G}, *, e\rangle$ and $\langle \underline{G}', *', e'\rangle$ as objects in Set.

(c) Next, the definition of products of categories is entirely predictable (I add this now for the record, though the idea won't be of interest to us until Part II):

Definition 23. If C and D are categories, then a product category C × D is such that

(1) Its objects are ordered pairs $\langle C, D\rangle$ where C is a C-object and D is a D-object;

(2) Its arrows from $\langle C, D\rangle$ to $\langle C', D'\rangle$ are all the pairs $\langle f, g\rangle$ where $f\colon C \to C'$ is a C-arrow and $g\colon D \to D'$ is a D-arrow.

(3) We define the identity arrow on $\langle C, D\rangle$ by putting $1_{\langle C,D\rangle} = \langle 1_C, 1_D\rangle$;

(4) Composition is then defined componentwise by putting $\langle f, g\rangle \circ_{\mathsf{C}\times\mathsf{D}} \langle f', g'\rangle = \langle f \circ_{\mathsf{C}} f', g \circ_{\mathsf{D}} g'\rangle$.[1] △

Of course, this definition requires us to have suitable pairing schemes in play, one for the relevant objects and one for the relevant arrows: but assuming those are available, it is immediate that this well-defines a sort of category.

(d) Thirdly, and more interestingly, consider quotients. Following closely what we said about quotients for groups in Defn. 7, we can say:

Definition 24. (1) If C is a category, then the relation $\sim$ is a *congruence* on its arrows iff it is an equivalence relation which respects composition.

That is to say, a congruence $f \sim g$ is an equivalence such that if, $f \sim g$, then (i) f and g share the same source and target (ensuring that equivalent arrows can compose in the same way), and (ii) $f \circ h \sim g \circ h$ and $j \circ f \sim j \circ g$ whenever the composites are defined.

(2) Suppose C is a category and $\sim$ is a congruence on its arrows. And suppose we have a quotient scheme for $\sim$, which sends an arrow f (and its $\sim$-equivalents) to $[f]$. Then $\mathsf{C}/\sim$ is the *quotient category* whose objects are the same as those of C and whose arrows are all the $[f]$ for f in C, with $[f]$ assigned the same source and target as an arrow in $\mathsf{C}/\sim$ as f has in C. △

We've defined the notion of congruence so that it becomes trivial to check that $\mathsf{C}/\sim$ actually is a category.

For a natural example, take the category Top. And consider the congruence $f \sim g$ which holds when f and g are arrows (i.e. continuous maps) between the same topological spaces which can be continuously deformed into each other (i.e. when there is a so-called homotopy between them). Then $\mathsf{Top}/\sim$ is the important homotopy category hTop.

(e) Recall from §5.4 that a single monoid M gives rise to a corresponding category M – this has some arbitrarily chosen thing as its sole object, while the category's arrows are the original monoid's objects, with the monoid-operation as composition. Now suppose $\sim$ is a congruence relation for M, i.e. an equivalence relation on the monoid objects which respects the monoid structure. This will then also be a congruence relation between arrows in the associated category M.

Starting from M, then, we can now quotient to get a category in two ways. We can form the category M corresponding to the monoid M, and then construct the quotient category $\mathsf{M}/\sim$ as per Defn. 24. Or we can first form a quotient monoid $M/\sim$ in exactly the same way as we form a quotient group, see §2.3(c), and then form the corresponding category to *that*.

Challenge: convince yourself that we end up with the same category – on some sensible understanding of 'same category' – either way.

[1] When more than one category is in play, it can occasionally be helpful to explicitly subscript '∘' to indicate which category we are composing arrows in!

7.2 Duality

(a) Now for an absolutely central new idea. A crucial way of getting a new category from old is by simply *reversing all the arrows*. More carefully, let's say:

Definition 25. Given a category C, then its *opposite* or *dual* C^{op} is the category such that

(1) The objects of C^{op} are the same as the objects of C.

(2) If f is an arrow of C with source A and target B, then f is also an arrow of C^{op} but there it is now assigned source B and target A.

(3) Identity arrows remain the same, i.e. $1_A^{op} = 1_A$.

(4) Composition-in-C^{op} is defined in terms of composition-in-C: put $f \circ^{op} g = g \circ f$. △

Here $\circ^{op}$ is, of course, composition in the new opposite category; and condition (4) is made transparent by the following linked diagrams:

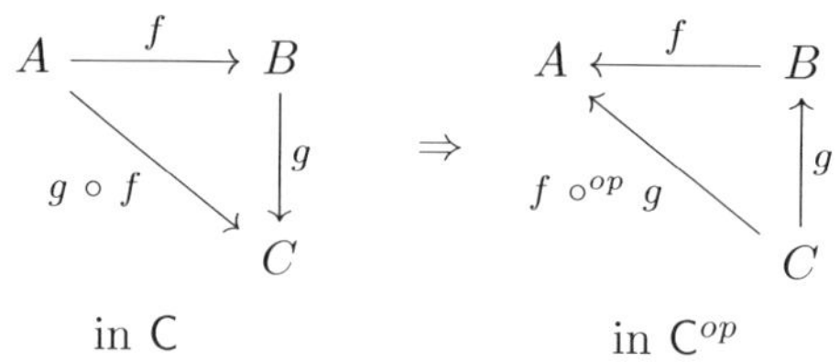

It is easy to see that our definition is in good order and that C^{op} is a category. It is also immediate that $(\mathsf{C}^{op})^{op}$ is C: this means *every* category is also the opposite of some other category.

In general, C^{op} will be a notably different category from C. An example: Set^{op} is tantamount to the category CABool of complete atomic Boolean algebras.[2]

(b) Note that according to our definition, if f is a C-arrow with the source A and the target B, then f also appears as a C^{op}-arrow, but now with the source B and target A. Hence the source and target of f in the one category are different from its source and target in the other.

There is a very important sense, then, in which we've set things up so that the source and target of an arrow f are *not* 'internal' to it. Rather, *you need the src and tar functions of the ambient category to tell you what the source and target of f are in that category.*

[2] We can glimpse the reason if we recall two elementary facts. (i) Given a set X, then the algebra of its subsets forms a complete atomic Boolean algebra $\mathcal{B}X$. And (ii) every complete atomic Boolean algebra is isomorphic to some such $\mathcal{B}X$.

So suppose we take a object X in Set^{op}, i.e. a set X, and associate it with the algebra $\mathcal{B}X$ in CABool. Then we can easily show that an arrow $f\colon X \to Y$ in Set^{op}, i.e. a set-function $f\colon Y \to X$, naturally generates a homomorphism $h\colon \mathcal{B}X \to \mathcal{B}Y$, i.e. an arrow in CABool (note directions of the arrows). And, without going into more details, this association of Set^{op} objects and arrows with CABool objects and arrows is enough to make Set^{op} equivalent in a good sense to CABool.

(c) Let's pause for a terminological aside. Given an arrow $f\colon A \to B$, many authors – as noted in §5.1(c) – follow Mac Lane in referring to A as the arrow's 'domain' and B as its 'codomain', irrespective of the ambient category.[3]

This way of talking is of course entirely apt when f really is a function or morphism. But it is less appropriate when we are dealing with a monoid-as-category M, or a category like Prop_L or Mat. And the terminology could positively lead us astray when we turn to a category like Set^{op}. For in this case the 'domain' and 'codomain' (Mac Lane's sense) of an arrow $f\colon A \to B$ are respectively the codomain and domain of the function f. This sort of tangle is best avoided by talking instead of 'sources' and 'targets', my preferred line.[4] End aside.

(d) What will matter most for us is not the construction of particular opposite categories, but rather the following *duality principle* which arises from the fact that every category is the opposite of another category.

Let's get a bit formal for a moment. Take L be a two-sorted first-order language with identity, with one sort of variable for objects, '$A, B, C \ldots$', and another sort for arrows '$f, g, h, \ldots$'. It has built-in function-expressions 'src' and 'tar' (denoting two operations taking arrows to objects), a built-in relation '$\ldots$ is the identity arrow for $\ldots$', and a two place function-expression '$\ldots \circ \ldots$' which expresses the function which takes two composable arrows to another arrow. We can regiment general propositions in the theory of categories in this language L.[5] The following is then a natural definition:

Definition 26. Suppose φ is a wff (well-formed formula) of L. Its *dual* φ^{op} is the wff you get by (i) swapping 'src' and 'tar' and (ii) reversing the order of composition, so '$f \circ g$' becomes '$g \circ f$', etc. △

And note, the duals of the axioms for a category are also instances of the axioms, as is quickly checked – which is why C^{op} is a category.

That last observation immediately gives us the duality principle we want:

Theorem 11. *Suppose φ is an L-sentence (a wff with no free variables) – so φ is a general claim about objects/arrows in an arbitrary category. Then if the axioms of category theory entail φ, they also entail the dual claim φ^{op}.*

Since we are dealing with a first-order theory, syntactic and semantic entailment come to the same, and we can prove the theorem either way:

---

[3] See for example Mac Lane (1997, p. 7), Awodey (2010, p. 4), Leinster (2012, p. 11), Riehl (2017, p. 3), Roman (2017, p. 2).

[4] As in, for example, Barr and Wells (1995, p. 14), Simmons (2011, p. 2). Some authors use both the source/target and domain/codomain idioms: see Borceux (1994, p. 4), Richter (2020, p. 8), Perrone (2023, p. 1).

[5] We in fact only need an identity predicate applying to objects in contexts like '$src(g) = tar(f)$': category theory, as we will see, otherwise only cares about distinguishing or equating objects 'up to isomorphism'. (We could avoid an identity predicate for objects altogether if we handle the way an arrow is 'typed' by its source and target by using the framework of a dependent type theory rather than a standard first order language. But that's a story for another day.)

Syntactic proof. If there's a first-order proof of φ from the axioms of category theory, then by taking the duals of every wff in the proof we'll get a proof of φ^{op} from the duals of the axioms.

But those duals of axioms are themselves axioms of category theory, so we have a proof of φ^{op} from the axioms. □

Semantic proof. If φ always holds, i.e. holds in every category C, then φ^{op} will hold in every C^{op} – but the C^{op}s comprise every category again, since every category is the opposite of some category, so φ^{op} also holds in every category. □

The duality principle might be simple but it is a hugely labour-saving result; we'll see this time and time again, starting in the next chapter.

7.3 Slice categories

(a) We will now look at another way of constructing new categories from old – or rather, we'll define a dual pair of constructions.

Suppose, then, that C is a category, and X is a particular C-object. We are first going to define a new category C/X whose *objects* are the pairs (A, f) for any C-object A and any C-arrow $f\colon A \to X$ (we will define the corresponding *arrows* of C/X in a moment).

And then there will be a dual construction, a new category X/C whose objects are again pairs (A, f), where A is any C-object but this time f goes in the opposite direction, i.e. is an arrow $f\colon X \to A$ in the original category C.

But why should we be interested in such constructions? Let's have a couple of very simple examples:

(1) Take an n-membered index set $I_n = \{c_1, c_2, c_3, \ldots, c_n\}$. Think of the members of I_n as 'colours'. Then a pair $(S, f\colon S \to I_n)$ gives us a set S whose members are coloured by f from that palette of n colours.

Hence with the arrows of the category appropriately defined – as I said, we'll come to that in a moment! – we can think of FinSet/I_n as the category of n-coloured finite sets, exactly the sort of structure that combinatorialists will be interested in.

(2) Pick a singleton set '1'. We have mentioned before that we can think of any element x of the set S as given by an arrow $\vec{x}\colon 1 \to S$.

Now think about a category $1/\mathsf{Set}$ whose objects are all the pairs of the form $(S, \vec{x}\colon 1 \to S)$. Each such object of $1/\mathsf{Set}$ provides us with a set and then a distinguished element of that set; in other words, the object works as a pointed set. Therefore, $1/\mathsf{Set}$ will be (or at least, in some strong sense to be explained later, comes to the same as) the category Set_* of pointed sets.

True, pointed sets aren't very exciting. But pointed topological spaces are. And, with 1 now some one-point topological space, $1/\mathsf{Top}$ similarly gives us the category of pointed topological spaces.

(b) OK, those examples gives us some initial motivation for being interested in slice categories, and we will meet more examples later. So let's explore further. But first a quick remark. Since, by definition, any of a category's arrows has a unique source in that category, we could without loss of information take the object-data of C/X to be not *pairs* of objects and arrows from C such as $(A, f\colon A \to X)$ but simply the C-arrows $f\colon A \to X$ *by themselves*. Many officially opt for this more economical definition for C/X-objects. Nothing hangs on this. And I'll occasionally talk in the economical idiom too when it makes for neatness.

The next question is: given C/X's objects $(A, f\colon A \to X)$ and $(B, g\colon B \to X)$ what's a sensible candidate for an *arrow* between them? If we want to construct C/X's data from ingredients available in C, what can we use to construct an arrow from (A, f) to (B, g)? The obvious candidate is a C-arrow j from A to B which interacts appropriately with the arrows f and g, so we get a commuting diagram in C.

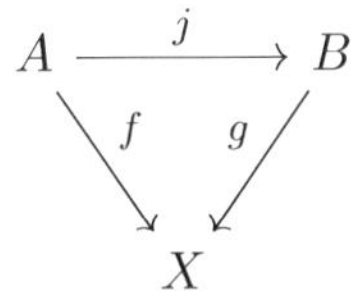

This thought prompts the following attempted definition:

Definition 27 (?). Let C be a category, and X be a C-object. Then the category C/X, the *slice category over* X, has the following data:[6]

(1) C/X-objects are pairs (A, f) for any C-object A and C-arrow $f\colon A \to X$.
(2) C/X-arrows from (A, f) to (B, g) are C-arrows $j\colon A \to B$ where $f = g \circ j$ in C.
(3) The identity arrow in C/X on the object (A, f) is the C-arrow 1_A.
(4) Given C/X-arrows j and k, where the target of j is the source of k, their composition is $k \circ j$ (where $k \circ j$ is the composite arrow in C).[7] △

Of course, we need to check that these data do satisfy the axioms for constituting a category. Let's make a start on doing that. In particular, we need to confirm that our definition of composition for C/X-arrows makes sense.

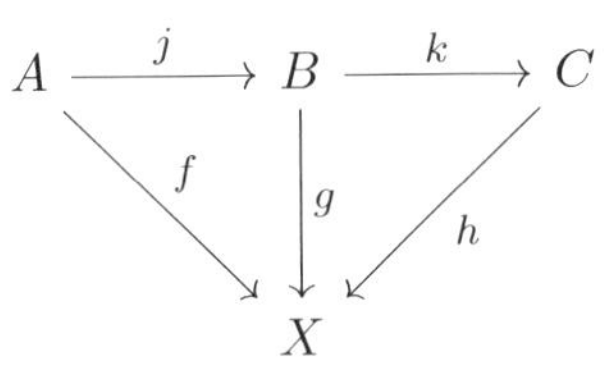

If $j\colon f \to g$ is a C/X-arrow, then $g \circ j = f$ and the left triangle commutes in C. If $k\colon g \to h$ is a C/X-arrow, then $h \circ k = g$ and the right triangle commutes in C. So pasting the triangles together, the whole resulting diagram commutes in C. Or in equations, we have $(h \circ k) \circ j = g \circ j = f$ in C, and therefore $h \circ (k \circ j) = f$. Hence $k \circ j$ will count as an arrow in C/X from f to h on our definition, as we require.

So far, so good!

(c) Unfortunately, however, there is a very annoying snag with our definition of a slice category C/X. It actually doesn't quite work as it stands. Why not?

[6] By the way, don't read anything into the entirely incidental similarity between the notation 'C/X' for the slice category over the object X and the notation '$\mathsf{C}/\!\sim$' for a quotient category with respect to the equivalence $\sim$.

[7] That is to say $k \circ_{\mathsf{C}/X} j$ is the same as $k \circ_{\mathsf{C}} j$.

Suppose $f\colon A \to X$ and $f'\colon A \to X$ are distinct arrows in C. Then (A, f) and (A, f') are distinct objects in C/X. But according to (3) these objects will both have 1_A as their identity arrows. However, by Theorem 10, we can't have distinct C/X-objects with the same identity arrow on them.

What to do? We could fiddle with the definition of a category so Theorem 10 no longer holds. But it is less disruptive to revise the definition of a C/X-arrow. So instead of saying that an C/X-arrow from (A, f) to (B, g) is a C-arrow $j\colon A \to B$ which makes the diagram commute, let's more carefully say:

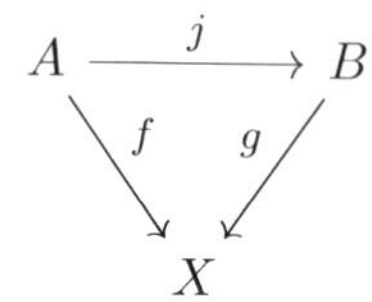

Definition 27. (2) The C/X-arrows from (A, f) to (B, g) are the complete commutative triangles in C formed by f, g together with an arrow $j\colon A \to B$ where $f = g \circ j$. △

In other words, a C/X-arrow is a *triple* of arrows (j, f, g) such that $f = g \circ j$.[8] Clauses (3) and (4) defining identity arrows and composition of arrows will now need adjusting to match (how?).

But it is irritating to have to fuss about this. So from now on, we'll cheat a tiny bit (as seems to be the standard way). When we talk of slice categories and the like, we'll continue to talk of an arrow from f to g as if it is simply a suitable j making the triangle commute. Since we can read off the triple when given an arrow specified as $j\colon f \to g$, no information at all is lost by just giving that single arrow. So that's what we'll do (crossing our fingers behind our backs).

(d) Now for the dual notion, namely the idea of a *co-slice category* X/C (or the slice category *under* X). As we said, the objects of this category are C-objects paired with C-arrows going in the opposite direction, i.e. they are of the form $(A, f\colon X \to A)$.[9] Then the rest of the definition is exactly as you would predict given our explanation of duality: simply go through the definition of a slice category reversing arrows and the order of composition. It is a useful exercise to check that this works.

7.4 Arrow categories, and categories of variable sets

(a) We have seen that an arrow in the slice category C/X is, strictly speaking, best seen as a complete commutative *triangle* in C. Here's another type of derived category where this time the arrows can be thought of as commutative *squares* in C.

Definition 28. If C is a category, then the corresponding *arrow category* $\mathsf{C}^{\to}$ has the following data:

---

[8] "The arrows of C/X from f to g are arrows j of C such that $f = g \circ j$, i.e. they are the same thing as commutative triangles from f to g." So say Lawvere and Rosebrugh (2003, p. 25), with lettering changed. But a single arrow j and a trio of arrows including j and forming a commuting triangle plainly *can't* be the same thing.

[9] Or, if you prefer, you could take the X/C-objects just to be arrows, in this case arrows of the form $f\colon X \to A$.

(1) The $\mathsf{C}^{\rightarrow}$-objects are all the C-arrows.

(2) The $\mathsf{C}^{\rightarrow}$-arrows from $f\colon X \to Y$ to $g\colon W \to Z$ are the commutative squares in C formed by arrows $j\colon X \to W$ and $k\colon Y \to Z$ such that $k \circ f = g \circ j$.

Composition is defined by amalgamating commuting squares in C to get another commuting square (how will this work?), and the identity arrow on a $\mathsf{C}^{\rightarrow}$-object is also defined in the obvious way (again, how?). △

(b) $\mathsf{Set}^{\rightarrow}$ in particular, then, will be the category whose objects are set functions, and whose arrows are suitable commutative squares. However, there's an alternative way of looking at this category that is well worth noting.

Each of $\mathsf{Set}^{\rightarrow}$'s objects is – in a different notation – some Set-arrow $u_X\colon X_0 \to X_1$; and we can think of this as representing a 'variable set' X which starts at stage 0 as the set X_0 and is then updated by u_X to become the set X_1 at stage 1. Then an arrow from the variable set X to the variable set Y can be regarded as a pair of arrows $j\colon X_0 \to Y_0$ and $k\colon X_1 \to Y_1$ making this square commute:

$$\begin{array}{ccc} X_0 & \xrightarrow{\;u_X\;} & X_1 \\ \big\downarrow{\scriptstyle j} & & \big\downarrow{\scriptstyle k} \\ Y_0 & \xrightarrow{\;u_Y\;} & Y_1 \end{array}$$

In other words, this map (j, k) is such that we can either start at the first stage of X, update to its second stage and then use our map to get to the second stage of Y; or we can again start at the first stage of X, map down to Y's first stage and *then* update to Y's second stage. We end up in the same place by either route.

Looked at this way, we can think of $\mathsf{Set}^{\rightarrow}$ as the *category of two-stage variable sets*. Indeed, Lawvere and Rosebrugh (2003, §6.2) *do* talk in this way of variable sets.

8 Kinds of arrows

We have defined the general notion of a category. Wse have met a lot of initial examples, and then seen how to construct yet more categories from old ones in various ways. Note again that our net is now cast very widely, going well beyond the initial motivating idea of a family of structures equipped with enough structure-respecting maps between them.

We are eventually going to want to impose some order on this proliferating universe of categories. And just as we organize groups by looking at the maps between them which respect group structure, we will want to introduce the key notion of *functors*, maps between categories which respect categorial structure. But not yet. Functors will be the fundamental organizing idea of much of Part II of these notes; however, for now, we are going to be continuing to look *inside* categories, before we proceed to look at relations *between* categories.

In this chapter, then, we make a start by characterizing a number of different kinds of arrows by the way they interact with other arrows. This will give us some elementary examples of categorial (re)definitions of familiar notions.

8.1 Monomorphisms, epimorphisms

(a) Let's begin with a simple definition, generalizing Defn. 11:

Definition 29. An arrow f in the category C is a *monomorphism* (for short, is *monic*) iff it is left-cancellable: in other words, whenever g and h are such that $f \circ g = f \circ h$, then $g = h$. △

Note that if the composites $f \circ g$ and $f \circ h$ are to exist and be equal, then g and h must be parallel arrows sharing the same source and target. So we can also put it this way: $f\colon Y \to Z$ is left-cancellable if whenever a fork of the form $X \underset{h}{\overset{g}{\rightrightarrows}} Y \xrightarrow{f} Z$ commutes, then $g = h$.[1]

Why is this notion of being monic/left-cancellable interesting? Well, first note that we have the following general result for categories where arrows are structure-respecting-functions (or set-proxies for functions):

[1]Annoyingly, in the diagrammatic version, f is 'cancelled' on the right. Bother! Recall the remark about notational inversion in §5.1(c)(vi). And recall Defn. 20* on what it takes for a fork to commute.

Theorem 12. *In a category where the arrows are functions, such as* Set, Pos *or* Grp, *if f is injective as a function, then f is a monomorphism.*

Proof. Suppose $f\colon C \to D$ is injective. Then in particular for any x, and any functions $g\colon A \to C$ and $h\colon A \to C$, we have $f(g(x)) = f(h(x))$ implies $g(x) = h(x)$. Hence in arrow-speak, if $f \circ g = f \circ h$ then $g = h$, and so f is left-cancellable. □

And in many categories where the arrows are functions, the reverse is true. For example,

Theorem 13. *In* Set, Pos *and* Grp, *if f is a monomorphism, it is injective as a function.*

Proof for Set. Suppose $f\colon C \to D$ is not injective. So, for some x, y we have $f(x) = f(y)$ but $x \neq y$. But x and y will be respectively picked out as the values (for the only inputs) of the functions $\vec{x}\colon 1 \to C$ and $\vec{y}\colon 1 \to C$, where 1 is your favourite singleton. Hence in Set we have $f \circ \vec{x} = f \circ \vec{y}$ but not $\vec{x} = \vec{y}$. So the non-injective f in Set isn't left-cancellable. Contraposing gives us our wanted result. □

Proof for Pos. The same argument works: just note that a singleton 1 can be equipped with a partial order to make a poset, and functions from 1 are automatically monotone, so also live in Pos. □

Proof for Grp. This takes a bit more work. Suppose that $f\colon C \to D$ is a group homomorphism between the groups $(\underline{C}, *, c)$ and $(\underline{D}, \star, d)$ but is not injective. So for some particular objects x, y we have $f(x) = f(y)$ but not $x = y$.

Now, note that for these objects,

$$f(x^{-1} * y) = f(x^{-1}) \star f(y) = f(x^{-1}) \star f(x) = f(x^{-1} * x) = f(c) = d.$$

Let $\underline{K}$ be the set of objects that f sends to D's group identity d – compare §2.7(b). Then $x^{-1} * y$ belongs to $\underline{K}$. And c is another *distinct* object that f sends to d (for if $x^{-1} * y = c$, then $x = x * c = x * x^{-1} * y = y$ contrary to hypothesis). Hence $\underline{K}$ has more than one member.

Theorem 8 tells us that $(K, *, c)$ forms a group K. Now define $g\colon K \to C$ to be the homomorphism which sends a K-object to the same object in C, while $h\colon K \to C$ sends everything to c. Since $\underline{K}$ has more than one member, $g \neq h$. But $f \circ g = f \circ h$ (both composites send everything in K to d). So the non-injective f in Grp isn't left-cancellable, isn't monic. Contraposing gives us our wanted result (and confirms our stated Theorem 5). □

So, putting our last two theorems together, we have now proved that in Set, Pos and Grp the monomorphisms are exactly the injective functions. And the same applies in most other 'naturally occurring' categories where arrows are functions.[2]

[2] But not all. For those who know about such things, an example is provided by the category of divisible groups.

(b) Now let's introduce the dual notion:

Definition 30. An arrow f in the category C is an *epimorphism* (for short, is *epic*) iff it is right-cancellable: in other words, whenever g and h are such that $g \circ f = h \circ f$, then $g = h$. △

Equivalently, $f\colon X \to Y$ is an epimorphism, is right-cancellable, if whenever a fork of the form $X \xrightarrow{f} Y \overset{g}{\underset{h}{\rightrightarrows}} Z$ commutes, then $g = h$.

Left and right cancellability are evidently dual properties – i.e. f is right-cancellable in C if and only if it is left-cancellable in C^{op}. And we easily get a companion result to Theorem 12:

Theorem 14. *In a category where the arrows are functions, such as* Set *or* Grp, *if f is surjective as a function, then f is an epimorphism.*[3]

Proof. Suppose $f\colon C \to D$ is surjective. And consider any two further functions onwards from the target of f, namely $g, h\colon D \to E$.

Suppose $g \neq h$. Then for some suitable y, $g(y) \neq h(y)$. But by the surjectivity of f, we know that $y = f(x)$ for some x in f's source domain, and therefore $g(f(x)) \neq h(f(x))$. So in arrow-speak, $g \circ f \neq h \circ f$.

Contraposing, if $g \circ f = h \circ f$, then $g = h$. Hence, in sum, the surjectivity of f entails that f is right-cancellable. □

There is an easy converse result in the special case of Set:

Theorem 15. *In* Set, *if f is an epimorphism, then it is surjective as a function.*

Proof. Suppose $f\colon C \to D$ is *not* surjective, so $f(C) \neq D$. Consider two functions $g, h\colon D \to E$ which agree on $f(C)$ but disagree on the rest of D. Then $g \neq h$, though $g \circ f$ and $h \circ f$ will agree everywhere on C; so f is not right-cancellable, not an epimorphism. Contraposing, if f is an epimorphism, it is surjective. □

We can also show e.g. that in Grp, the right-cancellable functions are surjective; but this is not so obvious.[4] And later in this chapter, §8.4, we'll meet an easy case where we have a right-cancellable arrow that *is* a function but which is *not* surjective.

---

[3]How are you supposed to remember which way round the labels 'monomorphism' and 'epimorphism' go? Well, you *could* try recalling that 'mono' means one, and the 'monomorphisms' are (we've seen) rather often the injective, one-to-one functions. While 'epi' is Greek for 'on' or 'over', and the 'epimorphisms' are (we've seen) fairly often surjective, onto, functions. But to be honest, what actually worked for me before the terms became embedded was going by the brute alphabetic proximity of *ML* and of *PR*: for a *M*onomorphism is *L*eft-cancellable, while an e*P*imorphism is *R*ight-cancellable.

[4]See e.g. Agore (2023, pp. 7–8). Why can't we recycle the proof of Theorem 15? Because while there may be such *functions* as the g and h there, that's not enough – we need functions-as-arrows, which in this case means functions which are *group homomorphisms*.

(c) Let's quickly prove a useful mini-theorem:

Theorem 16. *(1) Identity arrows are always monic. Dually, they are always epic too.*

(2) If f, g are monic, so is $f \circ g$ (assuming f and g compose). If f, g are epic, so is $f \circ g$.

(3) If $f \circ g$ is monic, so is g. If $f \circ g$ is epic, so is f.

The proofs are elementary but the results will be repeatedly invoked later.

Proof. (1) is immediate.

For (2), we need to show that if $(fg)j = (fg)k$, then $j = k$. So suppose the antecedent holds. By associativity, $f(gj) = f(gk)$. Whence, assuming f is monic, $gj = gk$. Whence, assuming g is monic, $j = k$.

Interchanging f and g, if f and g are monic, so is $(g \circ f)$. Being epic is dual to being monic. So applying the duality principle from §7.2, it follows that if f and g are epic, so is $(f \circ g)$.[5]

For (3) assume $f \circ g$ is monic. Suppose $gj = gk$. We need to show $j = k$. But $f(gj) = f(gk)$, hence $(fg)j = (fg)k$, hence since $f \circ g$ is monic we have $j = k$. The corresponding result for epics holds by duality. □

(d) Finally in this section, note that there is a notational convention that we use special styles of drawn arrows to represent cancellable arrows, and we will follow this convention, though not religiously:

$f\colon C \rightarrowtail D$ or $C \xrightarrow{f} D$ (with tail decoration) represents a monic, left-cancellable, f;

$f\colon C \twoheadrightarrow D$ or $C \overset{f}{\twoheadrightarrow} D$ represents an epic, right-cancellable f.

The convention is easy enough to remember: a left-cancellable arrow gets notated by an extra decoration on the tail of the arrow (i.e. on the left, when the arrow is drawn in the most common direction), and a right-cancellable arrow gets an extra decoration on the head (i.e. on the right).

8.2 Inverses

(a) We define some more types of arrow:

Definition 31. Given an arrow $f\colon C \to D$ in the category C,

(1) $g\colon D \to C$ is a *right inverse* of f iff $f \circ g = 1_D$.

(2) $g\colon D \to C$ is a *left inverse* of f iff $g \circ f = 1_C$.

(3) $g\colon D \to C$ is an *inverse* of f iff it is both a right inverse and a left inverse of f. △

Three remarks. First, on the use of 'left' and 'right' once again. Note that if we represent the situation in (1) with a commuting diagram like this

[5] Check this, as it is our first mini-application of the duality principle.

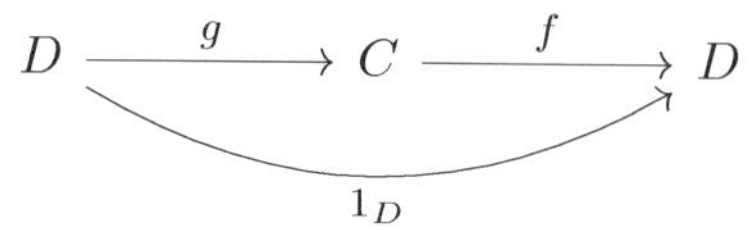

then f's right inverse g appears on the left. As with left/right cancellability, it is merely a matter of convention that we standardly describe the handedness of inverses by reference to the representation '$f \circ g = 1_D$' rather than by reference to our representing diagram.

Second, note that $g \circ f = 1_C$ in C iff $f \circ^{op} g = 1_C$ in C^{op}. So a left inverse in C is a right inverse in C^{op}. And vice versa. The notions of a right inverse and left inverse are therefore, exactly as you would expect, dual to each other; and the notion of an inverse is its own dual.

Third, if f *has* a right inverse g, then it *is* a left inverse (of g, of course!). Dually, if f *has* a left inverse, then it *is* a right inverse.

(b) Let's start by considering what happens in categories where arrows are common-or-garden functions. Here's a *very* easy result:

Theorem 17. *In a category where arrows are functions, if f has a left-inverse as an arrow, it is injective as a function. And if f has a right-inverse, it is surjective as a function.*

Proof. For the first part, we simply note that if $f(x) = f(y)$, then applying f's left inverse to both sides we can infer $x = y$.

For the second part, suppose $f : C \to D$ has a right inverse $g : D \to C$. Take any d in D. Then $f \circ g$ applied to d gives back d. In other words, there is an object c in C, where $c = g(d)$, such that $f(c) = d$. So f is surjective. □

So, putting together this last theorem with Theorems 12 and 14, the following hold for categories – typical 'concrete' categories – where arrows are functions (with '$\Rightarrow$' for 'implies', of course!).

$$f \text{ has a left inverse} \Rightarrow f \text{ is injective} \Rightarrow f \text{ is left-cancellable (monic).}$$
$$f \text{ has a right inverse} \Rightarrow f \text{ is surjective} \Rightarrow f \text{ is right-cancellable (epic).}$$

(c) What about categories where the arrows aren't functions, so the question of being injective or surjective doesn't arise? Well, having a left (right) inverse still implies being left (right) cancellable. Or to ring the changes on the terminology (both for inverses, and for cancellability), we have the first part of the following theorem:

Theorem 18. *(1) Every right inverse is monic, and every left inverse is epic.*

(2) But in general, not every monomorphism is a right inverse; and dually, not every epimorphism is a left inverse.

Proof of (1). Suppose g is a right inverse for f, which means that $f \circ g = 1$ (the identity arrow on the relevant object). 1 is monic by Theorem 16 (1). So g is monic by Theorem 16 (3). Similarly for the dual. □

Proof of (2). We can use a toy example. Take the two-object category **2**:

$$\circlearrowleft \bullet \xrightarrow{f} \star \circlearrowright$$

The non-identity arrow f can only compose with an identity arrow. So, for example, when we have $f \circ g = f \circ h$ it can only be because $g = h = 1_\bullet$. Hence f is monic. Similarly f is epic. But it lacks both a left and a right inverse. $\square$

Slightly more interesting proof of (2). Take the category Grp: consider its objects Z and $2Z$, respectively the additive groups $(\mathbb{Z}, +, 0)$, and $(2\mathbb{Z}, +, 0)$, where of course $2\mathbb{Z}$ is the set of even integers.[6] There is an injection homomorphism $i \colon 2Z \to Z$, and i is monic in Grp (why?).

But i is not a right inverse. That is to say, there is no $f \colon Z \to 2Z$ such that $f \circ i = 1_{2Z}$. For suppose otherwise. Then

$$f(1) +_{2Z} f(1) = f(1 +_Z 1) = f(2) = f \circ i(2) = 1_{2Z}(2) = 2$$

That's impossible since for any z, $f(z)$ is even. $\square$

(d) So monics need not in general be right inverses nor epics left inverses. But how do things pan out in the particular case of the category Set?

Theorem 19. *(1) In* Set*, every monomorphism is a right inverse apart from arrows of the form* $\emptyset \to D$.

(2) Also in Set*, the proposition that every epimorphism is a left inverse is (a version of) the Axiom of Choice.*

Proof of (1). Suppose $f \colon C \to D$ in Set is monic, hence one-to-one between C and $f(C)$. Consider a function $g \colon D \to C$ that reverses f on $f(C)$ and maps everything in $D - f(C)$ to some particular chosen object in C. Such a g is always possible to find in Set unless C is the empty set.

So by construction, $g \circ f = 1_C$, and f is a right inverse. $\square$

Proof of (2). Now suppose $f \colon C \to D$ in Set is epic, and hence a surjection. Assuming the Axiom of Choice, there will be a function $g \colon D \to C$ which maps each $d \in D$ to some chosen one of the elements c such that $f(c) = d$. Note that, in the general case, we *do* have to make an infinite number of choices, picking out one element among the pre-images of d for every $d \in D$: that's why Choice is involved. Given such a function g, $f \circ g = 1_D$, so f is a left inverse.

Conversely, suppose we have a partition of C into disjoint subsets indexed by (exactly) the elements of D. Let $f \colon C \to D$ be the function which sends an object in C to the index of the partition it belongs to; then f is surjective, hence epic. Suppose f is also a left inverse, so for some $g \colon D \to C$, $f \circ g = 1_D$. Then g is evidently a choice function, picking out one member of each partition.

So the claim that every epic is a left inverse in Set is equivalent to the Axiom of Choice. $\square$

[6] Well, not really – it's the set of set-implementations of the evens! But are we going to keep fussing about this sort of thing when it is not relevant to the local point?

(e) An arrow can have zero or one left inverse. It can also have more than one. For a mini-example in Set, consider $f\colon \{0,1\} \to \{0,1,2\}$ defined by $f(0) = 0$, $f(1) = 1$. Then $g\colon \{0,1,2\} \to \{0,1\}$ is a left inverse so long as $g(0) = 0$, $g(1) = 1$; we have two choices for $g(2)$, and hence two left inverses. By the duality principle, an arrow can also have zero, one, or many right inverses. However,

Theorem 20. *If an arrow has both a right inverse and a left inverse, then these are the same and are the arrow's unique inverse.*

Proof. Suppose $f\colon C \to D$ has right inverse $r\colon D \to C$ and left inverse $s\colon D \to C$. Then

$$r = 1_C r = (sf)r = s(fr) = s1_D = s.$$

Or, to put it diagrammatically, the following commutes:

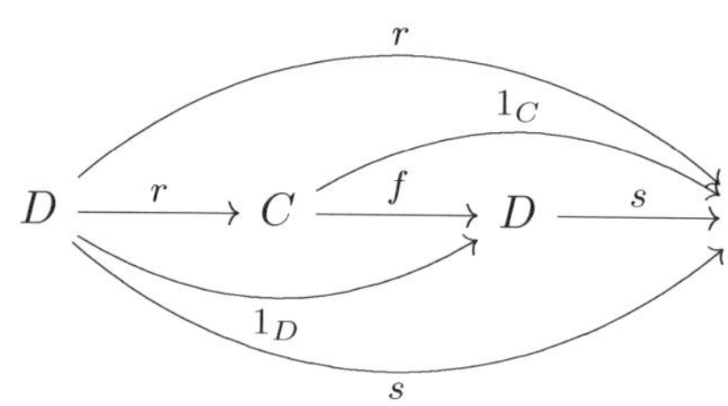

Hence $r = s$ and r is an inverse.

Suppose now that f has inverses, r and r'. Then r will be a right inverse and r' a left inverse for f, so as before $r = r'$. Therefore inverses are unique. □

(f) Our little example at the beginning of (e) shows us that we can, of course, have arrows $f\colon A \to B$ and $g\colon B \to A$ where $g \circ f = 1_A$ but $f \circ g \neq 1_B$.

Is there anything interesting that *can* always be said about $f \circ g$ when $g \circ f = 1$? Well, note we will then have

$$(f \circ g) \circ (f \circ g) = f \circ (g \circ f) \circ g = f \circ 1 \circ g = f \circ g$$

Suppose we say that an arrow j is *idempotent* when $j \circ j = j$ (for that to make sense, an idempotent arrow must have the same source and target). Then, when $g \circ f = 1$, the corresponding composite $f \circ g$ is an idempotent arrow.

8.3 Some more – less memorable? – terminology

There is an oversupply of alternative jargon used hereabouts. Unlike the quite essential 'monomorphism' and 'epimorphism', I won't be making much further use of these perhaps more opaque bits of terminology in these notes. But you'll come across them elsewhere, so I need at least to mention them.

Definition 32. Assume we have a pair of arrows $s\colon C \to D$, and $r\colon D \to C$ such that $r \circ s = 1_C$. Then r, which is a left inverse of s, is said to be a *retraction* of s. And s is a right inverse of r; but s is also called a *section* of r. △

In this usage, then, s *is* a section iff it *has* a retraction, etc.[7]

For a hint of the origin of this jargon, consider the following geometric example. Take P to be the plane minus a point as origin, and let S be the unit circle round the origin. Imagine P parameterized by polar co-ordinates r, θ centred at the origin, and S parameterized simply by θ (in each case, $0 \leqslant \theta < 2\pi$). Then consider the map $r\colon P \to S$ which sends a point (r, θ) on P to θ on S; this 'retracts' the whole plane onto the unit circle. While the map $s\colon S \to P$ which sends the point θ on the circle to $(1, \theta)$ in the plane locates, as it were, a 'section' of the plane. And trivially, $r \circ s\colon S \to P$ is an identity map. (But you can now forget that story!)

Definition 33. If f has a left inverse/is a right inverse, then f is also said to be a *split monomorphism*. If g has a right inverse/is a left inverse, then g is a *split epimorphism*. △

In this usage, we can say e.g. that the claim that every epimorphism splits in Set is the categorial version of the Axiom of Choice.

Note that Theorem 18 tells us that right inverses are monic, so a split monomorphism is indeed properly called a monomorphism. Dually, a split epimorphism is an epimorphism. But why 'split'? I haven't anything short and helpful to offer.

8.4 Isomorphisms

Before we ever encounter category theory, we are familiar with the notion of an isomorphism between groups, between metric spaces, between topological spaces, between orderings, etc. – it's a bijection between the underlying objects which preserves all the relevant structure.

How can we redefine this notion of isomorphism in arrow-theoretic, categorial, terms?

(a) First, what *doesn't* work.

In the extremal case, in the category Set of sets with no additional structure, the bijections are the arrows which are both monic and epic. Can we generalize from this case and define the isomorphisms of any category to be arrows which are monic and epic there?

No. Isomorphisms properly so called need to have inverses (if A and B have all the same relevant structure, then a map preserving all that structure should reverse). But being monic and epic *doesn't* always imply having an inverse. We can use again the toy case of the two-object category which has just one non-identity arrow. That non-identity arrow, as we saw in proving Theorem 18, is

[7] It could be said that there is a principled reason for preferring 'retraction' to 'left inverse', and 'section' to 'right inverse' – namely that the left/right terminology depends on a rather arbitrary decision about whether we notate the composition of s with r as '$r \circ s$' as opposed to something like 'sr'. But let that pass.

both monic and epic, but lacks an inverse. Or here's a generalized version of the same idea:

(1) Take the category P corresponding to some partially-ordered objects $(P, \preccurlyeq)$ – compare §5.4 (C4). Then there is at most one arrow f between any given objects of P. But for any f, if $f \circ g = f \circ h$, then g and h must share the same object as source and same object as target, hence $g = h$, so f is monic. Similarly f must be epic. But no arrows other than identities have inverses.

The arrows in that example aren't functions, however. So here's a case where the arrows *are* functions but where being monic and epic *still* doesn't imply having an inverse:

(2) Consider this artificial little example in Pos, the category of posets and monotone functions.

Suppose the posets A and B each involve just two objects x, y; and let $\leqslant_A$ be the empty relation, while $x \leqslant_B y$. Let $f\colon A \to B$ be the identity map, which is trivially monotone. Then f is monic and epic, but plainly doesn't have a monotone inverse.

And for a much more interesting case:

(3) Consider the category Mon of monoids. Among its objects are $N = (\mathbb{N}, +, 0)$ and $Z = (\mathbb{Z}, +, 0)$ – i.e. the monoid of natural numbers equipped with addition and the monoid of positive and negative integers equipped with addition. Let $i\colon N \to Z$ be the map which sends a natural number to the corresponding integer. This map doesn't have an inverse in Mon. But it is both monic and epic.

It is worth pausing to prove that last claim:

Proof. To prove i is monic, assume $i \circ g = i \circ h$. We need to show $g = h$. If those assumed composites are to exist and be equal, g and h must be parallel arrows from some monoid M to N. Suppose $g \neq h$. Then there is some M-object m such that the natural numbers $g(m)$ and $h(m)$ are different, which means that the corresponding integers $i(g(m))$ and $i(h(m))$ are different, so $i \circ g \neq i \circ h$. Contradiction. So $g = h$ as required.

Second, to prove i is epic, again take a monoid M and this time consider any two monoid homomorphisms $g, h\colon Z \to M$ such that $g \circ i = h \circ i$. Then g and h must agree on all integers from zero up. We'll now show that g and h agree on negative integers too, starting from -1. So note we have

$$\begin{aligned} g(-1) &= g(-1) \cdot 1_M = g(-1) \cdot h(0) = g(-1) \cdot h(1 + -1) \\ &= g(-1) \cdot h(1) \cdot h(-1) = g(-1) \cdot g(1) \cdot h(-1) \\ &= g(-1 + 1) \cdot h(-1) = g(0) \cdot h(-1) = 1_M \cdot h(-1) = h(-1). \end{aligned}$$

But if $g(-1) = h(-1)$, then

$$\begin{aligned} g(-2) &= g(-1+-1) = g(-1)\cdot g(-1) = h(-1)\cdot h(-1) \\ &= h(-1+-1) = h(-2), \end{aligned}$$

and the argument iterates, so we have $g(z) = h(z)$ for all $z \in \mathbb{Z}$, positive and negative. Hence $g = h$ and i is right-cancellable, i.e. epic. □

And note too, picking up a point from the end of §8.1(b), i in this example is an epic arrow that is a function but isn't surjective.

(b) The moral of our various examples? If we want our isomorphisms in general to be invertible, as we surely do, then it looks as though we'll have to build in that feature by definition.

So, at last, here's the official story:

Definition 34. An *isomorphism* in category C is an arrow that has an inverse. We conventionally represent isomorphisms by decorated arrows, thus: $\xrightarrow{\sim}$. △

That's a crucial definition. And from what we have already seen, we know or can immediately check that

Theorem 21. *(1) Identity arrows are isomorphisms.*

(2) An isomorphism $f\colon C \xrightarrow{\sim} D$ has a unique inverse which we can call $f^{-1}\colon D \xrightarrow{\sim} C$, such that $f^{-1} \circ f = 1_C$, $f \circ f^{-1} = 1_D$, $(f^{-1})^{-1} = f$, and f^{-1} is also an isomorphism.

(3) If f and g are isomorphisms, then $g \circ f$ is an isomorphism if it exists, whose inverse will be $f^{-1} \circ g^{-1}$. □

So let's quickly give some simple examples of isomorphisms in different categories:

(1) In Set, the isomorphisms are the bijective set-functions.

(2) In Grp, the isomorphisms are the bijective group homomorphisms.

(3) In Vect_k, the isomorphisms are invertible linear maps.

(4) But as we noted, in a poset category, i.e. a category P corresponding to some partially-ordered objects $(\mathrm{P}, \preccurlyeq)$, the only arrows with inverses are the identity arrows.

(c) Isomorphisms are monic and epic by Theorem 18. But as we have noted, arrows which are both monic and epic need not have inverses so need not be isomorphisms, e.g. in Pos and Mon. However, we do have this result:

Theorem 22. *If f is both monic and has a right inverse (or both epic and has a left inverse), then f is an isomorphism.*

Proof. If f has a right inverse, there is a g such that $f \circ g = 1$. Then $(f \circ g) \circ f = f$, whence $f \circ (g \circ f) = f \circ 1$. Hence, given that f is also monic, $g \circ f = 1$. So g is both a left and right inverse for f, i.e. f has an inverse. Dually for the other half of the theorem. □

Here's another easy result in the vicinity which we'll need later:

Theorem 23. *Suppose the following diagram commutes:*

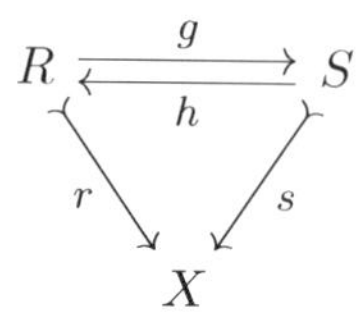

In other words, suppose r and s are both monic arrows with the same target, and there are g, h such that $r = s \circ g$ and $s = r \circ h$. Then g and h are isomorphisms and inverse to each other.

Proof. We have $r \circ 1_R = r = s \circ g = r \circ h \circ g$. Since r is monic, $h \circ g = 1_R$. Similarly, $g \circ h = 1_S$. So g and h are each other's two-sided inverse, and both are isomorphisms. □

(d) Finally, we should mention an often-used bit of terminology:

Definition 35. A category C is *balanced* iff every arrow that is both monic and epic *is* an isomorphism. △

We have seen that some categories like Set are balanced in this sense, while others like Pos and Mon are not. Top is another example of an unbalanced category.[8]

8.5 Isomorphic objects

(a) Having defined categorial isomorphisms, we can now introduce another absolutely key notion:

Definition 36. If there is an isomorphism $f\colon C \xrightarrow{\sim} D$ in C then the objects C, D are said to be *isomorphic* in C, and we write $C \cong D$. △

From the ingredients of Theorem 21, we immediately get

Theorem 24. *Isomorphism between objects in a category C is an equivalence relation.* □

For some examples:

(1) In Grp, any two Klein four-groups are isomorphic in the categorial sense (of course!).
(2) Similarly in Top, any two closed intervals of the real line equipped with the usual topology are isomorphic (of course!).
(3) While in FinOrd, the category of finite ordinals, no two distinct objects are isomorphic (such a category is called 'skeletal' – cf. §34.5).

[8] For the last case, consider that there is a continuous bijection (epic/monic) from the half-open interval $[0, 1)$ to S^1 but it is not an isomorphism in Top.

(4) In Set, any two singletons are isomorphic, since there will be a trivial bijection between them. More generally, any two objects in Set with the same cardinality are isomorphic.[9]

(b) As I remarked before, except when identifying the sources and targets of arrows, category theory doesn't tell us when objects are the same or different. The idea of isomorphic objects – which is itself defined in terms of the equality of some *arrows*! – is the nearest we get to a general categorial notion of equality for objects. And roughly speaking, just as group theory typically doesn't care about the distinction between isomorphic groups, category theory typically doesn't care about the distinction between isomorphic objects. For example, we'll see that categorially we only care about pinning down products and quotients 'up to isomorphism'.

Does this mean that, in some sense, we can simply identify isomorphic objects in category theory? Well, some care is needed here. We've noted that in Set any two singletons count as isomorphic; but wouldn't it strike us as rather odd to identify all singletons or to talk about *the* singleton in the way we talk about *the* Klein group? To be sure, there are contexts where any singleton will do: for example, in §5.6 we noted that there is a one-to-one correspondence between the elements x of a set X and the arrows $\vec{x}\colon 1 \to X$, and here any singleton 1 you care to choose will serve to make the point (also see §9.3(b)). But in other contexts, the pairwise distinctness of singletons would seem to be important, e.g. when we treat $\{\emptyset\}, \{\{\emptyset\}\}, \{\{\{\emptyset\}\}\}, \{\{\{\{\emptyset\}\}\}\}, \ldots$ as a sequence of *distinct* singletons in one possible construction (Zermelo's) for the natural numbers.

We can't delay to explore this sort of issue here: I am simply flagging up that there are questions that eventually need to be tackled around and about ideas of isomorphism-as-sameness. But these are best faced *after* getting a lot more category theory under our belt! – though see again e.g. Mazur 2008.

(c) Back to the technical basics. Let's quickly note that an isomorphism between objects in a category induces a bijection between the arrows to (or from) those objects:

Theorem 25. *If $C \cong D$ in* C, *then there is a one-to-one correspondence between arrows $X \to C$ and $X \to D$ for all objects X in* C, *and likewise a one-to-one correspondence between arrows $C \to X$ and $D \to X$.*

Proof. If $C \cong D$ then there is an isomorphism $j\colon C \xrightarrow{\sim} D$. Consider the map which sends an arrow $f\colon X \to C$ to $\overline{f} = (j \circ f)\colon X \to D$. This map $f \mapsto \overline{f}$ is injective (for $\overline{f} = \overline{g}$ entails $j^{-1}\overline{f} = j^{-1}\overline{g}$ and hence $f = g$). It is also surjective (for any $g\colon X \to D$, put $f = j^{-1}g$ then $\overline{f} = g$). That gives us a bijection, a one-to-one correspondence between arrows $X \to C$ and $X \to D$.

The dual claim is proved similarly. □

[9] In traditional set theory, we don't usually speak of bare sets as being 'isomorphic' (the notion doesn't appear in standard texts like Enderton (1977), Kunen (1980), Moschovakis (2006)).

8.6 Epi-mono factorization

(a) A simple definition and theorem:

Definition 37. An arrow $f: C \to D$ has an *epi-mono factorization* iff there is an epic arrow $e: C \twoheadrightarrow I$ and a monic arrow $m: I \rightarrowtail C$ such that f equals e composed with m, i.e. $f = m \circ e$. △

Theorem 26. *In* Set *every arrow has an epi-mono factorization.*

Proof. Immediate. In Set an arrow $f: C \to D$ is a function. Suppose $I = f[C]$ is the f-image of C. Then let the function $e: C \to I$ agree everywhere with f, and let $m: I \to D$ be the inclusion function which sends an element of I to itself as an element of D. Trivially e is surjective (so epic), m is injective (so monic), and $f = m \circ e$. □

Two observations:

(1) In Set, if f has an epi-mono factorization through both I and I', then $I \cong I'$. You might like to pause to think this through using informal pre-categorial reasoning. But I won't delay over this here, because later we will be able to show that the result holds in any rich enough category that is sufficiently Set-like – see Theorem 234.

(2) Also for future reference, let's have a particular example, still in Set. Suppose C and C' are both subsets of D, and let $C \sqcup C'$ be, as usual, the union of disjoint copies of C and C'. There is a simple function $j: C \sqcup C' \to D$ which sends each element of $C \sqcup C'$ to the original element of D it is a copy of. Then j has an epi-mono factorization $m \circ e$, where the monic $m: C \cup C' \hookrightarrow D$ sends every element of the union to itself.[10]

(b) We should note, however, that this kind of epi-mono factorization is not always available in other categories. For a toy example, consider the mini-category which has a single object o, its identity arrow 1_o, and a further non-identity arrow $f: o \to o$ such that $ff = f$. Then evidently f is neither epic nor monic (or else we could cancel from that equation to get $f = 1_o$). And so f doesn't have an epi-mono factorization.

Again, in some categories, more than one epi-mono factorization can be available which needn't go via isomorphic objects. Recall from §8.4 the situation in Mon where there is an arrow $i: N \to Z$ which is both epic and monic. Then we can have two epi-mono factorizations for i, as it equals both

$$N \xrightarrow{1_N} \!\!\!\!\rightarrow N \rightarrowtail^{i} Z \quad \text{and} \quad N \xrightarrow{i}\!\!\!\!\rightarrow Z \rightarrowtail^{1_Z} Z$$

and of course we don't have $N \cong Z$.

So it turns out that the kind of epi-mono factorization which we get in Set (universal for all arrows, and unique-up-to-isomorphism) is only found in a subfamily of categories.

[10] A hook at the start of an arrow is conventionally used to indicate an inclusion function.

8.7 Groups as categories, and groupoids

(a) Finally in this chapter, with the categorial notion of an isomorphism now in play, we can return to take up an earlier theme. For recall that we can consider a particular *monoid* as itself giving rise to a category – see §5.4 (C3). We can now say something about the categories which arise from those monoids which happen to be *groups*.

So take a group $(G, *, e)$ and – as we can do with any monoid – we define the associated category G to be the category whose sole object $\bullet$ is whatever you like, and whose arrows are simply the group objects G, with e the identity arrow. Composition of arrows in G is then defined as group multiplication of the group elements. Then, since every element in the group has an inverse, it follows immediately that every arrow in the corresponding G has an inverse. In other words, a group-as-a-category is a category with one object and whose every arrow is an isomorphism.

(b) What if we generalize to the case where there is perhaps more than one object but where the arrows all still have inverses? This sort of category is called a *groupoid*. There are significant examples. For instance: every topological space X gives rise to an associated groupoid $\Pi_1(X)$, the category whose objects are the points of X and whose arrows between two points are the homotopy classes of paths between them (arrows which have inverses defined in the obvious way).

9 Initial and terminal objects

When we defined an isomorphism in the previous chapter, we characterized an arrow of that type by reference to the way it relates to another arrow, its inverse. This is entirely typical of a category-theoretic (re)definition of a familiar notion: we look for comparable external, relational, characterizations of types of arrows and/or types of structured objects.

Here is Steve Awodey, offering some similar remarks about what he calls "category-theoretical definitions":

> These are characterizations of properties of objects and arrows in a category solely in terms of other objects and arrows, that is, just in the language of category theory. Such definitions may be said to be abstract, structural, operational, relational, or perhaps external (as opposed to internal). The idea is that objects and arrows are determined by the role they play in the category via their relations to other objects and arrows, that is, by their position in a structure and not by what they 'are' or 'are made of' in some absolute sense. (Awodey 2010, p. 29)

We proceed in this spirit to give some further examples of external category-theoretic definitions of a range of familiar notions. A prime exhibit will be the illuminating treatment of products, starting in the next chapter. In this chapter, however, we warm up by considering a particularly simple pair of cases.

9.1 Initial and terminal defined

(a) As we noted in §5.6, in Set,

(i) For any set X, there is one and only one set-function from the empty set $\emptyset$ to X – namely the empty function. Moreover, if the set S is such that for every X there is one and only one set-function from S to X, then S is the empty set.

(ii) For any set X, there is one and only one set-function from X to a singleton set $\{\star\}$ – namely the empty function if X is the empty set, or otherwise the function which maps every member of X to $\star$. Moreover, if the set S

is such that for every X there is one and only one set-function from X to S, then S is a singleton.

In category-speak, then: in Set the empty set is distinguished by being such that there is one and only one arrow *from* it to any object. And a singleton is distinguished by being such that there is one and only one arrow *to* it from any object.[1]

Let's now introduce a pair of quite natural concepts:

Definition 38. The object I is an *initial* object of the category C iff, for every C-object X, there is a unique arrow $!\colon I \to X$.

Dually, the object T is a *terminal* object of C iff, for every C-object X, there is a unique arrow $!\colon X \to T$.[2] △

On notation: the use of '!' to signal the unique arrows from an initial object or to a terminal object is quite common. If we want explicitly to indicate the target or source of such a unique arrow, we can write e.g. '$!_X$'. Evidently, an object is initial in C if and only if it is terminal in C^{op}.

Then, in summary, we've noted that

(1) The empty set is the unique initial object in Set, while any singleton is terminal.

Let's immediately have some more examples:

(2) In Pos too, the empty poset is initial, while any singleton equipped with the partial order that relates the singleton's member to itself is terminal.

(3) In the preordered natural numbers $(\mathbb{N}, \leqslant)$ thought of as a category, zero is the unique initial object and there is no terminal object. By contrast the preordered integers $(\mathbb{Z}, \leqslant)$ form a category that lacks both initial and terminal objects.

More generally, $(P, \preccurlyeq)$-treated-as-a-category has an initial object iff the preorder has a minimum, an object which $\preccurlyeq$-precedes all the others. Dually for terminal objects/maxima.

(4) $\mathsf{Set}_\star$, recall, is the category whose objects are non-empty sets equipped with a distinguished member and whose arrows are functions preserving distinguished members. Such a function from a singleton in $\mathsf{Set}_\star$ must map its (automatically distinguished) member to the distinguished member of its target X. And any such function from X to a singleton will be unique. Hence in $\mathsf{Set}_\star$ each singleton is both initial and terminal.

[1] Fine print. The official story is that we are working in a suitably capacious, though not-yet-fully-specified, universe of sets. So we haven't determinately pinned down Set. But familiar variants on the usual stories about sets do of course agree that there is an empty set and that there are singletons. I suppose I should, however, note for the record that there can be competent set theories *without* an empty set (Cantor's own, perhaps!) and/or *without* singletons distinguished from their members. For provocation on this topic, you might be diverted by Oliver and Smiley (2006).

[2] Some call terminal objects *final*; and then that frees up 'terminal' to mean *initial or final*. So when reading other treatments, you do need to check how 'terminal' is being used.

(5) What about $\mathsf{Set}^{\to}$, the category whose objects are the set functions, and where an arrow from $f\colon X \to Y$ to $g\colon W \to Z$ is a commutative square formed by arrows $j\colon X \to W$ and $k\colon Y \to Z$ such that $k \circ f = g \circ j$?

If 1 is a singleton, and hence a terminal object in Set, then its identity arrow 1_1 is terminal in $\mathsf{Set}^{\to}$. Why? Because there is a unique arrow from any $f\colon X \to Y$ to $1_1\colon 1 \to 1$. Why? Because there are unique arrows $j\colon X \to 1$, $k\colon Y \to 1$ since 1 is terminal, and these make the required square commutate because $k \circ f = 1_1 \circ j$ (both composites are arrows $X \to 1$ which must be equal, again since 1 is terminal).

Mini-exercise: does $\mathsf{Set}^{\to}$ have an initial object?

(6) Recall: in the slice category Set/X, an object is a pair (A, f) of an object A and arrow $f\colon A \to X$ from Set. And a Set/X-arrow from (A, f) to (B, g) is essentially a Set-arrow $j\colon A \to B$ such that $g \circ j = f$.

Consider then the Set/X-object $(X, 1_X)$. A Set/X-arrow from (A, f) to $(X, 1_X)$ is a Set-arrow $j\colon A \to X$ such that $1_X \circ j = f$, i.e. such that $j = f$ – which always exists and is unique! Hence $(X, 1_X)$ is terminal in Set/X.

But nothing in that argument depends on specific features of Set. We can generalize. In any slice category C/X, the C/X-object $(X, 1_X)$ is terminal.

(7) In Rel, the category of sets and relations, the empty set is both the sole initial and sole terminal object.

(8) What about M_2, the category whose objects are sets equipped with an idempotent function, (X, f), and where an arrow $j\colon (X, f) \to (Y, g)$ is an equivariant function $j\colon X \to Y$, i.e. one such that $j \circ f = g \circ j$? Evidently, if 1 is terminal in Set, then $(1, 1_1)$ is terminal in M_2. And if 0 is initial in Set, then $(0, 1_0)$ is initial in M_2.

(9) As in effect noted in §2.4, in Grp the one-element group is an initial object. The same one-element group is also terminal.

(10) The one-element ring is terminal in Ring too. But the initial object is more interesting – it's the ring of integers (why?).

(11) In Top, the empty set (considered as a trivial topological space) is the initial object. Any one-point singleton space is a terminal object.

(12) In the category Prop_L of propositions in the first-order language L, $\bot$ is initial and $\top$ is terminal.

(13) In the category Bool, the one-object algebra is terminal. While the two-object algebra on $\{0, 1\}$ familiar from propositional logic is initial – for a homomorphism of Boolean algebras from $\{0, 1\}$ to B must send 0 to the bottom object of B and 1 to the top object, and there's a unique map that does that.

(14) In the category Graph the empty graph is initial; the graph with one node and one edge looping from that node to itself is terminal (why?).

These various cases show that a category may have zero, one or many initial objects, and (independently of that) may have zero, one or many terminal objects.

Further, an object can be both initial and terminal. And there is, incidentally, a standard bit of jargon for this case:

Definition 39. An object O in the category C is a *null object* of the category C iff it is both initial and terminal. △

9.2 Uniqueness up to unique isomorphism

Evidently, the ideas of being initial and being terminal are dual, as they can be interrelated by reversing arrows. So for every general result about initial objects, there is a dual result about terminal objects.

Now, if a category C has any initial objects, they may be one or many. However, we have this key result:

Theorem 27. *Initial objects, when they exist, are 'unique up to unique isomorphism': i.e. if the C-objects I and J are both initial in the category C, then there is a unique isomorphism $f\colon I \xrightarrow{\sim} J$ in C. Dually for terminal objects.*

Proof. Suppose I and J are both initial objects in C. By definition there must be unique C-arrows $f\colon I \to J$, and $g\colon J \to I$. Then $g \circ f$ is an arrow from I to itself. But we know that one arrow from I to itself is the identity arrow 1_I. And since I is initial, there can only be one arrow from I to itself. Therefore $g \circ f = 1_I$. Exactly similarly, we can show $f \circ g = 1_J$.

Hence the unique arrow f has a two-sided inverse and is an isomorphism. □

Note this pattern of argument: versions get used a lot!

Next, a related result:

Theorem 28. *If I is initial in C and $I \cong J$, then J is also initial. Dually for terminal objects.*

Proof. Suppose (i) I is initial and (ii) $I \cong J$. By (i), for any X, there is a unique arrow $f\colon I \to X$. By (i) and (ii) the unique arrow $i\colon I \to J$ is an isomorphism.

Now take any arrow $g\colon J \to X$. Then $g \circ i\colon I \to X$, and so by the uniqueness of arrows from I, $g \circ i = f$. Hence g must be equal to $f \circ i^{-1}$. In other words, for any X there is a unique arrow g from J to X: thus J is also initial.

The dual of this line of argument delivers, of course, the dual result. □

It is standard to introduce notation for arbitrary initial and terminal objects (since categorially, we often won't care about distinctions among instances):

Definition 40. We use '0' to denote an initial object of C (assuming one exists), and likewise '1' to denote a terminal object.[3] △

[3]Note, we are overloading the symbol '1' again! Note also that null objects which are both initial and terminal are often alternatively called 'zero' objects. But that perhaps doesn't sit happily with the pretty standard practice of using '0' for an initial object. For 0 (in the sense of an initial object) typically isn't a zero (in the sense of null) object.

And here's a little theorem to help fix ideas:

Theorem 29. *In a category with a terminal object, any arrow* $f\colon 1 \to X$ *is monic.*

Proof. Suppose $f \circ g = f \circ h$; then, for the compositions to be defined and equal, both g and h must be arrows $Y \to 1$, for the same Y. Hence $g = h$ since 1 is terminal. □

9.3 Point elements

(a) A category can have a terminal object (so every object has a unique arrow *to* it), without having any arrows *from* that object (except to itself or to other terminal objects). For example, take a preordered collection viewed as a category, where the order has a maximum element.

By contrast, consider the category Set again. As we have remarked before, in this case arrows $\vec{x}\colon 1 \to X$ from a terminal object (a singleton!) correlate bijectively with elements $x \in X$. So, when working in Set, we can think of talk of such monic arrows $\vec{x}\colon 1 \to X$ as the categorial version of talking of elements of X.

Now imagine using $\vec{x}\colon 1 \to X$ to pick out an element x from X, and then applying a function $f\colon X \to Y$ to this element, to land at the element fx in Y. This element corresponds, of course, to the arrow $\overrightarrow{fx}\colon 1 \to Y$, and we get this commuting diagram:

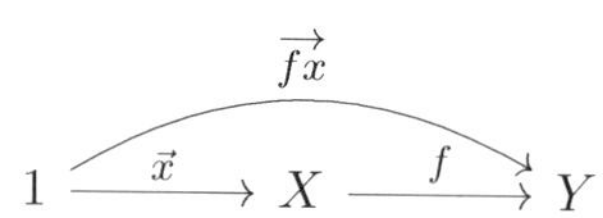

So: $\overrightarrow{fx} = f \circ \vec{x}$.[4]

(b) I said that in Set we can treat talk of arrows $\vec{x}\colon 1 \to X$ as the categorial version of talking of elements of the set X. We had better check, though, that it can't matter for us *which* terminal object 1 we use here in defining elements.

So suppose 1 and $1'$ are two terminal objects in Set. There is a unique isomorphism $j\colon 1' \xrightarrow{\sim} 1$. And if $\vec{x}\colon 1 \to X$ picks out a member x of X, then $\vec{x} \circ j\colon 1' \to X$ picks out the very same object. And of course conversely, if $\vec{x}'\colon 1' \to X$ picks out a member x' of X, $\vec{x}' \circ j^{-1}\colon 1 \to X$ does the same job.

Therefore – as far as general claims about elements are concerned – it can't matter whether elements are thought of as arrows from 1 or as arrows from $1'$.

(c) We now generalize and carry the idea over to other categories:

Definition 41. In a category C with a terminal object 1, a *point element* (or simply *element*) of the C-object X is a (monic) arrow $\vec{x}\colon 1 \to X$.[5]

[4] If we drop the over-arrow notation to mark elements-as-arrows, and drop the optional sign for composition, this would become, not very helpfully, $fx = fx$!

[5] Other standard terminology for such an arrow is, rather oddly, '*global* element'.

We can immediately see, however, that in categories C other than Set, these so-called point elements $1 \to X$ need not correspond nicely with the elements of X in the intuitive sense. In Grp, for example, a homomorphism from 1 (remember, that's a one-element group) to a group X has to send the only group element of 1 to the identity element e of X: so there is only one possible homomorphism $\vec{e}\colon 1 \to X$, irrespective of how many items there are forming the group X.

Put it this way. An arrow $1 \to X$ shines a very narrow beam into X. Still, varying through all the possible arrows $1 \to X$ (for a given fixed X) in Set will turn the spotlight through all the different elements (in the ordinary sense) of X. By contrast, an arrow $1 \to X$ in Grp can only spotlight the identity element of X.

(d) As is entirely familiar, a function f from the set X to the set Y is injective or surjective in the pre-categorial senses depending on how it behaves on elements of X and Y. We can now offer definitions that can be applied across categories:

Definition 42. In a category with a terminal object 1, an arrow $f\colon X \to Y$ is *point-injective* iff for any point elements $\vec{x}_1, \vec{x}_2\colon 1 \to X$, if $f \circ \vec{x}_1 = f \circ \vec{x}_2$ then $\vec{x}_1 = \vec{x}_2$.

While f is *point-surjective* iff for every point element $\vec{y}\colon 1 \to Y$, there is a point element of X, i.e. some $\vec{x}\colon 1 \to X$, such that $f \circ \vec{x} = \vec{y}$. △

In nice enough categories which are sufficiently Set-like (categories which, for a start, have enough point-elements) monic and point-injective arrows coincide, as do epic and point-surjective arrows. But this is far from always the case and, as we'll see, it is the notions monic/epic which will be much more frequently important for us.

9.4 Separators and well-pointed categories

(a) We can introduce a couple more useful bits of terminology. First,

Definition 43. The object S is a *separator* in category C iff for every pair of parallel C-arrows $f, g\colon X \to Y$, where $f \neq g$, there is an arrow $s\colon S \to X$ such that $f \circ s \neq g \circ s$. △

In other words, S is a separator if given two parallel arrows f and g (so we can't separate them merely by looking at their sources and/or targets), we can still always tell them apart 'looking from S' – we can probe their shared source using some arrow s from S, and find that f and g combine differently with that arrow.

Then there is the special case where a terminal object 1 is a separator:

Definition 44. Suppose the category C has a terminal object 1 which is a separator. Then C is said to be *well-pointed.* △

In other words, in a well-pointed category, whenever parallel arrows $f, g\colon X \to Y$ agree on all point elements – i.e. $f \circ \vec{x} = g \circ \vec{x}$ for all $\vec{x}\colon 1 \to X$ – then $f = g$. That's how things are in Set.

But compare again the situation in Grp. Take any two group homomorphisms $f, g\colon X \to Y$ where $f \neq g$. Still, for all possible $\vec{x}\colon 1 \to X$, $f \circ \vec{x} = g \circ \vec{x}$ (both composite arrows from 1 to Y must send the sole member of 1 to the identity element of the group Y). So to summarize, for the record:

Theorem 30. Set *is well-pointed. But* Grp*, for example, isn't.* □

(b) It is worth remarking, however, that Grp does have a separator (in fact it has many, though the terminal object isn't one):

Theorem 31. *Z, the additive group of integers, is a separator for* Grp.

Proof. Suppose we have parallel group homomorphisms $f, g\colon X \to Y$ in Grp, where $f \neq g$. Then choose some x such that $fx \neq gx$. Now consider the map $s\colon Z \to X$ that sends the integer j to x^j.[6] Then s is easily seen to be a group homomorphism, and by construction $f(s(1)) \neq g(s(1))$ so $f \circ s \neq g \circ s$. □

(c) An aside. It's always good to wonder about dual cases: so suppose that C were to have an *initial* object 0 which is a separator. What would this tell us about the category?

Well, if C has any parallel arrows $f, g\colon X \to Y$ such that $f \neq g$, then there will have to be a separating arrow $s\colon 0 \to X$ such that $f \circ s \neq g \circ s$. But both those composites are arrows from the initial 0 to Y, and so can't be distinct after all. Hence C can't have distinct parallel arrows $f, g\colon X \to Y$. Therefore it has to be a preorder category (and there is no real work for a separator to do!).

9.5 'Generalized elements'

(a) We have seen that, even when arrows in a category are functions, acting the same way on all point elements need not imply being the same arrow. An obvious question arises: can we generalize the notion of an element so that acting the same way on 'generalized elements' *does* always imply being the same arrow, whatever the category?

Well, suppose we say that

Definition 45. A *generalized element* (of shape S) of the object X in C is an arrow $s\colon S \to X$. △

'Generalized elements' give us more ways of interacting with the data of a category than the original point elements. And we immediately get

Theorem 32. *Parallel arrows in a category* C *are identical if and only if they act identically on* all *generalized elements.*

[6] The notation should be transparent. For positive j, $x^j = x{*}x{*}x{*}\ldots{*}x$, for j multiplicands, with $*$ the group operation in X. For negative j, $x^j = x^{-1} * x^{-1} * x^{-1} * \ldots * x^{-1}$, for $-j$ multiplicands. And $x^0 = e$, the group identity of X.

Proof. If $f, g \colon X \to Y$ act identically on *all* generalized elements of X, they in particular act identically on the generalized element $1_X \colon X \to X$: so $f \circ 1_X = g \circ 1_X$, and $f = g$.

And of course if $f, g \colon X \to Y$ are identical, they act identically on any generalized element. □

But that's far too trivial to be very exciting. More interesting will be the cases where there is some special class of generalized elements such that acting identically on *them* is enough to ensure arrows are equal. We saw, for example, that acting identically on 'generalized elements of shape Z' is enough to ensure equal arrows in Grp.

(b) In some respects, though, it is a bit odd to call any old arrow $S \to X$ an 'element' of X. For example, suppose we are in the category of topological spaces, and S^1 is a circle. Then a 'generalized element of shape S^1' in X is an arrow from $S^1 \to X$. In other words, it is a continuous map which yields a loop in X. But do we really want to think of this as in any sense an element of X? Doesn't this 'generalized element' correspond to, if anything, a subspace of X?

So I prefer largely to avoid the 'generalized element' jargon. And it is noticeable that quite a few writers on category theory do the same.[7] However, jargon aside, the *idea* is important – the idea of probing a target object X not just by arrows from a terminal object 1 but by arrows with other sources S.

9.6 And what about arrows to 0?

A terminal object 1 in C is defined as having a unique arrow from any C-object *to* it. And arrows *from* 1, when they exist, give us a notion of element which – at least in some categories – corresponds to the intuitive pre-categorial notion.

What about the dual case? An initial object 0 in C is defined as having a unique arrow *from* it to any C-object. And what about arrows *to* 0?

In some categories, there are no such arrows (other than the identity arrow). For a boring example, take any preordered-set-as-a-category where the order has a bottom element, so there is an initial object, but there are no 'downward' arrows. More importantly, there are no arrows $X \to 0$ in Set other than from 0 itself. In Grp where an initial object is a one-object group, arrows $X \to 0$ are the homomorphisms which send every group-member to the single object of 0.

You might like to think about some other examples in categories that you've now met. And we'll return to look again at arrows to 0 in the special context of so-called Cartesian closed categories in §18.2.

[7] To take three relatively recent examples of introductory books, none of Simmons (2011), Roman (2017) and Fong and Spivak (2019) have occasion to use the terminology. I believe it was introduced by Lawvere, so unsurprisingly we find it used in e.g. Lawvere and Rosebrugh (2003, pp. 15–17); but the remarks there about why talk of 'generalized elements' is apt to seem philosophically confused.

10 Pairs and products, pre-categorially

The discussion in the last chapter illustrates an absolutely central categorial theme. We defined initial objects and terminal objects not 'internally' but 'externally' in terms of the arrows for which they are source or target, and then we showed that the objects defined this way are 'unique up to unique isomorphism'.

This is a pattern which will keep on recurring, starting in the next chapter when we give a categorial definition of products. In this chapter, I set the scene by discussing pairings and products in pre-categorial terms.[1]

10.1 Ways of pairing numbers

(a) Suppose for a moment that we are working in a theory of arithmetic and we need to start considering ordered pairs of natural numbers. Perhaps we want to go on to use such pairs in constructing integers or rationals.

Then we can easily handle such ordered pairs of natural numbers without taking on any new commitments, by the simple trick of using *code-numbers*. For example, if we want a bijective coding between pairs of naturals and all the numbers, we could adopt the scheme of coding the ordered pair m, n by the single number $\langle m, n\rangle_B =_{\text{def}} \{(m+n)^2 + m + 3n\}/2$. Or, if we don't insist on every number coding a pair, we could instead adopt the policy of using powers of primes, setting $\langle m, n\rangle_P =_{\text{def}} 2^m 3^n$, which allows rather simpler decoding functions for extracting m and n from $\langle m, n\rangle_P$.

Relative to a coding scheme, we can call code-numbers such as $\langle m, n\rangle_P$ *pair-numbers*; and by a slight abuse of terminology we might even refer to m as the first element of the pair, and n as the second element.

(b) Now, you might be very tempted to protest that this coding trick is quite artificial compared with the set-theoretic way of dealing with ordered pairs of numbers. After all,

---

[1] I do think it is illuminating to take things slowly and to work up to the categorial story this way. Some authors introduce products much more briskly, later in the game and in a more sophisticated setting, after developing category theory much further than we have so far done: see for example Leinster (2014, p. 107), Riehl (2017, p. 77) and – with even less by way of intuitive motivation – Roman (2017, p. 98).

That approach, which is technically fine of course, does miss the opportunity to make the categorial treatment of products look as natural and uncontrived as it really is.

(i) a single pair-number $\langle m, n\rangle_P$ as just defined is neither ordered nor a twosome;
(ii) the number m is a member of (or is one of) the pair of m with n, but a number can't be a genuine member of a pair-number $\langle m, n\rangle_P$; and
(iii) such a coding scheme is quite arbitrary (e.g. we could equally well have used $3^m 5^n$ as a code for the pair m, n).

And that is all true, of course. But it is worth noting that we can lay *exactly* analogous complaints against, for example, the familiar Kuratowski implementation of ordered pairs that we all know and love. This treats the ordered pair of m with n as the set $\langle m, n\rangle_K = \{\{m\}, \{m, n\}\}$. But then:

(i′) $\langle m, n\rangle_K$ is not intrinsically ordered (after all, it is simply a *set*!), nor is it always two-membered (consider the case where $m = n$);
(ii′) even when it is a twosome, its members are not the members of the pair: in standard set theories, m cannot be a member of $\{\{m\}, \{m, n\}\}$; and
(iii′) the construction again involves quite arbitrary choices: thus $\{\{n\}, \{m, n\}\}$ or $\{\{\{m\}, \emptyset\}, \{\{n\}\}\}$ etc., etc., would have done equally as well as alternative implementations.[2]

On these counts, at any rate, coding pairs of numbers by using pair-numbers involves no worse a trick than coding them using Kuratowski's standard gadget.

There is indeed a rather ironic symmetry between the adoption of pair numbers as representing ordered pairs of numbers and another very familiar procedure adopted by the enthusiast for working in ZFC. For remember that standard ZFC knows only about pure sets. So to get natural numbers into the story at all – and hence to get Kuratowski pair-sets of natural numbers – the enthusiast for sets has to choose some convenient sequence of sets to implement the numbers (or to 'stand proxy' for numbers, 'simulate' them, 'play the role' of numbers, or even 'define' them – whatever your favourite way of describing the situation is). But someone who, for her particular purposes, has opted to play the game this way, treating pure sets as basic and dealing with natural numbers by selecting some convenient sets to implement them, is hardly in a position to complain about someone else who, for his purposes, goes in the opposite direction and treats numbers as basic and deals with ordered pairs of numbers by choosing some convenient code-numbers to implement *them*. Both theorists are in the implementation game.

(c) It might be retorted that the Kuratowski trick has the virtue of being a general-purpose device, available when you want to talk about pairs of objects other than numbers, while e.g. the powers-of-primes coding is of much more limited use. Again true. Similarly you can use a hammer to crack open all sorts of things, while nutcrackers are only useful for dealing with nuts. But that's not particularly to the point if it happens to be nuts you currently want to crack, efficiently and with lightweight resources. Likewise, if we want to implement

[2]The second of these is based on the original set-theoretic definition of an ordered pair, due to Norbert Wiener in 1914.

ordered pairs of numbers without ontological inflation – say in pursuing the project of 'reverse mathematics' (with its eventual aim of exposing the minimum commitments required for doing classical analysis, for example, as in Simpson 2009). Then pair-numbers are *exactly* the kind of thing we need.

10.2 Pairing schemes more generally

(a) So: pair-numbers $\langle m, n\rangle_P$ and Kuratowski-pairs $\langle m, n\rangle_K$ belong to two different schemes for pairing up numbers, each of which can work well enough in appropriate contexts. Let's now ask: what does it take to have such a workable scheme for pairing numbers with numbers? Or more generally, to have a scheme for pairing objects X with objects Y?

We've been here before, in §2.3. In essence, we need some *pair-objects* O to code up pairs; we need a binary *pairing function* that sends a given $x \mathrel{\varepsilon} \mathrm{X}$ and a given $y \mathrel{\varepsilon} \mathrm{Y}$ to a particular pair-coding object $o \mathrel{\varepsilon} \mathrm{O}$; and (of course!) we need a couple of *projection functions* which allow us to recover x and y from o. And the point illustrated by the case of rival pairing schemes for numbers is that we shouldn't care too much about the 'internal' nature of the pair-objects, so long as we can associate them 'externally' with suitable pairing and unpairing functions which fit together in the required way.

(b) Our earlier Defn. 3 which characterized pairing schemes was somewhat informally phrased. Let's now tidy things up – and for local typographical neatness, I'll now use '*pr*' generically for a pairing function rather than '$\langle\ ,\ \rangle$'.

Assume, as before, that X are some objects, as are Y, and as are O (these may or may not be all distinct), and assume that $x \mathrel{\varepsilon} \mathrm{X}$, $y \mathrel{\varepsilon} \mathrm{Y}$ and $o \mathrel{\varepsilon} \mathrm{O}$. Then:

Definition 3*. Let $pr\colon \mathrm{X}, \mathrm{Y} \to \mathrm{O}$ be a two-place function, while $\pi_1\colon \mathrm{O} \to \mathrm{X}$ and $\pi_2\colon \mathrm{O} \to \mathrm{Y}$ are one-place functions. Then $(\mathrm{O}, pr, \pi_1, \pi_2)$ *form a scheme for pairing* X *with* Y iff for all x, y and o, the following conditions hold:

(I) $\pi_1(pr(x, y)) = x$ and $\pi_2(pr(x, y)) = y$;

(II) $pr(\pi_1 o, \pi_2 o) = o$. △

Of course, (I) just records the desideratum that if we pair up objects using *pr*, and then unpair using the projection functions π_1 and π_2, we get back to where we started. While (II) records the complementary desideratum that if we take a pair-object o, extract the two paired objects by using π_1 and π_2, and then pair up the results again using *pr*, we also get back where we started.

It hardly needs to be said that, if O are all the natural numbers of the form $2^m 3^n$, and if we choose $pr(m, n) = \langle m, n\rangle_P = 2^m 3^n$ with $\pi_1 o$ and $\pi_2 o$ returning respectively the exponent of 2 and 3 in the factorization of o, then $(\mathrm{O}, pr, \pi_1, \pi_2)$ officially form a scheme for pairing naturals with naturals. And similarly for pairing based on Kuratowski pairs.

(c) Two points on notation. First, I could have used the familiar notation '$\mathrm{X} \times \mathrm{Y}$' rather than the blankly unhelpful 'O' to denote the pair-objects in our scheme for pairing X with Y. But avoiding that notation for a while will keep us honest and help loosen the hold of the idea that products must 'internally' be anything like Cartesian products.

Second, there is no need to over-interpret the brackets in '$(\mathrm{O}, pr, \pi_1, \pi_2)$': they are no more than punctuation, so you can read this as 'O together with the functions pr, π_1 and π_2'.

(d) We need to check some (very!) elementary facts about pairing schemes as defined. First,

Theorem 33. *If $(\mathrm{O}, pr, \pi_1, \pi_2)$ form a scheme for pairing* X *with* Y*, then (1) different pairs of objects are sent by pr to different pair-objects, i.e. $pr(x, y) = pr(x', y')$ iff $x = x'$ and $y = y'$. Also (2) pr, π_1 and π_2 are all surjective.*

Proof. For (1) suppose $pr(x, y) = pr(x', y')$. Then by condition (I) on pairing schemes, $x = \pi_1(pr(x, y)) = \pi_1(pr(x', y')) = x'$, and likewise $y = y'$.

We want (2) to be true so that O includes no more than we need, a condition which we built into our original Defn. 3. But it is immediate that pr is surjective by condition (II): any o is the value of pr for the corresponding inputs $\pi_1 o, \pi_2 o$.[3]

We also want every object x among X to be the first projection of a pair object o, etc. And it is again immediate that the projection function π_1 is surjective because, given x, we can take any y and put $o = pr(x, y)$, and then by (I), x is the value of π_1 for input o. Similarly for π_2. □

Second, as we'd also expect, for given candidate pair-objects, a pairing function fixes the two corresponding projection functions required for a pairing scheme, and vice versa, in the following sense:

Theorem 34. *(1) If $(\mathrm{O}, pr, \pi_1, \pi_2)$ and $(\mathrm{O}, pr, \pi'_1, \pi'_2)$ are both schemes for pairing* X *with* Y*, then $\pi_1 = \pi'_1$ and $\pi_2 = \pi'_2$.*

(2) If $(\mathrm{O}, pr, \pi_1, \pi_2)$ and $(\mathrm{O}, pr', \pi_1, \pi_2)$ are both schemes for pairing X *with* Y*, then $pr = pr'$.*

Proof. For (1), take any o, and suppose $o = pr(x, y)$ (there must be some such x and y since pr is surjective). Hence, applying (I) to both schemes, $\pi_1 o = x = \pi'_1 o$. Hence $\pi_1 = \pi'_1$. Similarly $\pi_2 = \pi'_2$.

For (2), take any x and y and let $pr(x, y) = o$. Then $\pi_1 o = x$ and $\pi_2 o = y$. Applying (II) to the second scheme, we have $pr'(x, y) = pr'(\pi_1 o, \pi_2 o) = o$. Whence $pr(x, y) = pr'(x, y)$. □

(e) Further, there is a sense in which all schemes for pairing X with Y are equivalent up to a unique isomorphism. More carefully,

Theorem 35. *If $(\mathrm{O}, pr, \pi_1, \pi_2)$ and $(\mathrm{O}', pr', \pi'_1, \pi'_2)$ are both schemes for pairing* X *with* Y*, then there is a unique bijection $f \colon \mathrm{O} \to \mathrm{O}'$ which respects pairing, i.e. which is such that for all x, y, $pr'(x, y) = f(pr(x, y))$.*

[3]Careful! We only need the binary function pr to be surjective *on the pair-objects.* The function $pr(m, n) = \langle m, n \rangle_P = 2^m 3^n$ in the pairing scheme for numbers which we considered a moment ago is not surjective *over all numbers.* Likewise a Kuratowski-style pairing function is not surjective over all sets.

Putting it another way, there is a unique bijection f such that, if we pair x with y using pr (in the first scheme), use f to send the resulting pair-object o to o', and then retrieve elements using π_1' and π_2' (from the second scheme), we get back to the original x and y.

Proof. Define f by putting $f(o) = pr'(\pi_1 o, \pi_2 o)$. Then it is immediate that $f(pr(x,y)) = pr'(\pi_1(pr(x,y)), \pi_2(pr(x,y))) = pr'(x,y)$.

To show that f is injective, suppose $f(o) = f(o^*)$. Then $pr'(\pi_1 o, \pi_2 o) = pr'(\pi_1 o^*, \pi_2 o^*)$. Apply π_1' to each side and then use condition (I) in Defn. 3*, and it follows that $\pi_1 o = \pi_1 o^*$. And likewise $\pi_2 o = \pi_2 o^*$. Therefore $pr(\pi_1 o, \pi_2 o) = pr(\pi_1 o^*, \pi_2 o^*)$. Whence by condition (II), $o = o^*$.

To show that f is surjective, take any o' among O′. Then put $o = pr(\pi_1' o', \pi_2' o')$. By the definition of f, $f(o) = pr'(\pi_1 o, \pi_2 o)$; plugging the definition of o twice into the right-hand side and simplifying using rules (I) and (II) confirms that $f(o) = o'$.

Hence f is a bijection with the right properties. And since any pair-object is $pr(x,y)$ for some x, y, the requirement that $f(pr(x,y)) = pr'(x,y)$ fixes f uniquely. □

(f) We've confirmed that pairing schemes as (re)defined in Defn. 3* do work exactly as we should want. Theorem 33 tells us that in a pairing scheme, different pairs get coded by different pair-objects, and also there are no redundancies, i.e. there are no more pair-objects in the scheme than we need. Theorem 34 tells us that, as we'd expect, pairing and unpairing functions fit together tightly – fix the first and that determines the second, and vice versa. Theorem 35 tells us that variant pairing schemes for pairing up X with Y will in a good sense all 'look the same'.

So far, then, so obvious. Should I have left those theorems as very elementary exercises? Quite possibly. But these things are rarely properly spelt out, so I thought it worth pausing over the details. Anyway, let's continue:

Theorem 36. *Given various pluralities of objects* O, X, Y *as before, and functions* $\pi_1\colon \mathrm{O} \to \mathrm{X}$, $\pi_2\colon \mathrm{O} \to \mathrm{Y}$, *suppose that there is a* unique *binary function* $pr\colon \mathrm{X}, \mathrm{Y} \to \mathrm{O}$ *such that*

(I) $\pi_1(pr(x,y)) = x$ *and* $\pi_2(pr(x,y)) = y$.

Then $(\mathrm{O}, pr, \pi_1, \pi_2)$ *form a scheme for pairing* X *with* Y.

Proof. We show (1) that uniqueness for pr implies it is surjective – or equivalently, that being non-surjective implies being non-unique. And then (2) the surjectivity of pr implies condition (II) for a pairing scheme, given (I).

(1) Suppose pr satisfies (I) but is *not* surjective. Then there will be at least one escapee o^*, such that there is no x and y such that $pr(x,y) = o^*$. In particular, $pr(\pi_1 o^*, \pi_2 o^*) \neq o^*$.

Consider then the function pr^* which agrees with pr on all inputs except that we stipulate $pr^*(\pi_1 o^*, \pi_2 o^*) = o^*$.

For all values of x and y other than $x = \pi_1 o^*, y = \pi_2 o^*$, (I) still holds. And by stipulation, for the remaining case $\pi_1(pr^*(\pi_1 o^*, \pi_2 o^*)) = \pi_1 o^*$ and $\pi_2(pr^*(\pi_1 o^*, \pi_2 o^*)) = \pi_2 o^*$.

Hence condition (I) always holds for pr^*, although $pr^* \neq pr$, thereby showing pr isn't unique.

(2) Since pr is surjective, for every o, there is some x and y such that $o = pr(x, y)$, and hence by (I) $\pi_1 o = x$ and $\pi_2 o = y$. But then $pr(\pi_1 o, \pi_2 o) = pr(x, y) = o$. Which gives us condition (II) for being a pairing scheme. □

10.3 Defining products, pre-categorially

Where have we got to?

We have introduced the general idea of a pairing scheme $(\mathrm{O}, pr, \pi_1, \pi_2)$. And note that we don't want to say that the relevant collection of pair objects O *by itself* forms a product of the relevant X with Y: it depends crucially on the rest of the pairing scheme whether the candidate pair-objects can play the right role.

But this definition chimes nicely with our last theorem:

Definition 46. $(\mathrm{O}, \pi_1, \pi_2)$ *form a product of* X *with* Y, where O are some objects, and $\pi_1 \colon \mathrm{O} \to \mathrm{X}$ and $\pi_2 \colon \mathrm{O} \to \mathrm{Y}$ are functions, if and only if (C) for any $x \in \mathrm{X}$ and $y \in \mathrm{Y}$, there is a *unique* $o \in \mathrm{O}$ such that (I*) $\pi_1 o = x$ and $\pi_2 o = y$. △

Why does this work? Think of defining pr by setting $pr(x, y) = o$. If there is an x and a y for which there are after all multiple choices for the value of o satisfying (I*), then there can be different candidates for pr which satisfy (I) in Theorem 36 but whose values $pr(x, y) = o$ peel apart at that x and y. While conversely, if there are different candidates for pr satisfying (I) whose values peel apart at some particular x and y, then at these x and y there will be different values of $o = pr(x, y)$ satisfying (I*).

11 Categorial products and coproducts

In the last chapter, starting from very elementary pre-categorial observations, we arrived at a characterization of a product of objects X with objects Y not in terms of the 'internal' make up of some pair-objects O, whatever they are, but 'externally' in terms of there being functions satisfying a certain uniqueness condition. Evidently, a categorially-flavoured definition!

Moreover, the only functions actually mentioned in Defn. 46 are unary, which is just what we want in a categorially-flavoured definition. Why? Because arrows in categories have single objects as sources – so when arrows are functions, they are always *monadic* functions.[1]

So let's see how we can generalize the story in the last chapter to get a arrow-theoretic definition of products which works across different kinds of category.

11.1 Products defined categorially

(a) Suppose we are initially working in a category where objects are structured sets and arrows are functions. Or more specifically, to fix ideas, suppose we are in Set. Then, instead of talking in the plural about some objects X and objects Y we can now talk in the singular about the set X and set Y of (proxies for) those objects.

But then, as we saw in §9.3, in the case of Set the point elements of an object X – in the sense of arrows from a terminal object to X – behave as elements intuitively should behave. Hence:

(1) In this setting, instead of talking of an object $x \varepsilon$ X and object $y \varepsilon$ Y, we can talk instead of two corresponding arrows $\vec{x}\colon 1 \to X$ and $\vec{y}\colon 1 \to Y$. Again, instead of talking of some object $o \varepsilon$ O, we can talk of an arrow $\vec{o}\colon 1 \to O$.

(2) Hence, instead of saying as in condition (C) in Defn. 46 that $\pi_1 o = x$ and $\pi_2 o = y$, we could say $\pi_1 \circ \vec{o} = \vec{x}$ and $\pi_2 \circ \vec{o} = \vec{y}$.

(3) And *that* is equivalent to saying that the following diagram commutes:

[1] Does it have to be this way? Definitely so, on the standard conception of a category. True, there does exist a contrasting notion of multicategory where the source of an arrow/morphism can be a list of objects. But this and related notions are *far* beyond our scope here.

(P)

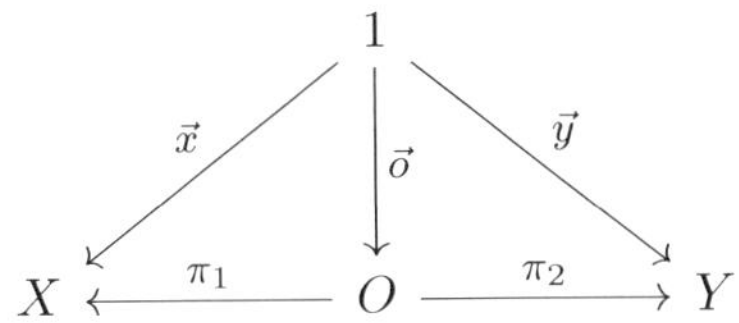

Hence we can transmute our Defn. 46 into a first-shot categorial definition applying to Set as follows: O equipped with the projection arrows $\pi_1\colon O \to X$, $\pi_2\colon O \to Y$ forms a product for X and Y in Set iff for each $\vec{x}\colon 1 \to X$ and $\vec{y}\colon 1 \to Y$ there is a *unique* arrow $\vec{o}\colon 1 \to O$ which makes our diagram (P) commute.

(b) So far, so good. But this won't give us what we want across *all* categories. For example, in Grp, our diagram (P) will always commute if O is a one-object group and π_1 and π_2 are the only possible arrows from it to X and Y (why?). But, trivial cases apart, such an O won't in any sense constitute a product of the groups X and Y.

In other words: in many categories, concentrating only on what happens with *point elements* $1 \to X$ and $1 \to Y$ isn't enough to give us a sensible notion for products of X with Y. What to do?

Following on from the discussion of §9.5, the obvious thing to try is this: move from considering only point elements to thinking about interactions with so-called *generalized elements* too. In other words, instead of thinking only about what happens when we probe X and Y with narrow-beam spotlights with source 1, we should also consider using wider-beam probes from other sources, i.e. using arrows $S \to X$ and $S \to Y$ more generally.

And this motivates proposing our official definition:[2]

Definition 47. In any category C, *a (binary) product* (O, π_1, π_2) *for* X *with* Y is an object O together with projection arrows $\pi_1\colon O \to X, \pi_2\colon O \to Y$, such that for any object S and arrows $f_1\colon S \to X$ and $f_2\colon S \to Y$ in C, there is always a *unique* 'mediating' arrow $u\colon S \to O$ such that the following diagram commutes:

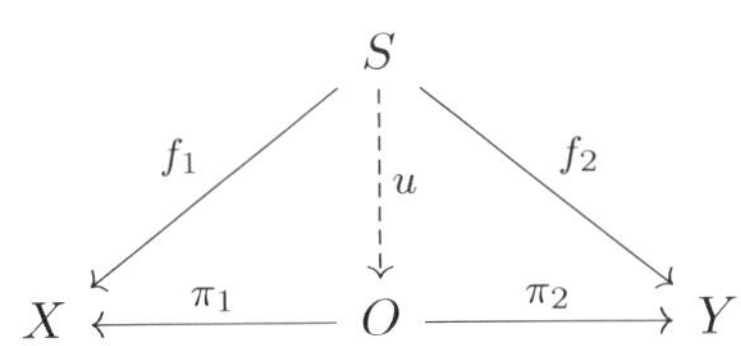

△

Note, by the way, we now adopt a very common convention: in a commutative diagram, we use a dashed arrow --→ to indicate an arrow that is to be uniquely fixed by the requirement that the diagram commutes.

[2]NB: 'motivates' does *not* mean 'forces'! (In Set, the first-shot definition of a product in terms of point elements is in fact equivalent to the official definition in terms of generalized elements. But that's a rather special case.)

(c) Our Defn. 47 is often served up 'neat', without any preceding ceremony. And I suppose it is true that you can stare hard at the definition and its accompanying diagram, and 'see' that it gives the sort of thing we need in a categorial context if O together with its projection arrows is to do the work of a product of X with Y.

Hand-waving more than a bit, the thought might go something along these lines. First, what is shown by the fact that – however we choose a vantage point S from which to probe X and independently probe Y (using some $f_1 \colon S \to X$ and some $f_2 \colon S \to Y$) – our diagram always commutes for *some* arrow u? That O packages together and preserves *enough* paired chunks of information about X and Y as probed from any vantage point we choose for us to be able tp retrieve that information again by going to O via u and then using the projection arrows. While second, the fact that the mediating arrow u is *unique* tells that O packages the information without redundancy, there is no slack for u to vary over, so O (so to speak) preserves *no more than enough.*

But that is not wonderfully transparent, is it? So I do think it has been well worth taking the longer route to our destination. We can already see that the pre-categorial Defn. 46 defines a product in an entirely natural way, given what we want from a pairing scheme. And this immediately gives us a categorial story about products in Set. The route from that to the cross-category Defn. 47 then involves a natural generalization (one that – as we note in a moment – still works as we'd want in Set).

11.2 Examples

Let's now check that our official definition behaves well in various categories. So, first,

(1) In Set, as you would most certainly hope, the usual Cartesian product treated as the set $X \times Y$ of Kuratowski pairs $\langle x, y \rangle$ of objects from X and Y, together with the projection functions $\langle x, y \rangle \overset{\pi_1}{\longmapsto} x$ and $\langle x, y \rangle \overset{\pi_2}{\longmapsto} y$, form a binary product.

For suppose we are given any set S and functions $f_1 \colon S \to X$ and $f_2 \colon S \to Y$. If, for $s \in S$, we put $u(s) = \langle f_1(s), f_2(s) \rangle$, the diagram evidently commutes. Now, for any pair $p \in X \times Y$, $p = \langle \pi_1 p, \pi_2 p \rangle$. Hence if $u' \colon S \to X \times Y$ is another candidate for completing the diagram, $u'(s)$ is a pair, so $u'(s) = \langle \pi_1 u'(s), \pi_2 u'(s) \rangle = \langle f_1(s), f_2(s) \rangle = u(s)$. Therefore u is unique.

And from now on, whatever the ambient category, we will typically default to the notation $X \times Y$ for the object forming a binary product of X with Y, thus $(X \times Y, \pi_1, \pi_2)$. By this stage in the game, you should hopefully be inoculated against the temptation to over-read this notation as *always* indicating something Cartesian-like (and see examples (5) and (7) below).

(2) We can similarly construct products in the category Pos which has posets as objects and order-respecting maps as arrows.

Suppose we take two posets $(X, \leqslant_X)$ and $(Y, \leqslant_Y)$ and form the usual Cartesian product $X \times Y$ of their underlying sets X and Y, and equip *this* with the component-wise product order: i.e., in the obvious notation, we define $\langle x, y\rangle \leqslant_{X\times Y} \langle x', y'\rangle$ to hold if and only if $x \leqslant_X x'$ and $y \leqslant_Y y'$. This gives us a poset $(X \times Y, \leqslant_{X\times Y})$.

Now, note that the obvious projection map from $(X \times Y, \leqslant_{X\times Y})$ to $(X, \leqslant_X)$ will be order-respecting, given our definition of the product order: so this projection map along with the companion projection map from $(X \times Y, \leqslant_{X\times Y})$ to $(Y, \leqslant_Y)$ will count as arrows in Pos. It is then easily confirmed that $(X \times Y, \leqslant_{X\times Y})$ equipped with these two projection maps forms a categorial product of our original two posets in Pos.

At the end of §2.7, I promised that it would turn out that product groups can be characterized by the existence of appropriate homomorphisms between groups. We now know how to do that:

(3) In a category of groups, a product of G_1 with G_2 is a group we'll call $G_1\times G_2$ equipped with homomorphisms $\pi_1\colon G_1 \times G_2 \to G_1$ and $\pi_2\colon G_1 \times G_2 \to G_2$ such that for any group H and homomorphisms $h_1\colon H \to G_1$ and $h_2\colon H \to G_2$, there is a unique homomorphism $u\colon H \to G_1\times G_2$ such that $h_j = \pi_j \circ u$ (for $j = 1, 2$).

For implementations of groups in the category Grp (so living in some universe of sets), we can use the usual Kuratowski construction for pairs. And then, relative to that pairing scheme, the standard direct product of the groups $(\underline{G}_1, *_1, e_1)$ and $(\underline{G}_2, *_2, e_2)$ will be the group $(\underline{G}_1 \times \underline{G}_2, \star, d)$, where $\star$ is defined component-wise, in other words $\langle x_1, x_2\rangle \star \langle y_1, y_2\rangle = \langle x_1 *_1 y_1, x_2 *_2 y_2\rangle$, and $d = \langle e_1, e_2\rangle$. The projection function π_1 which sends each $\langle x_1, x_2\rangle$ to x_1 is then easily checked to be a group homomorphism $\pi_1\colon G_1 \times G_2 \to G_1$. And we define π_2 similarly, of course.

We now need to confirm that, thus defined, the group $G_1 \times G_2$ equipped with the homomorphisms π_1 and π_2 really is a categorial product.[3] That's easy, following the same line of argument as in example (1).

So any two groups implemented in Grp have a product also to be found in Grp; for short, Grp has all products. Note, this relies on our 'category of all groups and their homomorphisms' having access to a scheme for pairing up any two groups: if we had thought of this category has (so to speak) free standing, not to

[3]To be extra clear: the group $G_1 \times G_2$ is an object living in Grp, and π_1 and π_2 are arrows also living in Grp. And that's all it takes for the product $(G_1 \times G_2, \pi_1, \pi_2)$ to count as existing in Grp. For the product is not an extra item over and above the product object and the projection arrows.

As already stressed, we shouldn't overinterpret our notation – the parentheses in the likes of '$(G_1 \times G_2, \pi_1, \pi_2)$' are mere punctuation, unlike say the curly brackets in '$\{G_1 \times G_2, \pi_1, \pi_2\}$' or angle brackets '$\langle G_1 \times G_2, \pi_1, \pi_2\rangle$' which would serve to introduce new sorts of set that don't themselves live in Grp.

be located in an arena already providing a pairing scheme, then the claim that it has all products would need an independent axiom.

(4) A product of topological spaces implemented in a universe of sets, defined in the usual way and equipped with the projection functions recovering the original spaces, is a categorial product of topological spaces in Top. So this category too has a binary product for any of its objects.

(5) Now revisit the category Prop_L introduced in §5.5, (C14). Its objects are propositions, closed sentences of a given first-order language L, and there is a unique arrow from X to Y iff $X \vDash Y$, i.e. iff X semantically entails Y.

In this case, consider the *logical* product of X with Y, i.e. their conjunction $X \wedge Y$. Take this together with the projections $X \wedge Y \to X$, $X \wedge Y \to Y$ (these are arrows because they encode entailments!). Then this gives us a *categorial* product of X with Y in Prop_L.

Why? Take any arrows in Prop_L from A to X and A to Y – i.e. assume $A \vDash X$ and $A \vDash Y$. Then of course we have $A \vDash X \wedge Y$, and we get the required and necessarily unique mediating arrow from A to $X \wedge Y$.

So far, then, so good: categorial products are beginning to line up nicely with products intuitively understood. Here's another example of a category (or rather a family of categories) which is well endowed with products:

(6) The slice category Set/X has a binary prduct for any pair of objects.

For suppose (A, f) and (B, g) are objects in the slice category. By definition, a product of those objects will be an object (O, o) equipped with projection arrows $\pi_1 \colon (O, o) \to (A, f)$ and $\pi_2 \colon (O, o) \to (B, f)$ satisfying our condition for being a product. In other words, for any (S, s) and arrows $j_1 \colon (S, s) \to (A, f)$ and $j_2 \colon (S, s) \to (B, g)$, we need there to be a unique mediating arrow u making this diagram commute in Set/X:

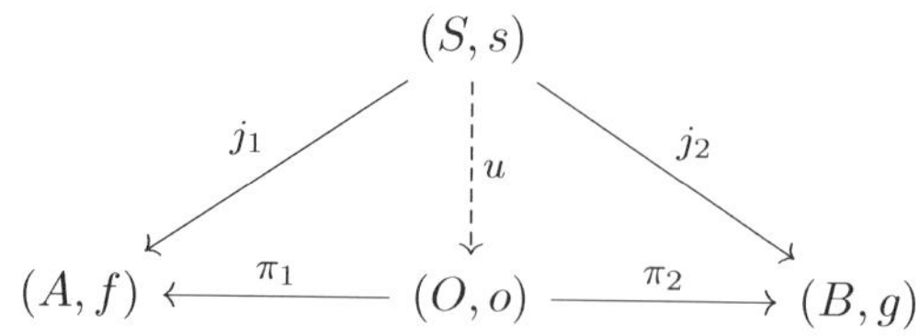

Recalling the definition of arrows in Set/X, that means we must have, in particular, $o = f \circ \pi_1$ in Set and likewise $o = g \circ \pi_2$.

Now, since we after a product, it is natural to try taking O to be some subset of $A \times B$ with π_1 being the Set-function which sends $\langle a, b \rangle$ in O to a, and with π_2 similarly sending $\langle a, b \rangle$ to b.

Suppose then that we put $O \subseteq A \times B$, where $\langle a, b \rangle \in O$ if and only if $fa = gb$, and define $o \colon O \to X$ by requiring $o\langle a, b \rangle = fa = gb$ for each $\langle a, b \rangle \in O$. As required, we'll get $o = f \circ \pi_1$ and $o = g \circ \pi_2$.

And so defined, $((O, o), \pi_1, \pi_2)$ will be a product of (A, f) and (B, g) as is now readily checked.

Let's have two more examples to be going on with:

(7) The category 2Set of sets with at most two members and functions between them does not have all binary products for the most trite of reasons – the product of a couple of two-member sets needs to have four members!

(8) Take preordered objects $(\mathrm{P}, \preccurlyeq)$ considered as a category P as in §5.4, (C4). Then, recall, there is an arrow $p \to q$ in the category iff $p \preccurlyeq q$.

What is a product of p and q in P? It will be an object $p \times q$ with projection arrows to p and q such that, for any pair of arrows from s to p and from s to q, there is a unique arrow from s to $p \times q$ making this diagram commute:

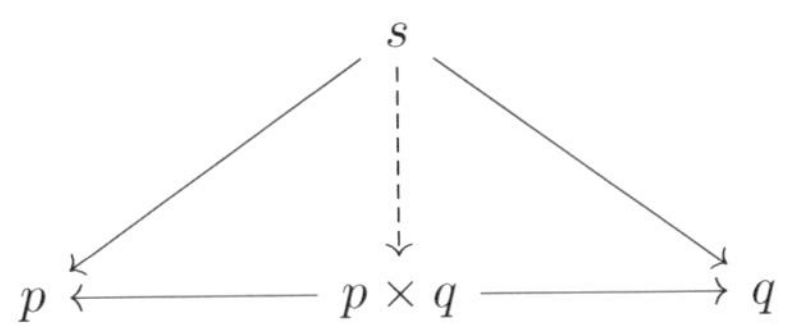

Which means that $p \times q \preccurlyeq p$ and $p \times q \preccurlyeq q$, and whenever $s \preccurlyeq p$ and $s \preccurlyeq q$, we have $s \preccurlyeq p \times q$. So the object $p \times q$ must be the *meet* or *greatest lower bound* of p and q in $(\mathrm{P}, \preccurlyeq)$.

Since pairs of objects in a preordering need not in general have greatest lower bounds, this – like the 2Set example – again shows that in general *a category may well not have all products* (or any other products than some trivial ones, as we shall see).

11.3 Products as terminal objects

(a) Defn. 47 defines the notion of *a* product of a pair of objects X and Y in a category. But in the next section we will prove that products are in fact unique up to unique isomorphism. First, though, it is helpful and illuminating to introduce a slightly different, though equivalent, way of defining products.

We'll need an auxiliary notion. Let's say[4]

Definition 48. A *wedge to X and Y* (in category C) is an object S and a pair of arrows $f_1 \colon S \to X$, $f_2 \colon S \to Y$. Call S the vertex of the wedge. △

Then a wedge $\begin{array}{l} \overset{\pi_1}{\nearrow} X \\ O \\ \underset{\pi_2}{\searrow} Y \end{array}$ is a product of X with Y iff, for any wedge $\begin{array}{l} \overset{f_1}{\nearrow} X \\ S \\ \underset{f_2}{\searrow} Y \end{array}$ to X and Y, there exists a unique arrow u making the following commute:

[4] I picked up 'wedge' from Harold Simmons (2011, cf. p. 25): 'span' is a less evocative but more conventional alternative. We'll see in §19.2 that wedges are simple instances of an indispensable more general construction.

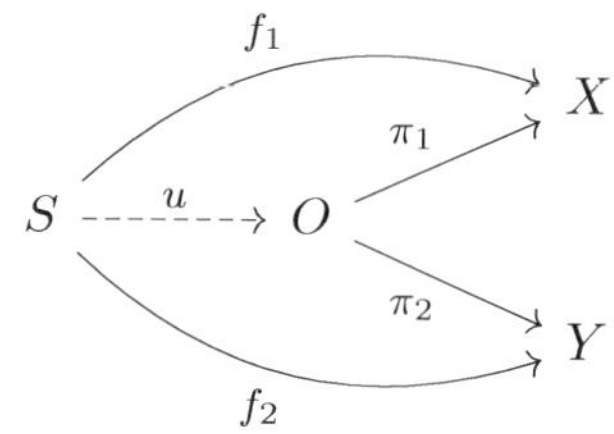

That's simply our previous definition put in different terms, with the diagram rotated! No mystery here.

In such a case where f_1 factors as $\pi_1 \circ u$ and f_2 as $\pi_2 \circ u$, we will say that the whole wedge $X \xleftarrow{f_1} S \xrightarrow{f_2} Y$ *factors through* the product wedge via the mediating arrow u.

(b) A quick terminological aside. It is a common categorial idiom to use the informal 'factors through' in a pretty relaxed spirit. Wikipedia's List of Mathematical Jargon puts it this way: "If for three objects A, B, and C a map $f\colon A \to C$ can be written as a composition $f = h \circ g$ with $g\colon A \to B$ and $h\colon B \to C$, then f is said to factor through any (and all) of B, g, and h." Talk of a wedge with its pair of arrows factoring through another wedge with its pair of arrows is a natural extension.

(c) Now for another definition involving wedges.

Recall, the category C/X, the slice category of C over X, has as its objects pairings of C-objects and C-arrows of the form $(S, f\colon S \to X)$.We are now going to introduce a new category C/XY, the *wedge category* of C over X and Y. Its objects are going to be triples of a C-object and *two* C-arrows of the form $(S, f\colon S \to X,\ g\colon S \to Y)$. In other words the objects of the wedge category C/XY are C-wedges to X and Y. And looking at the definition of slice categories, the corresponding definition for arrows in wedge categories should be predictable – just meditate on the following diagram:

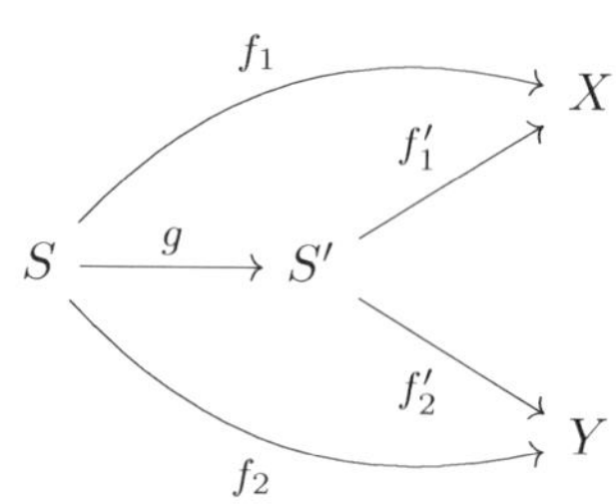

Thus, we will say:

Definition 49. Given a category C and C-objects X, Y, then the *wedge category* C/XY has the following data.

(1) Its objects are all the wedges (S, f_1, f_2) from any S to X, Y.
(2) And an arrow from (S, f_1, f_2) to (S', f'_1, f'_2) is a C-arrow $g\colon S \to S'$ such that the two resulting triangles commute: i.e. $f_1 = f'_1 \circ g$, $f_2 = f'_2 \circ g$.

Composition of two arrows in C/XY is defined as being the same as their composition as arrows of C.[5] △

(d) Finally, our new notion of the derived category C/XY to hand, we can revisit our previous definition of a product. A moment's more reflection shows that it is straightforwardly equivalent to

Definition 50. A product of X with Y in C is a terminal object of the wedge category C/XY. △

Think about it! – this really is rather cute.

11.4 Uniqueness up to unique isomorphism

(a) As noted, products need not exist for arbitrary objects X and Y in a given category C; and when they exist, they need not be strictly unique. However, when they do exist, then – as announced – they *are* 'unique up to unique isomorphism' (compare Theorems 27 and 35).

That is to say, we have:

Theorem 37. *If both* (O, π_1, π_2) *and* (O', π'_1, π'_2) *are products for* X *with* Y *in the category* C*, then there is a unique isomorphism* $f\colon O \xrightarrow{\sim} O'$ *commuting with the projection arrows (i.e. such that* $\pi'_1 \circ f = \pi_1$ *and* $\pi'_2 \circ f = \pi_2$*).*

Note the statement of the theorem carefully. It is *not* being baldly claimed that there is a unique isomorphism between any objects O and O' which are components of products for some given X, Y. That's false. For a very simple example, in Set, take the standard product object $X \times X$ comprising Kuratowski pairs of objects both taken from X. There are evidently two isomorphisms between $X \times X$ and itself, given by the maps $\langle x, x' \rangle \mapsto \langle x, x' \rangle$, and $\langle x, x' \rangle \mapsto \langle x', x \rangle$. The claim is, to repeat, that there is a unique isomorphism between the objects O of any two products (O, π_1, π_2) for X with Y *which commutes with the products' respective projection arrows.*

We are now going to prove our theorem twice over – rather ploddingly from first principles, and then more zippily using our redefinition of products as terminal objects.

Plodding proof. Since (O, π_1, π_2) is a product for X with Y in C, every wedge to X and Y factors uniquely through it, including itself. In other words, there is a unique u such that this diagram commutes:

[5] OK – we are cheating a bit again! For recall the irritating complication we mentioned in §7.3 when defining slice categories. We get the same irritating complication here, and really should define C/XY-arrows as whole commuting diagrams in C, not just as single C-arrows. I can perhaps leave it to pernickety readers to fuss about the details, but also think about why they don't matter for our purposes!

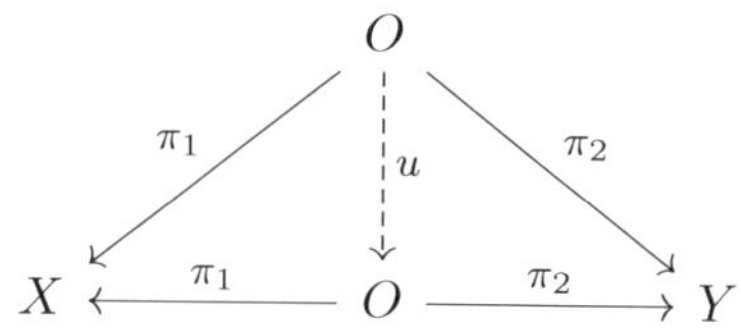

But evidently putting 1_O for the central arrow makes the diagram commute. So by the uniqueness requirement we know that

(i) Given a product (O, π_1, π_2) and an arrow $u: O \to O$, if $\pi_1 \circ u = \pi_1$ and $\pi_2 \circ u = \pi_2$ (so the product factors through itself via u), then $u = 1_O$.

Now, assuming (O', π_1', π_2') is also a product, (O, π_1, π_2) has to uniquely factor through it:

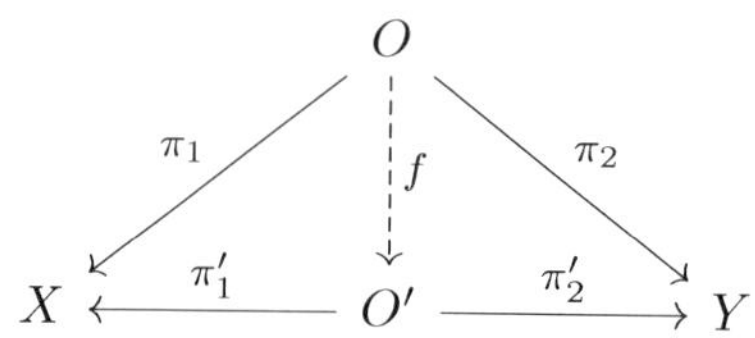

In other words, there is a unique $f: O \to O'$ commuting with the projection arrows, i.e. such that

(ii) $\pi_1' \circ f = \pi_1$ and $\pi_2' \circ f = \pi_2$.

And vice versa, since (O, π_1, π_2) is a product, (O', π_1', π_2') has to uniquely factor through *it*. That is to say, there is a unique $g: O' \to O$ such that

(iii) $\pi_1 \circ g = \pi_1'$ and $\pi_2 \circ g = \pi_2'$.

Whence,

(iv) $\pi_1 \circ g \circ f = \pi_1' \circ f = \pi_1$ and $\pi_2 \circ g \circ f = \pi_2$.

But $g \circ f$ is an arrow from O to O. So it follows by (i) that

(v) $g \circ f = 1_O$

The situation with the products is symmetric so we also have

(vi) $f \circ g = 1_{O'}$

Hence f has a two-sided inverse, i.e. is an isomorphism. □

You'll recognize the key proof idea here is closely akin to the one we used in proving Theorem 27, showing that initial objects are unique up to unique isomorphism. And we can in fact simply appeal to that earlier result:

Succinct proof using the alternative definition of products Both (O, π_1, π_2) and (O', π_1', π_2') are terminal objects in the category C/XY. Therefore by our earlier theorem there is a unique C/XY-isomorphism f between them.

By definition, this has to be a C-arrow $f: O \to O'$ commuting with the projection arrows. And it is immediate that an isomorphism in C/XY is also an isomorphism in C. □

(b) Here's a simple corollary of our last theorem.

Theorem 38. *In a category where the relevant products exist, $Y \times X \cong X \times Y$.*

Proof. Suppose $(X \times Y,\ \pi_1 \colon X \times Y \to X,\ \pi_2 \colon X \times Y \to Y)$ is a product of X with Y; then – applying the definition – $(X \times Y,\ \pi_2 \colon X \times Y \to Y,\ \pi_1 \colon X \times Y \to X)$ will count as a product of Y with X. Hence, by Theorem 37, there is an isomorphism between this particular product-object $X \times Y$ and the object $Y \times X$ of any other product of Y with X. □

(c) When discussing terminal objects, we not only showed that they are unique up to unique isomorphism (Theorem 27) but that any objects isomorphic to them are also terminal (Theorem 28).

Similarly for products. We've shown that they are unique up to unique isomorphism. We now prove that wedges that factor through a product via an isomorphism also give rise to products.

Theorem 39. *Suppose (O, π_1, π_2) is a product of X with Y and the wedge (O', π'_1, π'_2) factors through it by an isomorphism $o \colon O' \xrightarrow{\sim} O$; then (O', π'_1, π'_2) is also a product of X with Y.*

Proof. Take any wedge $X \xleftarrow{f_1} S \xrightarrow{f_2} Y$. We need to show (i) there is an arrow $v \colon S \to O'$ such that $f_j = \pi'_j \circ v$ (for $j = 1, 2$), and (ii) v is unique.

But we are given that (O, π_1, π_2) is a product of X with Y, so we know that there is a unique arrow $u \colon S \to O$ such that $f_j = \pi_j \circ u$. And by the assumption that the primed wedge factors through the unprimed wedge, we know $\pi'_j = \pi_j \circ o$, hence $f_j = \pi'_j \circ o^{-1} \circ u$. Therefore put $v = o^{-1} \circ u$, and that satisfies (i).

Now suppose there is another arrow $v' \colon S \to O'$ such that $f_j = \pi'_j \circ v'$. Then we have $o \circ v' \colon S \to O$, and also $f_j = \pi_j \circ o \circ v'$. Which makes the wedge with the apex S factor through the original product via $o \circ v'$. So by the uniqueness of mediating arrows, $o \circ v' = u$. Hence $v' = o^{-1} \circ u = v$. Which proves (ii). □

11.5 Notation for mediating arrows

Definition 51. Suppose (O, π_1, π_2) is a binary product for the objects X with Y, and the wedge (S, f_1, f_2) factors through it. We will now notate the unique mediating arrow from S to O as $\langle\!\langle f_1, f_2 \rangle\!\rangle$, thus:

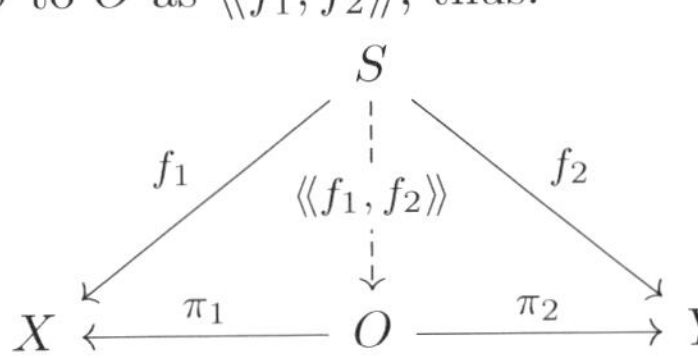

△

I should perhaps note that it is more conventional to use the simpler pair-style notation '$\langle f_1, f_2 \rangle$' here. However, I think it is well worth clearly signalling that a mediating arrow $\langle\!\langle f_1, f_2 \rangle\!\rangle$ is *not* in general to be thought of as an ordered pair of arrows (compare §7.1(c)). Though we do retain this much pair-like behaviour:

Theorem 40. *If* $\langle\langle f_1, f_2\rangle\rangle = \langle\langle g_1, g_2\rangle\rangle$, *then* $f_1 = g_1$ *and* $f_2 = g_2$.

Proof. Evidently, $f_1 = \pi_1 \circ \langle\langle f_1, f_2\rangle\rangle = \pi_1 \circ \langle\langle g_1, g_2\rangle\rangle = g_1$; similarly $f_2 = g_2$. □

A special case is also worth noting:

Definition 52. Suppose we are working in a category with the relevant products. Then the wedge $X \xleftarrow{1} X \xrightarrow{1} X$ must factor uniquely through a given product $X \times X$ via an arrow $\langle\langle 1_X, 1_X\rangle\rangle \colon X \to X \times X$.

That unique mediating arrow can also be notated δ_X, and is *the diagonal arrow* from X to $X \times X$. △

In Set, thinking of $X \times X$ in the usual way, δ_X sends an element $x \in X$ to $\langle x, x\rangle$. We can imagine elements $\langle x, x\rangle$ lying down the diagonal of a two-dimensional array of pairs: hence the label 'diagonal' and the notation 'δ'.

11.6 Two general comments

(a) We've seen that products, when they exist, are unique up to unique isomorphism. It is common, then, for category theorists to fall into talking loosely of *the* product of X and Y, etc.

Later, we'll meet other kind of widgets – equalizers, pullbacks, pushouts, and more – which when they exist are also unique up to unique isomorphism. Again, it is common to talk of *the* equalizer, *the* pullback, and so on.

For the sake of clarity, here in these notes, I'll try to largely avoid this tempting idiom. But you should be aware of its use elsewhere.

(b) We have defined a binary product for X with Y categorially as a wedge which has has a certain universal property – i.e. *any* other wedge to X and Y factors uniquely through it via a unique arrow. Since arrows are typically functions or maps, we can also naturally enough say more specifically that products are defined by a *universal mapping property.*

We've already seen universal mapping properties in action: terminal and initial objects are also defined by how any other object has a unique map/arrow to or from them. There will be lots more examples over coming chapters. We won't attempt yet a formal definition of what it is to be defined by a universal (mapping) property. But there's a common pattern of categorial definition here which you will start to recognize informally when you repeatedly come across it.

11.7 Coproducts

(a) We are going now to discuss the duals of products. But first, we should note a common terminological trope:

Definition 53. Duals of categorially defined widgets are very often called *co-widgets.* Thus a *co-widget* of the category C is a widget of C^{op}. △

For example, we have met co-slice categories, the duals of slice categories. True, there is a limit to this sort of thing – no one talks e.g. of 'co-monomorphisms' (instead of 'epimorphisms'). Still, the general convention is used widely. In particular, it is standard to talk of the duals of products as 'coproducts' – though in this case, as in some others, the hyphen is usually dropped.

(b) The definition of a coproduct is immediately obtained, then, by reversing all the arrows in our definition of products. Thus:

Definition 54. In any category C, *a coproduct* (O, ι_1, ι_2) *for the objects* X *with* Y is an object O together with two 'injection' arrows $\iota_1 : X \to O, \iota_2 : Y \to O$, such that for any object S and arrows $f_1 : X \to S$ and $f_2 : Y \to S$ there is always a unique 'mediating' arrow $v : O \to S$ such that the following diagram commutes:

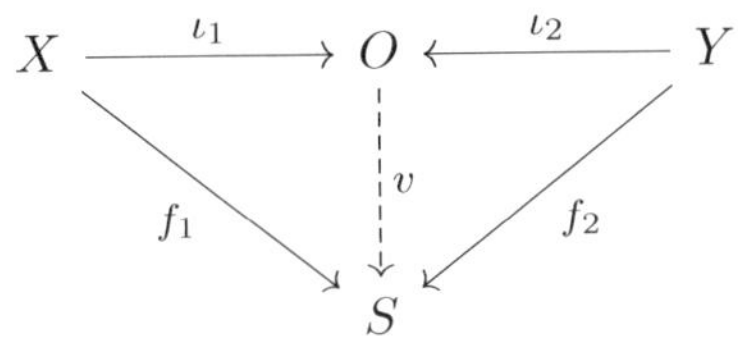

The object O in such a coproduct for X with Y is usually notated '$X + Y$' or '$X \amalg Y$'; and we can notate the mediating arrow v by '$[\![f_1, f_2]\!]$'. △

Do note, however, that the 'injections' in this sense need not be injective.

(c) It is useful to introduce another auxiliary notion. Let's say

Definition 55. A *corner from* X *and* Y (in category C) is an object S and a pair of arrows $f_1 : X \to S$, $f_2 : Y \to S$. Call S the vertex of the corner. △

Draw this situation, to see why corners are sensibly called corners![6] Then a coproduct of X with Y can be thought of as a corner from X and Y such that any corner from X and Y factors through it via a unique map v between the vertices of the corners.

We could now go on to define a category of corners from X and Y on the model of a category of wedges to X and Y, and define a coproduct of X with Y as an initial object of this category. It is a useful check on understanding to work through the easy details: just reverse arrows!

(d) Let's have some examples of coproducts. Start with easy cases:

(1) In Set, *disjoint unions* are instances of coproducts.

Given sets X and Y, let $X + Y$ be the set with members $\langle x, 0\rangle$ for $x \in X$ and $\langle y, 1\rangle$ for $y \in Y$. And let the injection arrow $\iota_1 : X \to X + Y$ be the function $x \mapsto \langle x, 0\rangle$, and similarly let $\iota_2 : Y \to X + Y$ be the function $y \mapsto \langle y, 1\rangle$. Then $(X + Y, \iota_1, \iota_2)$ is a coproduct for X with Y.

To show this, take any object S and arrows $f_1 : X \to S$ and $f_2 : Y \to S$, and then define the function $v : X + Y \to S$ as sending an element $\langle x, 0\rangle$ to $f_1(x)$ and an element $\langle y, 1\rangle$ to $f_2(y)$.

[6] Though if you prefer to call wedges 'spans', then you'll want to call corners 'cospans'.

By construction, this will make both triangles commute in the diagram in the definition above.

Moreover, if v' is another candidate for completing the diagram, then $v'(\langle x, 0\rangle) = v' \circ \iota_1(x) = f_1(x) = v(\langle x, 0\rangle)$, and likewise $v'(\langle y, 1\rangle) = v(\langle y, 1\rangle)$, whence $v' = v$, which gives us the necessary uniqueness.[7]

(2) We can think of the objects of FinOrd, the category of finite ordinals, as being the natural numbers (with the number n implemented by the n-membered set). Taking coproducts, i.e. disjoint unions, gives us addition of numbers (while taking products in FinOrd gives us multiplication). Looked at that way, we can think of addition and multiplication as *duals*, arising from the reversing of arrows. Who knew?!

(3) In Prop_L the *disjunction* $X \vee Y$ (with the injections $X \to X \vee Y, Y \to X \vee Y$) is a coproduct of X with Y.

(4) In the case of preordered objects $(\mathsf{P}, \preccurlyeq)$ considered as a category then a coproduct of p and q would be an object c such that $p \preccurlyeq c, q \preccurlyeq c$ and such that for any object d such that $p \preccurlyeq d, q \preccurlyeq d$ there is a unique arrow from c to d, i.e. $c \preccurlyeq d$.

Which means that the coproduct of p and q, if it exists, must be their *join* or *least upper bound* (equipped with the obvious two arrows, of course).

(5) The construction of coproducts can get markedly more complex. For example, in a category of groups, coproducts are (isomorphic to) the so-called 'free products' of groups.

Take the groups $G = (\mathrm{G}, \cdot, e)$ and $H = (\mathrm{H}, \odot, d)$. Assume that we have doctored the groups if necessary so that now $e = d$ while ensuring the objects G and H are otherwise disjoint. Form all the finite 'reduced words' $\mathrm{G} \star \mathrm{H}$ you get by juxtaposing objects from G and/or H, and then multiplying out neighbouring G-objects by $\cdot$ and neighbouring H-objects by $\odot$ as far as you can. Equip these objects $\mathrm{G} \star \mathrm{H}$ with the operation $\diamond$ of juxtaposition-of-words-followed-by-reduction. Then $G \star H = (\mathrm{G} \star \mathrm{H}, \diamond, e)$ is a group – the so-called free product of the two groups G and H – and there are 'injection' group homomorphisms $\iota_1 \colon G \to G \star H$, $\iota_2 \colon H \to G \star H$ (these send an object g or h respectively to itself as an object among $\mathrm{G} \star \mathrm{H}$).

Claim: $(G \star H, \iota_1, \iota_2)$ is a coproduct for the groups G and H. That is to say, for any group $K = (\mathrm{K}, *, k)$ and group homomorphisms $f_1 \colon G \to K$, $f_2 \colon H \to K$, there is a unique v such that this commutes:

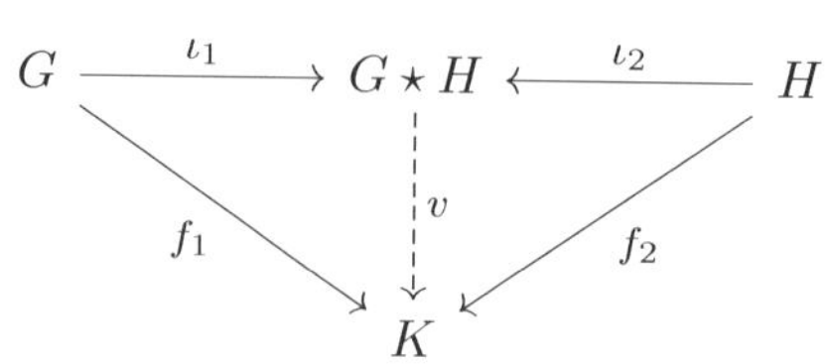

[7] A reality check: why won't the Cartesian product of X with Y (equipped with suitable injections) not also be a coproduct of X with Y?

Proof. Put $v\colon G \star H \to K$ to be the group homomorphism that sends a word such as $g_1 h_1 g_2 h_2 \cdots g_r$ (for g_i among G, and h_i among H) to $f_1(g_1) * f_2(h_1) * f_1(g_2) * f_2(h_2) * \cdots * f_1(g_r)$. By construction, $v \circ \iota_1 = f_1$, $v \circ \iota_2 = f_2$. That makes the diagram commute.

Let v' be any other candidate group homomorphism which makes the diagram commute. Then, to take a simple example, consider gh (one of the objects $\mathrm{G} \star \mathrm{H}$). We have $v'(gh) = v'(g) * v'(h) = v'(\iota_1(g)) * v'(\iota_2(h)) = f_1(g) * f_2(h) = v(\iota_1(g)) * v(\iota_2(h)) = v(g) * v(h) = v(gh)$. Similarly $v'(hg) = v(hg)$. By induction over the length of words w we can then go on to show quite generally $v'(w) = v(w)$. Hence, as required, v is unique. □

12 Products more generally

So we have arrived at a categorial definition of *binary* products. What, though, about products of three or more objects? What about products of an infinite collection of objects?

In fact, the story extending beyond the binary case more or less writes itself. But this chapter briskly spells things out.

12.1 Ternary products

(a) Here's a generalization of our previous definition, moving on from two-way to three-way products:

Definition 56. In any category C, *a ternary product* (O, π_1, π_2, π_3) *for the objects* X_1, X_2, X_3 is an object O together with three projection arrows $\pi_j \colon O \to X_j$ (for $j = 1, 2, 3$) such that for any object S and arrows $f_j \colon S \to X_j$ there is always a unique arrow $u \colon S \to O$ such that $f_j = \pi_j \circ u$. △

And then, exactly as we would expect, using the same proof ideas as in the binary case, we can prove that ternary products are unique up to a unique isomorphism, i.e.

Theorem 41. *If the ternary products* (O, π_1, π_2, π_3) *and* $(O', \pi'_1, \pi'_2, \pi'_3)$ *for* X_1, X_2, X_3 *both exist in* C, *then there is a unique isomorphism* $f \colon O \xrightarrow{\sim} O'$ *commuting with the projection arrows.* □

I can safely leave filling in the details as an exercise.

(b) We now note that if C has binary products for all pairs of objects, then it automatically has ternary products too:

Theorem 42. $(X_1 \times X_2) \times X_3$ *equipped in an obvious way with three projection arrows forms a ternary product of* X_1, X_2, X_3.

Proof. We hack through the details (sorry!).

Start by assuming $(X_1 \times X_2, \pi_1, \pi_2)$ is a product of X_1 with X_2, and also assume that $((X_1 \times X_2) \times X_3, \rho_1, \rho_2)$ is a product of $X_1 \times X_2$ with X_3. Now consider the following diagram.

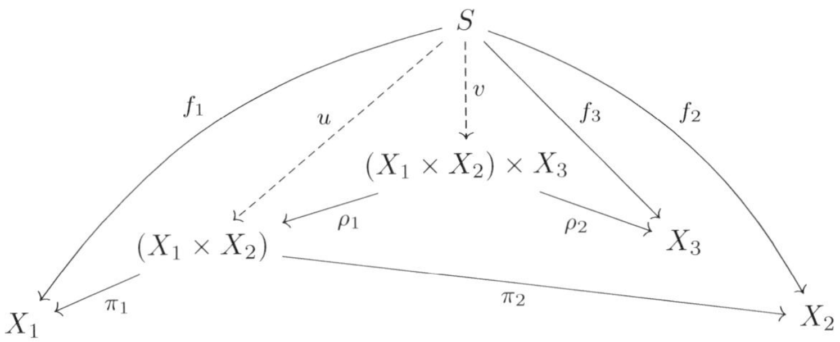

Take any object S and the three arrows $f_i \colon S \to X_i$. By our first assumption, (a) there is a unique $u \colon S \to X_1 \times X_2$ such that $f_1 = \pi_1 \circ u$, $f_2 = \pi_2 \circ u$. And by our second assumption, (b) there is a unique $v \colon S \to (X_1 \times X_2) \times X_3$ such that $u = \rho_1 \circ v$, $f_3 = \rho_2 \circ v$.

Therefore $f_1 = \pi_1 \circ \rho_1 \circ v$, $f_2 = \pi_2 \circ \rho_1 \circ v$, $f_3 = \rho_2 \circ v$.

Now consider the triple wedge $((X_1 \times X_2) \times X_3, \pi_1 \circ \rho_1, \pi_2 \circ \rho_1, \rho_2)$. This, we claim, is indeed a ternary product of X_1, X_2, X_3. And we've seen that the triple wedge with vertex S and arrows $f_i \colon S \to X_i$ factors through $(X_1 \times X_2) \times X_3$ via the arrow v. So it only remains to confirm v's uniqueness in this role.

Suppose that triple wedge also factors through $(X_1 \times X_2) \times X_3$ via the arrow w. In other words, suppose $w \colon S \to (X_1 \times X_2) \times X_3$ where $f_1 = \pi_1 \circ \rho_1 \circ w$, $f_2 = \pi_2 \circ \rho_1 \circ w$, $f_3 = \rho_2 \circ w$. Then $\rho_1 \circ w \colon S \to X_1 \times X_2$ is such that $f_1 = \pi_1 \circ (\rho_1 \circ w)$, $f_2 = \pi_2 \circ (\rho_1 \circ w)$. Hence by (a), $u = \rho_1 \circ w$. But since we also have $f_3 = \rho_2 \circ w$, it follows by (b) that $w = v$. □

(c) Of course, a similar argument will show that a product $X_1 \times (X_2 \times X_3)$ together with the obvious projection arrows will serve as another ternary product of X_1, X_2, X_3. So Theorem 41 entails

Theorem 43. *Assuming the products exist,* $(X_1 \times X_2) \times X_3 \cong X_1 \times (X_2 \times X_3)$.

Question: in what categories with binary products can we upgrade the claim that $(X_1 \times X_2) \times X_3$ is isomorphic to $X_1 \times (X_2 \times X_3)$ to an identity claim?

12.2 More finite products

Defn. 56 defines ternary, i.e. three-way, products: we can give exactly similar definitions for four-way, five-way, n-way products for any finite $n \geq 2$. And just as we can build a three-way product of X_1, X_2 and X_3 from two binary products, as in $(X_1 \times X_2) \times X_3$, we can build a four-way product of X_1, X_2, X_3 and X_4 from three binary products as in $((X_1 \times X_2) \times X_3) \times X_4$. More generally, if we can freely construct any binary products we like, we can also construct n-ary products for any finite $n \geq 2$.

So, to round things out, how do things go for the nullary and unary cases?

Following the same pattern of definition, a *nullary* product in C would be an object O together with *no* projection arrows, such that for any object S there is a unique arrow $u \colon S \to O$. Which tells us that a nullary product is a terminal object of the category.

And a *unary* product of X would be an object O and a single arrow $\pi \colon O \to X$ such that for any object S and arrow $f \colon S \to X$ there is a unique arrow $u \colon S \to O$ for which $\pi \circ u = f$. Putting $O = X$ and $\pi = 1_X$ evidently fits the bill. So the basic case of a unary product of X is X equipped with its identity arrow (and like any product, this is unique up to unique isomorphism). Unary products for all objects exist in all categories.

In sum, suppose we say

Definition 57. A category C *has all binary products* iff for all C-objects X and Y, there exists a binary product of X with Y in C. C *has all finite products* iff it has n-ary products for any n C-objects, for all $n \geq 0$. △

Then our preceding remarks establish

Theorem 44. *A category C has all finite products iff C has a terminal object and has all binary products.* □

Need we add that these theorems of course all dualize to coproducts? How?

12.3 Infinite products

We can now generalize still further, going beyond finite products to infinite cases:

Definition 58. Suppose that we are dealing with some C-objects X_j indexed by items j in some suite of indices J (not assumed finite). Then a product of the X_j, if it exists in C, is an object O together with a projection arrow $\pi_j \colon O \to X_j$ for each index j. It is required that for any object S and family of arrows $f_j \colon S \to X_j$ (one for each index), there is always a unique arrow $u \colon S \to O$ such that $f_j = \pi_j \circ u$. △

For the same reasons as before, such a generalized product will be unique up to unique isomorphism.

Now, we are only going to be really interested in cases where the suite of indices J is not *too* wildly large, so that J can be treated as a *set* in our favoured universe of sets (this means that, if we wanted, we could use our set theory to regiment theorizing about indexing by J). And here's some standard terminology:

Definition 59. A category C *has all small products* iff for any C-objects X_j, for $j \in J$ where J is some index set, these objects have a product. △

To emphasize, 'small' only indicates that we are taking products over collections of objects that are not too big to form a set. And note that smallness in this sense is relative to our favoured universe of sets. We'll be returning to such issues of size later, but here in Part I, we don't need to fuss.

13 Binary products explored

The conceptual basics about products are all in place. This chapter now reads into the record a variety of theorems, showing that categorial products have properties we naturally want them to have, and/or giving results which will be useful later.

To make things a bit more fun (if that's quite the right word!), I will state the chapter's theorems as a series of challenges for enthusiasts to prove. It is up to you how many you tackle.

13.1 Some elementary challenges!

Start by showing this easiest of lemmas, because we'll need it later:

Theorem 45. *Given a product* $(X \times Y, \pi_1, \pi_2)$ *and arrows* $S \underset{v}{\overset{u}{\rightrightarrows}} X \times Y$, *then, if* $\pi_1 \circ u = \pi_1 \circ v$ *and* $\pi_2 \circ u = \pi_2 \circ v$, *it follows that* $u = v$.

Now prove:

Theorem 46. *In a category with binary products, there is an order-swapping isomorphism* $o\colon X \times Y \xrightarrow{\sim} Y \times X$.

Next check a result which looks as though it ought to hold (but why?)

Theorem 47. *In a category that has a terminal object 1, the products* $1 \times X$ *and* $X \times 1$ *always exist, and* $1 \times X \cong X \cong X \times 1$.

Here, as always, when we mention only the object of a product, take this to be equipped with the natural projection arrows. Also show that, by contrast,

Theorem 48. *There are categories with initial objects where the products* $0 \times X$ *and* $X \times 0$ *exist but are not generally isomorphic to* 0.

(Hint: you needn't look beyond the very earliest examples of categories you met.)

Now for another lemma for later use:

Theorem 49. *If* $(1 \times X, !_{1 \times X}, \pi)$ *is a product of a terminal* 1 *with* X, *then* π *is an isomorphism. Similarly for the mirror image result.*

More interestingly, show that for any mediating arrow $\langle\!\langle f, g \rangle\!\rangle$ to some product,

Theorem 50. *Assuming* $\langle\!\langle f, g \rangle\!\rangle$ *and* e *compose,* $\langle\!\langle f, g \rangle\!\rangle \circ e = \langle\!\langle f \circ e, g \circ e \rangle\!\rangle$.

13.2 Six simple theorems, and a non-theorem

(a) We start with two warm-up exercises.

Theorem 45. *Given a product* $(X \times Y, \pi_1, \pi_2)$ *and arrows* $S \overset{u}{\underset{v}{\rightrightarrows}} X \times Y$, *then, if* $\pi_1 \circ u = \pi_1 \circ v$ *and* $\pi_2 \circ u = \pi_2 \circ v$, *it follows that* $u = v$.

Proof. The assumptions tell us that the same wedge $X \leftarrow S \rightarrow Y$ factors through the product both via u and via v:

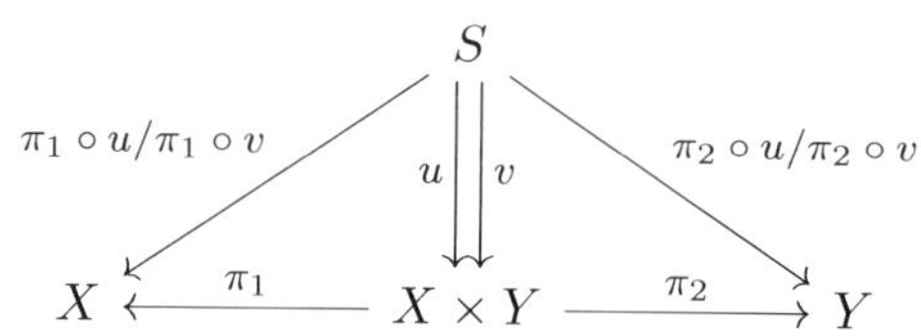

Hence $u = v$ by uniqueness of mediating arrows. □

Theorem 46. *In a category with binary products, there is an order-swapping isomorphism* $o\colon X \times Y \xrightarrow{\sim} Y \times X$.

Proof. Assume we have products $(X \times Y, \pi_1, \pi_2)$ and $(Y \times X, \rho_1, \rho_2)$. And now consider the left-hand diagram:

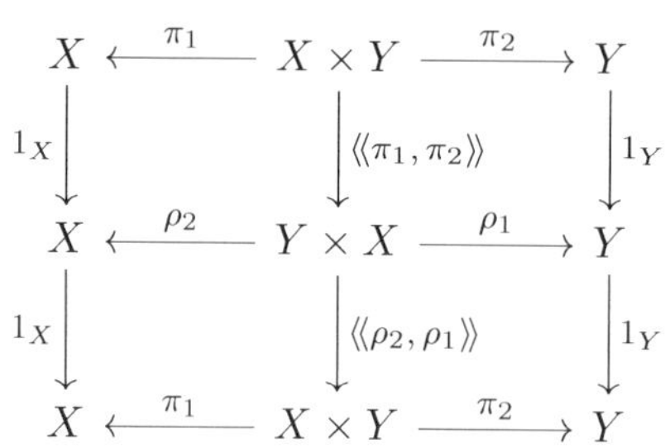

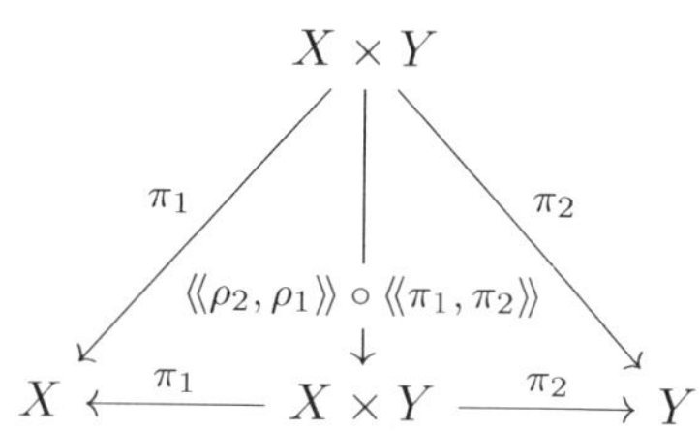

That diagram commutes by the definition of the mediating arrows, and hence the diagram on the right commutes. But evidently the unique mediating arrow must equal $1_{X \times Y}$. Whence $\langle\!\langle \rho_2, \rho_1 \rangle\!\rangle \circ \langle\!\langle \pi_1, \pi_2 \rangle\!\rangle = 1_{X \times Y}$.

Exactly similarly, $\langle\!\langle \pi_1, \pi_2 \rangle\!\rangle \circ \langle\!\langle \rho_2, \rho_1 \rangle\!\rangle = 1_{Y \times X}$.

So $o = \langle\!\langle \pi_1, \pi_2 \rangle\!\rangle$ has a two-sided inverse and is our required isomorphism. □

(b) Next, let's show:

Theorem 47. *In a category that has a terminal object 1, the products* $1 \times X$ *and* $X \times 1$ *always exist, and* $1 \times X \cong X \cong X \times 1$.

Proof. As before, we will use $!_X$ for the unique arrow from X to the terminal object 1, and 1_X is of course the identity arrow on X.

Consider then the wedge $1 \xleftarrow{!_X} X \xrightarrow{1_X} X$, and take any other wedge to 1 and X, namely $1 \xleftarrow{!_Y} Y \xrightarrow{f} X$. The following diagram then commutes:

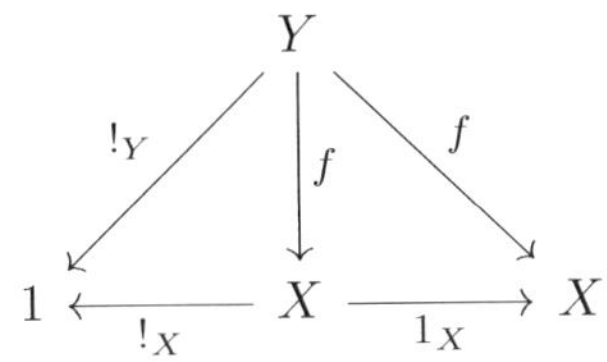

(the triangle on the left commutes because there can only be one arrow from Y to 1 which forces $!_X \circ f = !_Y$). And f is the only vertical arrow that makes the diagram commute. Hence $(X, !_X, 1_X)$ satisfies the conditions for being a product of 1 with X. So, by Theorem 37, given any product $(1 \times X, \pi_1, \pi_2)$, we have $1 \times X \cong X$. Exactly similarly, $X \times 1 \cong X$. □

Note that in a category with all finite products and hence a terminal object, we have both $(X_1 \times X_2) \times X_3 \cong X_1 \times (X_2 \times X_3)$ (as shown in the last chapter) and also $X \times 1 \cong X \cong 1 \times X$. If those isomorphism claims could be upgraded to claims of equality, then our category's objects equipped with a product-forming operation would give us a monoid with 1 as its unit. But in general we only get isomorphisms and thus a monoid-like or *monoidal* structure. We'll return to this thought.

(c) Question: do we also have $0 \times X \cong 0$ in categories with an initial object and the relevant product? Answer: Not always.

Theorem 48. *There are categories where the products $0 \times X$ and $X \times 0$ exist but are not generally isomorphic to* 0.

Proof. Take a category like Grp which has a null object, i.e. where $0 = 1$. Then $X \cong (1 \times X) = (0 \times X)$. But in general, it won't be the case that $X \cong 0$. Hence it won't be the case in general that $(0 \times X) \cong 0$. □

(d) To reduce clutter, we typically drop subscripts from (labels for) the unique arrows to terminal objects. So, in context, '!' might denote e.g. the unique arrow $!_X \colon X \to 1$, or might denote the unique arrow $!_{1 \times X} \colon : 1 \times X \to 1$, and so on. We also drop subscripts from identity arrows.

It then becomes a nice reality check to mentally replace the missing subscripts, for example in the following diagram and in the proof of our next little theorem:

Theorem 49. *If $(1 \times X, !, \pi)$ is a product of a terminal* 1 *with X, then π is an isomorphism. Similarly for the mirror image result.*

We know from Theorem 47 that there is an isomorphism between $1 \times X$ and X; but there could be other arrows between them. So it takes another argument to show that, in *any* wedge like (W) $1 \xleftarrow{\;!\;} 1 \times X \xrightarrow{\;\pi\;} X$, π has to be an isomorphism.

Proof. Consider, then, the following diagram:

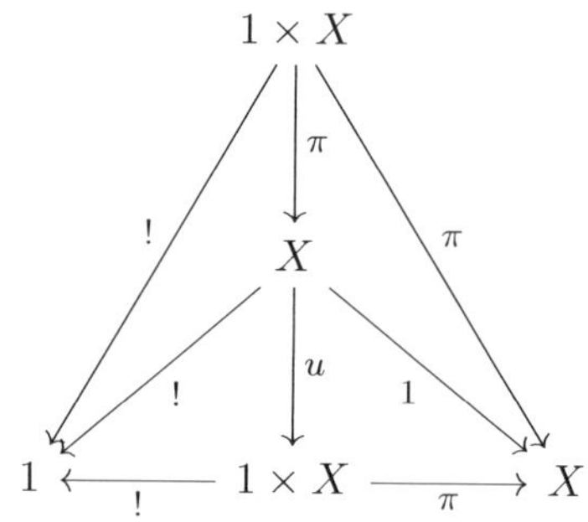

This commutes. Why?

First, there is a (unique) mediating arrow u making the bottom two triangles commute. In other words – and see §11.3 again for 'factors through' – the middle wedge (V) $1 \xleftarrow{\;!\;} X \xrightarrow{\;1\;} X$ factors through the bottom product (W) via a unique u, giving $\pi \circ u = 1$.

Similarly the top wedge, a copy of (W) again, factors through (V) as shown. (The top left triangle commutes, i.e. $!_X \circ \pi = !_{1 \times X}$ because arrows to the same terminal object are unique.)

Putting the upper and lower triangles together means that (W) factors through (W) via the mediating arrow $u \circ \pi$. But since (W) also factors through itself via 1, and such mediating arrows are unique by the definition of a product, it follows that $u \circ \pi = 1$.

Having inverses on both sides, π is therefore an isomorphism. □

(e) Before moving on to our next theorem, we should perhaps remark in passing on a non-theorem.

Suppose we have a pair of parallel composite arrows built up using the same projection arrow like this: $X \times Y \xrightarrow{\pi_1} X \overset{f}{\underset{g}{\rightrightarrows}} X'$. In Set, the projection arrow here 'throws away' the second component of pairs living in $X \times Y$, and all the real action then happens on X: so if $f \circ \pi_1 = g \circ \pi_1$, we should also have $f = g$. Generalizing, we might then suppose that, in any category, projection arrows in products are always right-cancellable, i.e. are epic. But this is false.

Consider the mini category with just four objects together with the following diagrammed arrows (labelled suggestively but noncommittally), plus all identity arrows, and the necessary two composites:

$$X' \overset{f}{\underset{g}{\leftleftarrows}} X \xleftarrow{\pi_1} V \xrightarrow{\pi_2} Y$$

If that is all the data we have to go on, we can consistently stipulate that in this mini-category $f \neq g$ but $f \circ \pi_1 = g \circ \pi_1$. Now, there is only one wedge of the form $X \longleftarrow ? \longrightarrow Y$, so trivially all wedges of that shape uniquely factor through it. In other words, the wedge $X \xleftarrow{\pi_1} V \xrightarrow{\pi_2} Y$ is a product and π_1 is a projection arrow. But by construction it isn't epic.

(f) Suppose $\langle\langle f, g \rangle\rangle$ is a mediating arrow from some wedge $X \xleftarrow{f} S \xrightarrow{g} Y$ to a product of X with Y, then

Theorem 50. *Assuming $\langle\langle f, g \rangle\rangle$ and e compose, $\langle\langle f, g \rangle\rangle \circ e = \langle\langle f \circ e, g \circ e \rangle\rangle$.*

Proof. Since we are assuming $\langle\langle f, g \rangle\rangle \circ e$ is defined, the target of e must be the source of the arrows f and g.

So this diagram commutes because each triangle commutes:

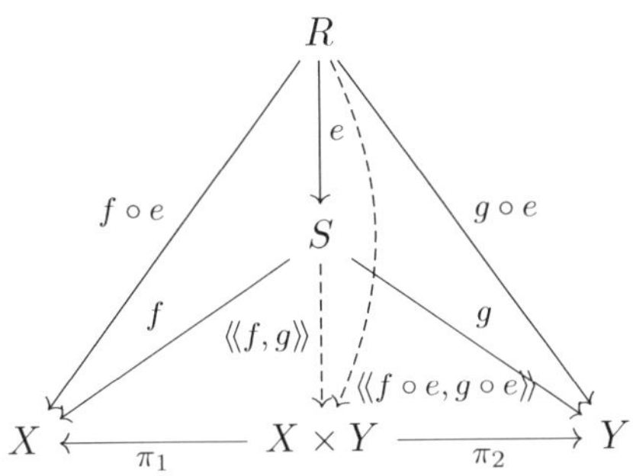

Hence in particular $\langle\langle f, g \rangle\rangle \circ e$ is a mediating arrow factoring the wedge with apex R through the product $X \times Y$.

But by definition, the unique mediating arrow is $\langle\langle f \circ e, g \circ e \rangle\rangle$. □

We should note a corollary. Defn. 52 defined δ_X as the arrow $\langle\langle 1_X, 1_X \rangle\rangle : X \to X \times X$. If f is an arrow whose target is X, then our theorem shows that $\delta_X \circ f = \langle\langle 1_X \circ f, 1_X \circ f \rangle\rangle = \langle\langle f, f \rangle\rangle$.

13.3 Arrows between two products

Suppose we have two arrows $f : X \to X'$, $g : Y \to Y'$. Then how can we characterize an arrow between products, $f \times g : X \times Y \to X' \times Y'$, which works component-wise? We want $f \times g$ to send the product of elements x and y (living in $X \times Y$) to the product of $f(x)$ and $g(y)$ (living in $X' \times Y'$). What is an appropriate categorial definition for $f \times g$?

Well, put in categorial terms, we require $f \times g$ to be such that the following diagram commutes:

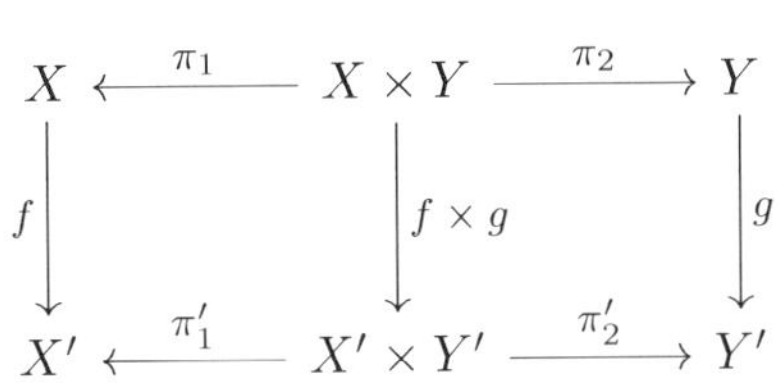

Note, however, that the vertical arrow is then a mediating arrow from the wedge $X' \xleftarrow{f \circ \pi_1} X \times Y \xrightarrow{g \circ \pi_2} Y'$ through the product $X' \times Y'$. Therefore $f \times g$ is

fixed uniquely by the requirement that that diagram commutes, and hence must equal $\langle\langle f \circ \pi_1, g \circ \pi_2 \rangle\rangle$.

Think through how this works in Set, and the following definition should now look a sensible one:

Definition 60. Given the arrows $f \colon X \to X'$, $g \colon Y \to Y'$, and the products $(X \times Y, \pi_1, \pi_2)$ and $(X' \times Y', \pi_1', \pi_2')$, then put $f \times g = \langle\langle f \circ \pi_1, g \circ \pi_2 \rangle\rangle \colon X \times Y \to X' \times Y'$. $f \times g$ then acts component-wise on the product $X \times Y$, like f on X and g on Y, and $\pi_1' \circ (f \times g) = f \circ \pi_1$ and $\pi_2' \circ (f \times g) = g \circ \pi_2$. △

13.4 More challenges!

Ask yourself: why *ought* the following theorems hold? Then prove them!

Theorem 51. *Assuming the arrows compose,* $(f \times g) \circ \langle\langle j, k \rangle\rangle = \langle\langle f \circ j, g \circ k \rangle\rangle$.

Theorem 52. *Suppose we have arrows* $f \colon X \to X$ *and* $g \colon Y \to Y$, *and an order-swapping isomorphism* $o \colon X \times Y \to Y \times X$. *Then* $o \circ (f \times g) = (g \times f) \circ o$.

Theorem 53. *Suppose we have parallel arrows* $f, g \colon X \to Y$ *in a category with binary products. Then the arrow* $\langle\langle f, g \rangle\rangle$ *is equal to the composite* $(f \times g) \circ \delta_X$.

Theorem 54. $1_X \times 1_Y = 1_{X \times Y}$

Theorem 55. *Assume that there are arrows*

$$\begin{array}{ccccc} X & \xrightarrow{\quad f \quad} & X' & \xrightarrow{\quad j \quad} & X'' \\ Y & \xrightarrow{\quad g \quad} & Y' & \xrightarrow{\quad k \quad} & Y'' \end{array}$$

Assume there are products $(X \times Y, \pi_1, \pi_2)$, $(X' \times Y', \pi_1', \pi_2')$ *and* $(X'' \times Y'', \pi_1'', \pi_2'')$. *Then we have the 'interchange law'* $(j \times k) \circ (f \times g) = (j \circ f) \times (k \circ g)$.

And finally, let's add an easy result which we'll need later and which provides a nice little reality check:

Theorem 56. *Suppose* $\pi_X : 1 \times X \to X$ *is the appropriate projection arrow for the product, and let* $\vec{x} \colon 1 \to X$ *be any point element of* X. *Then* $(\vec{x} \times 1_X) \circ \pi_X^{-1} \circ \vec{x} = \langle\langle \vec{x}, \vec{x} \rangle\rangle$.

13.5 Theorems about 'products' of arrows

(a) First, we want to show

Theorem 51. *Assuming the arrows compose,* $(f \times g) \circ \langle\langle j, k \rangle\rangle = \langle\langle f \circ j, g \circ k \rangle\rangle$.

Proof. Consider the following diagram:

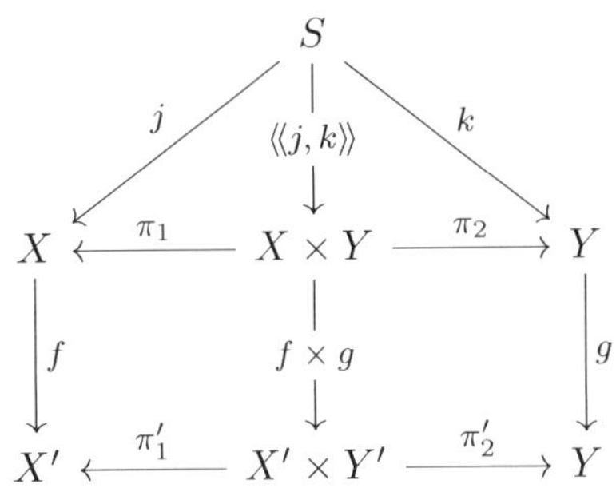

This commutes by assumption. So the composite $(f \times g) \circ \langle\langle j, k\rangle\rangle$ provides the mediating arrow in a product of $f \circ j$ and $g \circ k$. □

(b) Next, let's show

Theorem 52. *Suppose we have arrows $f\colon X \to X$ and $g\colon Y \to Y$, and an order-swapping isomorphism $o\colon X \times Y \to Y \times X$. Then $o \circ (f \times g) = (g \times f) \circ o$.*

This ought to hold because it shouldn't matter whether we first apply f and g component-wise to a product, and then swap the order of terms in the product, or alternatively first swap the order of terms, and then apply g and f component-wise.

Proof. Suppose we have products $(X \times Y, \pi_1, \pi_2)$ and $(Y \times X, \pi'_1, \pi'_2)$, and an order-swapping isomorphism $o\colon X \times Y \to Y \times X$. And now consider the following pair of diagrams (being very careful with the directions of the projection arrows!):

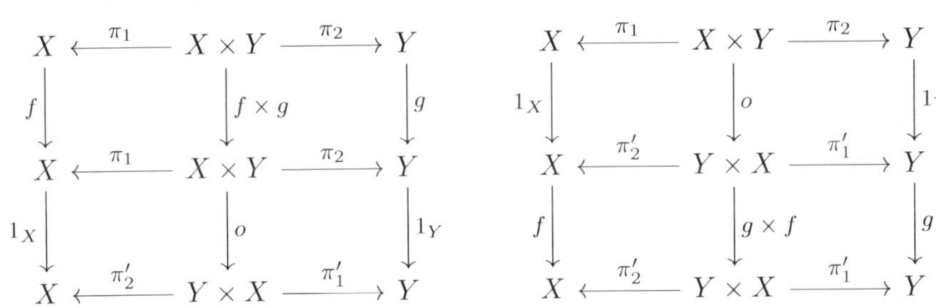

Both diagrams commute. Hence the wedge $X \xleftarrow{f \circ \pi_1} X \times Y \xrightarrow{g \circ \pi_2} Y$ factors through the bottom product via both $o \circ (f \times g)$ and $(g \times f) \circ o$. Those arrows must therefore be equal by the uniqueness of mediating arrows. □

(c) Do keep the distinction between $f \times g$ and $\langle\langle f, g\rangle\rangle$ firmly in mind! For a start, in a category with all products, $f \times g$ (for any arrows f, g) will always exist, and its source is the product of the typically different sources of f and g. While $\langle\langle f, g\rangle\rangle$ is only defined when f and g have the same source. However:

Theorem 53. *Suppose we have parallel arrows $f, g\colon X \to Y$ in a category with binary products. Then the arrow $\langle\langle f, g\rangle\rangle$ is equal to the composite $(f \times g) \circ \delta_X$.*

Think of the situation in Set, for example. The idea is that it should not matter whether we apply the functions f and g separately to some member x of X and then take the product of the results, or alternatively form the product of x with itself and then apply f and g to the resulting pair componentwise.

Proof. Take the diagram

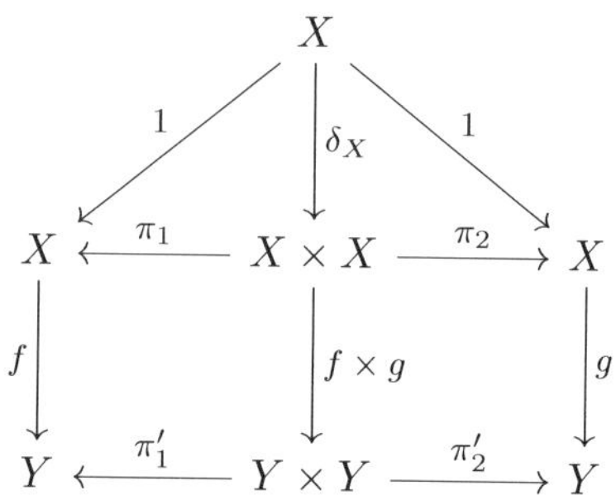

This commutes by the definitions of δ_X and $f \times g$. Hence this also commutes:

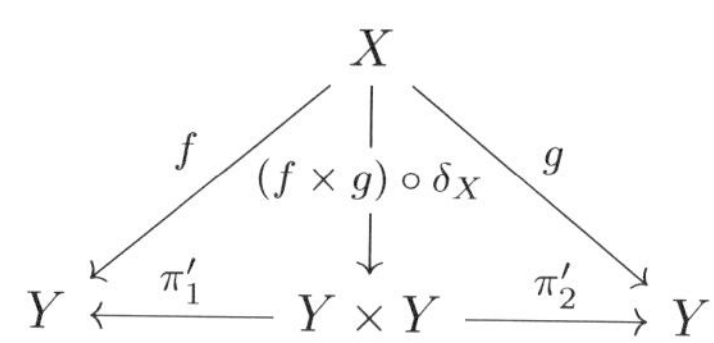

Which makes $(f \times g) \circ \delta_X$ the mediating arrow in a product diagram, so by uniqueness and the definition of $\langle\langle f, g\rangle\rangle$, we have $(f \times g) \circ \delta_X = \langle\langle f, g\rangle\rangle$. □

(d) Here's a special case: sometimes we have a function $f\colon X \to X'$ and we want to define an arrow from $X \times Y$ to $X' \times Y$ which applies f to the first component of a product and leaves the second alone. Then $f \times 1_Y$ will do the trick.

Now, it is tempting to suppose that if we have parallel maps $f, g\colon X \to X'$ and $f \times 1_Y = g \times 1_Y$, then $f = g$. But this actually fails in some categories. For example, consider again the toy category we met in §13.2, whose only arrows are as diagrammed

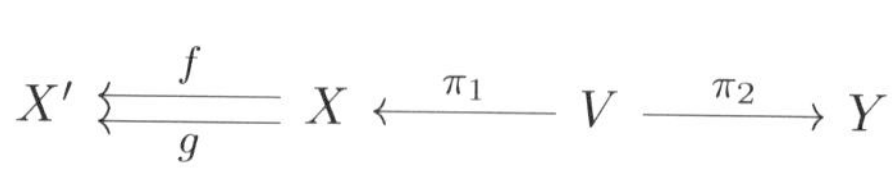

(together with the necessary identities and composites), and where $f \neq g$. Yet, by stipulation, $f \circ \pi_1 = g \circ \pi_1$, hence $\langle\langle f \circ \pi_1, 1_Y \circ \pi_2\rangle\rangle = \langle\langle g \circ \pi_1, 1_Y \circ \pi_2\rangle\rangle$, and hence $f \times 1_Y = g \times 1_Y$.

(e) Now for three more results we will need later.

Theorem 54. $1_X \times 1_Y = 1_{X \times Y}$

Proof. Consider the following diagram:

$$\begin{array}{ccccc}
X & \xleftarrow{\pi_1} & X \times Y & \xrightarrow{\pi_2} & Y \\
\downarrow{\scriptstyle 1_X} & & \downarrow{\scriptstyle 1_X \times 1_Y} & & \downarrow{\scriptstyle 1_Y} \\
X & \xleftarrow{\pi_1} & X \times Y & \xrightarrow{\pi_2} & Y
\end{array}$$

This commutes by the definition of $1_X \times 1_Y$.

So the wedge $X \xleftarrow{1_X \circ \pi_1} X \times Y \xrightarrow{1_Y \circ \pi_2} Y$ (i.e. $X \xleftarrow{\pi_1} X \times Y \xrightarrow{\pi_2} Y$) factors through itself (a product!) by $1_X \times 1_Y$. But of course it also factors through itself by $1_{X \times Y}$. Hence, by the uniqueness of mediating arrows for products, $1_X \times 1_Y = 1_{X \times Y}$. □

Theorem 55. *Assume that there are arrows*

$$\begin{array}{ccccc}
X & \xrightarrow{f} & X' & \xrightarrow{j} & X'' \\
Y & \xrightarrow{g} & Y' & \xrightarrow{k} & Y''
\end{array}$$

Assume there are products $(X \times Y, \pi_1, \pi_2)$, $(X' \times Y', \pi'_1, \pi'_2)$ *and* $(X'' \times Y'', \pi''_1, \pi''_2)$. *Then we have the 'interchange law'* $(j \times k) \circ (f \times g) = (j \circ f) \times (k \circ g)$.

Proof. By the defining property of arrow products applied to the three different products we get,

$$\pi''_1 \circ (j \times k) \circ (f \times g) = j \circ \pi'_1 \circ (f \times g) = j \circ f \circ \pi_1 = \pi''_1 \circ (j \circ f) \times (k \circ g).$$

Similarly

$$\pi''_2 \circ (j \times k) \circ (f \times g) = \pi''_2 \circ (j \circ f) \times (k \circ g)$$

The theorem then immediately follows by our warm-up lemma, Theorem 45. □

Theorem 56. *Suppose* $\pi_X : 1 \times X \to X$ *is the appropriate projection arrow for the product, and let* $\vec{x} \colon 1 \to X$ *be any point element of* X. *Then* $(\vec{x} \times 1_X) \circ \pi_X^{-1} \circ \vec{x} = \langle\!\langle \vec{x}, \vec{x} \rangle\!\rangle$.

Proof. The target equation makes sense because π is an isomorphism by Theorem 49, so it has an inverse π^{-1}. So now consider the following diagram:

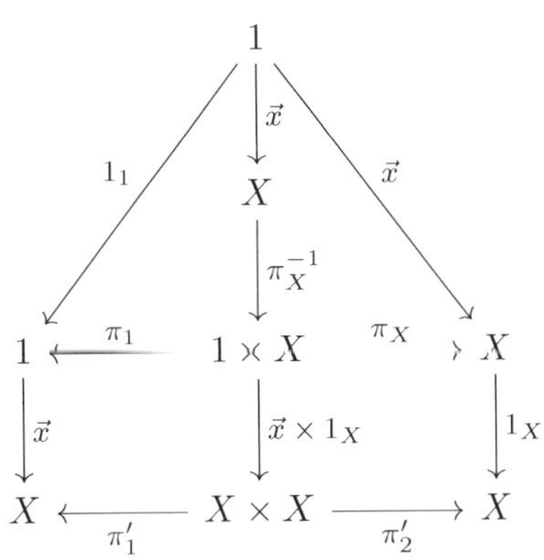

The top triangles commute trivially, and the bottom half of the diagram defines the product of arrows. Then the composite arrow down from 1 to $X \times X$ shows us that $(\vec{x} \times 1_X) \circ \pi_X^{-1} \circ \vec{x} = \langle\!\langle \vec{x} \circ 1_1, 1_X \circ \vec{x} \rangle\!\rangle = \langle\!\langle \vec{x}, \vec{x} \rangle\!\rangle$. □

(f) A (tantalizing?) remark. Given a category C with binary products, we've now seen that we can not only define an operation on *objects* that takes us from a pair of objects X, Y to an object $X \times Y$ but also define a companion operation on *arrows* that takes us from a pair of arrows f, g to a sort of 'product' arrow $f \times g$. Later, we'll see that we can package these definitions together nicely to give a map – officially, a functor – from $\mathsf{C} \times \mathsf{C}$ to C which operates on both the object and arrow data of $\mathsf{C} \times \mathsf{C}$ in a consistent way. More on that in due course.

13.6 Another category with all products

Double-check the definition of M_2, (C22) in §5.7. Then the final challenge in this chapter is to prove the following:

Theorem 57. *The category* M_2 *has all binary products.*

Proof. Suppose X, Y are objects of Set, and $f\colon X \to X$, $g\colon Y \to Y$ are idempotent functions. Then (X, f), (Y, g) are objects of M_2. We want to find their product. So consider the following diagram in M_2:

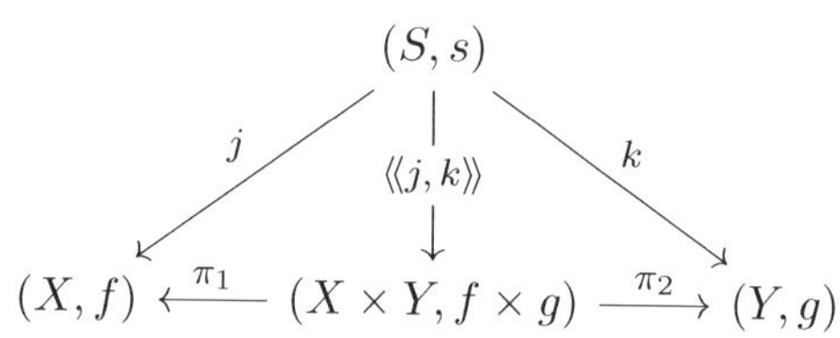

We want to find an M_2-object (O, o) with projection arrows to (X, f) and (Y, g) such that for any pair of M_2-arrows $j\colon (S, o) \to (X, f)$ and $k\colon (S, o) \to (Y, g)$ there will be a unique mediating arrow $u\colon (S, o) \to (O, o)$ making a commuting diagram in the usual way. And the obvious candidate to try first for (O, o) is $(X \times Y, f \times g)$ – what else?

We first need to check that $f \times g$ is idempotent, so that $(X \times Y, f \times g)$ really is an M_2-object. But that's easy: $(f \times g) \circ (f \times g) = (f \circ f) \times (g \times g) = f \times g$ with the first equation by the interchange law (Theorem 55), and the second by our initial assumption that f and g are idempotent.

Now, an M_2-arrow m from $(X \times Y, f \times g)$ to (X, f) is by definition a function such that $m \circ f \times g = f \circ m$. But by Defn. 60, $\pi_1\colon X \times Y \to X$ exactly fits the bill (where π_1 is the projection arrow for the product in Set). So that gives us our proposal for the product of (X, f) and (Y, g), as diagrammed.

Remembering that M_2-arrows are Set-arrows, our sole candidate for the mediating arrow completing the product diagram is then going to be $\langle\!\langle j, k \rangle\!\rangle$. But does this work? What's required is for $\langle\!\langle j, k \rangle\!\rangle$ to count as an M_2-arrow from (S, s) to $(X \times Y, f \times g)$. That holds so long as it is equivariant, which in this case means

(*) $\langle\!\langle j, k \rangle\!\rangle \circ s = f \times g \circ \langle\!\langle j, k \rangle\!\rangle$.

And it is now routine to check that identity.

So to finish, note that by Theorem 50

(i) $\langle\!\langle j, k \rangle\!\rangle \circ s = \langle\!\langle j \circ s, k \circ s \rangle\!\rangle$

While Theorem 51 tells us thar

(ii) $f \times g \circ \langle\!\langle j, k \rangle\!\rangle = \langle\!\langle f \circ j, g \circ k \rangle\!\rangle$.

But j is an M_2-arrow from (S, s) to (X, f), so $j \circ s = f \circ j$. Similarly, $k \circ s = g \circ k$. Hence from (ii) we have

(iii) $f \times g \circ \langle\!\langle j, k \rangle\!\rangle = \langle\!\langle j \circ s, k \circ s \rangle\!\rangle$.

Putting (i) and (iii) together gives us (*). □

14 Groups in categories

I'm going to pause at this point, before developing any more categorial apparatus. Because I want to show that we have already said enough to characterize so-called *internal groups* living in categories. I needn't take the discussion very far: the aim is simply to illustrate how we can begin to talk about one familiar type of mathematical structure in in category-theoretic terms.

14.1 Instead of binary functions

In category theory, arrows have single sources. So, as already noted in §10.3, in categories where arrows are functions, these are always monadic functions. But then how can we accommodate binary functions (or polyadic functions more generally)? In particular, how can we accommodate the binary operations that characterize groups?

Recall the orthodox set-theoretic approach: we model e.g. a two-place total function from numbers to numbers (addition, say) as a function $f\colon \mathbb{N}^2 \to \mathbb{N}$. Here, $\mathbb{N}^2$ is the Cartesian product of $\mathbb{N}$ with itself, i.e. is the set of ordered pairs of numbers. And an ordered pair is *one* thing, not two things. So a function $f\colon \mathbb{N}^2 \to \mathbb{N}$ is strictly speaking a *unary* function, a function that maps *one* argument, a pair-object, to a value: such an f is not a real binary function.

Of course, with the usual set-theoretic apparatus in play, we can arrange it that for any two things there is a pair-object that codes for them – we usually choose a Kuratowski pair. Hence we can unproblematically trade in a function from two objects for a related function from a single corresponding pair-object. And standard notational choices can make the move quite invisible. Suppose we adopt the modern convention of using '(m, n)', with common-or-garden parentheses, as our notation for the ordered pair of m with n. Then the notation '$f(m, n)$' invites being parsed either way, as representing a two-place function taking the two arguments m and n, or as a corresponding one-place function taking a single argument, the pair (m, n). But note: the fact that the trade between the two-place and the one-place function is now notationally disguised doesn't mean that it isn't being made!

In sum, a familiar procedure is to trade in an underlying binary function $\underline{f}\colon A, B \to C$ for a related unary function $f\colon A \times B \to C$. This same procedure is of course now available to us in any category with products.

14.2 Internal groups in Set, Top and Man

(a) So to our main theme. How can we characterize groups – OK, if you insist on being pernickety, implementations of groups – living in the category Set?

Dropping the underlining notation for underlying sets of groups, we need a set G in our category that collects together the elements of the relevant group, and we need three arrows:

(i) $m\colon G \times G \to G$ (this represents the group operation – so here we have traded the two-place group operation for an arrow from a corresponding single source, a Cartesian product);
(ii) $\vec{e}\colon 1 \to G$ (this element-as-arrow is going to pick out a particular group-element in G to be the group identity e);
(iii) $i\colon G \to G$ (this will send a group element to its inverse).

We then need to impose constraints on these arrows corresponding to the usual group axioms:

(1) We require the group operation represented by m to be associative. Categorially, consider the following diagram:

(G1)

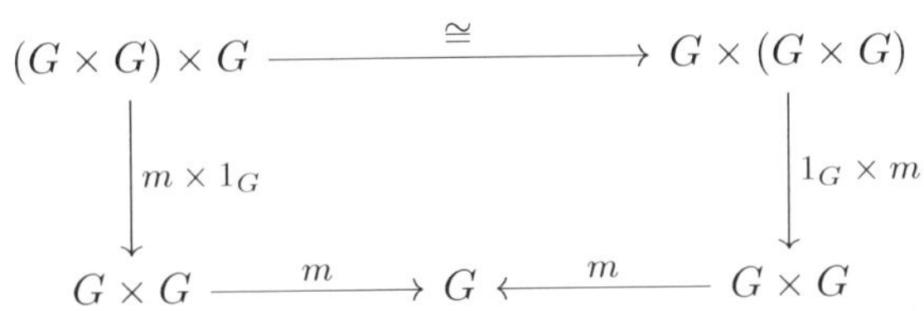

Here the arrow at the top represents the naturally arising isomorphism between the two triple products (cf. Theorem 43).

Remembering that we are working in Set, take an element $\langle\langle j, k\rangle, l\rangle \in (G \times G) \times G$. Going round on the left, that gets sent to $\langle m\langle j, k\rangle, l\rangle$ and then to $m\langle m\langle j, k\rangle, l\rangle$. Going round the other direction we get to $m\langle j, m\langle k, l\rangle\rangle$. So requiring the diagram to commute captures the associativity of the operation represented by m.

(2) We next require the distinguished object which $\vec{e}$ picks out to act as an identity for the group operation.

To characterize this condition categorially, start by defining the map $e!\colon G \to G$ by composing the unique map $!\colon G \to 1$ followed by $\vec{e}\colon 1 \to G$. In Set, $e!$ is then the function which sends anything in G to the group's identity element e, and we have the following product diagram:

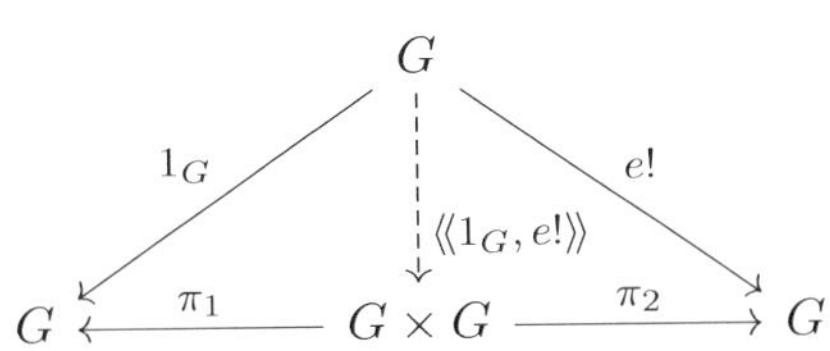

So we can think of the mediating arrow $\langle\!\langle 1_G, e! \rangle\!\rangle$ as sending an element $g \in G$ to the pair $\langle g, e \rangle$. And the element e then behaves like a multiplicative identity on the right just if m sends this pair $\langle g, e \rangle$ in turn back to g – i.e. if the top triangle in the following diagram commutes:

(G2)

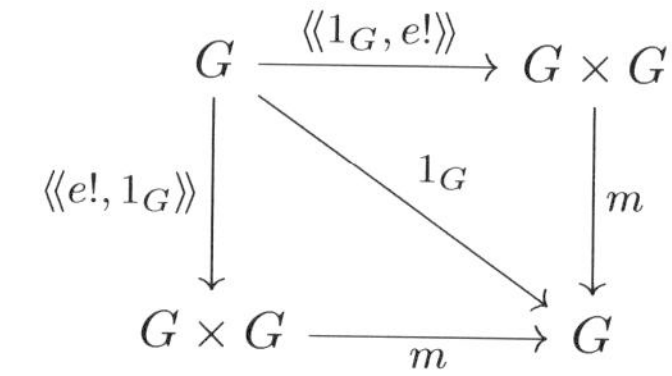

Similarly, the lower triangle commutes just if e behaves as an identity on the left. So, for e to behave as a two-sided identity, it is enough that the whole diagram commutes.

(3) Finally, we require that every element $g \in G$ has an inverse g^{-1} whose group product with g is e. Categorially, we can express this using an arrow $i \colon G \to G$ by requiring that the following commutes:

(G3)

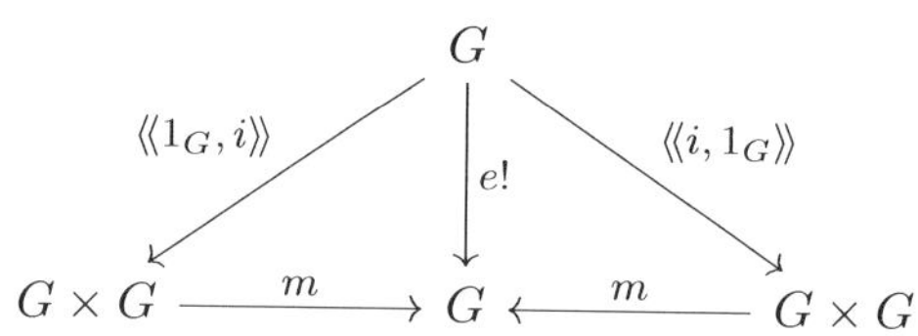

For take an element $g \in G$. Going left, the arrow $\langle\!\langle 1_G, i \rangle\!\rangle$ maps g to $\langle g, i(g) \rangle$ which is then sent by m to $m\langle g, i(g) \rangle$. The central vertical arrow meanwhile simply sends g to e. Therefore, the requirement that the left triangle commutes tells us that $m\langle g, i(g) \rangle = e$, as we want if i is to take an element to its inverse. Similarly the requirement that the right triangle commutes tells us that $m\langle i(g), g \rangle = e$.

In summary then, the usual group axioms correspond to the commutativity of our three diagrams (G1), (G2) and (G3).

Familiar elementary consequences of the axioms follow. For example, note that $e! \circ \vec{g} = \vec{e}$ (why?) Hence by Theorem 50 and (G2),

$$m \circ \langle\!\langle \vec{e}, \vec{g} \rangle\!\rangle = m \circ \langle\!\langle e!, 1_G \rangle\!\rangle \circ \vec{g} = 1_G \circ \vec{g} = \vec{g}.$$

and similarly

$$m \circ \langle\!\langle \vec{g}, \vec{e} \rangle\!\rangle = \vec{g}.$$

So suppose $\vec{e}_1$ and $\vec{e}_2$ both satisfy (G2). Then we'd have

$$\vec{e}_1 = m \circ \langle\!\langle \vec{e}_1, \vec{e}_2 \rangle\!\rangle = \vec{e}_2,$$

the categorial analogue of the basic fact that group identities are unique. Exactly as we want.

(b) Now note that this categorial treatment of groups in Set in fact makes sense when we are working in any category with binary products and a terminal object. So it is natural to generalize, as follows:

Definition 61. Suppose C is a category that has binary products and a terminal object. Let G be a C-object, and $m \colon G \times G \to G$, $e \colon 1 \to G$ and $i \colon G \to G$ be C-arrows. Then (G, m, e, i) is an *internal group* in C iff the three diagrams (G1), (G2), (G3) commute, where $e!$ in the latter two diagrams is the composite map $G \xrightarrow{!} 1 \xrightarrow{e} G$. △

An internal group is, alternatively, also called a 'group object'.

Then, if we don't fuss about the type-difference between an arrow $e \colon 1 \to G$ (in an internal group) and a designated element e (in a group), we have established the summary result

Theorem 58. *In the category* Set, *an internal group implements a group.* □

And conversely, every group living in our default world of sets can be implemented as an internal group in Set.

(c) Here are a couple more examples of internal groups in other categories:

Theorem 59. *(1) In the category* Top, *which comprises topological spaces with continuous maps between them, an internal group implements a topological group in the standard sense.*

(2) In the category Man, *which comprises smooth manifolds with smooth maps between them, an internal group implements a Lie group.*

The proofs of these two claims are pretty straightforward, at least if you know the usual definitions of topological groups and Lie groups. But I won't pause over the details here. I simply note that the categorial story here will very nicely bring out what is common between the various cases.

14.3 Groups in Grp

(a) A brain-teaser worth pausing over: what about the internal groups in our inclusive category of groups Grp?

In the spirit of §4.3, Grp is to comprise all groups-implemented-as-sets. So, on our chosen style of implementation as in e.g. §7.1, objects in the category will be, say, ordered triples $\langle \underline{G}, \cdot, \dot{1} \rangle$, where $\underline{G}$ is the underlying set of the group, $x \cdot y$ is a binary set-function from $\underline{G}$ to itself, and $\dot{1} \in \underline{G}$ is the group identity.

An 'internal group', by contrast, is a group $\langle \underline{G}, \cdot, \dot{1} \rangle$ together with homomorphisms m, e, i. If you want to package that up as a tuple, it's a quadruple $\langle \langle \underline{G}, \cdot, \dot{1} \rangle, m, e, i \rangle$ – and note that m as a set-function this time doesn't operate on the first component of the tuple, but on the first component of the first component. Thus regarded, the internal group has a different set-structure from the objects of Grp.

So, at least on our view of the category, an internal group of Grp is not a group in Grp!

(b) But that's getting a bit pernickety. The internal group can perfectly respectably be regarded as implementing a group, albeit in a different style to groups in Grp. And then the fun result is this:

Theorem 60. *In the category* Grp*, an internal group must be abelian.*

How strange! Yet the proof is relatively straightforward, quite cute, and a rather useful reality-check. So – at least for enthusiasts – here's the argument:

Proof. Suppose G, equipped with m, e, i is an internal group in Grp.

Then, since we are in the category Grp, the object G is itself *already* some group $\langle \underline{G}, \cdot, \dot{1} \rangle$.

Now note that the arrow $e \colon 1 \to G$ of the internal group must also pick out a distinguished element of $\underline{G}$, call it '$\tilde{1}$', an identity for m.[1]

Again, since we are in Grp, the arrows in the category are all group homomorphisms. In particular, m is a homomorphism from $G \times G$ to G.

Let $\times$ be the group operation of $G \times G$, which we can take to be fixed componentwise by the group operation of G in the standard way (see §11.2, Ex. (3)) – so $\langle x, y \rangle \times \langle z, w \rangle = \langle x \cdot z, y \cdot w \rangle$, for any elements $x, y, z, w \in \underline{G}$. Then,

$$m\langle x \cdot z, y \cdot w \rangle = m(\langle x, y \rangle \times \langle z, w \rangle) = m\langle x, y \rangle \cdot m\langle z, w \rangle$$

where the second equation holds because m is a homomorphism.

For elegance, let's rewrite $m\langle x, y \rangle$ as $x \star y$ (so $\tilde{1}$ is the unit for $\star$). Then we have established the interchange law

$$(x \cdot z) \star (y \cdot w) = (x \star y) \cdot (z \star w).$$

We will now use this law twice over (the proof from this point on uses what is standardly called the Eckmann–Hilton argument, a general principle applying when we have such an interchange law between two binary operations with units).

First, we have

$$\dot{1} = \dot{1} \cdot \dot{1} = (\tilde{1} \star \dot{1}) \cdot (\dot{1} \star \tilde{1}) = (\tilde{1} \cdot \dot{1}) \star (\dot{1} \cdot \tilde{1}) = \tilde{1} \star \tilde{1} = \tilde{1}$$

We can therefore just write 1 for the shared unit, and now show secondly that

$$\begin{aligned} x \cdot y &= (x \star 1) \cdot (1 \star y) = (x \cdot 1) \star (1 \cdot y) = x \star y \\ &= (1 \cdot x) \star (y \cdot 1) = (1 \star y) \cdot (x \star 1) = y \cdot x. \end{aligned}$$

By the end of the first line we have shown that $x \cdot y = x \star y$; so the binary operation represented by the internal group's arrow m is the same as G's own

[1] At this stage, note, the only assumption we've made is that everything fits together to make $\langle \underline{G}, \cdot, \dot{1} \rangle$ equipped with m, e, i an internal group in Grp. At this stage, it is up for grabs e.g. what the relation between $\dot{1}$ and $\tilde{1}$, if any, might be.

group operation. By the end of the second line we have shown that $x \cdot y = y \cdot x$, so G's own group operation commutes, making G abelian. Hence the operation m of the internal group (G, m, e, i) also commutes, and hence the internal group is indeed also abelian in the obvious sense. □

Not every group in the category Grp is abelian. Every internal group of that category *is* abelian. So not every group in Grp is equivalent to an internal group of that category.[2]

14.4 The story continues ...

(a) We can continue the story, defining further group-theoretic notions in categorial terms. For a start, we can categorially define the idea of a homomorphism between internal groups in a category.

Suppose (G, m, e, i) and (G', m', e', i') are internal groups in Set. Then a homomorphism between them is a C-arrow $h\colon G \to G'$ which 'preserves structure' by appropriately commuting with the group-objects' arrows. More precisely, a moment's reflection shows that h is a homomorphism if and only if the following three diagrams commute:

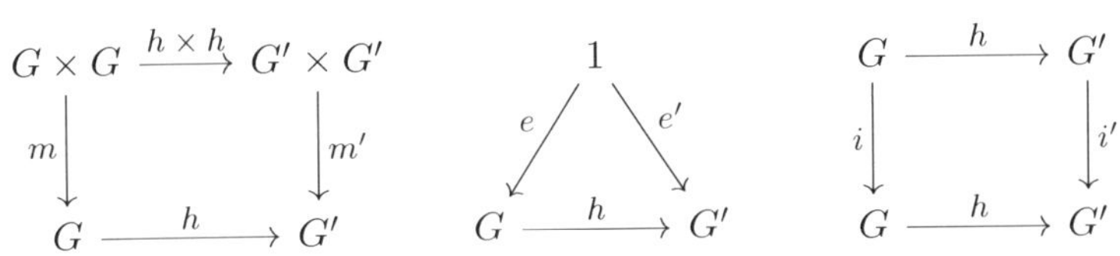

So we can in this way begin to recast core group-theoretic ideas into a categorial framework. And the richer the category we work in, the more group theory we can do: for example, if our category also has the resources for constructing quotients (see Chapter 15), then we can get quotient groups.

(b) The explorations we have gestured towards here can be continued in various directions. We can similarly define other kinds of algebraic objects and their morphisms within categories. And noting that we can now define group-objects and group-homomorphisms inside a given category, we could go on to categorially define whole *categories* of groups living in other categories. However, things do begin to get pretty abstract (and not in a way that is particularly helpful for us at this stage).

[2]Does that still seem strange? It shouldn't when we note that the group operations of members of Grp are any suitable binary operations satisfying certain conditions, while the group operations of internal groups are, specifically, group homomorphisms, which (as we have now seen) imposes extra constraints.

And note too that while we've shown that the internal group (G, m, e, i)'s structure as a group mirrors G's, this *doesn't* mean that the internal group 'contains itself' in any problematic way! Why not?

15 Quotients, pre-categorially

Forming product widgets is a ubiquitous procedure in pre-categorial maths. So too is forming new widgets by quotienting old widgets by suitable equivalence relations. The general idea, roughly put, is that we start with a given widget together with an equivalence relation on its objects, where equivalent objects behave in suitably congruent ways. We then, as it were, 'collapse' equivalent objects into a single object, and we arrive at a new widget formed from these 'collapsed' objects. In §2.3(c) we saw how this general idea begins to play out in one particular case, when we form a new group by quotienting an old one using a suitable congruence relation.

In this chapter, we will think a little more about how equivalence relations can be generated and about what is involved in quotienting. These pre-categorial reflections will shape our categorial account in the next chapter.

15.1 Equivalence relations

(a) To fix terminology:

Definition 62. For brevity, we'll say that the reflexive, symmetric, transitive closure of a relation R is (simply) its *closure*. △

The closure of R is, of course, the smallest equivalence relation containing R.

We now note two ways in which functions can generate equivalence relations.

Definition 63. Take any function $k\colon \mathrm{Y} \to \mathrm{Z}$. Then E_k – the *equivalence kernel* of k – is the equivalence relation on the objects Y such that yE_ky' if and only if $k(y) = k(y')$. △

Definition 64. Take any pair of functions $f, g\colon \mathrm{X} \to \mathrm{Y}$. Suppose P_{fg} is the relation on the objects Y projected by f and g acting on objects among X; i.e., $yP_{fg}y'$ if and only if there is some $x \in \mathrm{X}$ such that $f(x) = y$ and $g(x) = y'$. Then E_{fg} – the *equivalence projection* of f and g – is the closure of P_{fg}. △

Note, every equivalence relation $\sim$ on Y is the equivalence projection of some parallel functions into Y. Take some scheme for pairing Y with Y, and let X comprise the pairs $\langle y, y'\rangle$ where $y \sim y'$. Define $f, g\colon \mathrm{X} \to \mathrm{Y}$ so that f sends a pair $\langle y, y'\rangle$ to y and g sends $\langle y, y'\rangle$ to y'. Then of course $\sim$ is the equivalence projection of f, g.

(b) We will also need this notion:

Definition 65. The function $k\colon \mathrm{Y} \to \mathrm{Z}$ *respects the relation* R defined over the objects Y iff, whenever yRy', $k(y) = k(y')$. △

And here is an immediate lemma for future use:

Theorem 61. *If* $k\colon \mathrm{Y} \to \mathrm{Z}$ *respects the relation* R*, it also respects its closure.*

Proof. If k respects R, then the relation R is contained in the equivalence relation E_k. Therefore, since R's closure is the smallest equivalence relation containing R, that too must be contained in E_k.

It is then immediate that if y and y' are related by the closure of R, then $k(y) = k(y')$. □

(c) Suppose next that we have both a parallel pair of functions f, g into the objects Y, and also another function k onwards from Y. In other words, suppose we have a situation that we can depict like this, using a 'fork'-shaped diagram:

$$\mathrm{X} \underset{g}{\overset{f}{\rightrightarrows}} \mathrm{Y} \xrightarrow{\;k\;} \mathrm{Z}$$

The functions here generate two equivalence relations, the equivalence projection of f and g and the equivalence kernel of k. Question: what does it take for these to be the *same* relation?

Certainly, this much is required: in the same notation as before, when $yP_{fg}y'$, then yE_ky'. Hence, for any $x \in \mathrm{X}$, $k(f(x)) = k(g(x))$. So we need $k \circ f = k \circ g$; in other words, we need our fork diagram to commute.[1] Or putting it another way, k needs to respect the projected relation P_{fg}.

That's a necessary condition. It ensures that the equivalence relation E_k contains P_{fg} and hence contains its closure E_{fg}, the smallest equivalence relation containing P_{fg}.

But plainly the condition isn't sufficient. Suppose, for example, k sends every Y-object to the same target. Then $k \circ f = k \circ g$. In this case, however, the corresponding E_k is the *largest* equivalence relation on Y, relating any two Y-objects: and in the general case this won't be the same as the projected equivalence of f and g.

So the next question to ask is: what happens when we keep the prongs f and g of our fork fixed, but vary the handle k while still ensuring we get a commuting diagram? The resulting equivalence kernel E_k will then vary. And what we are after is the limiting case where the equivalence kernel is the *smallest* it can be and so is in fact E_{fg}. When does this happen?

We'll put this question on hold for the moment (though it is a nice mini-challenge to pause to think about), and first return to consider ...

[1] It needs to commute, that is to say, in the sense of our tweaked Defn. 20*.

15.2 Quotient schemes again

(a) Suppose we have an equivalence relation on some objects Y, and want to 'collapse' equivalent objects. We need some objects Q (which may or may not be some of Y) to play the role of the 'collapsed' objects, and a function $q\colon \mathrm{Y} \to \mathrm{Q}$ that sends equivalent objects among Y to the same target object. But we don't want q to be too indiscriminate and to collapse non-equivalent objects. And we also want to avoid redundant complications, so Q should only include the necessary target objects.

Wrapping these desiderata into a definition, let's say:

Definition 5* Given some objects Y and an equivalence relation $\sim$ defined over them, then the objects Q and the function $q\colon \mathrm{Y} \to \mathrm{Q}$ provide *a scheme* (Q, q) *for quotienting* Y *by* $\sim$ just when:

(i) if $y \sim y'$ then $q(y) = q(y')$;
(ii) if $q(y) = q(y')$ then $y \sim y'$; and
(iii) q is surjective, so for any $o \mathrel{\varepsilon} \mathrm{Q}$, there is some $y \mathrel{\varepsilon} \mathrm{Y}$ such that $o = q(y)$. △

Notation apart, this is of course the very same idea of a quotient scheme as more informally introduced in Defn. 5 right back in §2.3.

As we noted before, the canonical example of a scheme for quotienting Y by $\sim$ is provided by taking Q to be $\sim$-equivalence classes formed from Y, and q to be the function that sends an object among Y to the equivalence class it belongs to. But as we also emphasized, this is just one way of forming a quotient scheme. Exactly as with a pairing scheme, what actually matters about a quotient scheme is that it provides *some* (Q, q) which – viewed 'externally' – work together as described in Defn. 5*: the particular 'internal' nature of the quotient-objects Q is not of the essence. There is, in particular, no requirement that quotient-objects really *are* classes.

(b) Clause (i) of our definition tells us that q respects $\sim$. Clauses (ii) and (iii) together then tell us that q is a limiting case among the functions from Y that respect $\sim$. This thought can be captured by the following theorem:

Theorem 62. $(\mathrm{Q}, q\colon \mathrm{Y} \to \mathrm{Q})$ *is a scheme for quotienting* Y *by the equivalence relation* $\sim$ *if and only if (1)* q *respects* $\sim$ *and (2) for any function* $k\colon \mathrm{Y} \to \mathrm{Z}$ *that respects* $\sim$*, there exists a unique* $u\colon \mathrm{Q} \to \mathrm{Z}$ *such that* $k = u \circ q$*, i.e. such that this commutes:*[2]

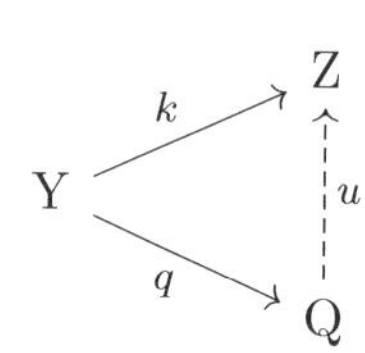

[2]Yes, we are for the moment still working pre-categorially; but we can of course still helpfully use diagrams in this chapter!

Proof ('only if'). Assume that (Q, q) form a scheme for quotienting Y by $\sim$. By condition (iii) in Defn. 5*, every object among Q is $q(y)$, for some y from Y. So we can define $u\colon \mathrm{Q} \to \mathrm{Z}$ by saying that, for each y, u sends $q(y)$ to $k(y)$.

We need to check that this does well-define a function. But by conditions (i) and (ii), and the assumption that k respects $\sim$, we can't have $q(y) = q(y')$ without having $k(y) = k(y')$.

And this is evidently the unique u such that $k = u \circ q$. □

Proof ('if'). We need to show that if (2) holds, so too do conditions (ii) and (iii) in Defn. 5*.

For (ii), suppose that q were to send objects in two different $\sim$-partitions of Y to the same q-value, while k (as could be the case) always sends objects in different $\sim$-partitions to different k-values. Then no u will make $k = u \circ q$. So the existence condition in (2) means q *can't* send objects in two different $\sim$-partitions to the same q-value. So (ii) is satisfied.

For (iii), suppose that, as well as all the requisite q-images of objects from Y, there are also one or more junk objects among Q. Then $k = u \circ q$ for any u that sends the q-values of objects in Y to their k-values but sends the junk objects wherever you like; so u wouldn't then in general be unique. Contraposing, if the uniqueness condition in (2) holds, so does (iii). □

(c) As with pairing schemes, we can similarly show that different schemes for quotienting Y by the equivalence relation $\sim$ will all 'look the same'. More carefully, we have

Theorem 63. *If* (Q, q) *and* (Q′, q') *are both schemes for quotienting* Y *by* $\sim$, *then there is a unique bijection* $f\colon \mathrm{Q} \to \mathrm{Q}'$ *that preserves the way objects from* Y *are 'collapsed together' by the schemes, i.e. such that* $q' = f \circ q$. *And* $\sim$ *will be the equivalence kernel of both* q *and* q'.

Proof. Since q is surjective onto Q, every object among Q is some $q(y)$ for y among Y. Likewise, every object among Q′ is some $q'(y)$. So define $f\colon \mathrm{Q} \to \mathrm{Q}'$ as sending $q(y)$ to $q'(y)$ for each y among Y. It is straightforward to check that this is our needed bijection. And the concluding claim is simply a consequence of our definitions. □

15.3 A key result about quotients to carry forward

We can return now to the question we left hanging at the end of §15.1. Given a commuting fork of the shape

$$\mathrm{X} \underset{g}{\overset{f}{\rightrightarrows}} \mathrm{Y} \xrightarrow{\;k\;} \mathrm{Z}$$

when is the equivalence projection of f and g the same as the equivalence kernel of k?

And here is the now-predictable answer:

Theorem 64. *Suppose that* $\mathrm{X} \underset{g}{\overset{f}{\rightrightarrows}} \mathrm{Y} \xrightarrow{q} \mathrm{Q}$ *commutes, and for each commuting fork* $\mathrm{X} \underset{g}{\overset{f}{\rightrightarrows}} \mathrm{Y} \xrightarrow{k} \mathrm{Z}$ *there is a unique* $u\colon \mathrm{Q} \to \mathrm{Z}$ *such that the following whole diagram commutes:*

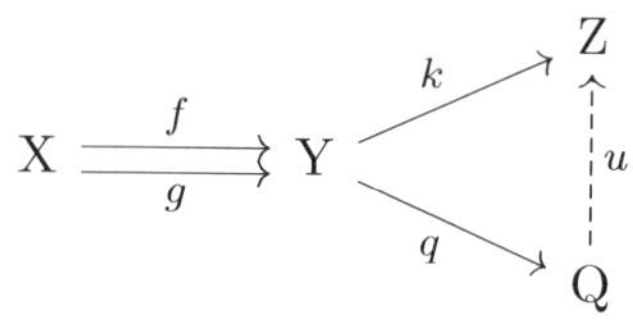

Then (Q, q) *is a scheme for quotienting* Y *by the equivalence projection of* f *and* g*, and this equivalence projection is also the equivalence kernel of* q.[3]

Proof. Since the forks commute, we know that q and k respect the relation P_{fg} and hence by Theorem 61 respect the equivalence relation E_{fg}. Hence by Theorem 62, (Q, q) is a scheme for quotienting Y by E_{fg}. But then the equivalence kernel of q is E_{fg} by Theorem 63. So we are quickly done! □

We can therefore relate claims about quotients with claims about limiting cases of commuting forks. And this looks to be *exactly* the kind of thing we can directly carry over into a categorial setting. Let's do that in the next chapter.

[3]See the previous footnote again!

16 Equalizers and co-equalizers

The informal story that we told in the last chapter suggests that we should be able to treat quotients in a categorial setting by invoking some particular commuting forks with unique arrows *from* them. In this way, the construction will be analogous to that for initial objects and coproducts.

However, starting out now on our official story, it is conventional to begin with the dual case. So we will first look at commuting fork diagrams with arrows in the opposite direction, and pick out special commuting forks with unique arrows going *to* them.

16.1 Forks and equalizers defined

For convenience, we'll start calling commuting fork diagrams simply 'forks' for short. With the direction of arrows reversed, then,

Definition 66. A *fork* (from W through X to Y) consists of an arrow $k\colon W \to X$ together with parallel arrows $f\colon X \to Y$ and $g\colon X \to Y$, such that $f \circ k = g \circ k$. In other words, to count as a fork, the resulting diagram must commute:[1]

$$W \xrightarrow{\quad k \quad} X \underset{g}{\overset{f}{\rightrightarrows}} Y \qquad \triangle$$

Now, when we were talking about products, we defined a product wedge from O to X and Y as a limiting case. It's a wedge such that any other wedge from W to X and Y uniquely 'factors through' it (in the sense of §11.3). Our next move is exactly analogous: we will define an equalizing fork starting from E and with the prongs $f, g\colon X \to Y$ as another limiting case. It's a fork such that any other fork sharing the same prongs f, g again uniquely 'factors through' it (in a closely related sense).

To spell that out:

Definition 67. Let $f, g\colon X \to Y$ be a pair of parallel arrows in the category C. Then the object E and arrow $e\colon E \to X$ form an *equalizer* (E, e) in C for those arrows if and only if (1) $f \circ e = g \circ e$ (making $E \xrightarrow{\ e\ } X \underset{g}{\overset{f}{\rightrightarrows}} Y$ a fork),

[1] Must commute, need I say again, in the sense of our tweaked Defn. 20*.

and (2) for any fork $W \xrightarrow{k} X \overset{f}{\underset{g}{\rightrightarrows}} Y$ there is a unique mediating arrow $u\colon W \to E$ making the following diagram commute:

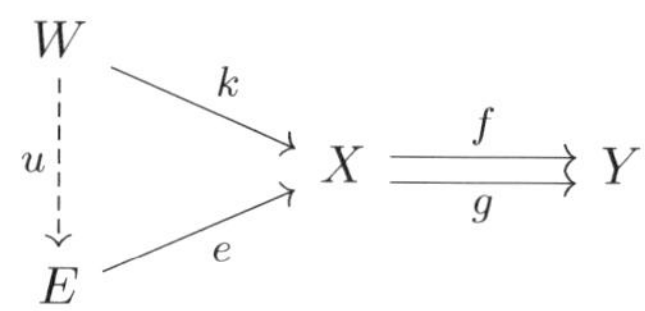

△

Definition 68. A category C *has all equalizers* iff all pairs of parallel C-arrows $f, g\colon X \to Y$ have an equalizer in C . △

16.2 Examples of equalizers

(1) Suppose in Set we have parallel arrows $f, g\colon X \to Y$ (in this case, the arrows are straightforwardly functions). Now consider the subset $X_{fg} \subseteq X$ that is the set of $x \in X$ such that $fx = gx$. Let $i\colon X_{fg} \hookrightarrow X$ be the simple inclusion map which sends an element of X_{fg} to the very same element of X. By construction, $f \circ i = g \circ i$. Therefore $X_{fg} \xhookrightarrow{i} X \overset{f}{\underset{g}{\rightrightarrows}} Y$ counts as a fork.

Claim: (X_{fg}, i) so defined is an equalizer for f and g. Why? Well, let's suppose $W \xrightarrow{k} X \overset{f}{\underset{g}{\rightrightarrows}} Y$ is another fork in Set. Can we make this diagram commute?

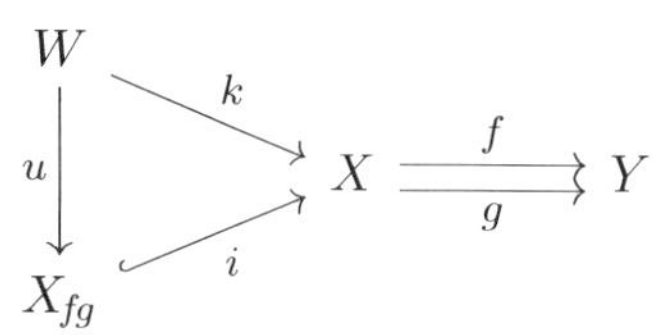

By the assumption that the top fork *is* a commuting fork, we know that for each $w \in W$, $f(k(w)) = g(k(w))$. Hence the k-images of objects in W must live in X_{fg}. Hence if we define u to be the arrow $\hat{k}\colon W \to X_{fg}$ that agrees with $k\colon W \to X$ for all $w \in W$, this will make the whole diagram commute. Moreover this is the unique possibility: in order for the diagram to commute, we need $k = i \circ u$, and since the inclusion i doesn't alter the value in X we reach, k and u must agree on all inputs (the functions just have different codomains).

Note that, since the described construction is always available, Set has all equalizers.

(2) As you would probably predict, equalizers in categories whose objects are sets-with-structure behave similarly.

Consider the category Mon. Given a pair of monoid homomorphisms $(X, *, e_X) \overset{f}{\underset{g}{\rightrightarrows}} (Y, \star, e_Y)$, take the subset E of X on which the functions agree. Evidently E must contain the identity element of X (since f and g agree on this element: being homomorphisms, both have to send e_X to the element e_Y). And suppose $a, b \in E$: then $f(a * b) = f(a) \star f(b) = g(a) \star g(b) = g(a * b)$, which means that E is closed under products of members.

So take E together with the monoid operation from $(X, *, e_X)$ restricted to members of E. Then $(E, *, e_X)$ is a monoid – for the shared identity element still behaves as an identity, E is closed under the operation, and the operation is still associative. And if we take $(E, *, e_X)$ and equip it with the injection homomorphism into $(X, *, e_X)$, this will evidently give us an equalizer for f and g.

Thus a parallel pair of monoid homomorphisms implemented in Mon always have an equalizer; for short, Mon has all equalizers. And note – to echo the remarks about groups and their products in §11.2 – we are here relying on our 'category of all monoids and their homomorphisms' having access to a separation principle which allows us to collect the elements on which parallel homomorphisms agree: if we had thought of this category has (so to speak) free standing, not to be located in an arena already providing a separation principle, then the claim that it has all equalizers would need some independent axiom.

(3) Next, take Top. What is the equalizer for a pair of continuous maps $X \overset{f}{\underset{g}{\rightrightarrows}} Y$? Well, take the subset of the underlying set of X on which the functions agree, and give it the subspace topology. This topological space equipped with the injection into X is then the desired equalizer.

(4) A more interesting case. Suppose we are in Grp and have a group homomorphism, $f\colon X \to Y$. There is automatically also a homomorphism $o\colon X \to Y$ that sends any element of the group X to the identity element in Y (this can be defined as the composite $X \to 1 \to Y$ of the only possible homomorphisms, where 1 is the one-object group which is both initial and terminal in the category of groups). Now consider what would constitute an equalizer for f and o.

Suppose K is the kernel of f, i.e. the subgroup of X whose elements are those that f sends to the identity element of Y, and let $i\colon K \hookrightarrow X$ be the inclusion map (trivially a homomorphism). Then $K \overset{i}{\hookrightarrow} X \overset{f}{\underset{o}{\rightrightarrows}} Y$ is a fork since $f \circ i = o \circ i$.

Let $W \xrightarrow{k} X \overset{f}{\underset{o}{\rightrightarrows}} Y$ be another fork. Now, $o \circ k$ sends every element of W to the unit of Y. By assumption, $f \circ k = o \circ k$, so $f \circ k$ also sends every element of W to the unit of Y; hence $k\colon W \to X$ must send any element of W to some element that lives in f's kernel K. Let $u\colon W \to K$ agree with $k\colon W \to X$ on all arguments. Then the following commutes:

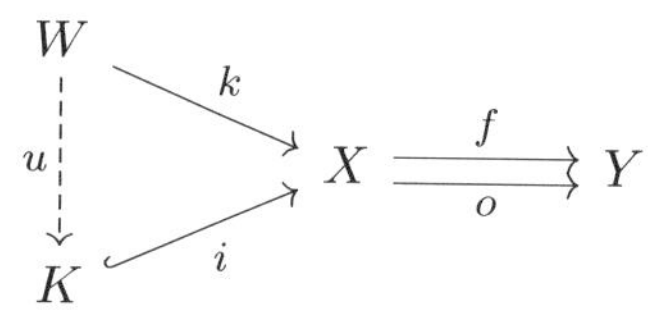

And u is the only possible homomorphism that will make the diagram commute. In sum, the equalizer of f and o is (up to isomorphism) f's kernel K equipped with the inclusion map into the domain of f.

Or putting it the other way about, we can define kernels of group homomorphisms categorially (up to isomorphism, as usual) in terms of equalizers. Which is rather nice.

16.3 Uniqueness up to unique isomorphism

(a) A quick terminological aside before proceeding further. I've defined an equalizer as an object E equipped with an arrow whose source is E, satisfying certain conditions. Since fixing the arrow fixes its source, we could without loss of information officially define an equalizer to be simply the relevant arrow. Many do this. Nothing hangs on the choice.

(b) To continue. Just as products are unique up to unique isomorphism, equalizers are too:

Theorem 65. *If (E, e) and (E', e') are both equalizers for $X \overset{f}{\underset{g}{\rightrightarrows}} Y$, then there is a unique isomorphism $v\colon E \xrightarrow{\sim} E'$ commuting with the equalizing arrows, i.e. such that $e' \circ v = e$.*

Plodding proof from first principles. We can use an argument that goes along very similar lines to the plodding proof we used to prove the uniqueness of products. This is of course no accident, given the similarity of the definitions of products and equalizers via a universal mapping property.

OK, if you want the details, assume (E, e) equalizes f and g. Then a fork from (E, e) on through f and g factors uniquely through *itself*, via some mediating arrow u, meaning that this commutes:

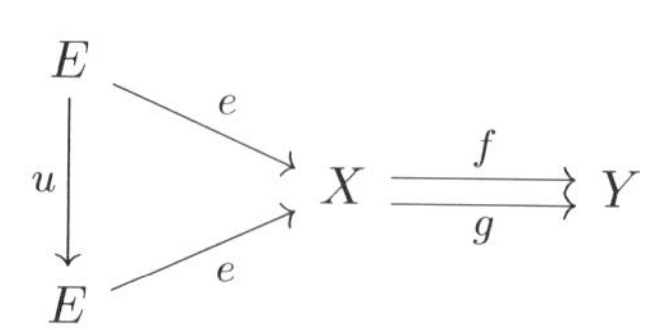

And this u (being unique) must be equal to 1_E.

But now note that for any u such that $e \circ u = e$ this diagram also commutes. Which gives us a first result

(i) If (E, e) is an equalizer, then $e \circ u = e$ implies $u = 1_E$.

Now suppose (E', e') is also an equalizer for f and g. Then the fork starting (E, e) must factor uniquely through this new equalizer via a (unique) mediating $v\colon E \to E'$ such that $e' \circ v = e$:

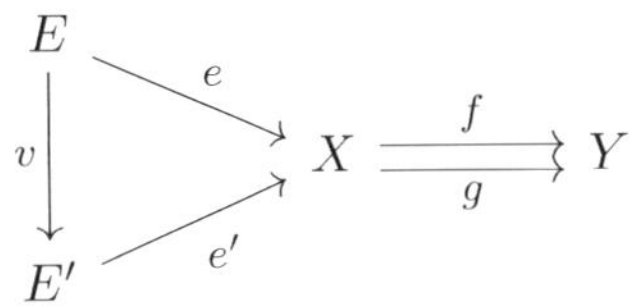

Similarly, swapping (E, e) and (E', e'), there is a unique w such that $e \circ w = e'$. Therefore $e \circ w \circ v = e$, and hence by (i) it follows that (with v and w as defined)

(ii) $w \circ v = 1_E$.

Since everything is symmetric in (E, e) and (E', e'), an exactly similar argument shows that

(iii) $v \circ w = 1_{E'}$.

Which gives the unique v a two-sided inverse, completing the proof that

(iv) v is an isomorphism. □

(c) We now quickly note that, as with products (see Defn. 49), we can give an alternative definition of an equalizer as a terminal object in a suitable category.

First we say

Definition 69. Given a category C and parallel arrows $f, g\colon X \to Y$, then the derived category of forks $\mathsf{C}_{f\|g}$ has as objects all forks $W \xrightarrow{k} X \underset{g}{\overset{f}{\rightrightarrows}} Y$.

And an arrow from $W \xrightarrow{k} \cdots$ to $W' \xrightarrow{k'} \cdots$ in $\mathsf{C}_{f\|g}$ is a C-arrow $v\colon W \to W'$ such that the resulting triangle commutes: i.e. such that $k = k' \circ v$.[2]

The identity arrow in $\mathsf{C}_{f\|g}$ on the fork $W \xrightarrow{k} \cdots$ is the identity arrow 1_W in C; and the composition of arrows in $\mathsf{C}_{f\|g}$ is defined as the composition of the arrows as they feature in C. △

It is easily checked that this does define a category,[3] and that our definition of an equalizer then comes to the following:

Definition 67*. An equalizer of $f, g\colon X \to Y$ in C is some (E, e), where E is a C-object, and e is a C-arrow $E \to X$, such that the fork $E \xrightarrow{e} X \underset{g}{\overset{f}{\rightrightarrows}} Y$ is terminal in $\mathsf{C}_{f\|g}$. △

But this redefinition immediately gives us

[2] Or rather, strictly speaking, we should take the arrow to be the entire commuting triangle, as we did for slice categories and for the same reason: see §7.3(c). But we won't fuss about this.

[3] Modulo that last footnoted tweak.

A slicker proof of Theorem 65. (E, e) and (E', e') are both terminal objects in the fork category $\mathsf{C}_{f\|g}$. So by Theorem 27 there is a unique $\mathsf{C}_{f\|g}$-isomorphism j between them. But, by definition, this has to be a C-arrow $j\colon E \xrightarrow{\sim} E'$ such that there is an arrow $j'\colon E' \xrightarrow{\sim} E$, where $j' \circ j = 1_E$ and $j \circ j' = 1_{E'}$. So j also has to be a C-isomorphism. □

16.4 Challenges!

Let's note the following simple results:

Theorem 66. *A preorder category has an equalizer for any parallel arrows. However, a group considered as a category has no equalizers for any distinct parallel arrows.*

Theorem 67. *If (E, e) is an equalizer, then e is a monomorphism. And if e is also epic, then it is an isomorphism.*

Theorem 68. *In* Set, *any subset S of a set X together with the inclusion map from S to X is an equalizer for a certain pair of parallel arrows from X to a two-member set Ω (which we can think of as $\{true, false\}$).*

Theorem 69. *The category* M_2 *has all equalizers.*[4]

Frankly, it is a bit of a stretch to announce some of these as 'challenges' to prove! But do pause to derive the results, before reading on.

Proof: A preorder category has all equalizers. Recall Defn. 17: in a preorder category, the only cases where we have parallel arrows $f, g\colon X \to Y$ are when $f = g$. But then it is easy to see that $(X, 1_X)$ equalizes f with itself, by thinking about the commuting diagram

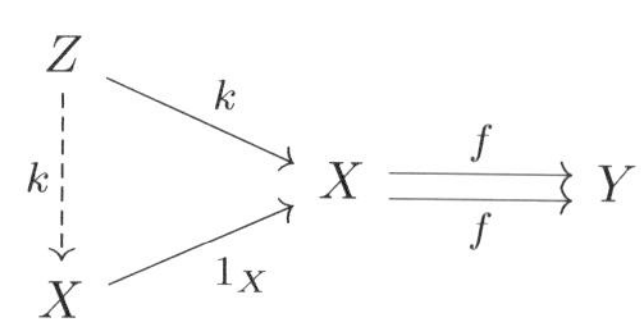

Since 1_X comes for free with any category containing X, the equalizer $(X, 1_X)$ always exists, as claimed. □

Proof: A category without any equalizers for distinct arrows. Recall §8.7: given a group $(G, *, e)$, we can define G to be the corresponding category whose sole object $\bullet$ is whatever you like, and whose arrows are simply the group objects G, with e the identity arrow. Composition of arrows in G is defined as group-multiplication $*$.

[4] For bonus points, you might also like to show that $\mathsf{Set}^{\to}$ and also Set/X (for any X) have all equalizers too.

Now, take distinct objects $g, h \varepsilon$ G. Then there will be no x such that $g * x = h * x$, or else we would have $g * x * x^{-1} = h * x * x^{-1}$ and hence $g = h$ after all. So correspondingly, in G, given distinct parallel arrows $g, h \colon \bullet \to \bullet$, there is no x such that $g \circ x = h \circ x$, and hence those arrows can't have an equalizer. □

Proof: Equalizing arrows are monic. Suppose (E, e) equalizes $X \overset{f}{\underset{g}{\rightrightarrows}} Y$, and also suppose $e \circ j = e \circ k$.

For the second supposition to make sense, j and k must be parallel arrows from some Z to E. And then the following diagram commutes:

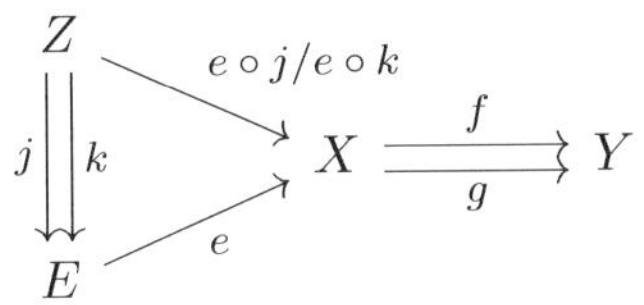

Therefore $Z \xrightarrow{e \circ j / e \circ k} X \overset{f}{\underset{g}{\rightrightarrows}} Y$ is a fork factoring through the equalizer. But by the definition of an equalizer, it has to factor uniquely, and hence $j = k$. In sum, e is left-cancellable in the equation $e \circ j = e \circ k$; i.e. e is monic. □

Proof: An epic equalizer is an isomorphism. Epic monomorphisms need not in general be isomorphisms; but they are iso in the special case when the monic is an equalizing arrow.

Assume again that (E, e) equalizes $X \overset{f}{\underset{g}{\rightrightarrows}} Y$, so that $f \circ e = g \circ e$. Hence if e is epic, it follows that $f = g$. Then consider the following diagram

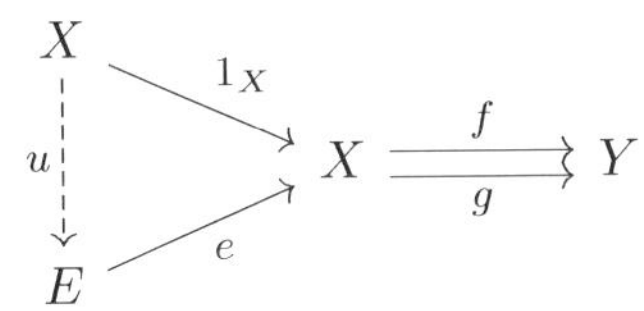

We know that the top fork commutes and uniquely factors through the equalizer, i.e. there is a unique u such that (i) $e \circ u = 1_X$.

But then also $e \circ u \circ e = 1_X \circ e = e = e \circ 1_E$. Hence, since equalizers are monic by the last theorem, (ii) $u \circ e = 1_E$.

Taken together, (i) and (ii) tell us that e has a two-sided inverse. Therefore e is an isomorphism. □

Our first example in §16.2 showed that in Set an equalizer of parallel maps $f, g \colon X \to Y$ will be provided by a suitable subset of X together with the inclusion map from that subset to X. Our next result, Theorem 68, shows the converse – a subset of X together with its inclusion map is an equalizer of a certain pair of well-chosen maps from X:

Proof: Subsets as equalizers. We are working in Set, and we'll use a very familiar device for thinking about subsets. So take a suitable two-object set we'll call Ω (whose members we might suggestively dub *true* and *false*). Then a subset $S \subseteq X$ has an associated characteristic function $\chi_S \colon X \to \Omega$, where $\chi_S(x) = true$ if and only if $x \in S$.

Now compare this with the boring function we'll dub $\top_X \colon X \to \Omega$ which indiscriminately sends *everything* in X to *true*. If $i \colon S \hookrightarrow X$ is the simple inclusion function, then

$$S \xhookrightarrow{\;i\;} X \underset{\top_X}{\overset{\chi_S}{\rightrightarrows}} \Omega$$

is a commuting fork. Moreover, (S, i) is an equalizer for $\chi_S, \top_X \colon X \to \Omega$. For consider the following diagram:

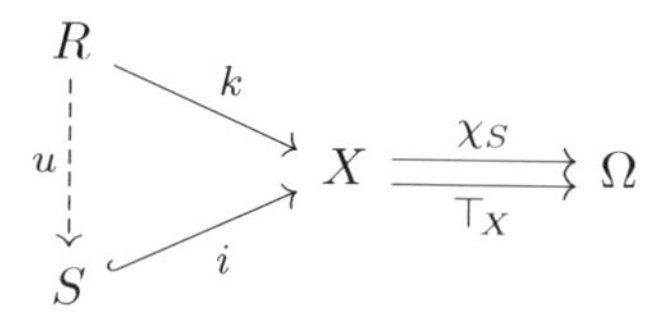

If we have a commuting fork at the top, the k-image of R must be contained in S (why?). In which case the unique way of making the whole diagram commute is to make u agree with k for all inputs.

As claimed, then, any subset S of a set X together with the inclusion map from S to X is an equalizer for a pair of parallel arrows from X to Ω. □

Proof: M_2 *has all equalizers.* Recall again that an object in this category is a set equipped with an idempotent function, and an arrow $f \colon (X, p) \to (Y, q)$ is a function such that $f \circ p = q \circ f$. Likewise for an arrow $g \colon (X, p) \to (Y, q)$. So we want to define an equalizer for any such f and g.

Suppose that $(W, w) \xrightarrow{\;k\;} (X, p) \underset{g}{\overset{f}{\rightrightarrows}} (Y, q)$ is a fork in M_2. In other words, suppose the following cube commutes in Set.

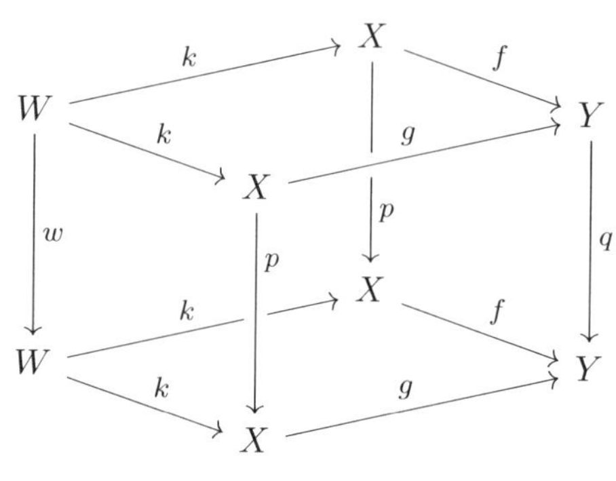

Then k must send an element of W into the subset $X_{fg} \subseteq X$ defined as before (x is a member iff $fx = gx$). And further, if the diagram is to commute, p must send a member of X_{fg} to a member of X_{fg}. This means we can define an idempotent function $\hat{p}\colon X_{fg} \to X_{fg}$ which is the restriction of p to X_{fg}.

And now we are flying! For consider the following diagram:

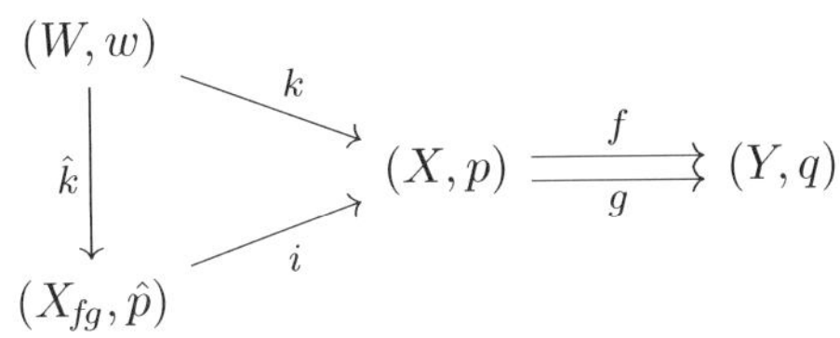

where i as before is the inclusion function from X_{fg} to X, and where $\hat{k}$ agrees with k for all members of W. It is now elementary to check that this commutes in M_2, with $\hat{k}$ as the only candidate to complete the diagram. So $(X_{fg}, \hat{p})$ equipped wirh i is our desired equalizer for f and g. □

16.5 Co-forks and co-equalizers defined

Now we dualize to get the notion of a co-equalizer. We simply reverse the arrows on forks as defined in Defn. 66 in §16.1 (and for convenience, swap back the labels 'X' and 'Y'). So we return back again to the sort of forks informally introduced in the last chapter. However, for clarity, we will now officially call commuting forks of that earlier kind 'co-forks', as in:

Definition 70. A *co-fork* (from X through Y to Z) consists of parallel arrows $f\colon X \to Y$, $g\colon X \to Y$ together with an arrow $k\colon Y \to Z$, such that $k \circ f = k \circ g$. In other words, to form a co-fork, this corresponding diagram must commute:

$$X \underset{g}{\overset{f}{\rightrightarrows}} Y \xrightarrow{\quad k \quad} Z$$ △

Dualizing Defn. 67, our definition of equalizers, we then get

Definition 71. Let $f, g\colon X \to Y$ be a pair of parallel arrows in the category C. Then the object C and arrow $c\colon Y \to C$ form a *co-equalizer* (C, c) in C for those arrows iff (1) $c \circ f = c \circ g$ (making $X \underset{g}{\overset{f}{\rightrightarrows}} Y \xrightarrow{\ c\ } C$ a co-fork), and (2) for any co-fork $X \underset{g}{\overset{f}{\rightrightarrows}} Y \xrightarrow{\ k\ } Z$ there is a unique mediating arrow $u\colon C \to Z$ making the following diagram commute:

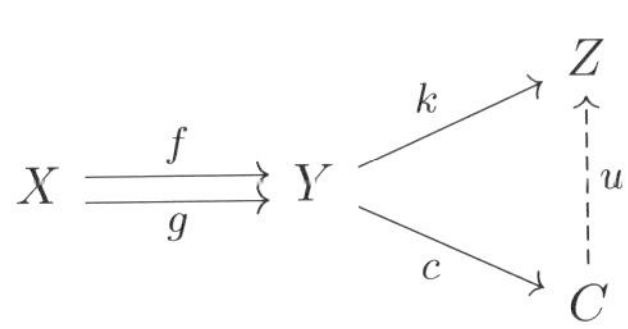

△

To stress the duality, in an equalizer (E, e) the object E is the *source* of the arrow e, in a co-equalizer (C, c) the object C is the *target* of the arrow c.

We need not spell out the dual arguments that co-equalizers are unique up to a unique isomorphism, or that the arrow of a co-equalizer is epic, etc.

16.6 Examples of co-equalizers

(a) Let's immediately turn to consider the key example.

What, then, do co-equalizers do in a category like Set? Central though the question is, we can be *very* brief, since we did all the necessary ground-work in the last chapter, which provided our motivation for thinking about equalizing forks/coforks. Simply trade in our previous plural talk about items Q for singular talk about some appropriate 'object' Q in a category that in some sense collects these items together.

So, take a pair of parallel arrows $f, g \colon X \to Y$ in Set. These set up a projected relation P_{fg} on Y, where $yP_{fg}y'$ if and only if there is some $x \in X$ such that $y = f(x)$ and $y' = g(x)$. Taking the closure of that relation gives us in turn an equivalence relation E_{fg} on Y. Then the pre-categeorial Theorem 64 becomes

Theorem 70. *In* Set, *a co-equalizer* (Q, q) *for the arrows* $f, g \colon X \to Y$ *provides a scheme for quotienting* Y *by* E_{fg}, *the equivalence projection of* f *and* g.

Conversely, if we form the set Q of equivalence classes for the relation E_{fg} in the conventional way, then Q equipped with the map that sends an object in Y to its equivalence class in Q will provide a categorial co-equalizer (Q, q). And since we can form equivalence classes ad libitum in a standard universe of sets, that tells us that

Theorem 71. Set *has a co-equalizer for any pair of parallel arrows.*

(b) Predictably enough, we get parallel results in other categories whose objects can be thought of as sets-equipped-with-structure. Take Grp, for example, and suppose (Q, q) is a co-equalizer for the parallel group homomorphisms $f, g \colon X \to Y$. Then $q \colon Y \to Q$ is a group homomorphism, and if we put $y_1 \sim y_2$ iff $q(y_1) = q(y_2)$, then by Theorem 9, $\sim$ is a congruence on the group Y, and Q will be a quotient of Y with respect to that congruence relation.

If you know enough topology, you might think about what a co-equalizer of two parallel arrows (continuous functions) $f, g \colon X \to Y$ should be in Top.

17 Exponentials

We have been thinking about how we form products, quotients, and a few other constructs, in 'ordinary mathematics'. And we've seen that what matters about product-widgets, quotient-widgets, and such like, is not their 'internal' make up, but how they 'externally' map to and from other widgets. This is the key insight which gets reflected in the categorial treatment of products, quotients, etc.

Now we move on to consider another kind of construction, namely exponentials. Again I'll start with some informal remarks; then we'll see how things play out in a categorial setting.

17.1 Instead of binary functions, again

(a) I have stressed more than once that, in categories where arrows are functions, arrows are always monadic functions. So as we asked before, how can we accommodate binary functions?

In fact, there are a couple of frameworks (already familiar before we get to category theory) that manage to do without genuine binary or multi-place functions by providing workable substitutes.

One of them we've met before:

(1) The default set-theoretic procedure is to trade in an underlying binary function $\underline{f}\colon A, B \to C$ for a related unary function $f\colon A \times B \to C$ – so $f\langle a, b\rangle = \underline{f}(a, b)$.

But now let's note that varieties of type theory usually deal with two-place functions in a quite different way.

To illustrate: addition – naively a binary function – is traded in for a function of the type $N \to (N \to N)$. This is a *unary* function that takes one number (of type N) and outputs something of a higher type, i.e. a unary function (of type $N \to N$). So we now get from two numbers as input to a numerical output in a couple of steps. We feed the first number to a function of type $N \to (N \to N)$, which delivers another function of type $N \to N$ as output; and then we can feed the second number to this second function.

This so-called 'currying' manoeuvre from type theory[1] is also perfectly adequate for certain formal purposes, and we can adopt the same device for use in a

[1]The trick of replacing the evaluation of a function that takes multiple arguments by the

set-theoretic framework. We can do the work of a binary function $\underline{f}\colon A, B \to C$ by a unary function that sends a member of A to a particular function from B to C. And where do functions from B to C live? In the 'exponential' C^B (which works as the collection of functions from B to C). Hence, in set-theoretic terms,

(2) Currying is essentially a matter of trading in a binary function $\underline{f}\colon A, B \to C$ for a related unary function $\widetilde{f}\colon A \to C^B$, i.e. the function that sends a to the function f_a which is the unary function whose value for input b is $\underline{f}(a, b)$.[2]

(b) The next question is how do these two substitutes f and $\widetilde{f}$ for the underlying binary function $\underline{f}$ fit together?

At a first shot, we want something like the following informal diagram to commute, where *eval* is a binary function that takes a function living in C^B (i.e. a function from B to C) and evaluates it for a given argument in B.

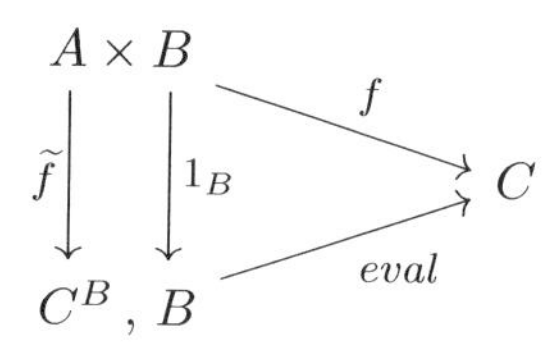

In other words, taking a pair $\langle a, b\rangle$ from $A \times B$, we can (1) use that pair as input to f. Or (2) we can use $\widetilde{f}$ to send a to a function $f_a\colon B \to C$ while carrying along b unchanged: and then *eval* takes f_a and b as its two inputs and outputs $f_a(b)$. By either route, we get the same result.

Now, that first shot gives us the core idea, except that it leaves us with a binary function *eval* still in play. So let's slightly revise. Let *ev* now be a *unary* function which takes an ordered pair of a function living in C^B and an argument from B, and still evaluates that function for that argument. Then, as a second shot, we'll say we need the following to commute:

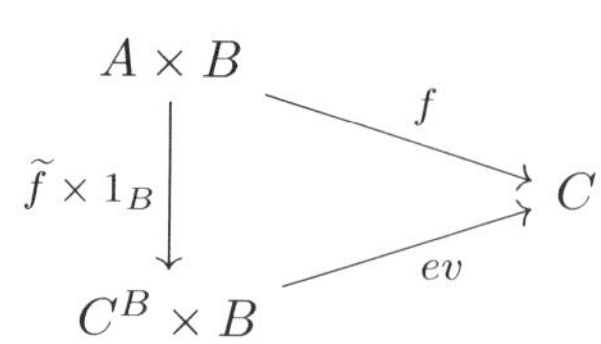

where $\widetilde{f} \times 1_B$ acts component-wise on $A \times B$, sending a pair $\langle a, b\rangle$ to $\langle f_a, b\rangle$, and *ev* takes the pair $\langle f_a, b\rangle$ and returns the value $f_a(b)$. Note: given f and given *ev* with its intended meaning, $\widetilde{f}$ will be the *unique* function from A to C^B that makes the diagram commute.

evaluation of a sequence of unary functions was developed by Haskell Curry: hence 'currying'. Moses Schönfinkel had the idea first, but somehow 'Schönfinkeling' never caught on!

[2] An alternative notation is also helpful. Suppose we use '$f(\cdot, \cdot)$' to explicitly mark how the function is waiting to be applied to two terms. Similarly, instead of 'f_a' we could write '$f(a, \cdot)$', marking how this function is to be applied to a single term. So $\widetilde{f}$ sends a to $f(a, \cdot)$.

17.2 Exponentials in categories

And now everything is nicely set up to carry over smoothly to our categorial framework.

We don't have native binary morphisms in category theory; we don't have binary arrows with two sources, $\underline{f}\colon A, B \to C$. But, as we've already seen, once we are working in a category that has products, we can use a version of the first set-theoretic trick and deploy corresponding arrows like $f\colon A \times B \to C$.

And we can also deploy an analogue of the currying trick, where we trade in our binary $\underline{f}\colon A, B \to C$ for the unary $\widetilde{f}\colon A \to C^B$. Or at least, we can do this if we have suitable exponential objects C^B and corresponding evaluation arrows ev available in our category. But which objects and arrows would these be? Given the discussion in the last section (when we were in effect looking inside the category Set), this is evidently the general story we want:

Definition 72. Assume C is a category with binary products. Then (C^B, ev), where C^B is an object and ev is an arrow $C^B \times B \to C$, constitute an *exponential of C by B* iff the following holds: for every object A and arrow $f\colon A \times B \to C$, there is a *unique* arrow $\widetilde{f}\colon A \to C^B$ making the following commute:

(Exp)

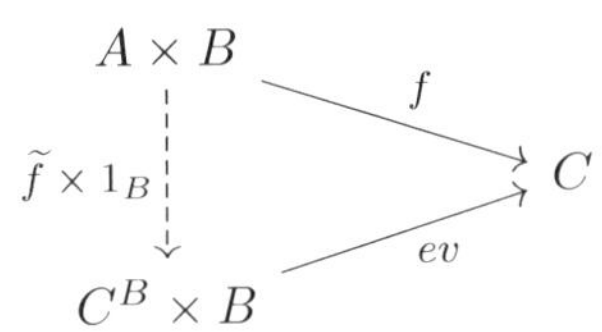

△

Here, all the objects and arrows are of course living in C. The product arrow $\widetilde{f} \times 1_B$, which acts componentwise on pairs in $A \times B$, is defined categorially in §13.3. And $\widetilde{f}$ – some write $curry(f)$ – is said to be f's *exponential transpose*.

Three quick comments.

(i) Note that, just as f uniquely fixes its exponential transpose, the converse is also true. Given any arrow $\widetilde{f}\colon A \to C^B$, then it must be the exponential transpose of $f = ev \circ \widetilde{f} \times 1_B\colon A \times B \to C$ (since then $\widetilde{f}$ makes the required diagram commute).

(ii) Putting $A = C^B$ and $f = ev$, we see that ev is the arrow whose exponential transform is 1_{C^B}.

(iii) A notational point. If we change the objects B, C the evaluation arrow $ev\colon C^B \times B \to C$ changes, since the source and/or target will change. Hence it might occasionally help to use more explicit notation and write the likes of 'ev_C' (as we have to do e.g. in §26.4).

And we can now add an obvious supplementary bit of terminology:

Definition 73. A category C *has all exponentials* iff for all C-objects B, C, there is a corresponding exponential (C^B, ev). △

17.3 Some categories with exponentials

(a) A category may have *no* exponentials, e.g. because there are not enough products in play (cf. the end of §11.2). Or it may have just a few trivial exponentials (cf. Theorem 74 below). But here are three initial examples of categories which *do* have all exponentials:

(1) Defns. 72 and 73 were of course purpose-built to ensure that Set counts as having all categorial exponentials – such an exponential of C by B is provided by the set C^B (the usual set of functions from B to C) equipped with the appropriate function-as-set ev.

Or at least, this is the case on most understandings of Set. Recall, however, we have so far left it open exactly how we are to conceive of our preferred universe of sets: and this is one point where details begin to matter. In particular, sets-according-to-Quine's-NF (or NFU, the version of the theory which allows for urelements) are not provided with a well-behaved ev function.[3] So from now on, then, we need to be at least a little more specific about the character of Set, and we will assume henceforth that it is sufficiently non-deviant to have all exponentials.

(2) We can note now that the construction of exponentials in Set as standardly understood applies equally in FinSet, the category of finite sets, since the set C^B is finite if both B and C are finite, and hence C^B is also in FinSet. Therefore FinSet has all exponentials.

(3) Prop_L is the category whose objects are sentences of a given first-order language L, and where there is a unique arrow from A to B iff $A \vDash B$.

Assuming L has the usual rules for conjunction and implication, then for any B, C, the conditional $B \to C$ provides an exponential object C^B, with the evaluation arrow $ev : C^B \times B \to C$ reflecting the modus ponens entailment $B \to C, B \vDash C$.

Why does this work? Recall that products in Prop_L are conjunctions. And note that, given $A \wedge B \vDash C$, then by the standard rules $A \vDash B \to C$ and hence – given $B \vDash B$ – we have $A \wedge B \vDash (B \to C) \wedge B$. We therefore get the required commuting diagram of this shape,

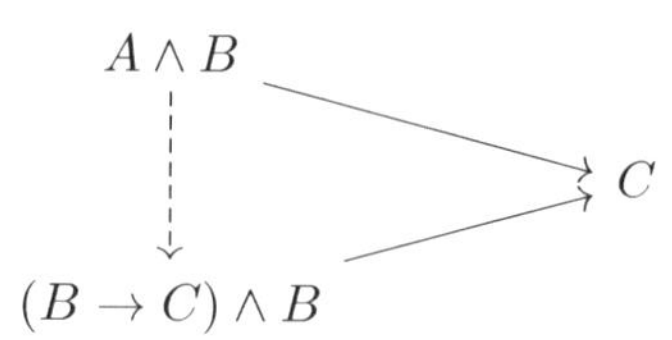

where the down arrow is the product of the implication arrow from A to $B \to C$ and the identity arrow from B to B.

[3] Quine's deviant set theory NF – which allows a universal set-of-all-sets – is explained at tinyurl.com/qui-nf. If you know a bit about stratification in NF, there is a very brief but clear explanation about why we can't there get a nice commuting diagram as in (Exp) at tinyurl.com/holmesev.

It can also be the case that a category has *some* non-trivial exponentials, though not *all* exponentials.

(4) Consider Count, the category of sets that are no larger than countably infinite, and of set-functions between them. If the Count-objects B and C are finite sets, then there is another finite set C^B that, with the obvious function ev, will serve as an exponential. But if B is a countably infinite set, and C has at least two members, then the set C^B is uncountable, so won't be available to be an exponential in Count – and evidently, nothing smaller will do.

(5) The standard example, however, of an interesting category that has some but not all exponentials is Top. If X is a space living in Top, then it is 'exponentiable', meaning that Y^X exists for all Y, if and only if it is so-called *core-compact* – and not all spaces are core-compact.

It would, however, take us far too far afield to explain and justify this example to non-topologists.

Back though to examples of categories that do have all exponentials!

(6) The category Pos has all exponentials. Why so?

We are looking for a general recipe for constructing an exponential of C by B, where those objects are posets (posets living in a non-deviant world of sets with all exponentials). Represent C's ordering by $\preccurlyeq_C$.

OK: take the set of monotone functions $f\colon B \to C$ (remember the arrows in Pos are order-respecting functions). And now equip this set with the order that puts $f \preccurlyeq_{C^B} f'$ iff for all $x \in B$, $f(x) \preccurlyeq_C f'(x)$. That gives us a poset C^B.

We can define products in Pos (as in §11.2), so we can in particular form $C^B \times B$. And now to get our categorial exponential, we need a suitable evaluation function $ev\colon C^B \times B \to C$. The natural candidate to chose is the function that takes $\langle f, b\rangle$ as input, applies the monotone function f from C^B to the element b from B, and outputs $f(b)$.

Remember, however, that ev is supposed to be an arrow living in Pos, so it needs to be an order-respecting map too. So we *do* have to check that ev as just defined is monotone. But actually that's easy. If $\langle f, b\rangle \preccurlyeq \langle f', b'\rangle$ in the product order for Pos, then by definition $f \preccurlyeq_{C^B} f'$ and $b \preccurlyeq_B b'$ (see §11.2 again). But then, as wanted, $ev\langle f, b\rangle = f(b) \preccurlyeq_C f'(b) \preccurlyeq_C f'(b') = ev\langle f', b'\rangle$.

So now the claim is that (C^B, ev) is our desired exponential in Pos. Well, is there always an exponential transpose for a monotone map $f\colon A \times B \to C$ that will get the required diagram to commute? As in Set, this transpose will need to be the function $\widetilde{f}\colon A \to C^B$ that maps $a \in A$ to the monotone function f_a which sends any $b \in B$ to $f\langle a, b\rangle \in C$.

But if this is to work, we need $\widetilde{f}$ to be available in Pos, i.e. it needs itself to be monotone. Again, we need to check! Suppose $a \preccurlyeq_A a'$. Then by definition of the order on products in Pos, $\langle a, b\rangle \preccurlyeq \langle a', b\rangle$ for any b from B. Hence, since f is monotone, $f\langle a, b\rangle \preccurlyeq_C f\langle a', b\rangle$ for all b, i.e. $f_a(b) \preccurlyeq_C f_{a'}(b)$ for all b. Hence by definition, $f_a \preccurlyeq_{C^B} f_{a'}$. So $\widetilde{f}$ is monotone as required.

(7) For the record, here's another result: the category M_2 also has all exponentials. This, recall, is the category whose objects are sets equipped with idempotent endofunctions, and where an arrow $j\colon (B, b) \to (C, c)$ is an equivariant function from B to C, i.e. is such that $j \circ b = c \circ j$.

In this case, the exponential object $(C, c)^{(B,b)}$ is not the set of *all* the functions from B to C but, predictably enough, just the set of *equivariant* functions from B to C. Then the rest of the story about exponentials will be adjusted to match, checking as we go along that the arrows ev and $\widetilde{f}$ that we define in the needed commuting digram are themselves equivariant and hence kosher M_2-arrows. As with Pos, the details again get a bit tedious; so I'll leave them for enthusiasts to explore.

17.4 Uniqueness up to unique isomorphism

(a) Defn. 72 talks of 'an' exponential of C with B. But – as you might expect by now, given that the definition is by a universal mapping property – exponentials are in fact unique, at least up to unique isomorphism:

Theorem 72. *Suppose the category* C *has two ways of forming an exponential of C by B, namely (E, ev) and (E', ev'): then there is a unique isomorphism between E and E' compatible with the evaluation arrows.*

Proof. Two commuting diagrams encapsulate the core of the argument, which parallels the proof of Theorem 37:

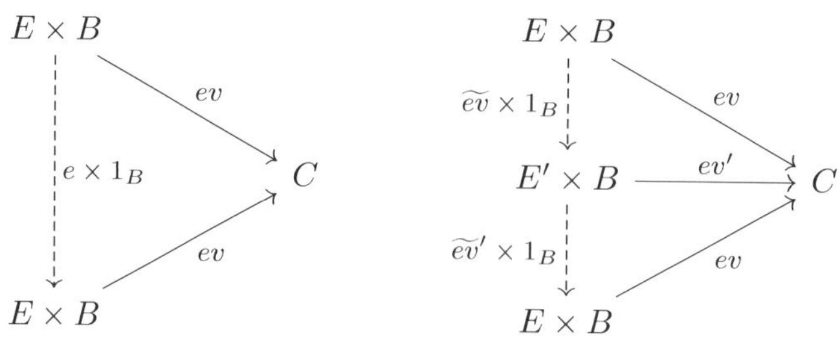

By definition, if (E, ev) is an exponential of C by B then there is a unique mediating arrow $e\colon E \to E$ such that $ev \circ e \times 1_B = ev$. But as the diagram on the left reminds us, 1_E will serve as the mediating arrow. Hence $e = 1_E$.

The diagram on the right then reminds us that (E, ev) and (E', ev') factor through each other, and putting the two commuting triangles together we get

$$ev \circ (\widetilde{ev}' \times 1_B) \circ (\widetilde{ev} \times 1_B) = ev.$$

Applying Theorem 55, we know that $(\widetilde{ev}' \times 1_B) \circ (\widetilde{ev} \times 1_B) = (\widetilde{ev}' \circ \widetilde{ev}) \times 1_B$, and hence

$$ev \circ (\widetilde{ev}' \circ \widetilde{ev}) \times 1_B = ev,$$

and now applying the uniqueness result from the first diagram

$$\widetilde{ev}' \circ \widetilde{ev} = 1_E.$$

Similarly, by interchanging E and E' in the second diagram, we get

$$\widetilde{ev} \circ \widetilde{ev}' = 1_{E'}.$$

Whence $\widetilde{ev}\colon E \to E'$ has a two-sided inverse and is an isomorphism. □

(b) When we were talking about e.g. products and equalizers, we gave two types of proof for their uniqueness (up to unique isomorphism). One was a direct proof from the definitions. For the other type of proof, we noted that products are terminal objects in a category of wedges, and equalizers are terminal objects in a category of forks, and then appealed to the uniqueness of terminal objects.

We have now given a proof of the first type, a direct proof, of the uniqueness of exponentials. Can we give a proof of the second type? We will expect so. And for the record, let's confirm this. Start with

Definition 74. Given objects B and C in the category C, then the category $\mathsf{C}_{E(B,C)}$ of parameterized maps from B to C has the following data:

1. Objects (A, f) comprising a C-object A, and a C-arrow $f\colon A \times B \to C$,
2. An arrow from (A, f) to (A', f') is any C-arrow $h\colon A \to A'$ that makes the following diagram commute:

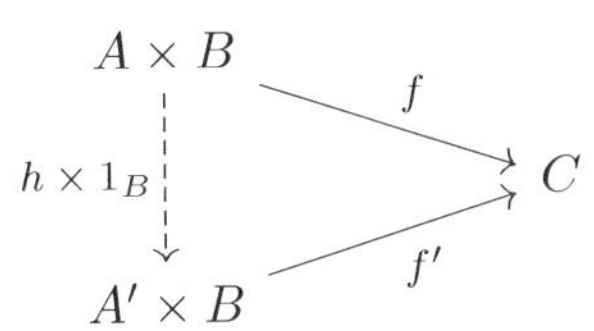

The identity arrows and composition are as in C. △

It is then easily checked that this does define a category, and evidently we have

Theorem 73. *An exponential (C^B, ev) in C is a terminal object in $\mathsf{C}_{E(B,C)}$.*

Since exponentials are terminal in a suitable category, that yields the second type of proof of their uniqueness.

17.5 Further general results about exponentials

(a) Let's next show that any category with binary products and a terminal object – has some trivial exponentials:

Theorem 74. *If the category* C *has binary products and a terminal object* 1, *then for any* C*-objects* B *and* C, *there exist exponential objects* 1^B *and* C^1.

Proof for 1^B. Consider this diagram:

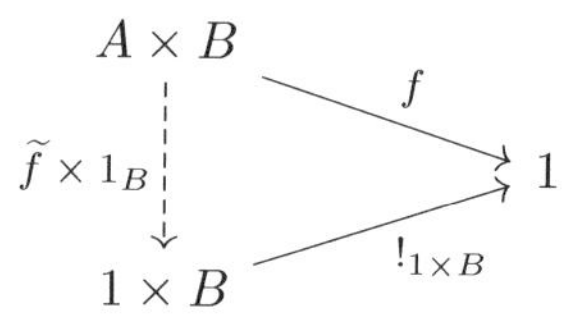

Since 1 is terminal, there is a unique arrow $\widetilde{f}$ from A to 1. And there is only one possible arrow from $A \times B$ to 1, so the diagram has to commute. So put $(1^B, ev) = (1, !_{1\times B})$ and by definition we get an exponential of 1 by B. (For another proof, using more sophisticated apparatus, see §43.5.) □

Proof for C^1. Suppose we can show that, given an arrow $f\colon A \times 1 \to C$, then there is always a unique $\widetilde{f}$ making this diagram commute,

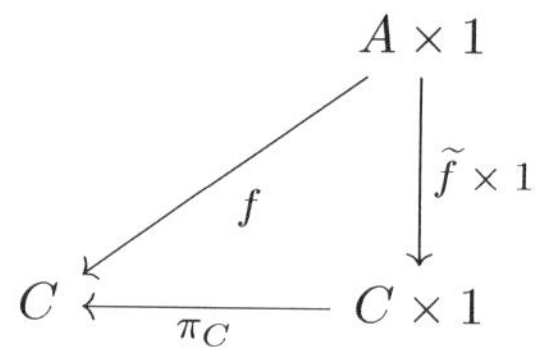

where π_C is the first projection from the product $C \times 1$. Then $(C^1, ev) = (C, \pi_C)$ will by definition serve as an exponential of C by 1.

So how do we construct $\widetilde{f}$, given f? Try brute force! By the definition of the product $C \times 1$, there is a unique mediating u making this next diagram commute:

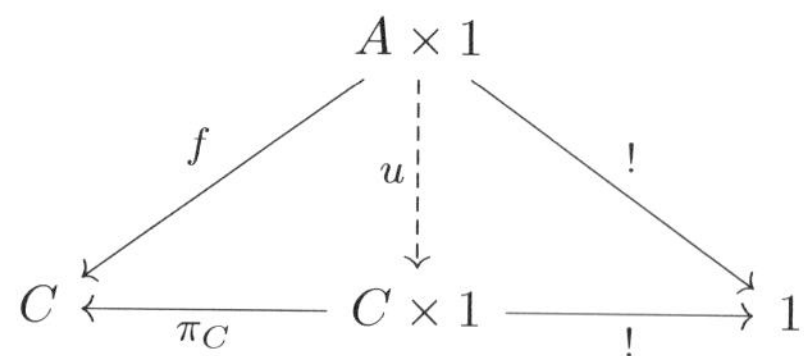

Now complete the diagram with the product wedge $A \xleftarrow{\pi_A} A \times 1 \xrightarrow{!} 1$:

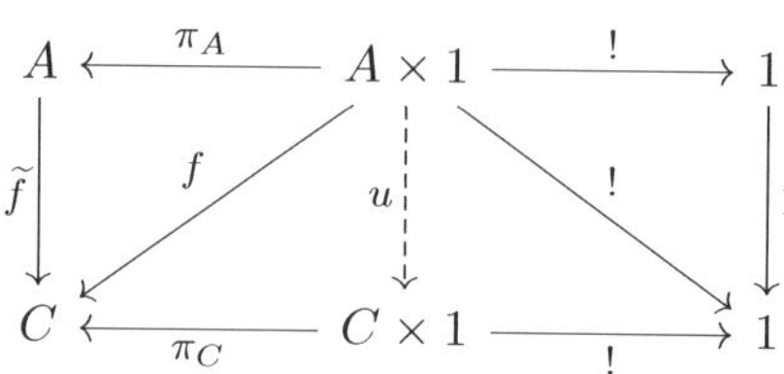

Both π_A and π_C must be isomorphisms by Theorem 49. So put $\widetilde{f} = f \circ \pi_A^{-1}$, and the whole diagram commutes. But this means that $u = \widetilde{f} \times 1$, by definition of the operation $\times$ on arrows in §13.3.

Hence, as we want, for each $f\colon A \times 1 \to C$ there is a corresponding $\widetilde{f}$ making the first of our three diagrams commute.

Moreover $\widetilde{f}$ is unique. For if $k \times 1\colon A \times 1 \to C \times 1$ also makes the third diagram commute then it must equal the unique u, i.e. $\widetilde{f} \times 1$. But if $k \times 1 = \widetilde{f} \times 1$, then by an easy exercise $k = \widetilde{f}$. □

(b) Next, we note

Theorem 75. *If there exists an exponential of C by B in the category C, then, for any object A in the category, there is a one-to-one correspondence between arrows $A \times B \to C$ and arrows $A \to C^B$.*

Proof. By definition of the exponential (C^B, ev), an arrow $f\colon A \times B \to C$ is associated with a unique 'transpose' $\widetilde{f}\colon A \to C^B$ making the diagram (Exp) commute.

The function we can notate $f \mapsto \widetilde{f}$, which sends an arrow to its transpose, is injective. For suppose $\widetilde{f} = \widetilde{g}$. Then $f = ev \circ (\widetilde{f} \times 1_B) = ev \circ (\widetilde{g} \times 1_B) = g$.

The function $f \mapsto \widetilde{f}$ is also surjective. Take any $k\colon A \to C^B$; then if we put $f = ev \circ (k \times 1_B)$, $\widetilde{f}$ is the unique map such that $ev \circ (\widetilde{f} \times 1_B) = f$, so $k = \widetilde{f}$.

Hence $f \mapsto \widetilde{f}$ is the required bijection making a one-to-one correspondence between arrows $A \times B \to C$ and arrows $A \to C^B$. □

This gives us a categorial analogue of the idea we met at the outset, where a two-place function of type $A, B \to C$ can get traded in for either a function of the type $A \times B \to C$ or alternatively for one of the type $A \to C^B$.

We also have:

Theorem 76. *Assuming the exponentials exist, there is also a one-to-one correspondence between arrows $A \to C^B$ and arrows $B \to C^A$.*

Proof. We simply note that arrows $A \times B \to C$ are in bijective correspondence with arrows $B \times A \to C$, in virtue of the isomorphism between $A \times B$ and $B \times A$ (see Theorems 25 and 38). We then apply the last theorem. □

(c) Note as a special case of Theorem 75 that, in a category with a terminal object and exponentials, there will be a bijection between arrows $1 \times B \to C$ and arrows $1 \to C^B$. But, since there is an isomorphism between B and $1 \times B$, we know that there is a bijection between arrows $B \to C$ and $1 \times B \to C$. And arrows $1 \to C^B$ in a category like Set are tantamount to elements of C^B. So, as we should want, in Set and similar categories, elements of C^B correspond one-to-one with arrows from B to C.

We will expand on this theme in quite a neat way in §18.4.

17.6 'And what is the dual construction?'

A good question to ask! After all, when introducing terminal objects, products, and equalizers, we more or less immediately went on to give the dual constructions of initial objects, coproducts, and co-equalizers, simply by turning arrows around. So, what happens if we turn around the arrows in our definition of exponentials?

Well, doing that won't by itself give us what we need, if we are to adhere to the mantra 'co-widgets in C are widgets in C^{op}'. To make things work properly, as well as reversing the arrows in Defn. 72, we need to replace the products with coproducts. Then we do get a coherent definition of co-exponentials in C which will be exponentials in C^{op}.

But is the concept actually of much immediate interest, at our level of enquiry? Not as far as I know! So I'll say no more about co-exponentials here.

18 Cartesian closed categories

Categories like Set, Prop_L and Pos which have all exponentials and which also have binary products and terminal objects (and hence all finite products) form an important class. This chapter briefly investigates.

18.1 A definition and some initial results

Definition 75. A category C is a *Cartesian closed category* (CCC) iff it has all finite products and all exponentials.[1] △

Such categories have some nice properties. In particular, exponentials in such categories behave as exponentials morally *ought* to behave. So we get:

Theorem 77. *If* C *is a Cartesian closed category, then for all* A, B, C *in* C

(1) If $B \cong C$*, then* $A^B \cong A^C$,

(2) $(A^B)^C \cong A^{B\times C}$,

(3) $(A \times B)^C \cong A^C \times B^C$.

I will here give a proof of (1), and outline a proof of (2). Enthusiasts can explore (3) now or wait to bring heavy-weight apparatus to bear in §43.5.

Proof that if $B \cong C$*, then* $A^B \cong A^C$. Here's the idea for a brute-force proof. We know that A^B comes along with an evaluation arrow we'll label $ev_B\colon A^B \times B \to A$. Since $B \cong C$, we can derive from that an arrow $g\colon A^B \times C \to A$. This has a unique associated transpose, $\widetilde{g}\colon A^B \to A^C$. Symmetrically, we can construct an arrow $\widetilde{h}\colon A^C \to A^B$. It remains to confirm that these arrows are (as you'd expect) inverses of each other, whence $A^B \cong A^C$.

[1] A terminological complexity: you need to be aware of a notable variation between what different authors count as a CCC.

Awodey (2010, p. 123), like many, follows the classic Mac Lane (1997, p. 97) in requiring only finite products and exponentials.

But e.g. Borceux (1994, p. 335) and Goldblatt (1984, p. 72) require all finite limits in the sense of Chapter 21's Defn. 87, rather than merely all finite products.

While Johnstone (2002, p. 46) notes that the weaker definition is the more embedded but rather deprecates that, and he calls categories satisfying the stronger condition 'properly Cartesian closed' (see Defn. 87).

Let's spell this out. So consider the following diagram (where $j\colon B \xrightarrow{\sim} C$ is an isomorphism witnessing that $B \cong C$):

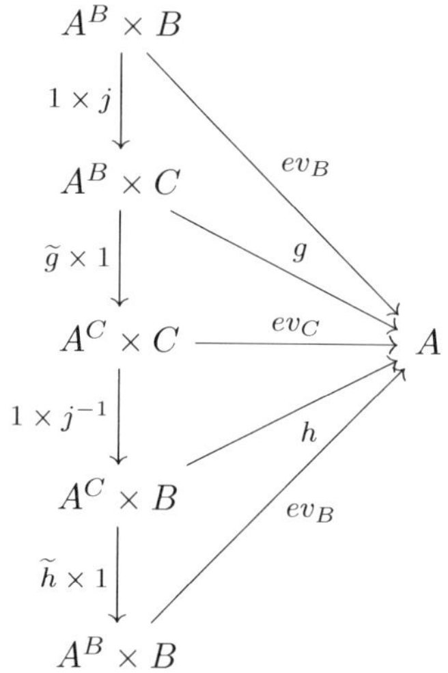

Here I've omitted subscripts on labels for identity arrows to reduce clutter. It is easy to see that since 1 and j are isomorphisms, so is $1 \times j$; hence $1 \times j$ has an inverse and it makes sense to put $g = ev_B \circ (1 \times j)^{-1}$. With g so defined the top triangle trivially commutes. The next triangle commutes by definition of the transpose $\widetilde{g}$; the third commutes if we now put $h = ev_C \circ (1 \times j^{-1})^{-1}$; and the bottom triangle commutes by the definition of the transpose $\widetilde{h}$.

Products of arrows compose componentwise, as shown in Theorem 55. Hence the composite vertical arrow reduces to $(\widetilde{h} \circ \widetilde{g}) \times 1$. However, by the definition of the exponential (A^B, ev_B) we know that there is a unique mediating arrow k such that this commutes:

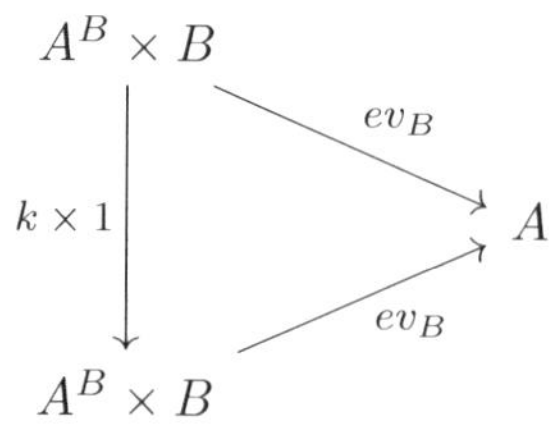

We now have two candidates for k which make the diagram commute, $\widetilde{h} \circ \widetilde{g}$ and the identity arrow. Hence by uniqueness, $\widetilde{h} \circ \widetilde{g} = 1$.

A similar argument shows that $\widetilde{g} \circ \widetilde{h} = 1$. We are therefore done. □

Outline proof that $(A^B)^C \cong A^{B \times C}$. We can give a similarly brute-force proof along the following lines. Start with the evaluation arrow $ev\colon A^{B \times C} \times (B \times C) \to A$. We can shuffle terms in the product to derive $ev'\colon (A^{B \times C} \times C) \times B \to A$. Transpose this once to get an arrow $A^{B \times C} \times C \to A^B$ and transpose again to get an arrow $A^{B \times C} \to (A^B)^C$. Then similarly find an arrow from $(A^B)^C \to A^{B \times C}$, and show the two arrows are inverses of each other. The devil, of course, is in all the details! (Alternatively, we can use a more sophisticated proof idea, as in §37.5.) □

18.2 Challenges!

A Cartesian closed category doesn't need to have an initial object. But when it does, we get a number of further results, some of which we will eventually need in Part III. Since they can be fairly smoothly established using what you already know, I'll group them together immediately as a sextet of challenges to prove:

Theorem 78. *If* C *is a Cartesian closed category that also has an initial object* 0, *then*

(1) $A \times 0 \cong 0 \cong 0 \times A$,

(2) $A^0 \cong 1$,

(3) *any arrow* $f\colon A \to 0$ *is an isomorphism,*

(4) *every arrow* $f\colon 0 \to A$ *is a monomorphism,*

(5) *there exists an arrow* $1 \to 0$ *iff all* C*'s objects are isomorphic to each other.*

And as a simple corollary,

(6) *the categories* Grp *and* Set$_*$ *are not Cartesian closed.*

So: pause to derive those results!

Proof that $A \times 0 \cong 0 \cong 0 \times A$. Since $A \times 0$ and $0 \times A$ exist by hypothesis, and are isomorphic by Theorem 38 we need only prove $0 \times A \cong 0$.

By Theorem 75, for all C, there is a one-to-one correspondence between arrows $0 \to C^A$ and arrows $0 \times A \to C$. But 0 is initial, so there is exactly one arrow $0 \to C^A$. Hence for all C there is exactly one arrow $0 \times A \to C$, making $0 \times A$ initial too. Whence $0 \times A \cong 0$. □

Proof that $A^0 \cong 1$. By Theorem 75 again, for all C, there is a bijection between arrows $C \to A^0$ and arrows $C \times 0 \to A$. And by the previous result and Theorem 25 there is a bijection between arrows $C \times 0 \to A$ and arrows $0 \to A$. Since 0 is initial there is exactly one arrow $0 \to A$, and hence for all C there is exactly one arrow $C \to A^0$, so A^0 is terminal and $A^0 \cong 1$. □

Proof that any arrow $f\colon A \to 0$ *is an isomorphism.* If there's an arrow $f\colon A \to 0$ then the wedge $A \xleftarrow{1_A} A \xrightarrow{f} 0$ exists and factors uniquely through $A \times 0$:

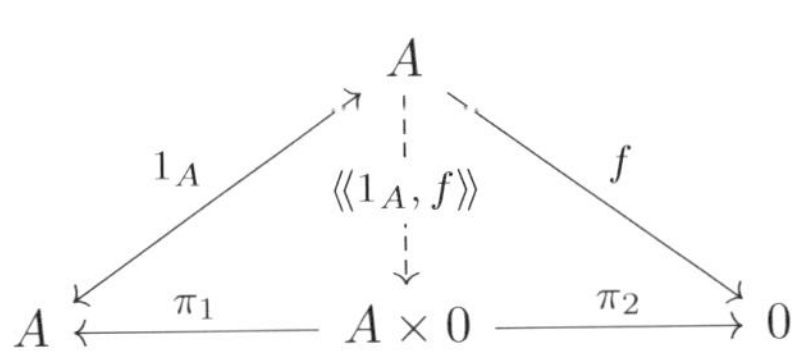

So $\pi_1 \circ \langle\langle 1_A, f \rangle\rangle = 1_A$. But $A \times 0 \cong 0$, so $A \times 0$ is an initial object, so there is a unique arrow $A \times 0 \to A \times 0$, namely $1_{A\times 0}$. Hence (travelling round the left triangle) $\langle\langle 1_A, f \rangle\rangle \circ 1_A \circ \pi_1 = 1_{A\times 0}$. Therefore $\langle\langle 1_A, f \rangle\rangle \circ \pi_1 = 1_{A\times 0}$, and $\langle\langle 1_A, f \rangle\rangle$ has a two-sided inverse. Whence $A \cong A \times 0 \cong 0$.

But then f is an arrow between two initial objects (since objects isomorphic to 0 are also initial by Theorem 28). And there can be only one such arrow and it will be an isomorphism (by Theorem 27). □

Proof that every arrow $f\colon 0 \to A$ is a monomorphism. Since 0 is initial, there is an arrow $f\colon 0 \to A$ for any target A.

Suppose we have arrows g, h such that $f \circ g = f \circ h$. Then for the composites to exist and be equal, g and h must be parallel arrows $g, h\colon X \to 0$ for some X. X will then be initial, as just shown, and hence $g = h$ by the final remark in the previous proof. □

Proof that there's an arrow $1 \to 0$ iff all C's objects are isomorphic. The 'if' direction is trivial. For 'only if', suppose there is an arrow $f\colon 1 \to 0$. Then, for any A there must be a composite arrow $A \xrightarrow{!_A} 1 \xrightarrow{f} 0$, and hence by our result a moment ago $f \circ !_A\colon A \to 0$ is an isomorphism and $A \cong 0$. So every object in the category is isomorphic to 0 and hence to each other. □

Proof that Grp *and* Set$_*$ *are not Cartesian closed.* Recall that the one-element group is both initial and terminal in Grp, so here $1 \cong 0$, and hence there is an arrow $1 \to 0$ in Grp. But not all groups are isomorphic. Therefore the category Grp cannot be Cartesian closed. (Since Grp has all finite products, it follows that this category must lack at least some exponentials.)

The same argument shows that Set$_*$ is not Cartesian closed, since the one-element set is both initial and terminal. □

18.3 Degeneracy

Our initial examples of Cartesian closed categories, Set, Prop$_L$ and Pos, are generously endowed with multiple non-isomorphic objects. At the other end of the scale, there is the example of the one-object one-identity-arrow instance 1 which – quite trivially – has a terminal object, all binary products, and exponentials (check that!).

And we can generalize from the one-object case:

Definition 76. A category where any object has exactly one isomorphism to itself and one isomorphism to any other object, and there are no other arrows, is a *degenerate* Cartesian closed category. △

That definition is in order because simply adding isomorphic copies of objects to the one-object one-identity-arrow case won't change a category in significant ways.

Let's have a theorem:

Theorem 79. *A Cartesian closed category with an initial object such that* $0 \cong 1$ *is degenerate.*

Proof. If $0 \cong 1$, then there is an arrow, in fact an isomorphism, $f\colon 1 \to 0$. Hence by part (5) of the last theorem, all the objects of our category are isomorphic, and hence all the objects are in particular isomorphic to 1, i.e. are terminal. So there is a *unique* isomorphism to it from itself and from any other object. In other words, our category is degenerate. □

18.4 'Naming' arrows

At the end of §17.5, I remarked that, in sufficiently nice categories, arrows from A to C correspond one-to-one with elements of C^A. We can say more.

Suppose we are in a Cartesian closed category. Then we can set up a one-to-one correlation between arrows $f\colon A \to C$ and arrows $\ulcorner f \urcorner\colon 1 \to C^A$ by considering the following commuting diagram where π_A is the projection arrow from the product object $1 \times A$ to A:

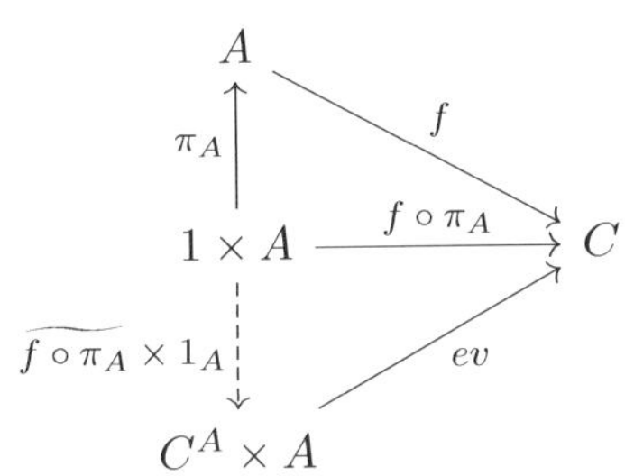

The top triangle trivially commutes, the bottom triangle commutes by definition of the exponential and the exponential transpose. Put $\ulcorner f \urcorner$ for $\widetilde{f \circ \pi_A}$. Then f uniquely fixes $\ulcorner f \urcorner$. For the converse, recall that π_A is an isomorphism by Theorem 49: then $\ulcorner f \urcorner$ fixes $f = (ev \circ \ulcorner f \urcorner \times 1_A) \circ \pi_A^{-1}$.

This inspires a definition:

Definition 77. In a Cartesian closed category, the *name* of an arrow $f\colon A \to C$ is the arrow $\ulcorner f \urcorner\colon 1 \to C^A$ which is the exponential transform of $f \circ \pi_A$ (where $\pi_A\colon 1 \times A \to A$ is the product's projection arrow). △

Why 'name'?[2] Because we have the following cute result which sort-of-says that if we evaluate the function with the name $\ulcorner f \urcorner$ when taken with the input x, we get the output fx (challenge! – prove the theorem before reading further):

Theorem 80. *If* $\vec{x}\colon 1 \to A$ *is a point element of* A *and* $f\colon A \to C$ *is an arrow, then* $ev \circ \langle\langle \ulcorner f \urcorner, \vec{x} \rangle\rangle = f \circ \vec{x}$.

[2]NB, there is some annoying variation in the terminology used hereabouts. I am following e.g. Goldblatt (1984, p. 78) and McLarty (1992, p. 58) in my restricted use of 'name' (for the transpose of an arrow from some $1 \times A$ to C). By contrast, e.g. Lawvere and Schanuel (2009, p. 313) use 'name' and the corner-quotes notation more generally to refer to the exponential transpose of an arrow from any $X \times A$ to C. You have been warned.

Proof. Consider these commutative diagrams:

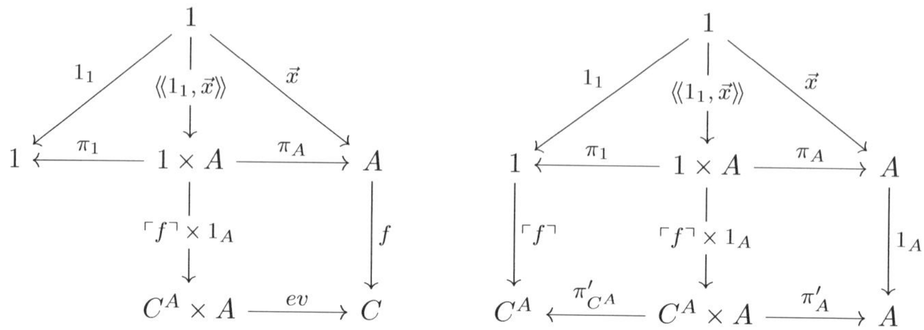

The diagram on the left pastes together a product diagram with the bottom triangle of the previous diagram redrawn as a square. It tells us that

$$ev \circ \ulcorner f \urcorner \times 1_A \circ \langle\!\langle 1_1, \vec{x} \rangle\!\rangle = f \circ \vec{x}.$$

The right-hand diagram pastes together the same product diagram with the diagram defining the arrow-product $\ulcorner f \urcorner \times 1_A$ (see §13.3). So it shows that the wedge $C^A \xleftarrow{\ulcorner f \urcorner \circ 1_1} 1 \xrightarrow{1_A \circ \vec{x}} C$ factors through the product $C^A \times A$ via the mediating arrow $\ulcorner f \urcorner \times 1_A \circ \langle\!\langle 1_1, \vec{x} \rangle\!\rangle$. Hence, by definition and the uniqueness of mediating arrows for products,

$$\ulcorner f \urcorner \times 1_A \circ \langle\!\langle 1_1, \vec{x} \rangle\!\rangle = \langle\!\langle \ulcorner f \urcorner, \vec{x} \rangle\!\rangle.$$

Putting those two equations together gives us our theorem. □

18.5 A fixed point theorem

(a) The assumption that there is a surjection from the set A to the set 2^A leads to contradiction, via a familiar diagonal argument (Cantor's Theorem). We can generalize: in Set, there is no surjection $A \to C^A$ if C has at least two members.

Inspired by the diagonal argument, we can now prove that in any Cartesian closed category the assumption that there is a surjective arrow $A \to C^A$ (now meaning *point-surjective* in the sense of Defn. 42) again puts a very tight constraint on C. Typically, there is no such surjection $A \to C^A$.

Here then is a version of the *Fixed Point Theorem* due to Lawvere (1969b):

Theorem 81. *In a Cartesian closed category, if there is a point-surjective arrow $A \to C^A$, then any arrow $j \colon C \to C$ has a fixed point, i.e. there is a point element $\vec{c} \colon 1 \to C$ such that $j \circ \vec{c} = \vec{c}$.*

Proof. Suppose $\widetilde{g} \colon A \to C^A$ is point-surjective. Then by definition, for any point element of C^A, as it might be $\ulcorner f \urcorner \colon 1 \to C^A$, there is a point element $\vec{a} \colon 1 \to A$ such that $\ulcorner f \urcorner = \widetilde{g} \circ \vec{a}$.

From comment (i) after Defn. 72 we know that our $\widetilde{g}\colon A \to C^A$ must be the exponential transpose of some $g\colon A \times A \to C$. So next consider the following composite arrow, where δ_A (i.e. $\langle\!\langle 1_A, 1_A \rangle\!\rangle$) is the diagonal arrow as in Defn. 52:

$$A \xrightarrow{\ \delta_A\ } A \times A \xrightarrow{\ g\ } C \xrightarrow{\ j\ } C$$

Call this composite $f\colon A \to C$. Its name is then $\ulcorner f \urcorner\colon 1 \to C^A$, which – as we said a moment ago – equals $\widetilde{g} \circ \vec{a}$ for some $\vec{a}\colon 1 \to A$.

Now note first that we have

$$\begin{array}{llll} (1) & f \circ \vec{a} & = j \circ g \circ \delta_A \circ \vec{a} & \text{(by definition of } f\text{)} \\ & & = j \circ g \circ \langle\!\langle \vec{a}, \vec{a} \rangle\!\rangle & \text{(by Thm 50, corollary)} \end{array}$$

But we also have[3]

$$\begin{array}{llll} (2) & f \circ \vec{a} & = (ev \circ \ulcorner f \urcorner \times 1_A \circ \pi_A^{-1}) \circ \vec{a} & \text{(comment pre Defn. 77)} \\ & & = ev \circ (\widetilde{g} \circ \vec{a}) \times 1_A \circ \pi_A^{-1} \circ \vec{a} & \text{(by choice of } \vec{a}\text{)} \\ & & = ev \circ (\widetilde{g} \circ \vec{a}) \times (1_A \circ 1_A) \circ \pi_A^{-1} \circ \vec{a} & \text{(by meaning of } 1_A\text{)} \\ & & = ev \circ (\widetilde{g} \times 1_A) \circ (\vec{a} \times 1_A) \circ \pi_A^{-1} \circ \vec{a} & \text{(by interchange law)} \\ & & = g \circ (\vec{a} \times 1_A) \circ \pi_A^{-1} \circ \vec{a} & \text{(by defn of } \widetilde{g}\text{)} \\ & & = g \circ \langle\!\langle \vec{a}, \vec{a} \rangle\!\rangle & \text{(by Theorem 56)} \end{array}$$

If we put $\vec{c} = g \circ \langle\!\langle \vec{a}, \vec{a} \rangle\!\rangle$, then (1) and (2) together tell us that $j \circ \vec{c} = \vec{c}$. Hence our arbitrarily selected endomorphism j has a fixed point. □

In a category like Set, for example, if an object C has more than one point element, we can find a permutation map $j\colon C \to C$ with no fixed point. So the fixed point theorem then tells us that there will be no point-surjection $A \to C^A$ – a generalized version of Cantor's Theorem.

(b) In a prefatory note to a reprint of this 1969 paper Lawvere writes that, as well as to recover Cantor's Theorem, "The original aim of this article was to demystify the incompleteness theorem of Gödel and the truth-definition theory of Tarski by showing that both are consequences of some very simple algebra in the cartesian-closed setting." This is perhaps doubly misleading. For a start, in so far as there is anything that needs 'demystifying' in e.g. Gödel's theorem, it isn't in the easy-once-spotted diagonalization trick but the whole idea of the arithmetization of syntax. And second, while various argumentative moves depending on diagonalizations in Cantor, Gödel and Tarski can seen as "special cases of a single theorem about a suitable kind of abstract structure" there is nothing especially categorial involved. As a nice paper by Noson Yanofsky (2003) shows, you can in fact squeeze the real goodness out of Lawvere's observations without essentially mentioning categories at all.

[3] I'm writing down a sequence of equations, but you might well want to draw a sequence of commutative diagrams to make it clearer what's going on. Some find that so-called string diagrams provide an even more helpful reasoning tool in cases like this. For an introduction to those, see e.g. Hinze and Marsden (2023).

18.6 CCCs and the lambda calculus?

Before moving on, I should very briefly mention a topic that we haven't discussed, namely the relationship between Cartesian closed categories and the typed lambda calculus. That calculus may well be familiar to you if you have interests in theoretical computer science; however, I judge that it would take us rather too far afield to pause here to set up a sufficiently accessible and meaningful account for those new to type theory.

The headline news, though, is that there are evident parallels between e.g. product and exponential constructions in a typed lambda calculus and products and exponentials in a CCC. And by treating objects in a CCC as the types of a lambda calculus – or, going in the opposite direction, by treating types of a type theory as objects in a category – we can set up a sort of equivalence between the two.

The classic account of this theme is in Part I of Lambek and Scott (1986). But that is at least a notch or two more sophisticated than anything in these present notes; so we won't pursue the topic here.[4]

[4]There is a supposedly introductory discussion in Awodey (2010, §2.5(c), §6.6); but this is not likely to be very helpful if you are new to type theories. If you *do* want an idea of what is going on by way of introduction to Lambek and Scott, try instead e.g. these nicely done notes by Kei Imada (2019) (once you know something about functors and equivalence of categories).

19 Limits and colimits defined

A terminal object is defined in terms of how *all* the other objects in the category relate to it (by each sending a unique arrow to it). A product wedge is defined in terms of how it relates to *all* the other wedges in a certain family (each factoring through it via a unique arrow to it). An equalizing fork is defined in terms of how *all* the other forks in a certain family relate to it (each factoring through it via a unique arrow to it). As noted before, then, terminal objects, products, and equalizers are *limiting cases*, defined in closely analogous ways using universal properties. Likewise, needless to say, for their duals.

Exponentials too are defined by a universal mapping property: '(C^B, ev) is an exponential iff for all f there is a unique $\widetilde{f}$ such that ...'. But intuitively, exponentials are not limiting cases of the same general sort as before. This chapter will confirm the intuition. We will capture what's common to terminal objects, products and equalizers by defining an official general class of *limits*. We also define a dual class of *colimits*, which includes initial objects, coproducts and co-equalizers. We'll see that exponentials are neither limits nor colimits.

Now, in giving a general categorial definition of products, we are already abstracting from various notions of products for different kinds of widgets, bringing out what they have in common. So in this chapter we are abstracting further, bringing out what is common between products, terminal objects, equalizers and more. Category theory does indeed stack layers of abstraction on layers of abstraction; but this can be revealing.

Having introduced the general idea of limits and colimits, in the next chapter we go on to explore a further pair of examples, so-called pullbacks and their duals. We meet some familiar constructions of 'ordinary' mathematics in this new guise. Then in Chapter 21 we prove an important general result of the following shape: if a category has certain basic limits then it will have *all* finite limits. (And this result will both dualize and extend to the infinite case in obvious ways.)

19.1 Cones over diagrams

We need to start by defining the notion of a *cone* over a diagram. Then in the next section we can use this to define the key notion of a *limit cone*.

Way back in Defn. 19, we loosely characterized a diagram D in a category C as what is represented by a representational diagram – i.e. as simply consisting in a bunch of objects with, possibly, (some) arrows between (some of) them. We'll now assume that the objects in D can handily be labelled by terms like 'D_j' where 'j' is an index from some suitable suite of indices J. We will also allow the limiting cases of diagrams where there are no arrows, and even the empty case where there are no objects. So, recasting our earlier definition:

Definition 19* A *diagram in a category* C consists in some (or no) C-objects D_j for indices j from the suite of indices J, together with some (or no) C-arrows between these objects. △

In Chapter 28, we will give a more abstract variant definition; but this current one will do for present purposes (and anyway, it motivates the abstract definition).

Definition 78. Let D be a diagram in category C. A *cone over* D comprises a C-object C, the *vertex* or *apex* of the cone, together with C-arrows $c_j : C \to D_j$ (often called the *legs* of the cone), one for each object D_j in D, such that *whenever* there is an arrow $d : D_k \to D_l$ in D, then $c_l = d \circ c_k$ – in other words the triangles from the vertex C always commute (for each k, l):

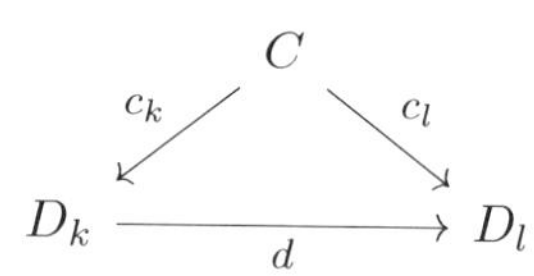

We use '(C, c_j)' as our notation for such a cone.[1] △

You can picture a typical case like this. Suppose the objects in the diagram D are arranged in a plane, along with whatever arrows D contains between those objects (as on the left). Now sit the object C above the plane, with a quiverful of arrows from C zinging down, one targeted at each object D_j in the plane (as on the right).

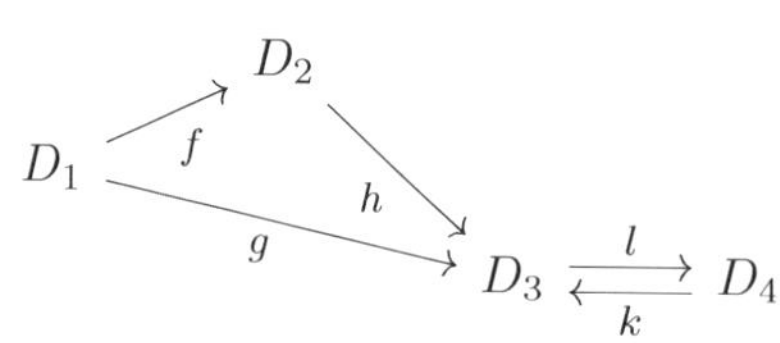

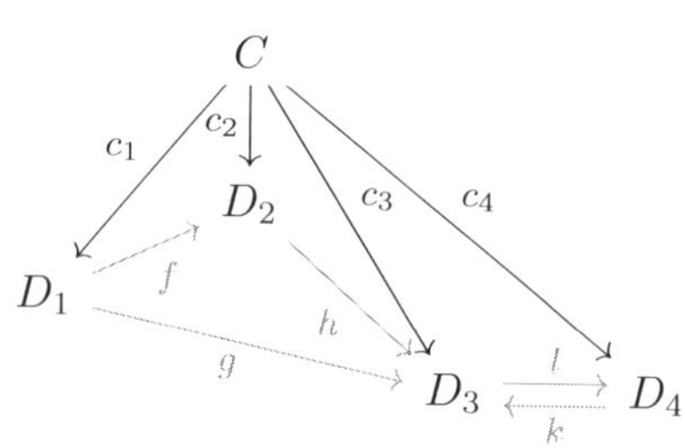

[1]The standard idiom '*cone over D*' rather strongly suggests that ingredients of the cone won't themselves be elements of D. But note that this *isn't* actually built into the definition.

I should also note, by way of aside, that some authors prefer to say more austerely that a cone is not a vertex-object-with-a-family-of-arrows-from-that-vertex, but simply a family of arrows from the vertex. Since we can read off the vertex of a cone as the common source of all its arrows, it is very largely a matter of convenience whether we speak austerely or explicitly mention the vertex. I'll here take the more explicit line (as I did when e.g. defining products or equalizers).

Those arrows $c_j\colon C \to D_j$ form the 'legs' of a skeletal cone. And the key requirement is that any new triangles formed, with C at the apex and some D_k, D_l at the base, must commute. So in our little example, we require $c_2 = f \circ c_1$, $c_3 = g \circ c_1$, $c_3 = h \circ c_2$, $c_3 = k \circ c_4$, $c_4 = l \circ c_3$.

19.2 Limits

(a) There of course can be many cones, with different vertices C, over a given diagram D. But, as with e.g. our earlier definition of products, we can define a limiting case, by means of a universal property:

Definition 79. A cone (L, λ_j) over a diagram D in C is a *limit (i.e. a limit cone) over* D iff *any* cone (C, c_j) over D uniquely factors through it, meaning that there is a unique mediating arrow $u\colon C \to L$ such that for each index j, $c_j = \lambda_j \circ u$.

And we will say (simply) that D *has a limit*, if there exists a limit cone over it. △

A picture will again help to make it clear what's going on. For (L, λ_j) to be a limit cone over D, there must for each cone (C, c_j) over D be a corresponding unique $u\colon C \to L$ which makes each pictured triangle commute:

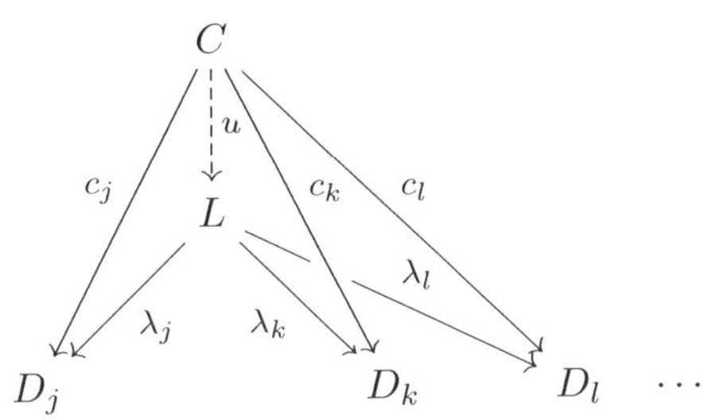

Here, for clarity's sake, I've left out of the picture any arrows there are between the D-objects.

In sum: you can think of a limit cone as one of the *shallowest* cones over D, which other cones will factor through via a unique mediating arrow.

(b) Hopefully that last diagram makes the general concept clear. But to fix ideas, let's immediately confirm that three announced examples of 'limiting cases' – namely terminal objects, products, equalizers – are (or are tantamount to) limits in the sense defined.

(1) We start with the easy null case. Take the empty diagram in C – *zero* objects and so, necessarily, no arrows.

Then a cone over the empty diagram is simply an object C, a lonely vertex (there is no further condition to fulfil), and an arrow between such minimal cones C and L is simply an arrow $C \to L$. Hence L is a limit cone over the empty diagram if and only if there is a unique arrow to it from any other object – i.e. if and only if L is a terminal object in C!

(2) Consider now a diagram that consists in *two* naked objects D_1 and D_2, still with no arrow between them.

Then a cone over such a diagram is a wedge with vertex C and arrows to D_1, D_2 (compare Defn. 48). And hence a limit cone is simply a product of D_1 with D_2.

(3) Next consider a diagram that again has two objects D_1 and D_2, but now with two parallel arrows f and g between them.

A cone over this is a commuting diagram of this shape:

$$\begin{array}{ccc} & C & \\ {\scriptstyle c_1}\swarrow & & \searrow{\scriptstyle c_2} \\ D_1 & \underset{g}{\overset{f}{\rightrightarrows}} & D_2 \end{array}$$

If there is such a cone, then we must have $f \circ c_1 = c_2 = g \circ c_1$. Which means that $C \xrightarrow{c_1} D_1 \underset{g}{\overset{f}{\rightrightarrows}} D_2$ is a fork. Conversely, of course, given such a fork, we can turn it into a cone by adding the arrow $c_2 = f \circ c_1 \colon C \to D_2$. Since c_1 fixes what c_2 has to be to complete the cone, we can without loss focus on the part of the cone consisting of (C, c_1).

What is the corresponding part of a limit cone over $D_1 \underset{g}{\overset{f}{\rightrightarrows}} D_2$? It consists in some (E, e) such that there is a unique u such that $c_1 = e \circ u$. Hence (E, e) is an equalizer of the parallel arrows (compare Defn. 67). So equalizers are (parts of) limits.

By contrast, exponentials are evidently *not* limits in the sense of limit cones.

19.3 Uniqueness up to unique isomorphism

(a) You know what comes next! We have the predictable result:

Theorem 82. *A limit over a given diagram D, if one exists, is unique up to a unique isomorphism commuting with the cones' arrows. That is to say, if (L, λ_j) and (L', λ'_j) are both limit cones over D, then there is a unique isomorphism between the vertices $v \colon L' \to L$ such that $\lambda_j \circ v = \lambda'_j$ for all j.*

And as with comparable uniqueness theorems like Theorems 37, 65 and 72, we can prove this in two ways (you shouldn't need me to fill in the details).

Firstly, then, we can use brute force:

Plodding proof from first principles. Suppose (L, λ_j) is a limit cone over D. Note that if $\lambda_j \circ u = \lambda_j$ for all indices j, then (L, λ_j) would factor through itself via u. But the limit cone factors through itself via 1_L. Hence, since mediating arrows are unique,

(i) if $\lambda_j \circ u = \lambda_j$ for all indices j, then $u = 1_L$.

Now suppose (L', λ'_j) is another limit cone over D. Then (L', λ'_j) uniquely factors through (L, λ_j), via some v, so

(ii) $\lambda_j \circ v = \lambda'_j$ for all j.

And likewise (L, λ_j) uniquely factors through (L', λ'_j) via some w, so

(iii) $\lambda'_j \circ w = \lambda_j$ for all j.

Whence

(iv) $\lambda_j \circ v \circ w = \lambda_j$ for all j.

Therefore by (i),

(v) $v \circ w = 1_L$.

And symmetrically we can show

(vi) $w \circ v = 1_{L'}$.

Whence v is not only unique (by hypothesis, the only way of completing the relevant diagrams to get the arrows to commute) but is an isomorphism. □

(b) We have already seen that

(1) A terminal object in C is ... wait for it! ... terminal in the category C.
(2) A product of X with Y in C is a terminal object in the derived category C/XY of wedges to X and Y.
(3) An equalizer of parallel arrows $f, g\colon X \to Y$ in C is (part of) a terminal object in the derived category $\mathsf{C}_{f\|g}$ of forks through X to Y.

Predictably, limit cones more generally are terminal objects in appropriate categories.

To spell this out, we first note that the cones (C, c_j) over a given diagram D in C form a category in a very natural way:

Definition 80. Given a diagram D in category C, the derived category $Cone(D)$ – *the category of cones over D* – has the following data:

(1) Its objects are the cones (C, c_j) over D.
(2) An arrow from (C, c_j) to (C', c'_j) is any C-arrow $f\colon C \to C'$ such that $c'_j \circ f = c_j$ for all indices j. In other words, for each D_j, D_k, D_l, ..., in D, the corresponding triangle with remaining vertices C and C' commutes:

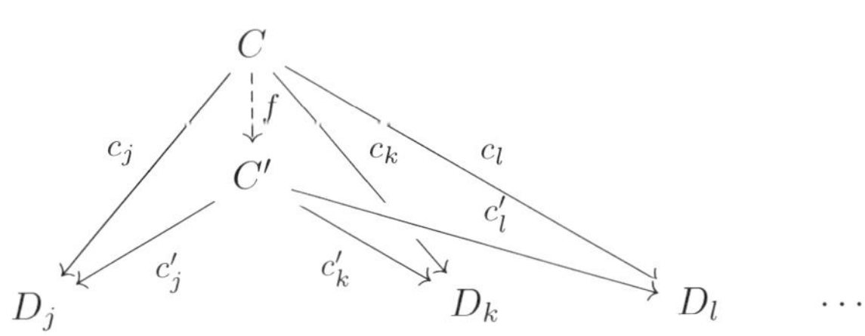

The identity arrow on a cone (C, c_j) is the C-arrow 1_C. And composition for arrows in *Cone*(D) is composition of the corresponding C-arrows. △

It is entirely routine to confirm that *Cone*(D) is a category. Our earlier definition of a limit cone is then immediately equivalent to this:

Definition 81. A *limit for D* in C is a terminal object in *Cone*(D). △

And we now have an alternative proof of our desired uniqueness result:

Succinct proof of Theorem 82. Since a limit cone over D is terminal in the category *Cone*(D), it is unique in *Cone*(D) up to a unique isomorphism. But such an isomorphism in *Cone*(D) must be an isomorphism in C commuting with the cones' arrows. □

19.4 Challenges!

We now want to prove some further general results about limits, some of which we'll need later. And to make things a bit more interesting, let's present the three theorems as a series of challenges to prove.

But first, some quick questions. We have seen that limit cones in C over the null diagram are terminal objects of C, those over arrowless two-object diagrams are products, and those over two-object two-arrow diagrams are in effect equalizers. So it is natural to ask, for example:

Queries *What is a limit over a two-object diagram that is arrowless except that each object comes with its identity arrow? What is a limit over an arrowless* one-*object diagram? How about limits over arrowless* three-*object diagrams? What are limits over diagrams with two objects and a single arrow between them? What about limits over wedges?*

Now for our needed theorems:

Theorem 83. *Suppose (L, λ_j) is a limit cone over the diagram D in* C, *and (L', λ'_j) is another cone over D which factors through (L, λ_j) via an isomorphism $o \colon L' \xrightarrow{\sim} L$. Then (L', λ'_j) is also a limit cone.*

Theorem 84. *Again let (L, λ_j) be a limit cone over a diagram D in* C. *Then the cones over D with vertex C correspond one-to-one with* C-*arrows from C to L.*

Theorem 85. *If (C, c_j) is a cone over a diagram D in* C *it is also a cone over the smallest O of* C *which contains D.*

Hint: the smallest O which contains D will have the same objects, with all the necessary identity arrows, and all compositions of D's arrows (and then compositions of *these* arrows, etc.).

19.5 Responses

(a) Let's start with our five queries. A limit over an arrowless diagram with two objects is a product of those objects. But what if we decorate those initial objects with their identity arrows? Nothing changes!

The underlying point is this. To make a cone over a diagram D, if there is any arrow $d\colon D_k \to D_l$, then the legs of the cone $c_k\colon C \to D_k$ and $c_l\colon C \to D_l$ must form a commuting triangle making $c_l = d \circ c_k$. Whatever the shape of D, if we add to the diagram an identity arrow $1\colon D_k \to D_k$, then we will have $c_k = 1 \circ c_k$, and so any cone over D is still a cone over the augmented diagram.

(b) Next, what is a limit cone over a diagram in C that consists in a single object D? A cone (C, c) over D is any object C together with an arrow $c\colon C \to D$. So – mindlessly applying the definition! – a limit cone will comprise an object L and arrow $\lambda\colon L \to D$ such that for any $c\colon C \to D$ there is a unique $u\colon C \to L$ such that $\lambda \circ u = c$.

One candidate such limit is evidently given by $L = D$ and $\lambda = 1_D$.[2] But you know that limits are usually only unique up to isomorphism. You should expect the same here. So you'd expect any L that is isomorphic to D, when equipped with that isomorphism, gives us another limit over D.

And that's right. Here's a slow-motion argument for the same conclusion. All cones over D need to factor uniquely through a candidate limit cone (L, λ). In particular, this applies (i) to the cone $(D, 1_D)$ and also (ii) to the cone (L, λ) itself.

From case (i) we know that there is a unique u such that $\lambda \circ u = 1_D$. From case (ii) we know there is a unique v such that $\lambda \circ v = \lambda$, and evidently $v = 1_L$. But $\lambda \circ u \circ \lambda = \lambda$, so we now know that $u \circ \lambda = 1_L$. Which means that λ has a two-sided inverse, i.e. has to be an isomorphism between L and D.

(c) To answer our next query, we note that a limit cone over a diagram that consists of three isolated objects is a limiting case of a three-way wedge over those objects, i.e. is a ternary product. See Defn. 56.

(d) What is the shallowest cone over $D_1 \xrightarrow{f} D_2$? Evidently, the cone with vertex D_1 and legs $1\colon D_1 \to D_1$ and $f\colon D_1 \to D_2$. But remembering the point in (b), a cone whose vertex L is isomorphic with D_1 and with legs $\lambda\colon L \xrightarrow{\sim} D_1$ and $f \circ \lambda\colon L \to D_2$ will do equally well.

Similarly, the shallowest cone over the wedge $D_1 \xleftarrow{f} D_2 \xrightarrow{g} D_3$ is, up to isomorphism, the cone with vertex D_2 and legs $f, 1_{D_2}, g$, so is the wedge we started off with decorated with the identity arrow on D_2.

(e) Next, Theorem 83 should be very reminiscent of Theorem 39. It is proved in exactly the same way (simply generalize from the case where we have two 'legs' π_j to the case where there are possibly many legs λ_j).

So let's move on to

[2] "Ahah! – so the vertex of a limit '*over*' a diagram D can be an object *already in* D?" Yes! Our definition of a cone over D allows for this, as noted in fn. 1.

Theorem 84. *Suppose (L, λ_j) is a limit cone over a diagram D in C. Then the cones over D with vertex C correspond one-to-one with C-arrows from C to L.*

Proof. Take any arrow $u\colon C \to L$. If there is an arrow $d\colon D_k \to D_l$ in the diagram D, then since (L, λ_j) is a cone, $\lambda_l = d \circ \lambda_k$, whence $(\lambda_l \circ u) = d \circ (\lambda_k \circ u)$. Since this holds generally, $(C, \lambda_j \circ u)$ is a cone over D.

But since (L, λ_j) is a limit, every cone over D with vertex C is of the form $(C, \lambda_j \circ u)$ for unique u.

Hence there is a one-to-one correspondence between arrows $u\colon C \to L$ and cones over D with vertex C. Moreover, the described correspondence is a natural one, involving no arbitrary choices. □

(f) Let's introduce another natural idea:

Definition 82. The (reflexive, transitive) *closure* of a diagram D in a category C is the smallest diagram that includes all the objects and arrows of D, but that also (i) has an identity arrow on each object, and (ii) for any two of its composable arrows, it also contains their composition. △

In other words, the closure of a diagram D in C is what you get by adding identity arrows where necessary, forming composites of any composable arrows you now have, then forming composites of what you have at the next stage, and so on and so forth. Since the associativity of the composition operation will be inherited from C, it is immediate that the closure of a diagram D in C is itself a category, the smallest subcategory of C that contains D.

And now we can prove

Theorem 85. *If (C, c_j) is a cone over a diagram D in C it is also a cone over the smallest O of C that contains D.*

Proof. The closure of D has no additional objects, so (C, c_j) still has a leg from the vertex C to each object in the closure. As noted before, it is trivial that, given an identity arrow $1_k\colon D_k \to D_k$, we have $c_k = 1_k \circ c_k$. Therefore we only need to show that a cone over composable arrows in D is still a cone when their composite is added to D.

Suppose we have a cone over a diagram including the arrows $d\colon D_k \to D_l$ and $d'\colon D_l \to D_m$. By the definition of a cone, that means $c_l = d \circ c_k$ and $c_m = d' \circ c_l$. Hence $c_m = (d' \circ d) \circ c_k$. In other words, the new triangle with apex C and base arrow $d' \circ d\colon D_k \to D_m$ also commutes.

Hence the (commuting-where-required) cone remains a (commuting-where-required) cone if we add the composite arrow $d' \circ d\colon D_k \to D_m$. Iterating gives us our theorem. □

This result will prove rather significant when we turn to give our more abstract account of limits in Chapter 28.

19.6 Cocones and colimits

(a) Now let's dualize! Reverse the relevant arrows and you get definitions of cocones and colimits. So, dualizing Defns. 78 and 79 we get:

Definition 83. Let D be a diagram in category C. Then a *cocone under* D is a C-object C, together with an arrow $c_j\colon D_j \to C$ for each object D_j in D, such that whenever there is an arrow $d\colon D_k \to D_l$ in D, the following triangle commutes:

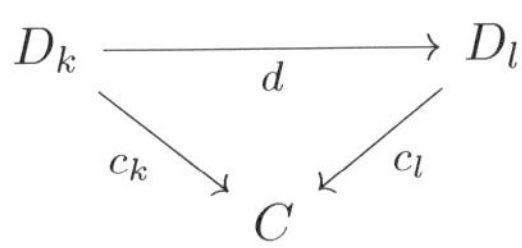

We again use '(C, c_j)' as our notation for such a cocone, letting context settle whether we are talking of cocones rather than cones. △

Definition 84. A cocone (L, λ_j) under a diagram D in C is a *colimit* (i.e. a limit case of a cocone) under D iff it factors uniquely through *any* cocone (C, c_j) under D, meaning that there is a unique mediating arrow $u\colon L \to C$ such that for each index j, $u \circ \lambda_j = c_j$.

And we will say (simply) that D *has a colimit*, if there exists a colimit cocone under it. △

Here's another picture, to fix ideas. For (L, λ_j) to be colimit under D, for any cocone (C, c_j) under D, there must be a corresponding unique $u\colon L \to C$ that makes each pictured triangle commute:

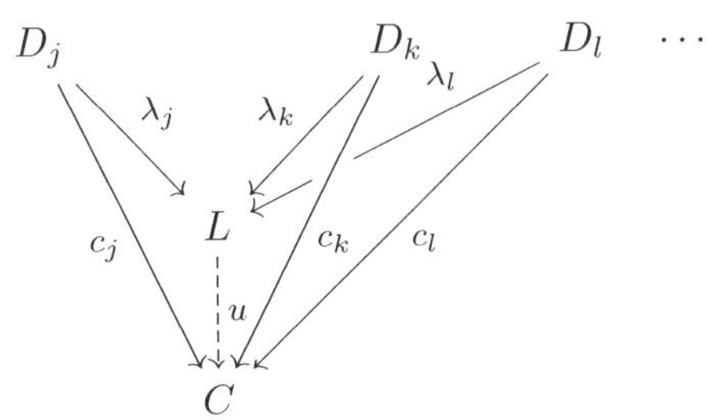

Again, for clarity's sake, I've left out of the picture any arrows there are between the D-objects.

(b) The cocones under D form a category with objects the cocones (C, c_j) and an arrow from (C, c_j) to (C', c'_j) being any C-arrow $f\colon C \to C'$ such that $c'_j = f \circ c_j$ for all indices j. So here's an equivalent definition: a colimit for D is an initial object in the category of cocones under D. (Check that!)

(c) It is now routine to confirm that our earlier examples of initial objects, coproducts and co-equalizers do count as colimits.

(1) The null case where we start with the empty diagram in C gives rise to a cocone that is simply an object in C. So the category of cocones over the empty diagram is the category C we started with, and a limit cocone is just an initial object in C.

(2) Consider now a diagram with only *two* objects D_1 and D_2, still with no arrow between them. Then a cocone over such a diagram is simply a corner from D_1, D_2 (in the sense we met in §11.7); and a limit cocone in the category of such cocones is simply a coproduct.

(3) And if we start with the diagram $D_1 \underset{g}{\overset{f}{\rightrightarrows}} D_2$ then a limit cocone over this diagram gives rise to a co-equalizer.

Evidently, exponentials are no more colimits than they are limits.

(d) Which of course is not to deny that there is *something* in common between limits/colimits and exponentials – after all, both are defined by universal mapping properties in an intuitive sense. A limit cone over a diagram is one such that every cone over that diagram factors through it via a unique map. Likewise for colimits. An exponential is essentially an arrow $ev\colon C^B \times B \to C$ such that any arrow $f\colon A \times B \to C$ factors through it by a unique map $\hat{f} \times 1_B\colon A \times B \to C^B \times B$. And relatedly, for both limits and exponentials in C we can cook up derived categories – categories of cones over diagrams in C, categories of parameterized maps in C – such that limits/exponentials in a C are terminal objects in the derived categories. Likewise colimits are initial in a derived category of cocones.

In §39.5 we will find a way of defining a more general categorial construction that subsumes both limits/colimits and exponentials. But it is questionable whether this adds much extra illumination.

20 Pullbacks and pushouts

In this chapter, we explore one more kind of limit, so-called pullbacks. Predictably, we will also (although more briefly) consider their duals, so-called pushouts.

With products and quotients, we first thought about these pre-categorially, and then gave a treatment in categorial terms. This time we'll go the other way about. We will first define the notion of a pullback in abstract categorial terms, and then think about how this relates to various already-familiar concrete constructions. For example, we'll make links with forming intersections of sets, inverse images, and kernels of group homomorphisms.

20.1 Pullbacks defined

(a) As we saw in the last chapter, taking a limit over a wedge simply returns the same wedge. But what about taking a limit over a co-wedge or *corner*, i.e. a diagram D like $D_1 \xrightarrow{d_1} D_3 \xleftarrow{d_2} D_2$?

Things now get much more interesting! Let's draw this *as* a corner, and then a cone over it can be pictured as a commutative diagram like this:

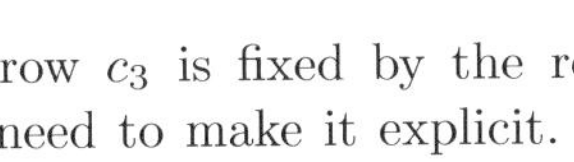

But note, the diagonal arrow c_3 is fixed by the requirement that $d_1 \circ c_1 = c_3 = d_2 \circ c_2$. So we don't need to make it explicit. And from now on, to keep things uncluttered, I'll follow convention and leave out such diagonal arrows when drawing cone diagrams over corners.

So, what is the *limit* case for this type of cone? It will be a cone with vertex L and three projections $\lambda_j\colon L \to D_j$ such that any cone (C, c_j) over D factors uniquely through (L, λ_j). In other words, for any (C, c_j) there is a unique $u\colon C \to L$ such that this diagram commutes (again leaving out the diagonal arrows):

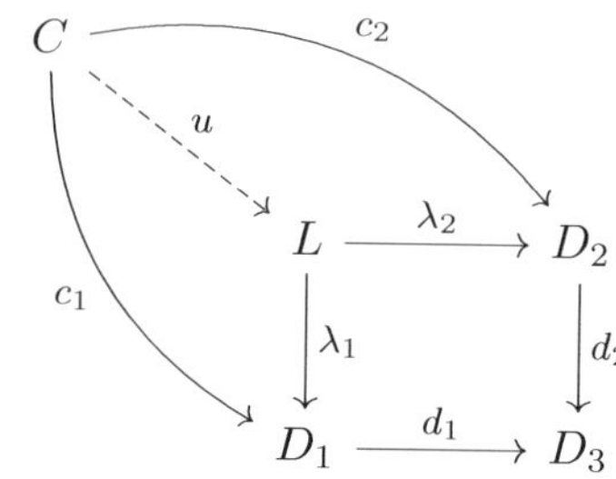

There is standard terminology for such a limit:

Definition 85. A limit for a corner diagram is its *pullback.*

The commuting square formed by a corner and its limit, with or without its diagonal, is a *pullback square* (alternatively, *Cartesian square*).

Further, in the illustrated square, the arrow λ_1 is said to arise by *pulling back* d_2 along d_1; likewise λ_2 is said to arise by pulling back d_1 along d_2. △

As with limits generally, pullbacks are not unique. However, applying Theorem 82, we have

Theorem 86. *If both $\lambda_a\colon L_a \to D_1$ and $\lambda_b\colon L_b \to D_1$ arise from pulling back d_2 along d_1, then there is a unique isomorphism such that $L_a \cong L_b$, and λ_a and λ_b factor through each other via that isomorphism.* □

(b) A pullback for a corner, then, is a limit wedge (W), $D_1 \xleftarrow{\lambda_1} L \xrightarrow{\lambda_2} D_2$ (forgetting about the diagonal arrow we've left undrawn). And, in a sense, the notion of a pullback is a generalization of the notion of a product. Compare:

(i) If (W) is to be an ordinary product, *every* wedge $D_1 \xleftarrow{c_1} C \xrightarrow{c_2} D_2$ must factor uniquely through that limit wedge (W).

(ii) But, for a pullback, only those wedges $D_1 \xleftarrow{c_1} C \xrightarrow{c_2} D_2$ *that form a commuting square with the opposite corner* $D_1 \xrightarrow{d_1} D_3 \xleftarrow{d_2} D_2$ need to factor uniquely through the limit wedge (W).

Evidently, then, not all pullbacks need be ordinary products.

(c) On notation. (i) It is conventional to indicate that a diagrammed square is a pullback by marking the vertex of the limit over the opposite corner with a little corner-symbol. (ii) The fact that pullbacks are a restricted version of products encourages a product-like notation for the vertex of a pullback, as here:

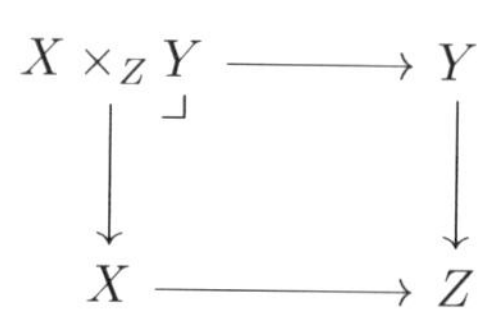

Though NB: the limit object essentially depends on the particular *arrows* from X and Y and not only on the object Z that is their shared target.

20.2 Examples

(a) Consider then what happens in the category Set.

Start with a corner formed by two set functions f, g as in $X \xrightarrow{f} Z \xleftarrow{g} Y$. Put $X \times_Z Y = \{\langle x, y\rangle \in X \times Y \mid f(x) = g(y)\}$. And let the projection function $\pi_1 \colon X \times_Z Y \to X$ send $\langle x, y\rangle$ to x, while $\pi_2 \colon X \times_Z Y \to Y$ sends $\langle x, y\rangle$ to y. Then it is easy to see that this is a pullback square:

$$\begin{array}{ccc} X \times_Z Y & \xrightarrow{\pi_2} & Y \\ \big\downarrow{\scriptstyle \pi_1} & \lrcorner & \big\downarrow{\scriptstyle g} \\ X & \xrightarrow{f} & Z \end{array}$$

What about the special case where X and Y are subsets of Z, with f and g inclusion functions? Then $X \times_Z Y = \{\langle x, y\rangle \in X \times Y \mid x = y)\} = \{\langle z, z\rangle \mid z \in X \cap Y\}$. But that's isomorphic to $X \cap Y$ and so, equivalently, we get a pullback square

$$\begin{array}{ccc} X \cap Y & \hookrightarrow & Y \\ \big\downarrow & \lrcorner & \big\downarrow \\ X & \hookrightarrow & Z \end{array}$$

And note, in particular, that this is a pullback when $Z = X \cup Y$.

In short: up to isomorphism, we can define intersections as pullback objects.

(b) Again in a category like Set, suppose we start from a corner of this kind:

$$\begin{array}{ccc} & & Y \\ & & \big\downarrow{\scriptstyle 1_Y} \\ X & \xrightarrow{f} & Y \end{array}$$

Then a pullback object is provided by $\{\langle x, y\rangle \in X \times Y \mid f(x) = y\}$. But that's isomorphic to $\{x \in X \mid \exists y f(x) = y\} = f^{-1}[Y]$. So we have the following square:

$$\begin{array}{ccc} f^{-1}[Y] & \longrightarrow & Y \\ \big\downarrow & \lrcorner & \big\downarrow{\scriptstyle 1_Y} \\ X & \xrightarrow{f} & Y \end{array}$$

Hence we can also think of forming the inverse image as a pullback construction, where the arrow $f^{-1}[Y] \to X$ arises by pulling back the arrow 1_Y along f.

Now generalize. Suppose we are given an inclusion function $i \colon Y' \hookrightarrow Y$. Then the inverse image of Y' under the function $f \colon X \to Y$ is obtained by pulling back i along f like this, where f' is the restriction of f to $f^{-1}[Y']$:

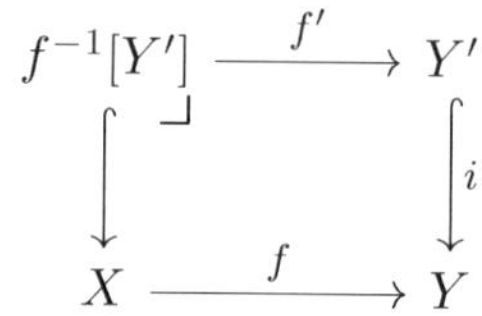

To check that this *is* a pullback square, consider any wedge $X \xleftarrow{c} C \xrightarrow{d} Y'$ such that $f \circ c = i \circ d$:

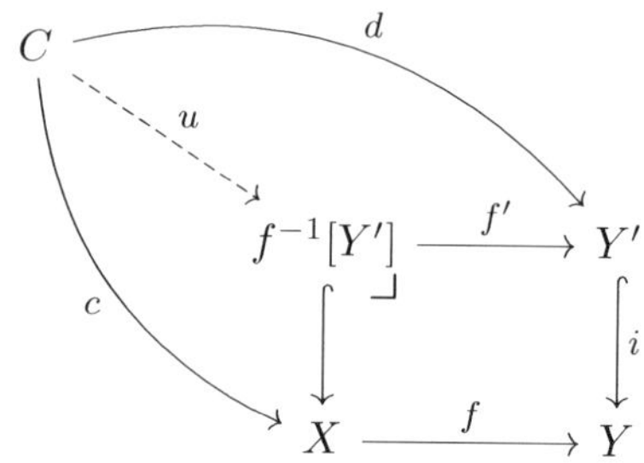

By assumption, $f \circ c$ sends any element of C to $Y' \subseteq Y$; hence c must send that element to an element of $f^{-1}[Y]$. So define u to agree with c on all elements of C, the only option, and the left triangle commutes. Then $f' \circ u$ will agree with $f \circ c$ on all elements of C and hence agree with d, making the top triangle commute.

(c) Let's turn now to look at a nice example in Grp.

Remember, the trivial group 1 is both initial and terminal in Grp. So now consider the corner $1 \xrightarrow{!} Z \xleftarrow{f} Y$. Does it have a pullback? Well, a cone over this corner will look like this (omitting the diagonal as usual):

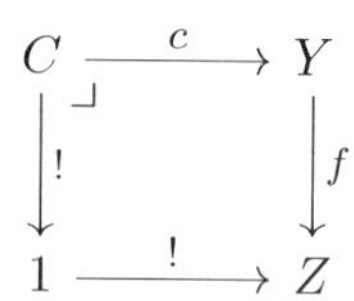

And if this diagram is to commute, then $f \circ c$ has to send every object in the group C to the group identity of Z. So c has to send every object of C somewhere in the kernel of f.

We'll expect, then, that we get a limiting case – a pullback cone – when its vertex is the kernel K of f, and the non-trivial 'leg' is simply the inclusion map $i\colon K \hookrightarrow Y$. Let's confirm that. Consider the diagram

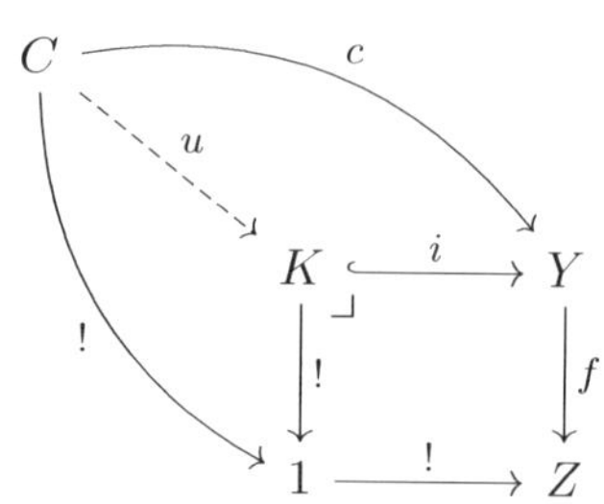

For the outer wedge to make a commuting square with the corner, as we said, c needs to map C into the kernel of f. But if $u\colon C \to K$ agrees everywhere with $c\colon C \to Y$, we'll get a commuting upper triangle (and in the only way possible). So K with the appropriate 'legs' gives us our limit cone.

In short, the kernel of a group homomorphisms is a pullback object.

20.3 Pullbacks, products, equalizers

(a) In general, wedges forming a pullback square won't be products. However, we do have the following result in the converse direction:

Theorem 87. *In a category with a terminal object, all binary products are pullbacks.*

Proof. Consider the diagram

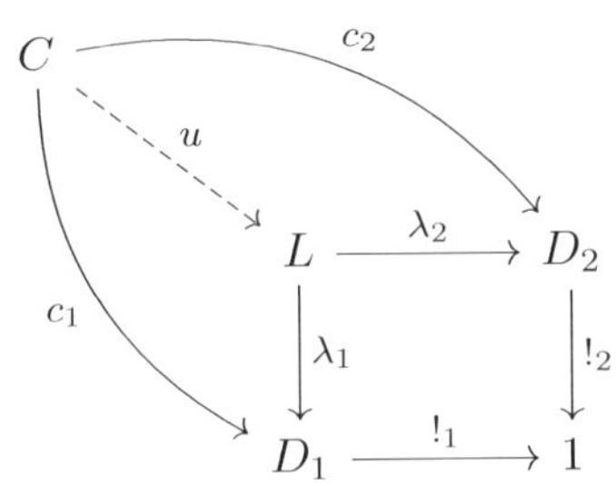

By definition, if the wedge with vertex L is a product for D_1 with D_2 then, for any wedge $D_1 \xleftarrow{c_1} C \xrightarrow{c_2} D_2$, there will be a unique arrow $u\colon C \to L$ making the outer triangles commute. But of course, adding the only possible arrows from D_1 and D_2 to the terminal object 1 will make the resulting square commute. Therefore that unique u will make the whole diagram commute.

Hence the product wedge for D_1 with D_2 with vertex L is a pullback limit for the corner $D_1 \xrightarrow{!} 1 \xleftarrow{!} D_2$. □

Here is an easy corollary that we'll need later, but can state right away:

Theorem 88. *A category that has a terminal object and a pullback for every corner also has all binary products.*

Proof. Given objects D_1 and D_2 in the category, then by the definition of the terminal object 1 there will be a corner $D_1 \xrightarrow{!} 1 \xleftarrow{!} D_2$. This corner, by assumption, will have a pullback. But, as the last proof shows, the wedge that forms the pullback is a product of D_1 with D_2. □

(b) Now for another very easy theorem, this time relating equalizers to pullbacks:

Theorem 89. *(E, e) is an equalizer for the parallel arrows $f, g\colon X \to Y$ if the following is a pullback square:*

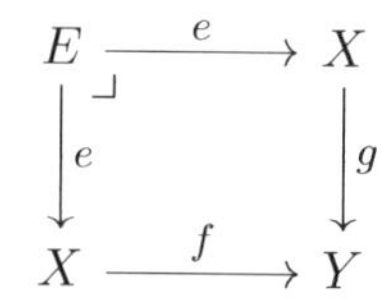

Proof. Suppose the given square is a pullback. Then whatever object W and arrow $k\colon W \to X$ we take, there will be a unique u making the following diagram commute:

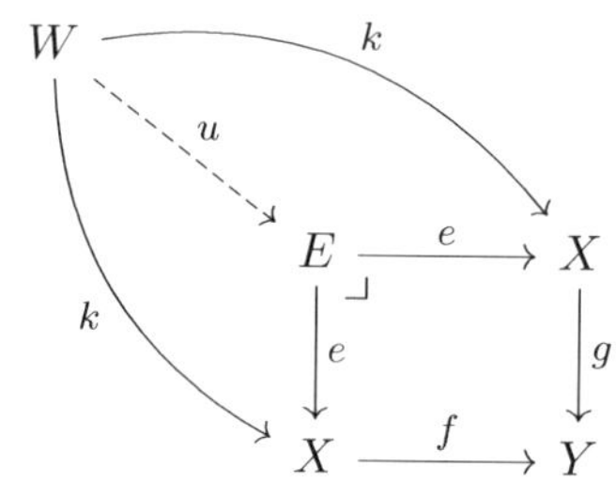

But that is simply a re-drawn version of our Defn. 67 of (E, e) as an equalizer. □

20.4 Challenges!

(a) Do at least note the statements of the following theorems. And why not fix ideas by finding the proofs for yourself (not exactly serious challenges)?

Theorem 90. *Pulling back an isomorphism yields an isomorphism.*

Start with a corner, namely $X \xrightarrow{f} Z \xleftarrow{g} Y$, where g is an isomorphism. In this case, whatever category we are in, we can always make a pullback square where g is pulled back along f, and any pullback of g will be an isomorphism.

Proof. Consider the diagram

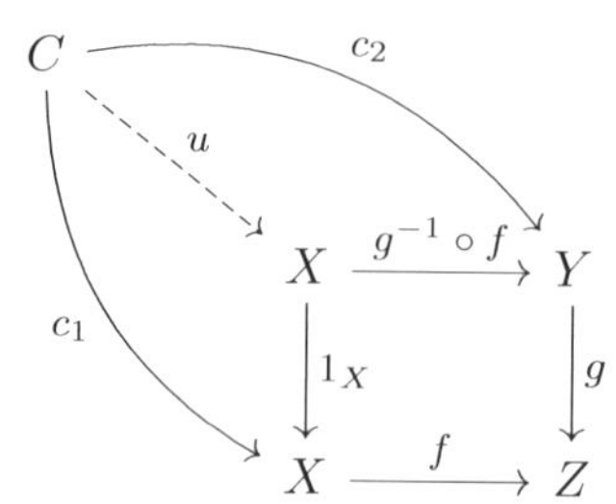

By assumption the inverse g^{-1} exists and the inner square commutes.

And now suppose a wedge with vertex C makes a commuting square with the opposite corner, so $f \circ c_1 = g \circ c_2$, which means $g^{-1} \circ f \circ c_1 = c_2$. Then the unique arrow $u = c_1$ makes the diagram commute. Hence the inner square is a pullback.

Therefore, by Theorem 86, any result of pulling f back along g will factor through 1_X by an isomorphism – i.e. will be an isomorphism. □

(b) Now two theorems about pullbacks and monomorphisms:

Theorem 91. *Pulling back a monomorphism yields a monomorphism.*

In other words, start with a corner $X \xrightarrow{f} Z \xleftarrow{g} Y$ with g monic. Then if we *can* pull back g along f – and that will depend on the category we are in – the resulting arrow will be monic. (NB, this again doesn't depend on the character of f.)

Proof. Suppose that we have the pullback square

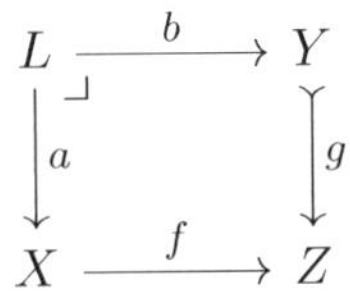

And suppose too that, for some arrows $j, k\colon C \to L$, $a \circ j = a \circ k$. To show that a is monic, we need to prove $j = k$.

WE have $g \circ b \circ j = f \circ a \circ j = f \circ a \circ k = g \circ b \circ k$. Hence, given that g is monic, $b \circ j = b \circ k$. Now consider the two diagrams

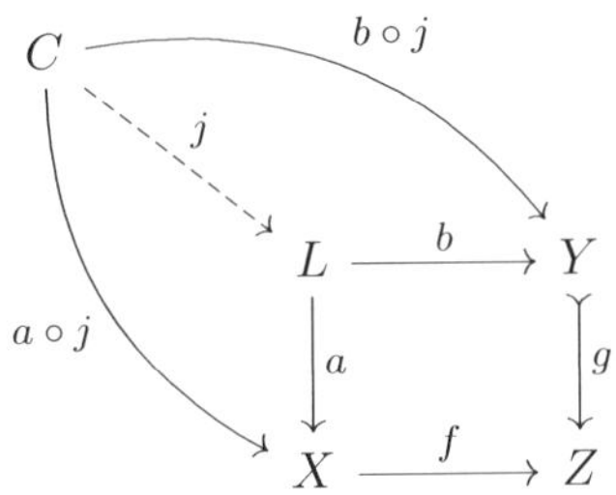

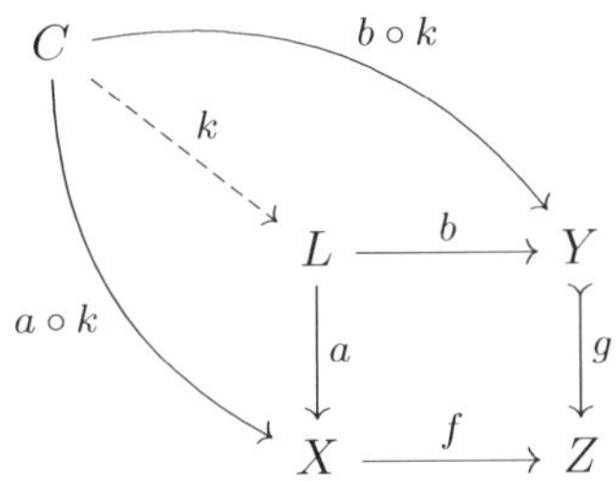

Since $f \circ a \circ j = g \circ b \circ j$, the wedge $X \xleftarrow{a \circ j} C \xrightarrow{b \circ j} Y$ is a cone over the original corner, so uniquely factors through the limit via j. Likewise the wedge $X \xleftarrow{a \circ k} C \xrightarrow{b \circ k} Y$ is also a cone over the original corner, factoring uniquely through the limit via k. But we've just shown that those two cones are the same. Hence $j = k$, as was to be proved. □

Theorem 92. *The arrow $f\colon X \to Y$ is a monomorphism in C if and only if the following is a pullback square:*

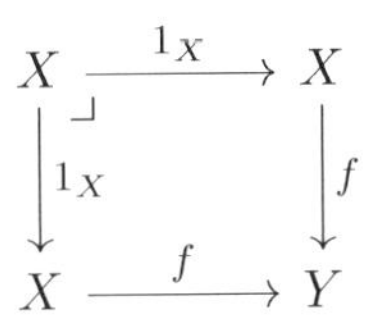

Proof. Suppose this is pullback square. Then any cone $X \xleftarrow{a} C \xrightarrow{b} X$ over the corner $X \xrightarrow{f} Y \xleftarrow{f} X$ must uniquely factor through the limit

with vertex X. In other words, if $f \circ a = f \circ b$, then there is a u such that $a = 1_X \circ u$ and $b = 1_X \circ u$, hence $a = b$ – so f is monic.

Conversely, suppose f is monic. Then given any cone $X \xleftarrow{a} C \xrightarrow{b} X$ over the original corner, $f \circ a = f \circ b$, whence $a = b$. But that means the cone factors through the cone $X \xleftarrow{1_X} X \xrightarrow{1_X} X$ via the unique a, making that cone a limit and the square a pullback square. □

(c) Now for a *very* useful result, often called simply *the pullback lemma*:

Theorem 93. *Suppose we have two joined commuting squares like this:*

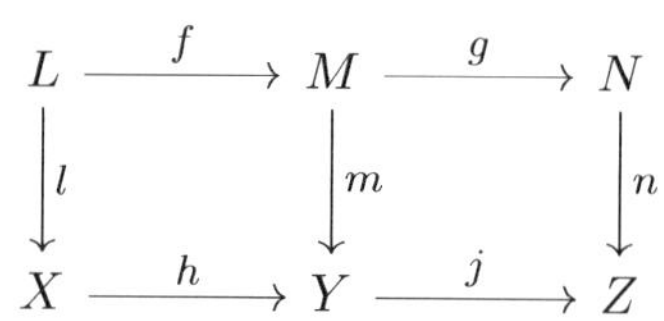

Then (1) if the two inner squares are pullback squares, the outer rectangle is a pullback square(!) too. And (2) if the right-hand square and the outer rectangle are both pullback squares, so is the left-hand square.

Proof of (1). If the outer rectangle is to be a pullback square, we need to show that a wedge $X \xleftarrow{c} C \xrightarrow{d} N$ over the corner $X \xrightarrow{j \circ h} Z \xleftarrow{n} N$ factors uniquely through L. Suppose then that we assume that this next diagram commutes:

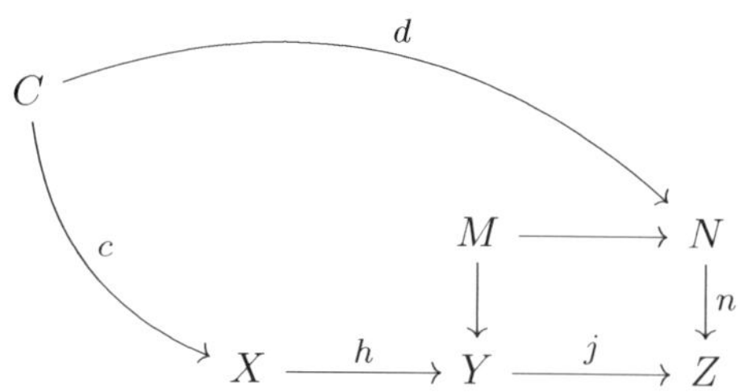

Since the right-hand square is a pullback, there must be a unique $v \colon C \to M$ from the vertex of the wedge $Y \xleftarrow{h \circ c} C \xrightarrow{d} N$ making this commute:

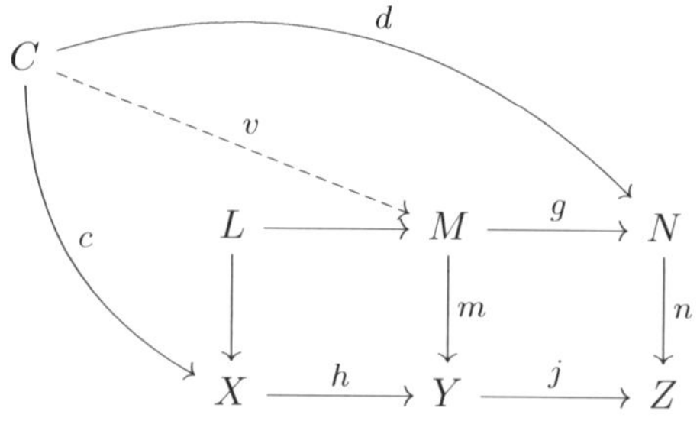

Since the left-hand square is also a pullback, the wedge $X \xleftarrow{c} C \xrightarrow{v} M$ must factor through L via a unique arrow $u \colon C \to L$, making all this commute:

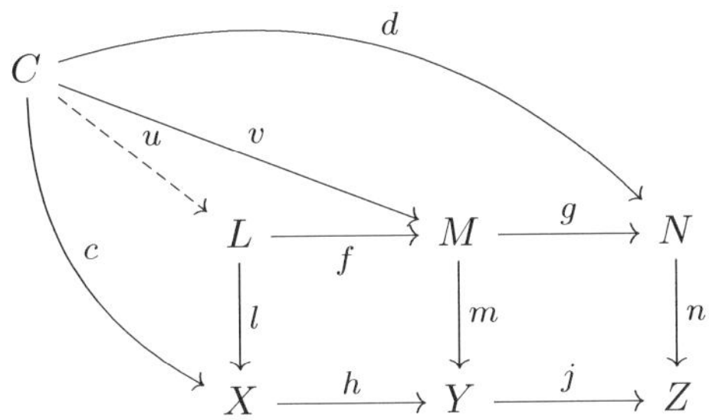

So the wedge $X \xleftarrow{c} C \xrightarrow{d} N$ factors through L via a unique map $u\colon C \to L$. Hence the whole rectangle *does* form a pullback. ☐

I can perhaps leave you to similarly prove part (2) of the pullback lemma.

(d) We saw that there is a sense in which pullbacks are like modified products. Here is another less obvious way of looking at how pullbacks and products are related.

Theorem 94. *A pullback square for a corner $D_1 \longrightarrow V \longleftarrow D_2$ in the category* C *is a product of $D_1 \to V$ and $D_2 \to V$ as objects of the slice category* C/V.

Proof. Let's consider these next two diagrams, while keeping our wits about us!

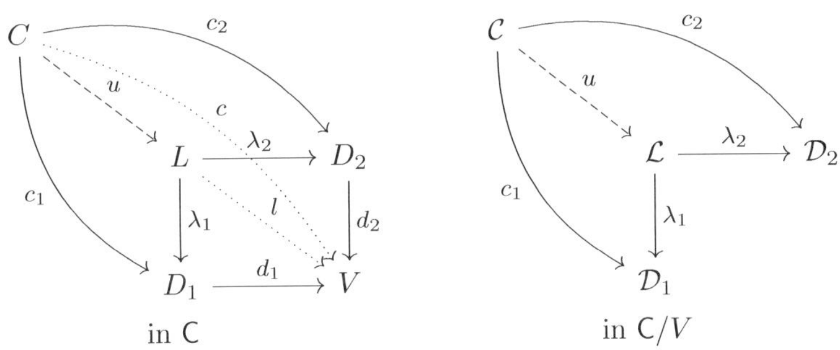

So, we have a pullback in C over the corner $D_1 \longrightarrow V \longleftarrow D_2$. Being a pullback square, for any wedge with vertex C that forms a commuting square with that corner, there will be a unique u making the whole diagram commute. And this time I have indicated the two diagonals $l\colon L \to V$ and $c\colon C \to V$.

Next, recall from §7.3 that both objects and arrows in C/V are arrows in C. Using a different font to label C/V-objects for clarity, let $\mathcal{D}_j$ be the C-arrows $d_j\colon D_j \to V$, and let $\mathcal{L}$ be $l\colon L \to V$ and $\mathcal{C}$ be $c\colon C \to V$.

Now, since $c = d_j \circ c_j$ in C, we have (by definition) C/V-arrows $c_j\colon \mathcal{C} \to \mathcal{D}_j$. And since $l = d_j \circ \lambda_j$, we have C/V-arrows $\lambda_j\colon L \to D_j$. There they are, diagrammed on the right.

Finally note that, as we vary in C over any wedge $D_1 \xleftarrow{c_1} C \xrightarrow{c_2} D_2$ that makes a commuting square with the opposite corner, we vary in C/V over every wedge $\mathcal{D}_1 \xleftarrow{c_1} \mathcal{C} \xrightarrow{c_2} \mathcal{D}_2$. But in C, however we vary the wedge, there is a unique arrow u in C such that $\lambda_j \circ u = c_j$ (by the definition of the pullback); which means that for every corresponding wedge in C/V, there is again the same unique arrow such that $\lambda_j \circ u = c_j$.

Therefore, lo and behold, the wedge $\mathcal{D}_1 \xleftarrow{\lambda_1} \mathcal{L} \xrightarrow{\lambda_2} \mathcal{D}_2$ is a product of $\mathcal{D}_1$ and $\mathcal{D}_2$. But the wedge is just the whole pullback square in C (with the diagonal drawn in), and the $\mathcal{D}_j$ are the arrows $D_j \to V$.

So we are done. □

(e) We know from Theorem 82 that if we have two limits over the same diagram, then there is a unique isomorphism between their vertices. Applied to pullbacks, if we have two pullbacks over the same corner diagram, there will be an isomorphism between their vertices making the pullbacks factor through each other – that was Theorem 86.

Likewise, we know from Theorem 83 that if a cone factors through a limit cone over a diagram via an isomorphism, then it is also a limit over that diagram. Let's now apply this to pullbacks, to get a useful lemma:

Theorem 95. *Suppose that, in the given diagram, the inner square with vertex L is a pullback square, and that $i\colon L \to L'$ is an isomorphism making the whole diagram commute:*

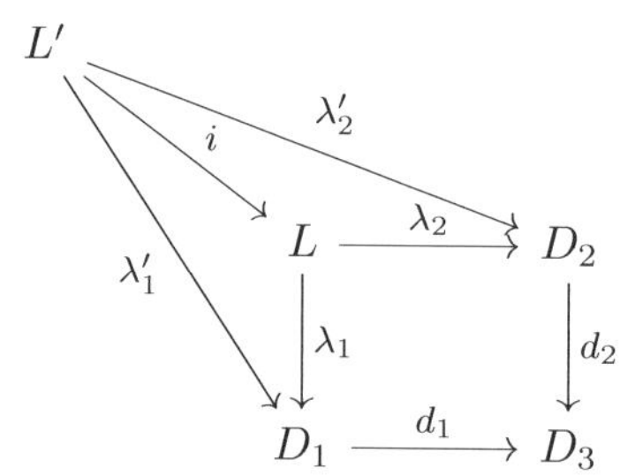

Then the outer square with vertex L' is also a pullback.

Proof. This follows from the general theorem of course. But we can easily prove it directly by chasing arrows round some diagrams.

We want to prove that, for any wedge $D_1 \xleftarrow{c_1} C \xrightarrow{c_2} D_2$ such that $d_1 \circ c_1 = d_2 \circ c_2$, there is a unique $u\colon C \to L'$ making the left-hand diagram commute.

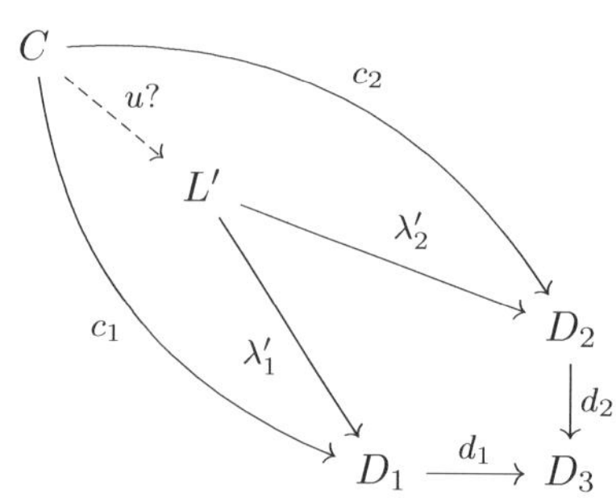

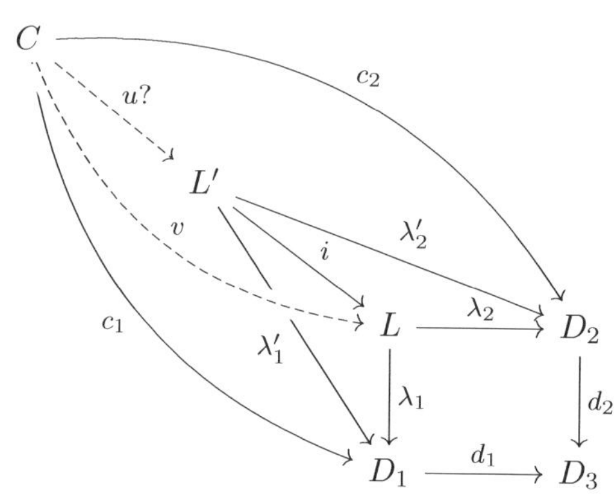

Since the original square is a pullback, the wedge with vertex C (the C-wedge for short) must factor through the L-wedge via a unique $v\colon C \to L$ as shown in the right-hand diagram. But we can then read off the commuting diagram that the C-wedge factors through the L'-wedge via the arrow $u = i^{-1} \circ v$.

And that's the unique possibility. For suppose the C-wedge also factors through the L'-wedge by some u'. Then again we can read off the commuting diagram that the C-wedge factors through the L-wedge via the arrow $i \circ u'$.

However, we know that v is the unique arrow that can play this role, so $i \circ u' = v$, whence $u' = i^{-1} \circ v = u$. □

20.5 Pushouts

(a) A pullback is the *limit* of a *corner*. What is a colimit for a corner? Check the relevant diagram and it is obviously the corner itself. That's uninteresting. But the real dualization of the notion of a pullback is – of course – provided when we take the colimit of a *'co-corner'*, i.e. of a wedge. Let's explore.

Suppose then we take a wedge D, i.e. a diagram $D_1 \xleftarrow{d_1} D_3 \xrightarrow{d_2} D_2$. A cocone under this diagram makes a commutative square (omitting again the diagonal arrow which is fixed by the others).

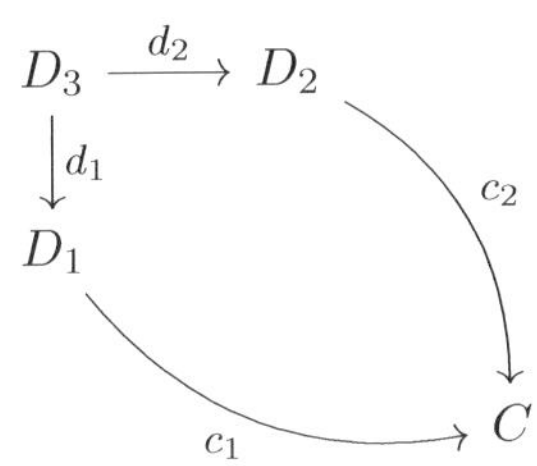

And a limit cocone of this type will be a cocone with apex L and projections $\lambda_j\colon D_j \to L$ such that for any cocone (C, c_j) under D, there is a unique $u\colon L \to C$ such that the dual of the whole pullback diagram in §20.1 commutes, as here:[1]

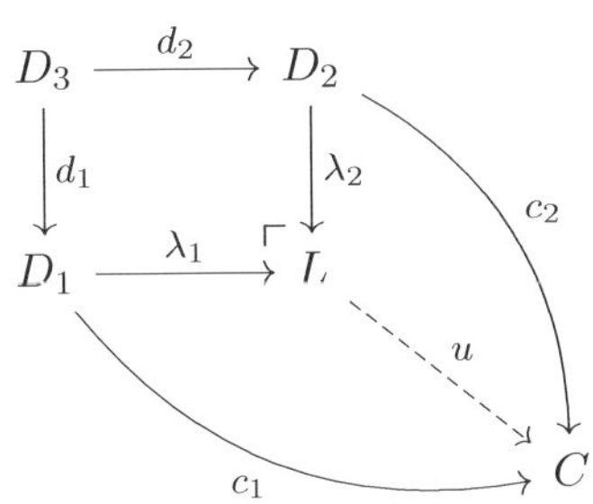

So we say:

Definition 86. A colimit for a wedge diagram is a *pushout*. △

[1] It's not conventional, but I have added a corner symbol to mark the vertex of the dual limit, by analogy to the usual use of a corner symbol to mark the vertex of a pullback.

As you would expect, we then have dual theorems about pushouts – e.g. they are unique up to unique isomorphism, pushing out an epimorphism yields an epimorphism, etc.

(b) What about examples of pushouts? Let's start with a nice example in Set. We noted before that this square is a pullback: but is also a pushout:

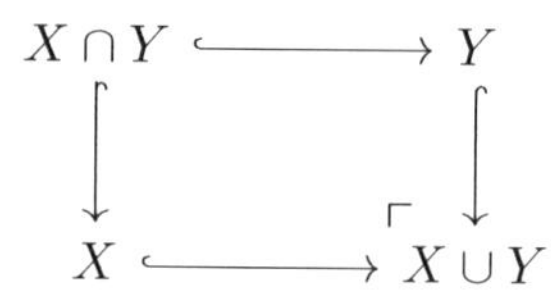

Why so? We need to show that for any cocone (C, c_j) under the wedge $X \leftarrow X \cap Y \rightarrow Y$ there is a unique arrow $u\colon X \cup Y \rightarrow C$ making this next diagram commute (I've now labelled the arrows for convenience):

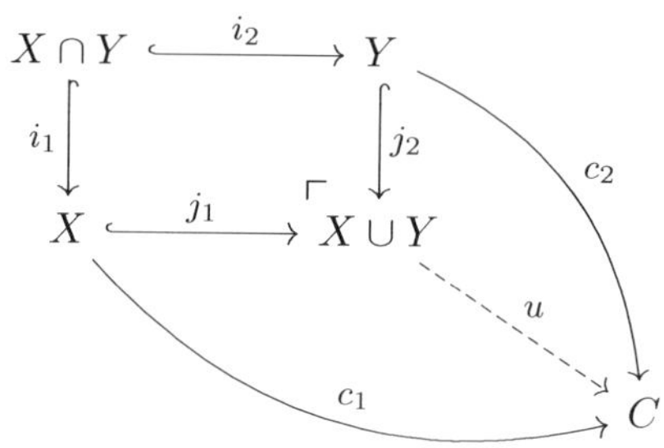

Let u agree with c_1 on members of X and c_2 on members of Y. That gives us a well-defined function because, by the assumption that (C, c_j) is a cocone for the wedge, $c_1 \circ i_1 = c_2 \circ i_2$, and by the assumption that i_1 and i_2 are inclusions (so don't affect values), that means c_1 and c_2 do agree on $X \cap Y$. Thus defined, u makes the added triangles commute because j_1 and j_2 are inclusions too, and this is evidently the unique possibility.

(c) What about pushouts more generally in Set?

In this category, as we've seen, the object of the pullback limit cone over a corner diagram $X \xrightarrow{f} Z \xleftarrow{g} Y$ is not in general the whole product $X \times Y$ but a subset of that, namely the subset of the product consisting of pairs $\langle x, y \rangle$ where $f(x) = g(y)$. Dually, we'll expect the object of the colimit pushout cocone under a wedge diagram $X \xleftarrow{f} Z \xrightarrow{g} Y$ to be not the whole coproduct (disjoint union) $X + Y$ but ...?

Well, how *do* we need to tinker with the coproduct (thought of as constructed in the usual way by tagging members of X with 0 and members of Y with 1, and combining the results)? It turns out that we need to quotient by the smallest equivalence relation generated by the relation that holds between each $(f(z), 0)$ and $(g(z), 1)$.

For our purposes, however, I probably don't need to pause to spell out the details why (though it is a nice challenge to think things through). Nor will I pause to explain why in Grp, pushouts produce so-called free products with amalgamation. I merely note that pushouts often give rise to less familiar constructions than pullbacks.

(d) There is a pushout for any corner in Set. So consider the square formed by two copies of an arrow $f\colon X \to Y$ and their pushout, as on the left:

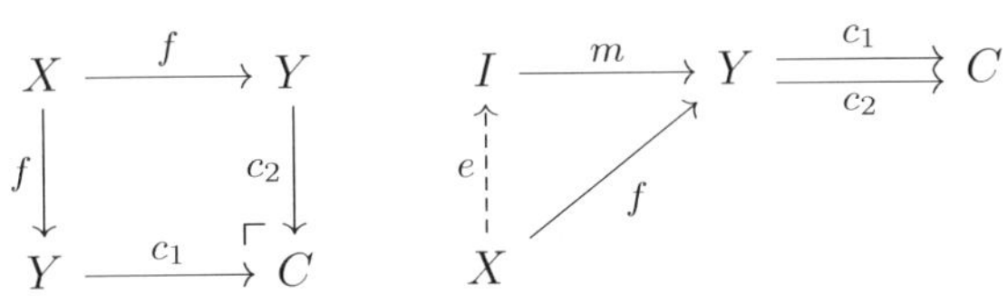

So f followed by c_1 and c_2 forms a commuting fork, which in Set must factor through the equalizer (I, m) of c_1 and c_2, as shown in the diagram on the right.

A claim to think about. Still in Set, the natural candidate for I is $f[X]$, the image of X under f. Then m – which must be monic, since it is an equalizer – can be the inclusion function. And e will then be the surjective and hence epic function that agrees with f on all elements of X. So $m \circ e$ is an epi-mono factorization of f.

In Part III, we'll be able to show that this construction (via a pullback and an equalizer) always works in nice enough categories to give us an epi-mono factorization of any arrow.

21 The existence of limits

We have now seen that a whole range of very familiar constructions from various areas of ordinary mathematics can be regarded as instances of taking limits or colimits of (tiny) diagrams in appropriate categories. Examples so far include: forming Cartesian products or logical conjunctions, taking disjoint unions or free products, quotienting out by an equivalence relation, taking intersections, and taking inverse images.

Not every familiar kind of construction in a category C involves taking a limit cone or cocone in C. We have already met one important exception, namely exponentials: we meet another exception in Chapter 22, so-called subobjects. But plainly we are mining a very rich seam here – and we are already making good on the promise to show how category theory helps reveal recurring patterns across different areas of mathematics. So what more can we say about limits?

It would get tedious to explore what it takes for a category to have limits for other kinds of diagrams, case by case. But fortunately we don't need to do such a piecemeal examination. In this chapter we prove a major theorem: that if a category has certain basic limits of kinds that we have already met, then it has *all* finite limits. Similarly, needless to say, for the dual case.

21.1 The key theorems stated

(a) Start with a package of definitions:

Definition 87. The category C *has all finite limits* iff for any finite diagram D in C – i.e. for any diagram with a finite number of objects and arrows – C has a limit over D. A category with all finite limits is said to be *finitely complete.*

A category with all finite limits and exponentials too is *properly Cartesian closed.*[1] △

So now the obvious question is: what does it take for a category to be finitely complete?

We'll work up to an answer by initially establishing (in §21.2)

Theorem 96. *If a category has all binary products and has equalizers for every pair of parallel arrows, then it has a pullback for every corner.*

[1]See footnote to Defn. 75.

Then (in §21.3) we'll adapt the proof-strategy for that initial result to establish our first main result:

Theorem 97. *If a category has a terminal object, and has all binary products and equalizers, it is finitely complete.*

And we could leave it at that. But there is some interest in also proving a variant completeness result. Recall, we have already shown the easy

Theorem 88. *A category that has a terminal object and a pullback for every corner also has all binary products.*

But we can also show (in §21.4)

Theorem 98. *If a category has a terminal object and has a pullback for every corner, then it will have an equalizer for every pair of parallel arrows.*

And from those two theorems and our first main result we can immediately derive this variant completeness theorem:

Theorem 99. *If a category has a terminal object and has a pullback for every corner, it is finitely complete.*

Given ingredients from our previous discussions, since the categories in question have terminal objects, binary products and equalizers, we can immediately conclude

Theorem 100. Set *and* FinSet *are finitely complete, as are categories of algebraically structured sets such as* Mon, Grp, Ab, Ring. *Similarly* Top *is finitely complete.*[2]

Mutilating such categories by zapping some of their objects can boringly result in incomplete categories. Rather less artificially, a poset-as-a-category (for example) may lack many products and hence not be finitely complete.

(b) Theorem 97 and its companion Theorem 99 are perhaps the first unobvious Big Results we have met. Their proofs are not exactly difficult, but get a bit intricate. You need to know the Results; however, nothing later depends on your knowing the proof-details explained over the next three sections. So by all means – at least on a first reading – skip forward to §21.5 where I briefly outline the move from the finite to the infinite case, and to §21.6 where everything gets snappily dualized.

[2] It is worth noting, to echo points made in §11.2 and §16.2, that these finite completeness results fall out because we are taking our categories of 'all' monoids, groups, rings, topological spaces to be implemented in a universe of sets which provides all the constructions of the products and equalizers that we need. If we instead had taken these inclusive categories to be, so to speak, free standing, then in each case we'd need to invoke special axioms to justify finite completeness claims.

21.2 Products plus equalizers imply pullbacks

So, as announced, we are going to begin by proving

Theorem 96. *If a category has all binary products and has equalizers for every pair of parallel arrows, then it has a pullback for every corner.*

Proof. Start with an arbitrary corner $X \xrightarrow{f} Z \xleftarrow{g} Y$.

There is nothing to equalize yet. Our only option, then, for constructing a pullback is to begin by constructing some product. So: let's take a product $X \times Y$ with the usual projections $\pi_1 \colon X \times Y \to X$ and $\pi_2 \colon X \times Y \to Y$. This immediately gives us parallel arrows $X \times Y \underset{g \circ \pi_2}{\overset{f \circ \pi_1}{\rightrightarrows}} Z$.

We are assuming that we can always equalize parallel arrows, so there is some (E, e) for which $f \circ \pi_1 \circ e = g \circ \pi_2 \circ e$. We can picture the situation like this:

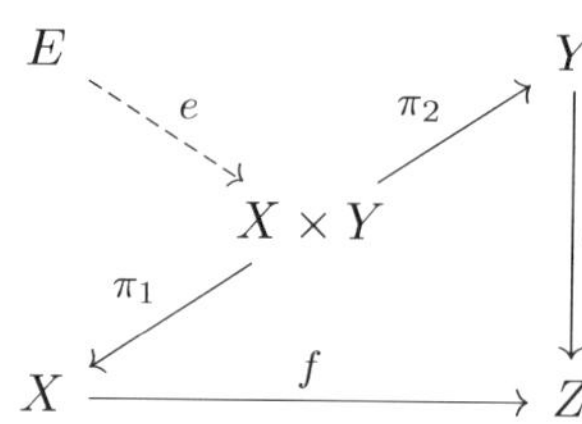

But be careful! This is *not* a fully commuting diagram. We are *not* assuming that the two composite arrows from $X \times Y$ to Z are equal – after all, we are in the business of equalizing those arrows!

Now add the arrows which give us two new commuting triangles like this:

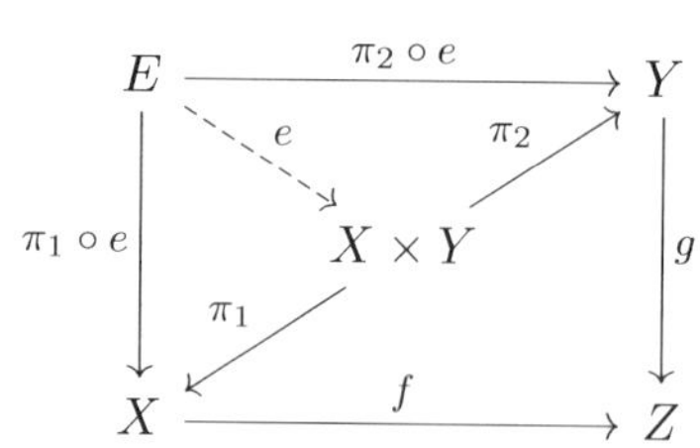

And note that the wedge formed by E with the arrows $\pi_1 \circ e$, $\pi_2 \circ e$ looks as if it should be a sort of limiting case among wedges completing a commuting square with the original corner (after all, E is part of a limit). So hopefully that wedge is a pullback for the corner.

Fairly routine checking confirms that conjecture. For consider any other cone over the original corner. In other words, leaving the diagonals to take care of themselves, consider any wedge $X \xleftarrow{c_1} C \xrightarrow{c_2} Y$ with $f \circ c_1 = g \circ c_2$: we need to show that this factors uniquely through E.

So stare at the following diagram!

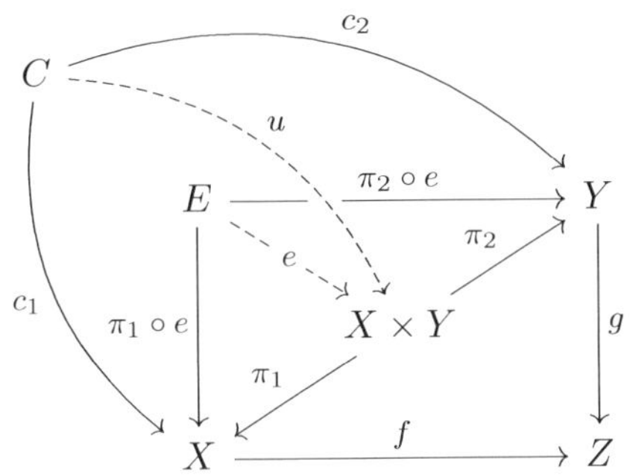

Our new wedge will also factor through the product $X \times Y$, which is why we have added to the diagram a unique $u\colon C \to X \times Y$ such that $c_1 = \pi_1 \circ u$, $c_2 = \pi_2 \circ u$.

Hence $f \circ \pi_1 \circ u = g \circ \pi_2 \circ u$. Therefore $C \xrightarrow{u} X \times Y \overset{f \circ \pi_1}{\underset{g \circ \pi_2}{\rightrightarrows}} Z$ is itself a fork, which must factor uniquely through the equalizer E via some v, like this:

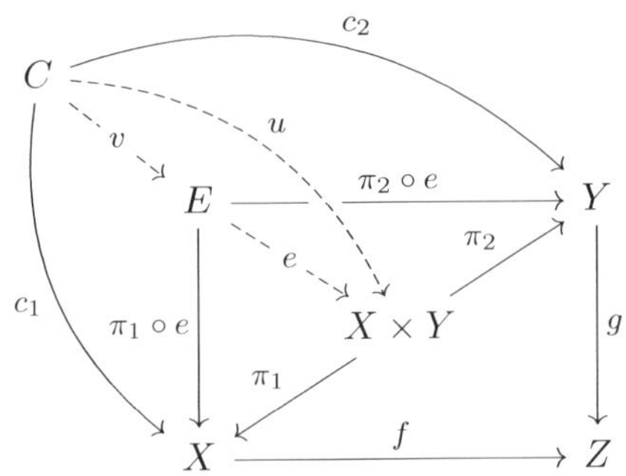

That is to say, there is a $v\colon C \to E$ such that $e \circ v = u$. Hence $\pi_1 \circ e \circ v = \pi_1 \circ u = c_1$. Similarly $\pi_2 \circ e \circ v = c_2$. Therefore our wedge with vertex C factors through the wedge with vertex E, as we need.

To finish, we have to establish that v is the only way that the wedge with vertex C can factor through E. Suppose then that $v'\colon C \to E$ also makes $\pi_1 \circ e \circ v' = c_1$, $\pi_2 \circ e \circ v' = c_2$. Then the wedge $X \xleftarrow{c_1} C \xrightarrow{c_2} Y$ factors through $X \times Y$ via $e \circ v'$; but the wedge factors uniquely through the product $X \times Y$ by u. Therefore $e \circ v' = u = e \circ v$. But equalizers are monic by Theorem 67, so $v' = v$, and we are done. □

That's neat! So just one comment about this line of proof. We did the obvious thing in starting the proof by considering a binary product $X \times Y$. But a moment's reflection shows that the argument would have gone equally well if we had instead taken a product of every object in sight, i.e. $X \times Y \times Z$. We know that if we have all binary products we get ternary products too. And then we can (so to speak) carry Z along for the ride.

21.3 Deriving the finite completeness theorem

(a) Our goal now is to prove the announced main result:

Theorem 97. *If a category has a terminal object, and has all binary products and equalizers, it is finitely complete.*

And we are going to generalize the basic strategy pursued in proving the cut-down version in the previous section, where we showed that having binary products and equalizers implies at least having pullbacks. So, here's the outline four-stage plan:

(i) Given a finite diagram D, we again start by forming a multi-product (P, p_i) of objects from D – and we'll take the easy option of forming a product of every object in sight (which we can do since C has all finite products if it has terminal objects and binary products – see §12.2).

(ii) We then aim to find some appropriate parallel arrows out of this product P, arrows that we are going to equalize. But what will be the target of these parallel arrows? In the previous proof we were looking at a mini-diagram whose arrows had the same target Z. In the general case we'll be looking at a diagram D whose arrows have multiple targets. We'll need to package up those targets together by forming another multi-product, (Q, q_j).

(iii) On the model of the proof of Theorem 96, we next define our two parallel arrows P to Q.

(iv) Finally, we construct an equalizer (E, e) for these arrows (which we can do since C has all equalizers by assumption). And then we show that we can again use E as the vertex of the desired limit cone over the diagram D pretty much as before.

The devil, of course, is in the details! And to be frank, you won't lose anything if you skip right past them.

(b) Still with me? We want to show that given a finite diagram D, with objects D_i and arrows $d_j : D_k \to D_l$, we can construct a limit over D.

Stage (i): We form a multi-product of all the diagram objects D_i. So this product is (P, p_i), where $P \cong D_1 \times D_2 \times \ldots \times D_n$, and the p_i are the various projection arrows from P to the objects D_i. Simple!

Stage (ii): We now look at all the various arrows $d_j : D_k \to D_l$ in D, and form the multi-product of their *targets*, where a particular object is to feature in the multi-product as many times as it is the target of some arrow d_j. So this product is (Q, q_j), where Q might be (for example) $D_1 \times D_1 \times D_3 \times D_4 \times D_4 \times D_4 \times \ldots \times D_n$, and each q_j is the projection arrow aimed at an instance of the target of the corresponding arrow d_j.

Stage (iii): The name of the game is now to define a pair of parallel arrows $v, w : P \to Q$ which we are then going to equalize.

There are two arrows from P to Q which arise quite naturally:

(1) First, take the vertex P together with the projection arrows $p_j \colon P \to D_j$, one for each D_j occurring in the product Q. These arrows will form a 'multi-wedge' over the same objects as the multi-product (Q, q_j) is over. Hence this multi-wedge must factor through (Q, q_j) by a unique mediating arrow $v \colon P \to Q$, so that $p_j = q_j \circ v$ for each arrow $p_j \colon P \to D_j$.

(2) Second, take an arrow $d_j \colon D_k \to D_l$ and compose it with the projection arrow $p_k \colon P \to D_k$, and we get an arrow $d_j \circ p_k \colon P \to D_l$. This arrow has the same target as the corresponding q_j. So consider the 'multi-wedge' with vertex P and all those arrows $d_j \circ p_k$. This 'multi-wedge' too must again factor through the multi-product (Q, q_j) by a unique mediating arrow $w \colon P \to Q$, so that $d_j \circ p_k = q_j \circ w$ for each arrow $d_j \colon D_k \to D_l$.

Since we are assuming that all parallel arrows have equalizers in C, we can take the equalizer of v and w, namely (E, e).

And now the key claim, modelled on the key claim in our proof of Theorem 96: $(E, p_i \circ e)$ will be a limit cone over the given diagram D.

(c) Let's state this as a more specific Theorem 97* that immediately implies the less specific Theorem 97:

Theorem 97*. *Let D be a finite diagram in a category C that has a terminal object, binary products and equalizers. Let (P, p_i) be a product of the objects D_i in D, and (Q, q_j) be a product of the objects D'_j (one occurrence for each arrow of the kind $d_j \colon D_k \to D_l$ with $D'_j = D_l$). So there are arrows*

$$P \underset{w}{\overset{v}{\rightrightarrows}} Q$$

such that the following diagrams commute for each $d_j \colon D_k \to D_l$:

$$\begin{array}{ccc} P & \xrightarrow{\ v\ } & Q \\ & \searrow^{p_l} & \downarrow{q_j} \\ & & D_l \end{array} \qquad\qquad \begin{array}{ccc} P & \xrightarrow{\ w\ } & Q \\ {p_k}\downarrow & & \downarrow{q_j} \\ D_k & \xrightarrow{\ d_j\ } & D_l \end{array}$$

Let (E, e) be an equalizer of v and w. Then $(E, p_i \circ e)$ will be a limit cone over D in C.

Proof. We've already shown that (P, p_i) and (Q, q_j) exist, and that v and w exist such that the given diagrams always commute. By assumption, an equalizer (E, e) for v and w must exist.

Next we confirm $(E, p_i \circ e)$ is a cone over the diagram D. Suppose then that there is an arrow $d_j \colon D_k \to D_l$. For a cone, we require $d_j \circ p_k \circ e = p_l \circ e$.

But we have $d_j \circ p_k \circ e = q_j \circ w \circ e = q_j \circ v \circ e = p_l \circ e$, where the inner equation holds because e is an equalizer of v and w and the outer equations are given by the commuting diagrams above.

It remains to show that $(E, p_i \circ e)$ is a *limit* cone. So suppose (C, c_i) is any other cone over D. Then there must be a unique $u\colon C \to P$ such that every c_i factors through the product cone (P, p_i) over D and we have $c_i = p_i \circ u$.

Since (C, c_i) is a cone over D, for any $d_j : D_k \to D_l$ in D we have by assumption that $d_j \circ c_k = c_l$. Hence $d_j \circ p_k \circ u = p_l \circ u$, and hence for each q_j from the product (Q, q_j), $q_j \circ w \circ u = q_j \circ v \circ u$. But then we can apply a generalized version of Theorem 45 about products, and conclude that $w \circ u = v \circ u$. Which means that

$$C \xrightarrow{\ u\ } P \overset{v}{\underset{w}{\rightrightarrows}} Q$$

is a fork, which must therefore uniquely factor through the equalizer (E, e). That is to say, there is a unique $s\colon C \to E$ such that $u = e \circ s$, and hence for all i, $c_i = p_i \circ u = p_i \circ e \circ s$.

But that is to say, (C, c_i) factors uniquely through $(E, p_i \circ e)$ via s. Therefore $(E, p_i \circ e)$ is indeed a limit cone. And we are done! □

21.4 Deriving the variant completeness theorem

Our first completeness result in the bag! Our next target is to prove a companion to Theorem 96 which told us that products and equalizers give us pullbacks:

Theorem 98. *If a category has a terminal object and has a pullback for every corner, then it will have an equalizer for every pair of parallel arrows.*

Then, as we noted at the beginning of the chapter, this – together with Theorems 88 and 97 – immediately implies our variant completeness theorem

Theorem 99. *If a category has a terminal object, and has a pullback for every corner, it is finitely complete.*

Proof of Thm. 98. Take the parallel arrows we want to equalize, say $f, g\colon X \to Y$, but now think of these as forming a wedge $Y \xleftarrow{\ f\ } X \xrightarrow{\ g\ } Y$.

Like any wedge, this wedge will factor uniquely through the appropriate product of the outside objects of the wedge – in this case $Y \times Y$. And by Theorem 87, this particular product is available in our category with all pullbacks and a terminal object. Denote its unique mediating arrow $\langle\!\langle f, g\rangle\!\rangle\colon X \to Y \times Y$.

So now consider the corner $X \xrightarrow{\langle\!\langle f,g\rangle\!\rangle} Y \times Y \xleftarrow{\ \delta_Y\ } Y$, where δ_Y is the 'diagonal' arrow $\langle\!\langle 1_Y, 1_Y\rangle\!\rangle$ (see Defn. 52). This is nice to think about since (to hand-wave a bit!) the first arrow packages up the parallel arrows we want to equalize. While the second sort of arrow is always available in a category with products, and in effect does some sort-of-equalizing – e.g. in **Set** it sends something from Y to an ordered pair of two equal objects.

Now take this corner's pullback (the only thing to do with it!):

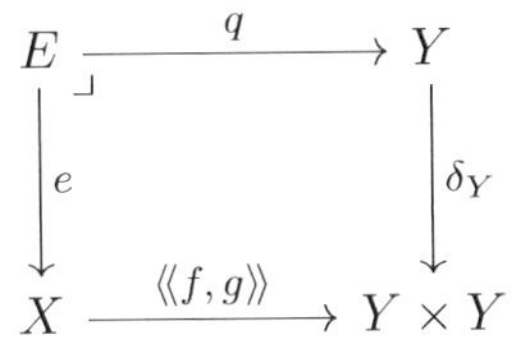

Intuitively, $E \xrightarrow{e} X \xrightarrow{\langle\langle f,g\rangle\rangle} Y \times Y$ sends something from E to what is, according to the other route round the commuting square, a pair of equals. So, morally, (E, e) ought to be an equalizer for $X \overset{f}{\underset{g}{\rightrightarrows}} Y$.

And, from this point on, it is a routine proof to check that (E, e) is the equalizer we want. Here goes …

By the commutativity of the pullback square, $\delta_Y \circ q = \langle\langle f, g\rangle\rangle \circ e$. Appealing to the definition of δ_Y and Theorem 50, it follows that $\langle\langle q, q\rangle\rangle = \langle\langle f \circ e, g \circ e\rangle\rangle$. Hence, by Theorem 40, $f \circ e = q = g \circ e$. Therefore $E \xrightarrow{e} X \overset{f}{\underset{g}{\rightrightarrows}} Y$ is a fork. It remains to show that it is a limit fork.

Take any other fork $C \xrightarrow{c} X \overset{f}{\underset{g}{\rightrightarrows}} Y$. So we have $f \circ c = q' = g \circ c$, and hence $\langle\langle f, g\rangle\rangle \circ c = \langle\langle q', q'\rangle\rangle = \delta_Y \circ q'$.

Hence the outer square in this next diagram commutes. So the wedge with vertex C factors through E (because E is the vertex of the pullback) via a unique mediating arrow v:

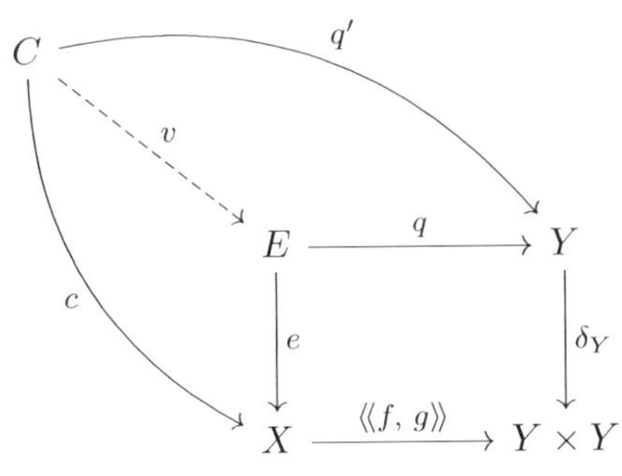

It follows that v makes this next diagram commute:

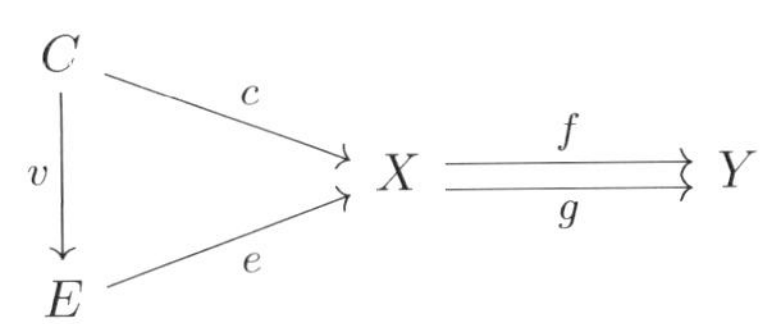

And any $v' \colon C \to E$ that makes the latter diagram commute will also be a mediating arrow making the previous diagram commute, so $v' = v$ by uniqueness of mediators in pullback diagrams. Hence (E, e) is an equalizer. □

21.5 Infinite limits

We can extend our key Theorem 97 to reach beyond the finite case. Let's say that a diagram D is 'small' if it has only a set's worth of objects and set's worth of arrows. In other words, the objects are D_i for $i \in I$ where I is some set of indices; similarly the arrows are d_j for $j \in J$ where J is again a set. Then

Definition 88. The category C *has all small limits* if C has a limit over any small diagram D. A category with all small limits is also said to be *complete.* △

Again, as in talking of so-called 'small' products in §12.3, this usage is a conventional: 'small' limits can be huge constructions – the only requirement is that they are taken over diagrams that are no-bigger-than-set-sized.

An easy inspection of the proof of Theorem 97 shows that the argument will continue to go through when applied to infinite diagrams, so long as we assume that we can sensibly handle as many objects and arrows as needed. Suppose then we are still dealing with a category that has products for all collections of objects O indexed by set-sized suites of indices, etc. Then, without further ado, we can state:

Theorem 101. *If* C *has all small products and has equalizers, then it has all small limits, i.e. is complete.*

We can similarly extend Theorem 100 to show that

Theorem 102. Set *is complete – as are the categories of structured sets* Mon, Grp, Ab, Ring. Top *too is complete.*

We have already met a category that, by contrast, is finitely complete but is evidently not complete, namely FinSet.

21.6 Dualizing again

Our definitions and results in this chapter dualize (of course!). Thus:

Definition 89. The category C *has all finite colimits (is finitely cocomplete)* iff for any finite diagram D in C, the category has a colimit over D.

C *has all small colimits* if for any diagram D whose objects and arrows are not too many to be indexed by a set, then C has a colimit over D. A category with all small colimits is also said to be *cocomplete.* △

Theorem 103. *If a category has initial objects, binary coproducts, and co-equalizers, then it has all finite colimits, i.e. is finitely cocomplete.*

If a category has all small coproducts and has co-equalizers, then it is cocomplete.

Theorem 104. Set *is cocomplete – as are the categories of structured sets* Mon, Grp, Ab, Ring. Top *too is cocomplete.*

Note that a category can be (finitely) complete without being (finitely) cocomplete and vice versa. For a generic source of examples, take again a poset $(P, \preccurlyeq)$ considered as a category. This automatically has all equalizers and coequalizers (see Theorem 66). But it will have other limits (colimits) depending on which products (coproducts) exist, i.e. which sets of elements have suprema (infima). For a simple case, take a poset with a maximum element and such that every pair of elements has a supremum: then considered as a category it has all finite limits (but maybe not infinite ones). But it need not have a minimal element and/or infima for all pairs of objects: hence it can lack some finite colimits despite having all finite limits.

22 Subobjects

We have discussed categorial treatments of various constructions of ordinary mathematics, such as products, quotients, exponentials, inverse images, and the like. One perhaps surprising omission: we haven't yet said anything about surely the most basic of such constructions, namely the simple business of forming subobjects (as in subsets, subgroups, subspaces). It is time to fill the gap.

22.1 Subsets revisited

Look again at Theorem 68 and its proof in §16.4. We were working in the category Set. Then

(i) We chose some two-membered set Ω to be, so to speak, a *truth-value object*, treating its two members – call them 'T' and 'F' – as coding for *true* and *false*.

(ii) A subset $S \subseteq X$ has an associated *characteristic function* $\chi_S \colon X \to \Omega$ that sends $x \in X$ to T if $x \in S$ and sends x to F otherwise.

(iii) Fix on a singleton set 1. Let $!_X \colon X \to 1$ be the function that sends every member of X to the sole member of 1, and let $\top \colon 1 \to \Omega$ be the function that sends the sole member of 1 to T. Then $\top_X = \top \circ !_X$ is the composite map that sends *every* element of X to T.

(iv) The proof of Theorem 68 then showed that the subset S together with the inclusion function $i \colon S \hookrightarrow X$ forms a kind of limit, because (S, i) is an equalizer for the pair of parallel arrows $\chi_S, \top_X \colon X \to \Omega$.

We might now reasonably wonder whether there is a cross-category generalization of this story. Can we always treat subobjects as in effect limits like this?

Immediately, however, we face two questions:

(Q1) First, regarding subobjects as limits means we can only pin them down up to isomorphism: is this acceptable?

(Q2) Second, when generalizing from Set, can we find a 'truth-value object' like Ω in other categories, in a way that enables us then to go on to define subobjects in terms of limits involving the local Ω?

Answers:

(A1) Yes, we can live with identifying subobjects only up to isomorphism.

(A2) No, things have to go the other way about. In the general case, we need to get a prior notion of subobject into play first. Only *then* can truth-value objects Ω, together with characteristic functions as arrows to Ω, get defined by their interplay with subobjects as already understood.

This chapter elaborates on (A1); the next chapter explains (A2).

22.2 Subobjects and monic arrows

(a) By now you are entirely used to the idea that, at least using the resources of category theory, we don't get a direct handle on the constituents or 'innards' of an object in a category. We only get to see how an object is related 'externally' to others by the arrows of the category. Categorially we come to know an object not by digging inside (so to speak) but by finding the company it keeps: as the old adage has it, "by their friends ye shall know them" (cf. §37.6).

Now ordinarily, when we ask whether S is a part of X, we are asking about the innards of X: do they already comprise all of S? Categorially, however, we can't ask straight out whether one object S in a category is a part of another object X in *that* sense. Rather, for S to count as a subobject of X, we will need instead for there to be the right sort of arrow $S \to X$ (what else?).

"Yes, yes, of course: we want an inclusion arrow!" But hold on! How can we specifically define an *inclusion* arrow in categorial terms?

The non-categorial definition, looking e.g. at objects in Set, is that an inclusion arrow $i\colon S \hookrightarrow X$ sends every constituent element of S to the very same element of X. However, the best that category theory has to offer by way of an account of 'elements of S' and 'elements of X' are, respectively, arrows $1 \to S$ and $1 \to X$, and *they* can't be the same. What to do?[1]

(b) Let's meditate a bit on the next two diagrams, still in Set, and still assuming that we have a grip on the 'internal' idea of an inclusion map i:

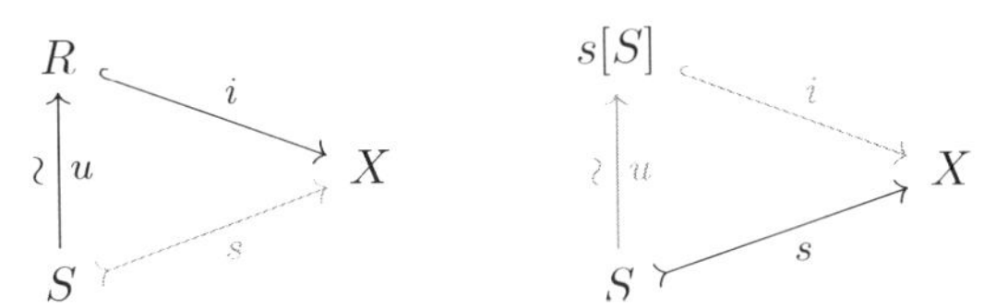

Suppose, in the left-hand diagram, that we are given a subset R included (in the ordinary sense) in X, with i the associated inclusion function. And suppose we are also given an isomorphism u between S and R. Categorially, i will be monic (being injective), and u will be monic (as any isomorphism is); hence their composition $s = i \circ u\colon S \rightarrowtail X$ will be monic by Theorem 16.

[1] At some earlier points we relied on our pre-categorial understanding of the idea of an inclusion function in order to illustrate some categorial idea; *now* we are noting a problem about defining such an inclusion function in purely categorical, arrow-theoretic, terms.

Now alternatively suppose, in the right-hand diagram, that we are given a monic arrow $s\colon S \rightarrowtail X$. Being injective in Set, this sets up a derived isomorphism u from S to its s-image $s[S]$ which then can be sent by the obvious inclusion map into X, so that again $s = i \circ u$.

Hence, *at least up to isomorphism*, the object S equipped with a monic arrow $s\colon S \rightarrowtail X$ will be tantamount to a subobject of X defined by inclusion.

(c) Now we throw away the ladder we've just climbed up.

Let's accommodate ourselves to the idea that (as with other categorial notions) we only want to characterize subobjects up to isomorphism. In this spirit, and dropping any reliance on the 'internal' notion of inclusion, we can propose the following definition:

Definition 90. (S, s) is a *subobject* of an object X in the category C when S is some object and $s\colon S \rightarrowtail X$ is a monomorphism. △

For example, in the category Grp, we take a subobject of a group G living in the category to be any group S which has a monic homomorphism s into G. We don't care whether S is 'really' a subgroup (as traditionally understood), we don't care whether the constituent elements of S are 'really' to be found inside G – because, in the spirit of group theory itself, we mostly only care about distinguishing groups up to isomorphism. So we will only be interested in whether S has, so to speak, the right shape to map injectively into G via a monic. Or so goes the story.

You can see, however, why I didn't introduce this categorial treatment of subobjects much earlier. For you do need to be softened up by quite a bit of exposure to some attractive up-to-isomorphism categorial treatments of other informal mathematical ideas to be primed to find the perhaps initially surprising Defn. 90 half-way sensible!

(d) OK: we have defined a subobject of X as an object S equipped with a monic arrow $s\colon S \rightarrowtail X$.

Of course, fixing the monic fixes its source. Therefore we could without any loss of information tersely specify a subobject (S, s) by simply giving the relevant monic s. In fact, for reasons of aesthetics and/or brevity, I'll later often fall into the snappier idiom, and talk as if subobjects simply are monics.

But for now, let's stick with our more elaborated way of specifying subobjects. Which is the natural one. Compare: we defined a product of X and Y as an object we can denote $X \times Y$ equipped with projection arrows $\pi_1\colon X \times Y \to X$ and $\pi_2\colon X \times Y \to Y$ (all satisfying certain conditions). Now, since the projection arrows fix their shared source, we could without any loss of information have specified a product by just giving those arrows. But that would surely have been unhappy: the key idea we want is that a product of two objects is indeed another object, but one coming equipped with the projections which (pre-categorially) we might ordinarily take for granted. Similarly here, when talking about subobjects: the key idea is that a subobject of a given object is another object, but one equipped with something as good as the injection which (pre-categorially) we might ordinarily take for granted.

22.3 Images

One familiar way of getting subobjects in Set is as images. Thus $f\colon X \to Y$ generates $I = f[X]$, the image of X under f, where $f[X] \subseteq Y$. And f has an epi-mono factorization through I, where the surjective and hence epic $e\colon X \twoheadrightarrow I$ is defined as agreeing everywhere with f, and where $m\colon I \rightarrowtail Y$ is the monic inclusion function, so is a categorial subobject of Y (see §8.6).

And now suppose that f also factors through another subobject of Y, i.e. through another monic arrow $m'\colon I' \rightarrowtail Y$, so the two large triangles here commute:

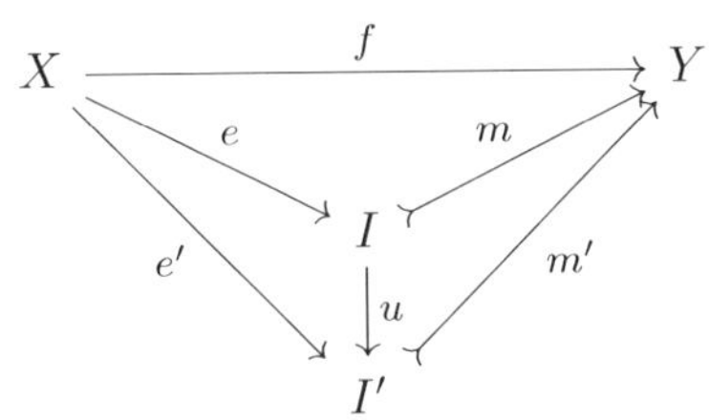

(note, though, that there is no assumption that e' is also epic).

Still working in Set, it is easy to see that there will be an arrow $u\colon I \to I'$ which makes the whole diagram commute. Just let u send $f(x) \in I$ to the unique object in I' that the injective m' sends to $f(x) \in Y$, and both inner triangles will then commute. However, there won't in general be an arrow in the other direction, from I' to I which makes the triangles commute (why not?).

Which suggests that this is a sensible categorial definition:

Definition 91. An *image* of an arrow $f\colon X \to Y$ is a subobject of Y, namely $(I, m\colon I \rightarrowtail Y)$, such that (i) f factors through the monic arrow m, i.e. for some $e\colon X \to I$, $f = m \circ e$, and (ii) for any monic m', if f also factors through m' so does m (through some mediating arrow u). △

An arrow f need not have an image. But if it does, e is unique given m: if $f = m \circ e_1 = m \circ e_2$, then $e_1 = e_2$ since m is monic.

Further, a mediating u is also unique: if $m = m' \circ u_1 = m' \circ u_2$ then $u_1 = u_2$ since m' is monic.

Finally, in Set, the source of the subobject $m\colon I \to Y$ will be at least isomorphic to f's image of X in the conventional sense, i.e. to $f[X]$. Much later, in Chapter 45, we'll prove that in nice enough categories like Set, images are unique up to isomorphism.

22.4 Ordering subobjects

(a) Evidently, we *can't* immediately re-apply Defn. 90 to give us subobjects of subobjects. There are, however, very natural definitions of inclusion and equivalence between subobjects of X, as follows:

Definition 92. If (R, r) and (S, s) are subobjects of X, then we say (R, r) *is included in* (S, s) – or in symbols $(R, r) \preccurlyeq (S, s)$ – if and only if r factors through s, i.e. when there is an arrow $h\colon R \to S$ such that $r = s \circ h$.

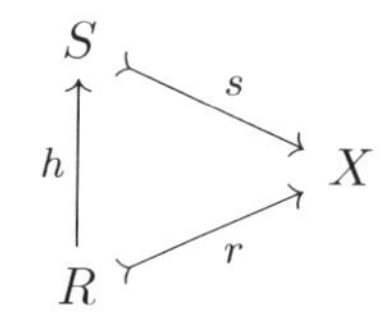

If both $(R, r) \preccurlyeq (S, s)$ and $(S, s) \preccurlyeq (R, r)$ then we say the subobjects are *equivalent*, in symbols $(R, r) \equiv (S, s)$. △

Question: Wouldn't it be even more natural to also require the mediating arrow h in our inclusion diagram to be monic too? Answer: We don't need to write that into the definition because h *is* monic by Theorem 16(3). And note that if $(R, r) \preccurlyeq (S, s)$ because there is an arrow $h\colon R \to S$ such that $r = s \circ h$, then h is unique. For suppose $r = s \circ h = s \circ h'$: then $h = h'$ because s is monic. So we don't have to write uniqueness into the definition either.

Theorem 105. *(1) The relation $\preccurlyeq$ on subobjects of X is a preorder.*

(2) The relation $\equiv$ on subobjects of X is an equivalence.

(3) Moreover, $(R, r) \equiv (S, s)$ if and only if r and s factor through each other by an isomorphism and so $R \cong S$.

Proof. (1) Inclusion is easily seen to be reflexive and transitive. Consider:

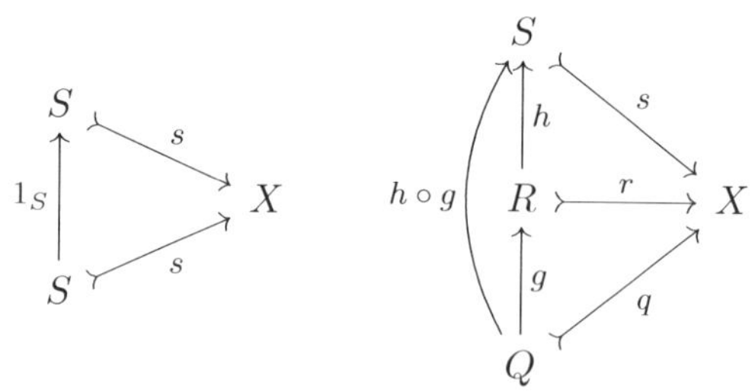

(2) Given (1), then (2) is immediate.

(3) One direction of the biconditional again is immediate. For the other direction, we appeal to Theorem 23 which tells us that if monics factor through each other, they do so via isomorphisms. □

(b) Here is another easy theorem:

Theorem 106. *(1) In any category, $(X, 1_X)$ is a maximum in the preorder $\preccurlyeq$ on subobjects of X.*

(2) In a Cartesian closed category with an initial object, $(0, 0_X)$ is a minimum, where 0_X is the unique arrow $0 \to X$.

Proof. (1) Since $1_X : X \to X$ is monic (by Theorem 16), $(X, 1_X)$ is a subobject of X. If (R, r) is any subobject of X, then r factors through 1_X since there is an arrow h such that $r = 1_X \circ h$ (trivially, put $h = r$). Hence $(R, r) \preccurlyeq (X, 1_X)$.

(2) Since $0_X : 0 \to X$ is monic (by Theorem 78(4)), $(0, 0_X)$ is a subobject of X. If (R, r) is any subobject of X, then 0_X factors through r since there is an arrow h such that $0_X = r \circ h$ (put $h = 0_R$, and rely on the fact that there is only one arrow from 0 to X). Hence $(0, 0_X) \preccurlyeq (R, r)$.[2] □

22.5 How many subobjects?

(a) In Set, a singleton set $\{\bullet\}$ will of course have *just two subsets* in the ordinary sense; however it will have *infinitely many subobjects* in the categorial sense. For any singleton $\{\star\}$ equipped with (the unique possible) arrow $\{\star\} \rightarrowtail \{\bullet\}$ will count as a subobject of $\{\bullet\}$. Indeed our singleton will have more than set-many such subobjects, since in standard set theory there are too many singletons to form a set! Isn't that quite desperately awkward?

No! For note that all those singleton subobjects of $\{\bullet\}$ will evidently be equivalent, leaving the one outlier subobject, the empty set equipped with the empty function. So our singleton does have just two subobjects up to isomorphic equivalence. And the result generalizes:

Theorem 107. *In* Set, *the subsets of X correspond one-to-one with equivalence classes of subobjects of X.*

Proof. Given a subset S of X (in the ordinary sense), there is a monic inclusion function $s \colon S \hookrightarrow X$, and hence a corresponding subobject (S, s). So consider the map m that sends each subset S to the equivalence class of subobjects of X containing the corresponding (S, s).

Evidently, m is onto. So it remains to confirm m is one-to-one. So suppose S_1 and S_2 are different subsets. Then the inclusions $s_1 : S_1 \hookrightarrow X$ and $s_2 : S_2 \hookrightarrow X$ will have different images so, by Theorem 105, $(S_1, s_1) \not\equiv (S_2, s_2)$. Therefore m sends S_1 and S_2 to different equivalence classes. □

We get parallel results in other categories too. For example, equivalence classes of subobjects in Grp correspond one-to-one to subgroups, and similarly equivalence classes of subobjects in Vect_k correspond to vector subspaces, and so on. (But topologists might like to work out why in Top the equivalence classes of subobjects don't straightforwardly correspond to subspaces.)

(b) Since monics with the target X typically don't line up one-to-one with subobjects of X as ordinarily understood, some theorists want to get things back in sync, by officially defining subobjects not as objects equipped with monics,

[2] Note that '0_X', so defined, denotes an arrow from 0 to X. However, '1_X' of course doesn't denote an arrow from 1 to X, but the identity arrow from X to itself. So, the neatness of the parallel between results (1) and (2) is exaggerated by a slightly sneaky choice of notation!

nor as simply the monics themselves, but as equivalence classes of monics;[3] while others oscillate between our original definition and that alternative definition.[4]

On balance, however, I do prefer to keep things simple and cleave to our Defn. 90. After all, by this stage of the game we are used to the idea that, categorially, we only define products (say) up to isomorphism. And this is to be welcomed, we said: for we shouldn't really care about the implementation details for different schemes for forming products, so long as the gadgetry works in the right way. Moreover, there is no point in trying to wrap up the different equivalent ways of forming the product of X and Y into a single package, using an equivalence class of some sort. Likewise for the other category-theoretic widgets we've encountered: definition up to isomorphism is actually what we want.

So let's have the courage of our categorial convictions and be content to initially define subobjects of X as objects equipped with appropriate monics, $(S, s\colon S \rightarrowtail X)$, letting equivalent subobjects do the same job without feeling a need to package them together from the start. (But we'll return to this issue in the next chapter, §23.2 fn. 2.)

22.6 Looking forward: an algebra of subobjects?

(a) Recall the discussion of §20.2 where we characterized the intersection of two subsets of a set X (up to isomorphism, at least) using a pullback diagram. So, now generalizing that earlier idea beyond the category Set, the following looks an inviting definition to try:

Definition 93. In a category with pullbacks, if (R, r) and (S, s) are subobjects of X, then any $(R \cap S, r \cap s)$ is an *intersection* of them, where $R \cap S$ is the vertex of a pullback over the corner formed by r and s,

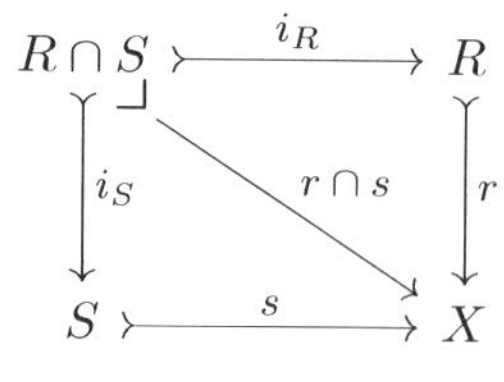

and $r \cap s\colon R \cap S \to X$ is the resulting diagonal, equalling the composite arrow round the square on either path. △

The labels 'i_R' and 'i_S' are intended to suggest injections of $R \cap S$ into R and S respectively – and being pullbacks of monics, these two arrows are monics too.

[3] See, for example, Arbib and Manes (1975, p. 35), Agore (2023, p. 12), Barr and Wells (1995, p. 43), Leinster (2014, p. 125).

[4] See Goldblatt (1984, pp. 76–78). Similarly Awodey (2010, p. 90): "We shall use both notions of subobject, making clear when monos are intended, and when equivalence classes thereof are intended." And compare Johnstone (2002, p. 18) who says that "like many writers on category theory" he will be deliberately ambiguous between the two definitions in his use of 'subobject'.

Hence the diagonal arrow, being a composite of monics is another monic, so we have well-defined $(R \cap S, r \cap s)$ as a subobject of X.

Note: defining an arrow as being the result of a pullback doesn't fix it uniquely. That's why our definition talks about 'an intersection'. However, we can apply Theorems 86 and 105, which immediately give us:

Theorem 108. *Any two intersections of (R, r) and (S, s) are equivalent as subobjects.* □

(b) We will see that intersections as we've defined them do behave nicely in various predictable ways. Here's one illustration of the point:

Theorem 109. *In a category with pullbacks, if (R, r) and (S, s) are both subobjects of some X, they have an infimum, a greatest lower bound, namely their intersection $(R \cap S, r \cap s)$.*

Proof. Using the terse idiom for subobjects (where we just mention the relevant monic), it is immediate from the defining pullback square in Defn. 158* that $r \cap s$ $\preccurlyeq$-precedes both r and s.

Now suppose that q is any other subobject which $\preccurlyeq$-precedes both r and s. Then by definition there is an arrow $i\colon Q \to R$ such that $q = r \circ i$ and an arrow $j\colon Q \to S$ such that $q = s \circ j$. So the bent outer square here commutes,

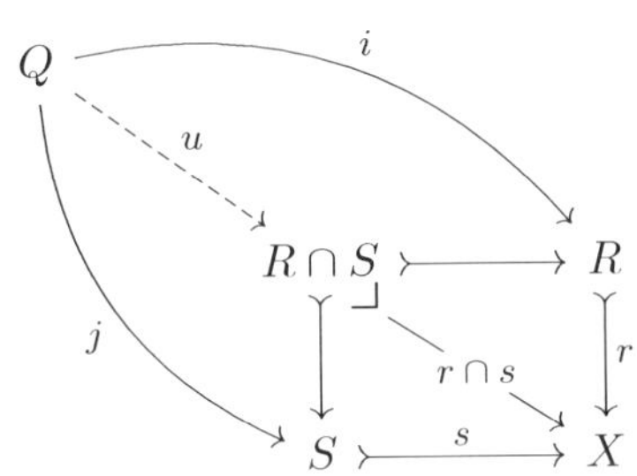

with composites along both outer paths equalling q. And hence, because the lower square is a pullback, there is an arrow u making the whole diagram commute. But then, $q = (r \cap s) \circ u$ and therefore $q \preccurlyeq r \cap s$.

Hence $r \cap s$ is a greatest lower bound as advertised. □

(c) Elementary set theory recounts how the family of subsets of a given set X form a nice algebraic structure when we take intersections, unions, and complements, and the natural inclusion order on that family interacts with the algebra so that intersections are infima etc. And now we seem to have the beginnings of a parallel story to tell about algebras of subobjects in categories rather more generally. This is intriguing, and we'll want to know more. The story does get a little complicated, however: so let me put the topic on hold for now, while promising to pick up the theme again in Part III.

23 Subobject classifiers

In the previous chapter, we took an already-familiar idea, the idea of a monic arrow, and showed how it provides a sensible enough categorial notion of subobject. In this chapter we introduce a new idea: this will enable us to define a categorial generalization of the notion of a characteristic function.

23.1 Motivation

(a) We'll start by taking an argument from §16.4 about inclusion functions and applying it to monics in Set more generally.

So again suppose Ω is a two-object set we can think of as $\{true, false\}$. And let $\top_X$ be the function that sends everything in X to *true*.

Now suppose that (S, s) is a subobject of X in the sense of Defn. 90: in other words, $s\colon S \rightarrowtail X$ is any monic arrow into X. Then let $\chi_s\colon X \to \Omega$ be the characteristic function that sends $x \in X$ to *true* iff $x \in s[S]$.[1]

Evidently $S \overset{s}{\rightarrowtail} X \underset{\top_X}{\overset{\chi_s}{\rightrightarrows}} \Omega$ is a commuting fork. Moreover, any other fork through $\chi_s, \top_X$ factors uniquely through it.

Why? Consider the diagram

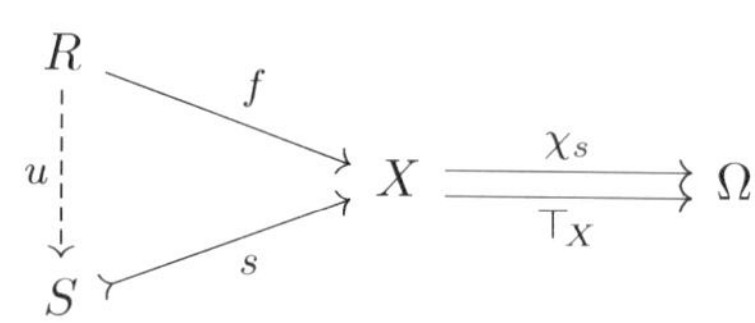

For (R, f) to produce a fork through $\chi_s, \top_X$, members of $f[R]$ must be sent to *true* by χ_s. Hence $f[R] \subseteq s[S] \subseteq X$. If we define u to send $x \in R$ to the pre-image of $f(x)$ under s (which is unique since s is monic), then the diagram commutes. Moreover, this u is the only arrow to give us a commuting diagram.

Therefore the subobject $(S, s\colon S \rightarrowtail X)$ in Set is an equalizer for the corresponding χ_s and $\top_X$.

---

[1]When we subscript a characteristic arrow χ to indicate what it is characteristic for, it is the specific monic we want to record, not just its source. Why? Because the same source S can be sent by different monics $s\colon S \rightarrowtail X$ to different 'parts' of X, giving us different subobjects; the characteristic arrow is telling us about which particular 'part' is the target of s.

(b) That was a *very* minor tweak to what we had before in §16.4. But the point of rehearsing it is to set the scene for an equivalent (and not-so-obvious) way of redescribing this situation, still in Set.

Start with the observation that the map $\top_X \circ s \colon S \to \Omega$, which sends everything in S to the value T, is equal to the composite map $S \xrightarrow{!_S} 1 \xrightarrow{\top} \Omega$ where 1 is a terminal object in the category. Similarly for the map $\top_X \circ f \colon R \to \Omega$: this is equal to the composite map $R \xrightarrow{!_R} 1 \xrightarrow{\top} \Omega$.

Hence, replacing composite arrows in the previous diagram by equal arrows and then re-arranging, the claim that (S, s) equalizes the arrows χ_s and $\top_X$ in Set is equivalent to saying that for any $f \colon R \to X$ such that $\chi_s \circ f = \top \circ !_R$ there is a unique $u \colon R \to S$ which makes the following diagram commute:

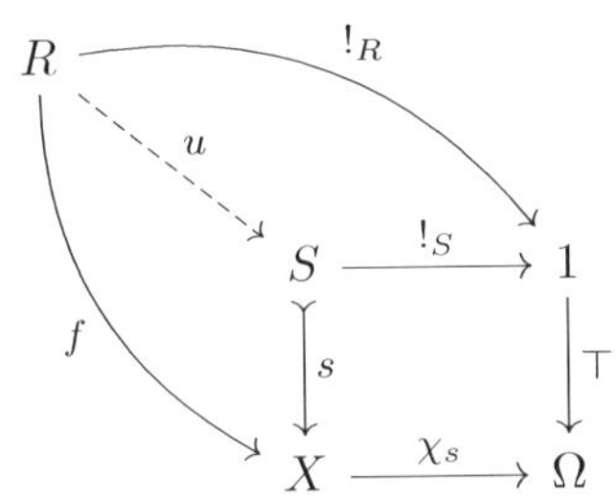

But that is to say that the lower square is a pullback square.

(c) In summary: in Set, any subobject arrow $s \colon S \to X$ is a pullback from $\top \colon 1 \to X$ along a suitable $\chi_s \colon X \to \Omega$.

Fine! And now we should be able to carry that kind of linkage between subobjects and pullbacks across to other categories which have a suitable Ω which works like a set of truth-values and an associated 'true'-selecting map $\top \colon 1 \to \Omega$.

However, to pick up the thought I trailed at the end of §22.1, we *can't* parlay this linkage into an alternative definition of a subobject in terms of a pullback limit – since that would presuppose we *already* have a handle on a general notion of 'truth-value object' applicable in other categories. And we don't.

Instead, we have to look at things exactly the other way around. In fact, what we can extract here is a general specification of a 'truth-value object' Ω and an associated 'true'-selecting map $\top \colon 1 \to \Omega$ (which, recall, must be monic by Theorem 29). *We pin these down across categories by requiring that they interact in the described way with subobjects as we previously defined them.*

23.2 Defining a subobject classifier $(\Omega, \top)$

(a) Our discussion, then, motivates a new and distinctively categorial idea:

Definition 94. In a category C with a terminal object 1 and pullbacks, an object Ω and arrow $\top \colon 1 \rightarrowtail \Omega$ provide a *subobject classifier* $(\Omega, \top)$ if and only if for any $(S, s \colon S \rightarrowtail X)$, i.e. for any subobject s of any X, there is a unique *characteristic* arrow $\chi_s \colon X \to \Omega$ making this a pullback square:

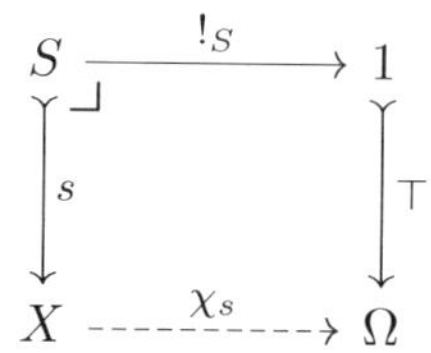

△

We can think of the classifying object Ω as a 'truth-value object', and the arrow $\top: 1 \to \Omega$ as its point element *true*.

Note: the characteristic arrow χ_s is, so to speak, the *most discriminating* arrow making the square commute. Think again of how things go in Set. Then any arrow $\chi_{s'}: X \to \Omega$ will give us a commuting square, if $\chi_{s'}$ is the characteristic function for S' where $S \subseteq S' \subseteq X$. But if S' contains elements not in S, the square won't be a limiting case, won't be a pullback (why not?).

(b) Let's immediately check that this works as intended, and subobjects which categorially come to the same, i.e. are equivalent, have the same characteristic arrow:

Theorem 110. *Assuming we are in a category with a subobject classifier, (R, r) and (S, s) are equivalent subobjects of X if and only if $\chi_r = \chi_s$. Further, any arrow $\chi: X \to \Omega$ is the characteristic arrow of some subobject of X, unique up to equivalence.*

Proof of 'if'. Suppose $\chi_r = \chi_s$ and consider this diagram:

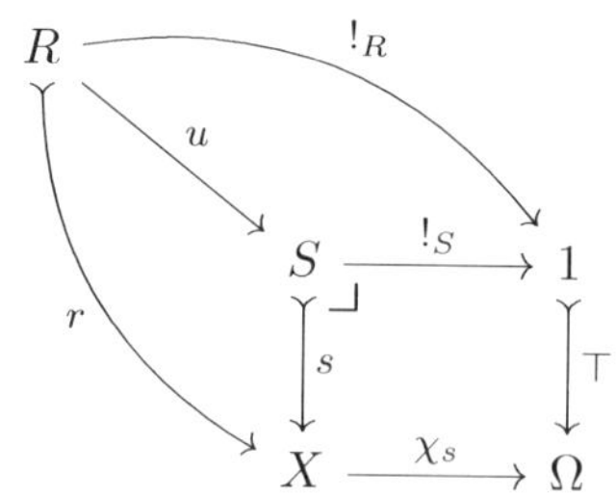

The outer 'square' commutes by definition of χ_r and the assumption $\chi_r = \chi_s$. The inner square is a pullback, so the wedge with vertex R has to factor through the wedge with vertex S via some u. So we have $r = s \circ u$. But that means $(R, r) \preccurlyeq (S, s)$. Similarly we can prove $(S, s) \preccurlyeq (R, r)$. Therefore $(R, r) \equiv (S, s)$. □

Proof of 'only if'. Suppose $(R, r) \equiv (S, s)$ and consider the same diagram. By Theorem 105(3), the equivalence assumption entails that there is an isomorphism u between R and S such that the lower triangle commutes. And the top triangle must commute. So Theorem 95 applies: it shows that the outer square with vertex R is a pullback.

But then by definition of the subobject classifier $(\Omega, \top)$, the bottom arrow from X to Ω must be equal to χ_r. Therefore $\chi_r = \chi_s$. □

Proof that every arrow $\chi\colon X \to \Omega$ is a characteristic arrow. We just note that pulling back $\top$ along χ will give us some arrow $s\colon S \to X$ which will be monic by Theorem 91, and χ will be characteristic for it. Further Theorem 86 tells us that every pullback of $\top$ along χ will factor through that s by an isomorphism and will therefore be equivalent as a subobject.[2] □

23.3 $\top$, $\bot$, and $\neg$

(a) If I say that we can think of the arrow $\top\colon 1 \to \Omega$ as the point-element *true* of Ω, an obvious question will immediately occur to you: can we similarly characterize a *false* element $\bot\colon 1 \to \Omega$?

Well, think how things work in **Set** with its classifying object $\{true, false\}$. How can we categorially define an arrow $\bot\colon 1 \to \Omega$ which sends the sole member of 1 to *false*? **Set** has an initial object 0 (namely the empty set), and there is a unique arrow $0 \to 1$ (the empty function, as noted in §9.1). This is trivially a monomorphism. So like any monic arrow in **Set**, this arrow gives us a subset of its target. Which one? The empty set (what else?). And what is the characteristic arrow $1 \to \Omega$ associated with the empty set? It has to be the function $\bot$ which maps the sole member of the singleton 1 to *false*.

That's quite neat. So let's now generalize that line of thought:

Definition 95. In a category with a subobject classifier and an initial object 0, $\bot\colon 1 \to \Omega$ is the characteristic arrow of $0 \to 1$, the unique arrow that makes this a pullback diagram:

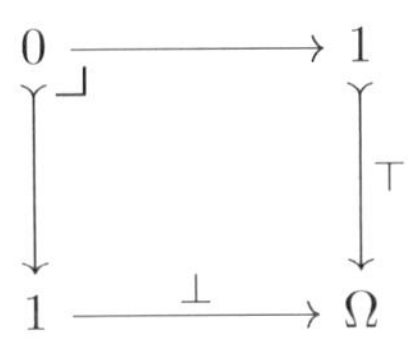

△

(b) Plainly we want 'truth' and 'falsehood' to be distinct. And they will be in nice enough categories. In particular we should note

Theorem 111. *In a non-degenerate Cartesian closed category with a subobject classifier, $\bot \neq \top$.*

Proof. Suppose $\bot = \top$ in a Cartesian closed category. Then this supposition makes the inner square below a pullback square:

[2] We noted in §22.5 that some want to identify subobjects not with individual monics but with equivalence classes of monics. However – a reasonable question – where are such equivalence classes to be found? Will appropriate classes themselves be among the data of C? If we are working in an impoverished category, it seems not necessarily so. However, if we are working in a category with a subobject classifier and hence with characteristic arrows, we needn't fret. We've seen that subobjects of X thought of as equivalence classes of monics into X line up one-to-one in a natural way with corresponding arrows from X to Ω. So if you *really* want to package subobjects into equivalence classes, you could therefore get much the same effect by officially redefining subobjects of X as characteristic arrows $X \to \Omega$.

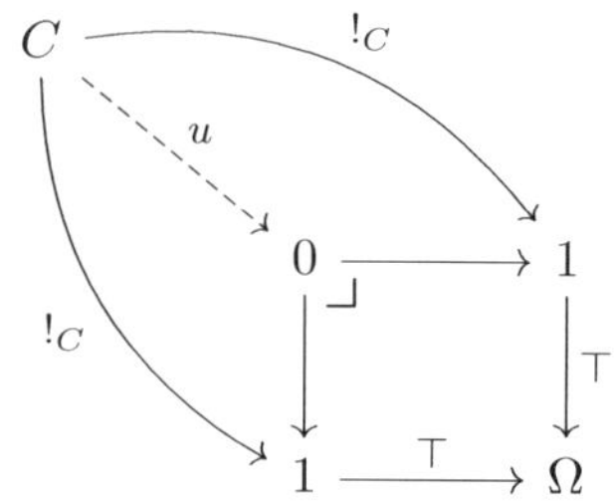

Hence, for any C, since the composite arrows from C to Ω along the two outer routes are identical, there must be an arrow $u\colon C \to 0$ completing the diagram since this is a pullback. But C was arbitrary, and so there will in particular be such an arrow when $C = 1$.

But we know from the proof of Theorem 79 that the existence of an arrow $1 \to 0$ in a Cartesian closed category leads to degeneracy. □

(c) Like any arrow from 1, the 'falsehood' arrow $\bot\colon 1 \to \Omega$ is monic (see Theorem 29). So, assuming we are in a nice enough category, $(1, \bot)$ is itself a subobject of Ω with its own characteristic arrow. Here it is:

Definition 96. In a category with a subobject classifier and initial object (so $\bot$ is defined), the arrow $\neg\colon \Omega \to \Omega$ is the characteristic arrow of $(1, \bot)$, making this a pullback:

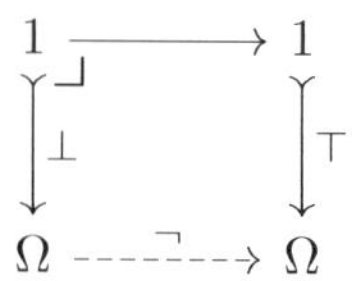

△

The negation-like notation '$\neg$' here seems pretty appropriate, for we have the following simple result (dropping explicit composition signs):

Theorem 112. *$\neg\bot = \top$ and $\neg\top = \bot$.*

Proof. Again assume we are in a category with pullbacks, and $\top$, $\bot$, and $\neg$ are defined. Then our first claim is true by definition. For the second, consider the left-hand diagram below:

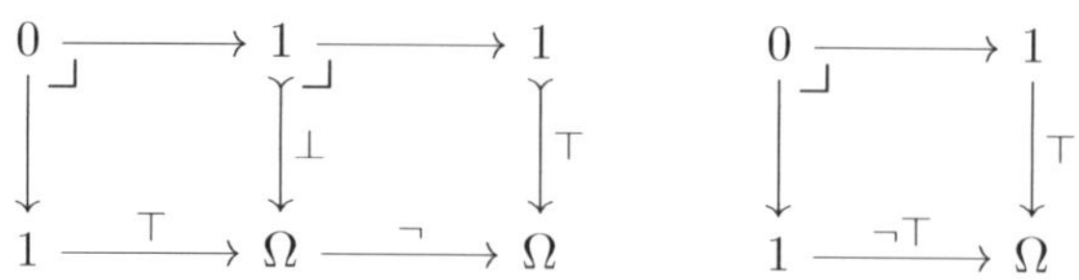

One square is the pullback defining $\bot$, reflected about the diagonal. The other square is the pullback which introduces $\neg$. Therefore the overall rectangle in the left diagram, equivalently the right-hand diagram, commutes and is also a pullback by the pullback lemma Theorem 93.

But $\bot$ is by definition the sole arrow $1 \to \Omega$ which completes the pullback from $\top\colon 1 \to \Omega$ to $!\colon 0 \to 1$. Hence, $\neg\top = \bot$. □

Later, we will be introducing other 'logical' arrows (a conjunction, disjunction, and conditional) to live alongside the negation arrow $\neg$, and these interact to give us an 'internal logic', at least in sufficiently rich categories. But that's a topic for Part III. Negation is making an early appearance here because it will be useful to have it in play in §25.1.

23.4 Three instructive examples

We know that Set has a subobject classifier whose truth-value object Ω is a simple two-element set. The same two-object classifier works in Finset, and even in 2set, the category of sets which have no more than two members. But as we remarked back in §11.2, the impoverished category 2set doesn't have all binary products. So a category can (quite radically!) fail to be Cartesian closed in the sense of Defn. 75 yet still have a subobject classifier.

Things can also go the other way around. In other words, a category can be rich enough to be Cartesian closed yet lack a subobject classifier: Pos is an example. Why? Because Pos is not a balanced category as we proved in §8.4, and we are later going to prove that a category can only have a subobject classifier if it is balanced (that's Theorem 116 in the next section). Top is also unbalanced, and so the same applies.

The category Grp by contrast is balanced, but still doesn't have a subobject classifier, though I'm not going to pause to prove that. Some other negative cases are easier to see. The category of rings doesn't have any ring homomorphisms from its terminal object, the one-element ring, to any other non-trivial ring: so we can't get a non-trivial Ω and arrow $\top\colon 1 \to \Omega$. So the possibility of having a subobject classifier in Ring is immediately squashed.

But here are three instructive positive cases:

Theorem 113. *These categories have subobject classifiers:*

(1) Set^2, *i.e.* $\mathsf{Set} \times \mathsf{Set}$,
(2) Graph,
(3) M_2.

(1) *Proof sketch:* Set^2 *has a subobject classifier.* Recall Defn. 23. Let's use undecorated notation for items in Set (thus $1, \Omega, \top$), and superscripted notation for items in Set^2 (so $1^2, \Omega^2, \top^2$).

In Set, the subobject classifier is, for short, a suitable $\Omega = \{T, F\}$ equipped with an arrow $\top\colon 1 \to \Omega$ (with $\top$ sending the sole member $*$ of the chosen terminal object 1 to T). Set^2 doubles everything up. So the pair $1^2 = \langle 1, 1\rangle$ is terminal. And $\Omega^2 = \langle \Omega, \Omega\rangle$ equipped with the Set^2-arrow $\top^2 = \langle \top, \top\rangle\colon 1^2 \to \Omega^2$ provides a subobject classifier for Set^2. Simple exercise: check this claim.

There are then *four* point elements of Ω^2, the four arrows $\langle \top, \top\rangle$ plus $\langle \top, \bot\rangle$, $\langle \bot, \top\rangle$, $\langle \bot, \bot\rangle$. Which illustrates in a particularly easy way that subobject classifiers need not be merely 'two valued' structures. □

(2) *Proof sketch:* Graph *has a subobject classifier.*[3] The category has a terminal object 1, a graph with a single node and one loop. And the classifying object Ω is a graph as on the right:

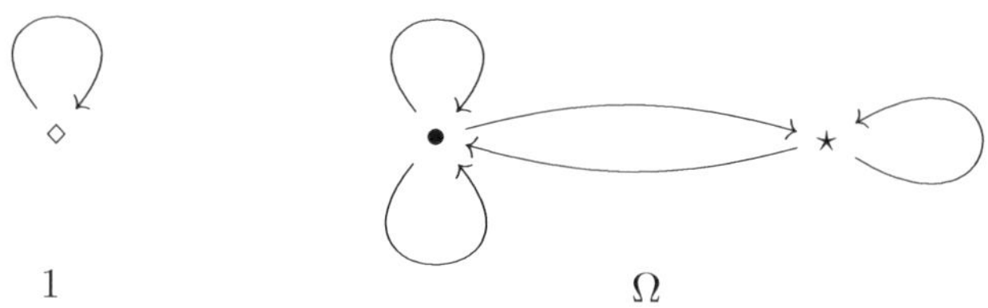

The required arrow $\top\colon 1 \rightarrowtail \Omega$ needed to complete the subobject classifier is the graph homomorphism which sends the terminal graph 1's single node to Ω's node $\bullet$ and 1's loop to one of $\bullet$'s loops.

This specification of Ω no doubt initially appears entirely mystifying! But suppose we start with a graph X and take a subgraph $(S, s\colon S \rightarrowtail X)$, where s is a monic graph homomorphism of course (so s acts on both nodes and edges). And for brevity's sake, we will slightly abuse language and say that a node n from X is 'in' S when there is a node m in S such that $s(m) = n$; similarly an edge e from X is 'in' S when there is an edge d in S such that $s(d) = e$. In other words, let's speak as if the monic s is simple inclusion.

Now think how S looks from the point of view of X:

(i) As far as nodes are concerned, a node n in X can be *in* S or *out*.
(ii) As far as edges are concerned, an edge e in X can be in S, along with its nodes of course, i.e. *everything* about e is also in S.
(iii) Otherwise the edge e from X is not in S: but that leaves four possibilities as far as e's nodes are concerned – either *both* its nodes are in S, or just e's *source* is in S, or just e's *target*, or else *nothing* of e's is in S.

Ah! Compare the situation in Set: looking at one of the subsets S of X from the point of X, we ask 'is this element of X still in S?'. That's a simple yes/no, two possibility, question – which is why we only need a two-member truth-value set to encode the possibilities. Here in Graph there are more possibilities to classify: when we look at our subgraph S from the point of view of X, there are two possibilities for nodes, and five for edges. Which demystifies why our classifying object Ω is more complex, a graph with two nodes and five edges, reflecting the different modes of 'inclusion'.

Suppose then that we label the nodes and arrows of Ω like this, reflecting those different possibilities for nodes and arrows in subgraphs:

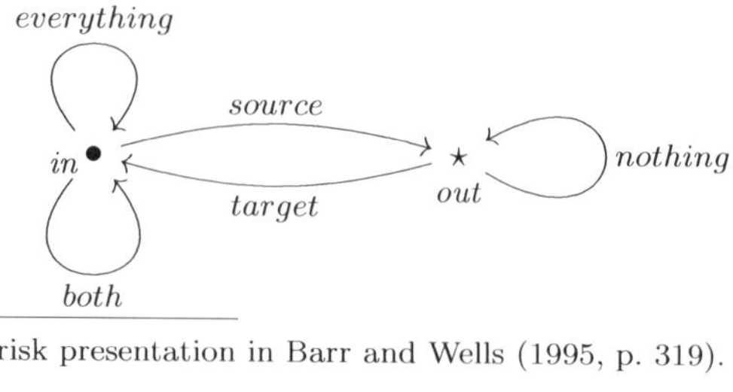

[3]Inspired by the brisk presentation in Barr and Wells (1995, p. 319).

Our arrow $\top\colon 1 \rightarrowtail \Omega$ now more specifically sends 1's node to Ω's node *in* and 1's loop to Ω's edge *everything*.

Similarly, $\bot\colon 1 \to \Omega$ should send 1's node to the 'out' node $\star$ and send 1's loop to the 'nothing' loop.

And note: that leaves the third option of an arrow sending 1's node to the 'in' node and 1's loop to the 'both' loop. So in this case, *there are three point elements of* Ω, three 'truth values'.

OK: so returning to our subgraph $(S, s\colon S \rightarrowtail X)$, now consider the graph homomorphism $\chi_s\colon X \to \Omega$ which sends each node n of the graph X to *in* or *out* depending on whether n is in S, and sends each edge e from X to the appropriate arrow in Ω which represents how much of e is in S. Then χ_s will make the required square commute – and moreover it is the most discriminating graph homomorphism which will do this.

Hence, hand-waving a bit, this gives us a pullback square and we are done. $(\Omega, \top)$ as just defined is our desired subobject classifier in Graph, and χ_s is the characteristic arrow for the subobject (S, s). □

I won't pause to tighten up the details of the last stage of that proof sketch, as I have said enough to get across the basic idea, and this is one of those cases where playing with a few more diagrams will convince.

(3) *Proof sketch:* M_2 *has a subobject classifier.*[4] Recall that back in §5.7 (C22) we defined objects of M_2 as sets equipped with idempotent functions, thus (X, f). Arrows in M_2 are 'equivariant' functions: in other words, the Set-function $j\colon X \to Y$ counts as an M_2-arrow $j\colon (X, f) \to (Y, g)$ so long as $j \circ f = g \circ j$.

Let's continue to use undecorated notation for items in Set, with superscripted notation for items in M_2 (thus $1^{\mathrm{M}}, \Omega^{\mathrm{M}}, \top^{\mathrm{M}}$). 1^{M} is simply $(1, 1_1)$ where 1 is initial in Set and 1_1 is its identity arrow. A classifying object Ω^{M} will be some (Ψ, ψ) (a set equipped with an idempotent function ψ), and the truth-seeking arrow $\top^{\mathrm{M}}$ will be an equivariant set function $\top^{\mathrm{M}}\colon (1, 1_1) \to (\Psi, \psi)$.

Now, if $(\Omega^{\mathrm{M}}, \top^{\mathrm{M}})$ is actually to work as a subobject classifier, it must be the case that, for every monic $j\colon (X, f) \to (Y, g)$, there is a unique χ_j making the diagram on the left a pullback in M_2:

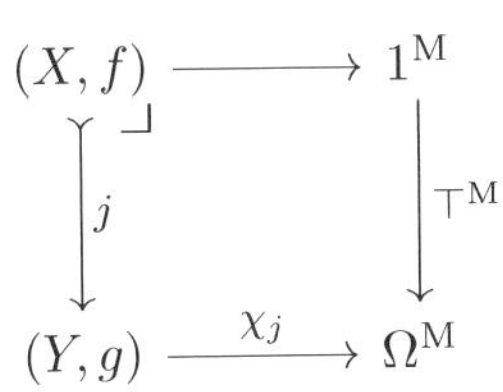

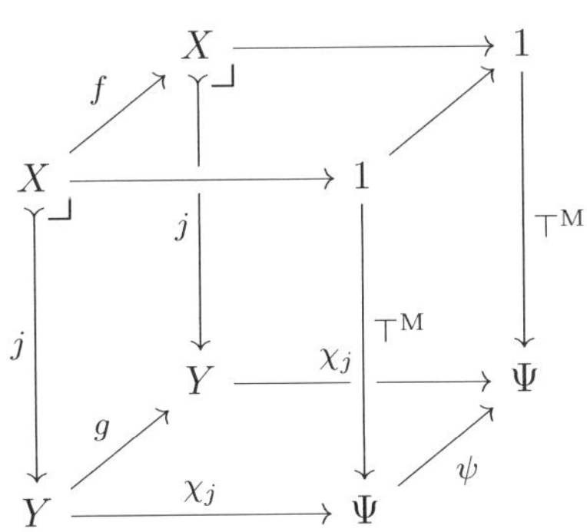

[4] I am indebted here to Daniel Schepler.

Which requires Ψ, ψ and $\top^{\mathsf{M}}$ to be such that for any monic $j\colon X \rightarrowtail Y$ there is a unique χ_j making the diagram on the right commute in Set.

The top and the left-hand faces of the cube commute by definition. Dropping the superscript, assume $\top$ sends the sole member of the singleton 1 to an object T in Ψ: then the right-hand face of the cube will commute so long as ψ sends T to T. Note, by the way, that these assumptions ensure that $\top\colon (1, 1_1) \to (\Psi, \psi)$ is a kosher M_2-arrow (since $\top \circ 1_1 = \psi \circ \top$). Further, assume that χ_j sends an object $y \in Y$ to T precisely when $y \in j[X]$.

Now pick an object y in Y. There are three ways things might go for it and its 'descendant' $g(y)$, depending on whether they are in the j-image of X.

(i) Suppose $y \in j[X]$, i.e. there is some x such that $y = j(x)$. Then $g(y) = j(f(x))$, so $g(y) \in j[X]$. In this case, given our assumption about χ_j, everything will commute as we chase x round the diagram.

But what happens if we've picked an object y which isn't some $j(x)$?

(ii) Suppose neither y nor $g(y)$ is in $j[X]$. Let χ_j send any such y to F (a member of Ψ distinct from T). Then, as we chase round y, the bottom face will commute so long as ψ sends F to F.

(iii) Suppose lastly that y isn't but $g(y)$ is in $j[X]$. Then, by (i), on the back face of the cube χ_j will send $g(y)$ to T. On the front face, χ_j can't send y to T, nor can it send y to F and give us a commuting base to the cube (given what (ii) says about ψ). Hence χ_j will need to send y to some third member of Ψ which we'll dub $\frac{1}{2}$. Then, for such a y, the bottom face will commute so long as ψ sends $\frac{1}{2}$ to T.

So that dictates a limiting-case solution, giving us a subobject classifier for M_2. In sum, put Ψ to be a three-member set, $\{T, \frac{1}{2}, F\}$, and define ψ by $T \mapsto T, \frac{1}{2} \mapsto T, F \mapsto F$ (which evidently, as required, is idempotent).

It remains to officially check that this works. But let that stand as enough by way of motivation. □

An important remark about this case before proceeding. There is an M_2-arrow $\top\colon (1, 1_1) \to (\Psi, \psi)$. We can also define $\bot\colon (1, 1_1) \to (\Psi, \psi)$ as sending the sole member of 1 to F: again, that is evidently a kosher M_2-arrow since $\bot \circ 1_1 = \psi \circ \bot$.

Can we define a third arrow $I\colon (1, 1_1) \to (\Psi, \psi)$ which sends the sole member of 1 to $\frac{1}{2}$ in Ψ? No. That wouldn't make the grade as an M_2-arrow because (given how ψ is defined) $I \circ 1_1 \neq \psi \circ I$.

So, looked at externally, M_2's classifying object Ψ is a set with three elements. But the category itself discerns only two point elements.[5]

[5] Our three examples are enough for the purposes of these notes. If you want some more illustrations of how classifying objects can be significantly more complex than in Set, see e.g. Goldblatt (1984, §§4.4–4.5).

23.5 Four general theorems about subobject classifiers

(a) Let's start by noting that, even after we've fixed on a 'truth-value object' Ω, it can be arbitrary which arrow $1 \rightarrowtail \Omega$ is chosen as $\top$ to form a subobject classifier. This is evidently so in Set: if Ω is e.g. $\{0, 1\}$, which element is to be selected to encode *true*?

And this is the case in Graph too – because, given a graph of the right shape to be the classifying object, it is still up for grabs which of the loops on the two-loop node we decide to be the value of $\top$ when applied to 1's loop.

However, if a category has a subobject classifier at all, the truth-value object Ω itself will be determinate up to isomorphism:

Theorem 114. *(i) If a category has subobject classifiers* $(\Omega, \top)$, $(\Omega', \top')$, *then* $\Omega \cong \Omega'$. *Conversely, (ii) if* $(\Omega, \top)$ *is a subobject classifier, and there is an isomorphism* $j\colon \Omega \xrightarrow{\sim} \Omega'$, *then* $(\Omega', \top')$ *is a subobject classifier where* $\top' = j \circ \top$.

Proof for (i). Theorem 29, to repeat, tells us arrows with source 1 are monic, so $(1, \top)$ counts as a subobject of Ω.

Hence, given $(\Omega', \top')$ is a subobject classifier, there is a unique χ such that the left square in diagram (i) below is a pullback:

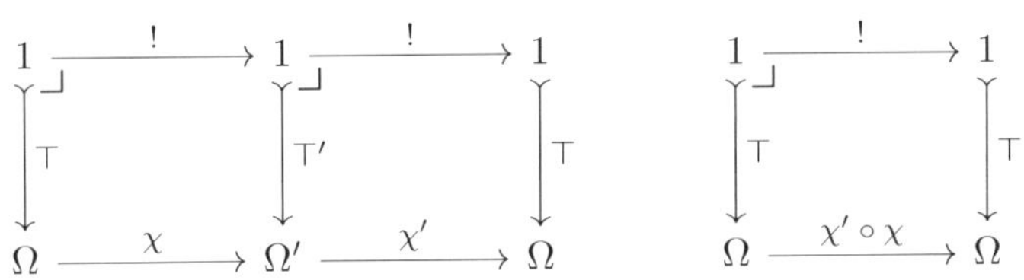

(i) (ii)

Further, since $(1, \top')$ counts as a subobject of Ω' and $(\Omega, \top)$ is a subobject classifier, there is a unique χ' such that the right square in (i) is a pullback. Hence by Theorem 93 the whole rectangle, equivalently (ii), is a pullback.

But by the very definition of the subobject classifier $(\Omega, \top)$, given the particular subobject $(1, \top)$ there must be a *unique* arrow $\Omega \to \Omega$ that will make a pullback square like (ii), and 1_Ω will do the trick. Hence we must have $\chi' \circ \chi = 1_\Omega$. Exactly similarly, of course, $\chi \circ \chi' = 1_{\Omega'}$.

Therefore χ and χ' are mutually inverse – and hence we have found our required isomorphism between Ω and Ω'. □

Proof for (ii). Consider the following diagram. The right-hand square is easily checked to be a pullback square, so the whole rectangle is a pullback by the pullback lemma.

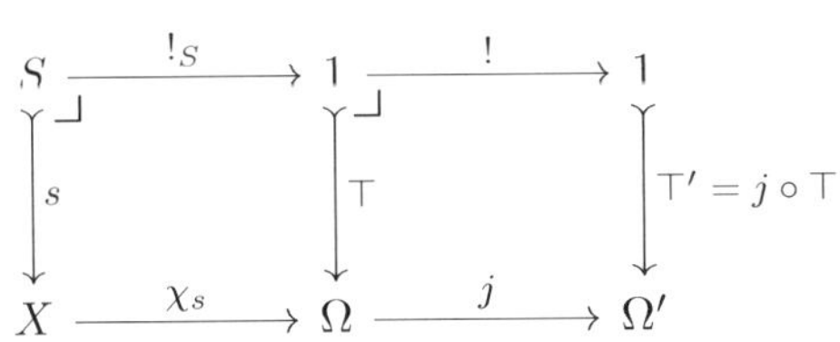

So that shows there is at least one arrow, namely $j \circ \chi_s$, along the bottom of the rectangle giving us a pullback square. And to show it is unique, suppose that $\psi\colon X \to \Omega'$ also completes a pullback. Then by the obvious argument $j^{-1} \circ \psi\colon X \to \Omega$ would complete the left-hand square as a pullback, entailing that $j^{-1} \circ \psi = \chi_s$ and so $\psi = j \circ \chi_s$ again. □

(b) Next, let's show that it was actually unnecessary to stipulate that our subobject classifier involves an arrow from a *terminal* object to the truth-value object Ω. In fact, we get that for free from the following theorem:

Theorem 115. *Suppose in* C *that there is an object* Ω *and monic* $\top\colon Z \rightarrowtail \Omega$ *such that for any* $(S, s\colon S \rightarrowtail X)$ *there is a unique arrow* $\chi\colon X \to \Omega$ *which (with some arrow* $S \to Z$*) makes a pullback square of this form:*

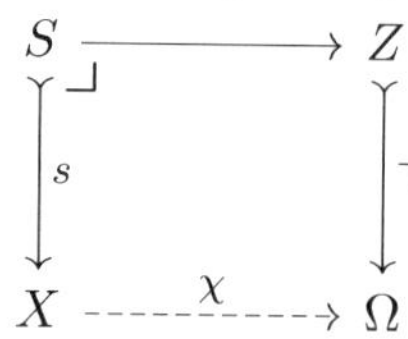

Then Z *is terminal in* C.

Proof. Take (S, s) to be, in particular, some $(X, 1_X)$. Then, by assumption, there must be at least one arrow $f\colon X \to Z$ making the left-hand diagram a pullback.

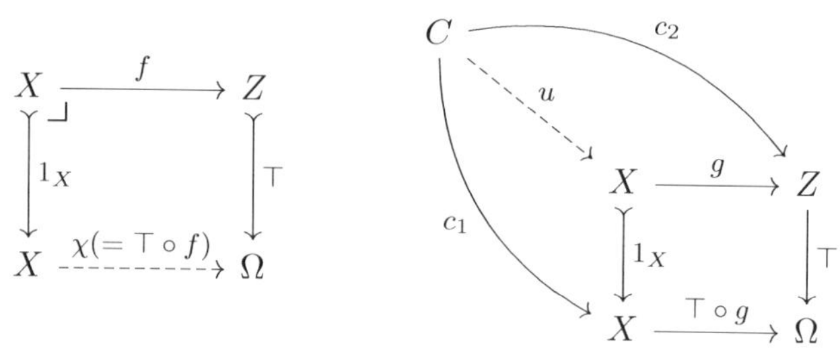

Suppose g is also an arrow from X to Z. Trivially the inner square on the right commutes. Suppose that the arrows $c_1\colon C \to X$, $c_2\colon C \to Z$ also make a commuting square with the opposite corner, so that $\top \circ g \circ c_1 = \top \circ c_2$. Then since $\top$ is monic, $g \circ c_1 = c_2$. So that makes the arrow $u = c_1$ the unique arrow completing the diagram. Hence the inner square is a pullback. But by assumption there is a unique arrow χ along which we can pull back $\top$ to 1_X. That means $\top \circ g = \top \circ f$, and therefore $g = f$, again since $\top$ is monic.

Hence there is one and only one arrow from any X to Z; therefore Z is terminal. □

(c) We will next recycle an earlier argument to show

Theorem 116. *In a category with a subobject classifier, any monic* $s\colon S \rightarrowtail X$ *is an equalizer, equalizing the parallel arrows* $\chi_s, \top_X\colon X \to \Omega$. *Hence a category with a subobject classifier is balanced.*

Proof. Take the pullback square defining the characteristic arrow of (S, s): and suppose that $f \colon R \to X$ is any arrow such that $\chi_s \circ f = \top \circ !_R$.

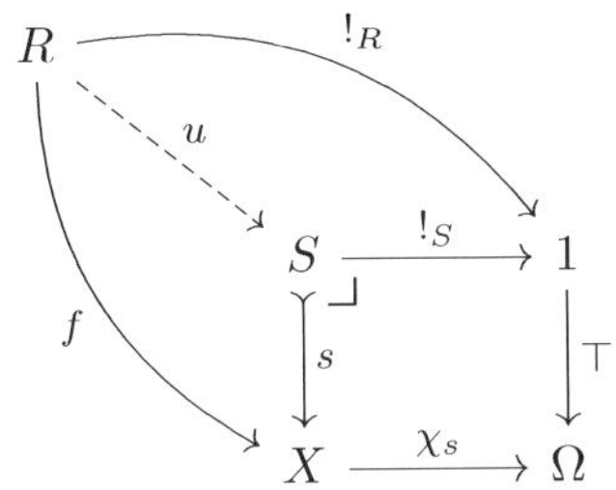

Then by the universal property of a pullback square there is a unique $u \colon R \to S$ making the diagram commute. Reversing the argument of §23.2, it then follows that – equivalently – for any fork with handle f and prongs χ_s and $\top_X$ will factor uniquely via the corresponding u through the fork with handle s and the same prongs. So the monic s is an equalizer.

Hence, in a category with a subobject classifier, any epic monic will be an epic equalizer, and hence is an isomorphism by Theorem 67. So the category is balanced. □

(d) We've seen a number of examples before where we can revamp a definition of the form (i) 'so-and-so is a widget in category C iff there is a unique arrow doing such-and-such' into a crisp definition of the form (ii) 'a widget in C is a terminal (or initial) object of the associated category C*'. For the record, let's note that we can do the same for subobject classifiers. Start with

Definition 97. Assume C is a category with pullbacks. Then Monic(C) is the derived category whose objects are monic arrows of C and where an arrow from $m \colon X \rightarrowtail Y$ to $m' \colon X' \rightarrowtail Y'$ is a pullback square

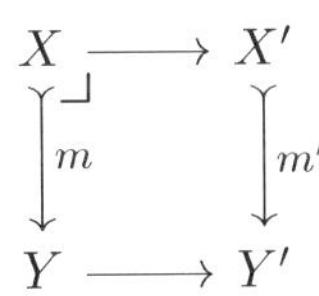

△

It is easily checked that Monic(C) is a category (essentially because pasting together pullback squares gives another pullback). And then

Theorem 117. *$(\Omega, \top \colon 1 \rightarrowtail \Omega)$ is a subobject classifier for* C *iff $\top$ is a terminal object for* Monic(C).

Proof. It is immediate from the definition that if $(\Omega, \top \colon 1 \rightarrowtail \Omega)$ is a subobject classifier for C, then $\top$ is terminal in Monic(C).

For the converse, Theorem 115 tells us that any monic $\top \colon Z \rightarrowtail \Omega$ which is terminal in Monic(C) will require Z to be a terminal object in C, and hence that $(\Omega, \top)$ will indeed be a subobject classifier in C. □

Using the fact that terminal objects are unique up to isomorphism we get another proof of Theorem 114.

23.6 A brisk aside about duals

By now, you should have got into the habit, whenever presented with a new categorial gadget, of asking 'what's the dual'?

If we take subobjects to be equivalence classes of monics, then their duals are equivalence classes of epics. In fact, some call an equivalence class of epics with source C a *quotient* of C (see e.g. Agore 2023, p. 15). But it is more natural, I think, to introduce quotients as certain co-equalizers as in §16.6.

As for a 'quotient classifier', i.e. the dual construct to a subobject classifier $(\Omega, \top\colon 1 \to \Omega)$, what would that be? Dualizing everything, it would be an object and epic arrow $(\Psi, V\colon \Psi \twoheadrightarrow 0)$ such that, for any epic $e\colon X \twoheadrightarrow S$, there exists a unique $\varphi_e\colon \Psi \to X$ making this a pushout:

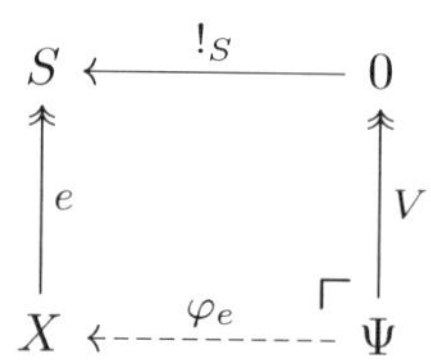

But that cannot happen e.g. in Set and similar categories: for a start, the only arrow into 0 is from 0 itself (a corollary: Set^{op} can't have a subobject classifier). No surprise, then, that the idea of duals to subobject classifiers doesn't get much air-time. We move on.

24 Power objects

We have seen how to handle e.g. products, quotients, exponentials, and much more, in a categorial setting. But there's another 'ordinary' mathematical construction which we haven't mentioned yet – the one involved in forming powersets. We will be returning to say more about subobjects (and how to form their intersections, unions, and complements). But first, in this short chapter, we'll show how in a sufficiently rich category that has both exponentials and a subobject classifier we can get a categorial analogue of powersets.

24.1 Power objects

(a) Think again about sets. Pick one, say Y. As is entirely familiar, subsets $S \subseteq Y$ then correspond one-to-one to the characteristic functions $\chi_S \colon Y \to \Omega$ (where Ω is our favoured set $\{\mathit{true}, \mathit{false}\}$). Hence we can take the powerset $\mathcal{P}Y$ to be tantamount to the set of functions from Y to Ω. And we can think of that set in turn as the exponential Ω^Y.

So we are now going to generalize wildly, define a cross-category notion of power-object, and show that – in a rich enough category which has a classifying object like Ω and exponentials – a power-object of Y is again provided by Ω^Y.

(b) Pre-categorially, we define a powerset $\mathcal{P}Y$ by reference to its 'internal' constitution (we say what its *members* are). Can we give a more 'external', more categorially flavoured, characterization of $\mathcal{P}Y$ (by talking about *morphisms* to or from it)? Let's take things in stages.

(i) Consider any function f from some set X to Y's powerset $\mathcal{P}Y$. This sends an element $x \in X$ to a (possibly empty) set $f(x) \subseteq Y$. And there will be a unique set of pairs $R \subseteq X \times Y$ such that $\langle x, y \rangle \in R$ iff $y \in f(x)$.

(ii) Or, looking at things the other way about, suppose we start with an inclusion function $r \colon R \hookrightarrow X \times Y$. Then there will be a unique function $f_r \colon X \to \mathcal{P}Y$ such that $f_r(x) \ni y$ (i.e. $y \in f_r(x)$) iff $\langle x, y \rangle \in R$.

(iii) Let $\ni_Y \subseteq \mathcal{P}Y \times Y$ be the extension of the converse of the membership relation $\in$ as restricted to the sets in $\mathcal{P}Y$. Then we've just seen that a pair $\langle x, y \rangle$ is in R if and only if $\langle f_r x, y \rangle$ is in $\ni_Y$ – in other words, $\langle x, y \rangle$ is in R if and only if it is sent by the map $f_r \times 1_Y$ into $\ni_Y$. Which makes R the inverse image of $\ni_Y$ under $f_r \times 1_Y$.

(c) OK: so now we want to 'categorify' that last idea, which tells us how the powerset $\mathcal{P}Y$ and (the converse of) the membership relation restricted to $\mathcal{P}Y$ together interact with morphisms. Instead of talking about an inclusion function, we talk of a monic arrow r. Likewise, instead of talking of $\ni_Y$ as included as a subset in $\mathcal{P}Y \times Y$, let's say that there is a monic arrow $\varepsilon\colon \ni_Y \rightarrowtail \mathcal{P}Y \times Y$. And instead of talking of R as an inverse image of $\ni_Y$ under $f_r \times 1_Y$, we will specify (R, r) as a pullback of ε along $f_r \times 1_Y$ (compare §20.2(b)). We also want to generalize to categories other than the category of sets. We then land on the following perhaps unexpected but standard definition:

Definition 98. A *power object* for an object Y is an object $\mathcal{P}Y$ together with some monic arrow $\varepsilon\colon \ni_Y \rightarrowtail \mathcal{P}Y \times Y$ such that for any object X and any monic arrow $r\colon R \rightarrowtail X \times Y$ there is a unique arrow $f_r\colon X \to \mathcal{P}Y$ (depending on r) making a pullback square

$$\begin{array}{ccc} R & \longrightarrow & \ni_Y \\ {\scriptstyle r}\big\downarrow & \lrcorner & \big\downarrow{\scriptstyle \varepsilon} \\ X \times Y & \xrightarrow{\ f_r \times 1_Y\ } & \mathcal{P}Y \times Y \end{array}$$

△

Two comments. First, in the general case, the continued use of the notation '$\ni_Y$' or something similar for the source of the monic ε is mere convention, and needn't any longer have anything directly to do with membership.

Second, consider the case where $X = 1$. If we take any subobject of Y, (R, r), we can associate the arrow $r\colon R \rightarrowtail Y$ with a unique arrow $r'\colon R \rightarrowtail 1 \times Y$ (via the isomorphism between R and $1 \times R$). Then this derived arrow r' will have a unique corresponding arrow $f_{r'}\colon 1 \to \mathcal{P}Y$. So we have nicely associated subobjects of Y with point elements of $\mathcal{P}Y$ (compare the way that subsets of a set Y are elements of the powerset $\mathcal{P}Y$).

(d) For the record, note that our definition determines the object data of power objects up to isomorphism:

Theorem 118. *If a category has two power objects for the object Y, namely $(\mathcal{P}Y, \varepsilon)$, $(\mathcal{P}Y', \varepsilon')$, then $\mathcal{P}Y \cong \mathcal{P}Y'$.*

Proof. Use the same idea as in the proof of Theorem 114! I can leave details as an exercise. □

(e) And with our shiny new definition to hand, we can now state the main result that I trailed at the beginning of the chapter:

Theorem 119. *Any properly Cartesian closed category with a subobject classifier has a power object (Ω^Y, ε) for each object Y.*

Being properly Cartesian closed gives us the products, pullbacks and exponentials we need for the official proof in the next section.

There is also a theorem in the opposite direction which is worth at least noting (though I won't fully prove it):

Theorem 120. *If a category with finite limits has a power object for every object, then it also has an exponential for every object and a subobject classifier.*

24.2 Proving that Ω^Y is a power object for Y

(a) We want to show that Ω^Y suitably equipped with a monic ε always provides a power object for Y (if we have products, pullbacks, exponentials and a subobject classifier all available):

Proof. Start by considering any subobject of $X \times Y$, i.e. an object R and monic $r\colon R \rightarrowtail X \times Y$.

With a subobject classifier in play, there must be a unique $\chi_r\colon X \times Y \to \Omega$ making the following commute:

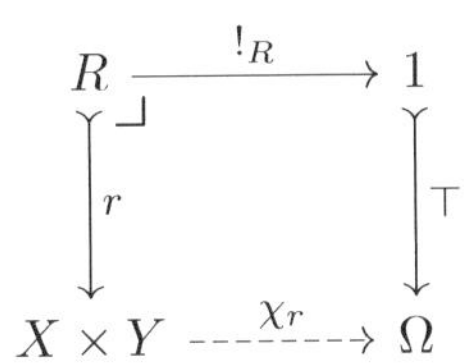

Now assume we also have exponentials available. Then given the definition of Ω^Y it follows that there will be a unique transpose $\widetilde{\chi_r}\colon X \to \Omega^Y$ which makes the following commute:

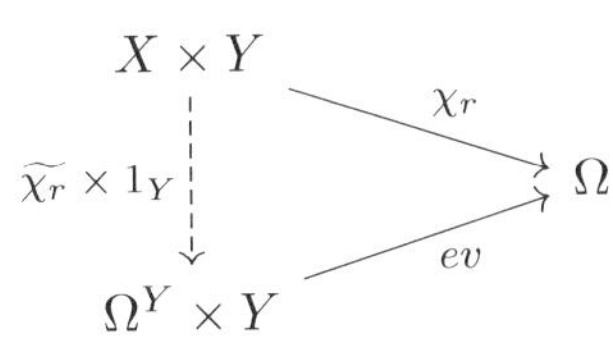

Again, given that a pullback is available for any corner, we can take the corner $\Omega^Y \times Y \xrightarrow{ev} \Omega \xleftarrow{\top} 1$ and form a pullback square, borrowing suggestive notation for the resulting pullback of $\top$:

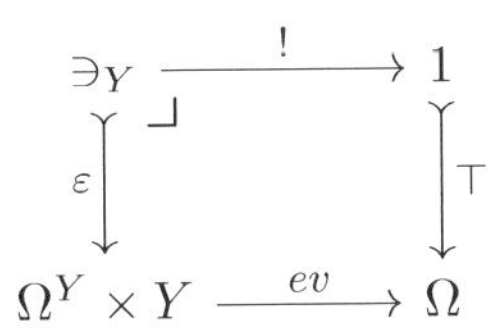

Note, since this *is* a pullback, and the arrow $\top$ is monic, the arrow from $\ni_Y$ down to $\Omega^Y \times Y$ is also monic (by Theorem 91).

We can now combine our three diagrams into a single commuting diagram, with $\widetilde{\chi_r}$ uniquely fixed:

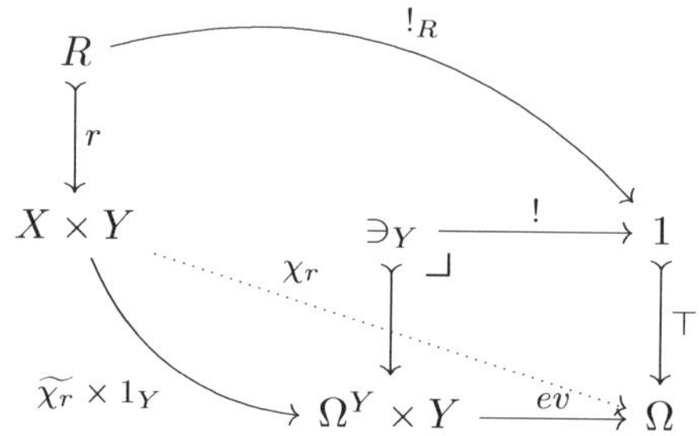

Ignoring the dotted diagonal, we have a wedge $\Omega^Y \times Y \xleftarrow{\widetilde{\chi_r} \times 1_Y \circ r} R \xrightarrow{!_R} 1$ forming a commuting square with the opposite corner with vertex Ω. But then, since the inner square with vertex $\ni_Y$ is a pullback, there must by definition be a unique arrow $u \colon R \to \ni_Y$ making the diagram commute. Therefore, rearranging, we have the following commuting diagram:

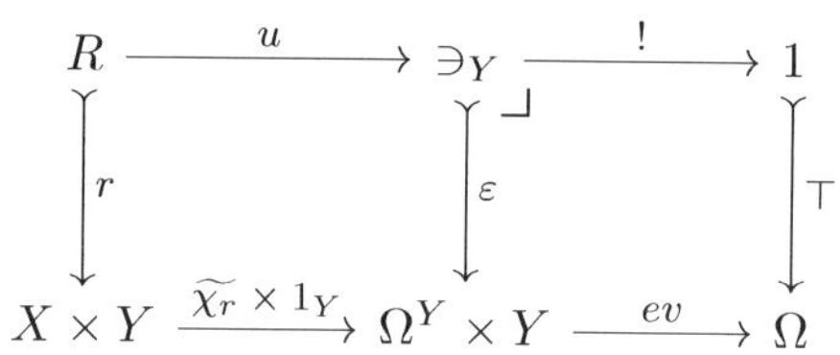

Since $!_{\ni_Y} \circ u = !_R$ and $\chi_r = ev \circ (\widetilde{\chi_r} \times 1_Y)$, the outer rectangle is the original pullback square. The right-hand square is a pullback. Hence by Theorem 93 we can conclude that the left-hand square is a pullback square

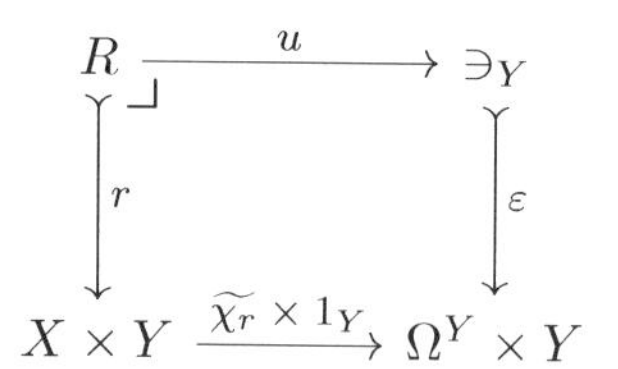

with $\widetilde{\chi_r}$ uniquely fixed to make it so.

Hence (Ω^Y, ε) is our desired power object! □

(b) In the previous section, I also stated another theorem, claiming that if a category with finite limits has a power object for every object, then it also has an exponential for every object and a subobject classifier. It is perhaps worth giving a proof-sketch for the easier part:

Proof sketch: limits plus power objects give a subobject classifier. We have seen that Ω^Y can be traded for $\mathcal{P}Y$; so $\mathcal{P}1$ should give us our desired Ω. So let's follow up that hint.

If a category with a terminal object 1 and products has a power object for every object, then for any $r \colon R \rightarrowtail X$ there will be a unique f making the left-hand diagram a pullback:

$$\begin{array}{ccc} R & \longrightarrow & \ni_1 \\ {\scriptstyle \langle\langle r, !\rangle\rangle}\downarrow & \lrcorner & \downarrow{\scriptstyle \varepsilon} \\ X \times 1 & \xrightarrow{f \times 1_1} & \mathcal{P}1 \times 1 \end{array} \qquad \begin{array}{ccc} R & \longrightarrow & \ni_1 \\ {\scriptstyle r}\downarrow & \lrcorner & \downarrow{\scriptstyle \varepsilon'} \\ X & \xrightarrow{f} & \mathcal{P}1 \end{array}$$

Of course, that first diagram isn't quite what we want. But we know $X \times 1 \cong X$ and $\mathcal{P}1 \times 1 \cong \mathcal{P}1$. Hence we should be able to massage away the idle complications introduced by forming products with 1. In this way, then, we will be able to show that f also uniquely makes a diagram of the second kind a pullback, and (in the light of Theorem 115) we will be done! □

25 An axiom of infinity: NNOs

Categories can have limits, colimits, exponentials and subobject classifiers too, yet still be decidedly tame affairs. FinOrd is an example: yes, it has infinitely many objects, but each of them is a finite entity (in effect, a natural number). So let's ask: what condition can we put on a category which will require it to contain some intuitively infinitary object? Or as we might put it: *what can serve as a categorial axiom of infinity*?

A neat answer was proposed by F. W. Lawvere (1964). The idea is to require our category to include a so-called 'natural numbers object', which is an object N equipped with two related arrows $z\colon 1 \to N$ and $s\colon N \to N$ satisfying certain conditions. As we'll see, N behaves in a crucial way like the – infinite! – collection of all natural numbers.

To avoid distracting details, I'll state most of the theorems in this chapter as applying in a 'nice enough category'. You can take that to mean a non-degenerate, properly Cartesian closed, category with a subobject classifier. However, that's strictly speaking more than is required for some results; it might be a useful exercise, then, to work out how much needs to be assumed for each theorem.

25.1 Natural numbers objects defined

(a) We'll start with the familiar notion of a sequence. A sequence has a first member. And each member is followed by a unique successor.

In general, a sequence of successors can circle round and repeat itself. But of course, the natural numbers form a special kind of sequence, an ω*-sequence*. No distinct two numbers have the same successor – so the sequence of successors, starting from zero, plods on for ever without repetition. And there are no stray numbers, living outside that sequence of successors of zero. This suggests that, in a categorial spirit, we might first define a family of sequences and then hope to locate the natural numbers (up to isomorphism, in the usual way) as forming a limiting case.

Now, to handle sequences in arrow-speak, we need an arrow-as-point-element which we can think of as picking out the first member of the sequence, and then we need an arrow-as-operation which takes a member to its 'successor'. Let's officially say, then:

Definition 99. If C is a category with a terminal object, then a C-object X equipped with C-arrows $i\colon 1 \to X$ and $f\colon X \to X$ is a *sequence object* in C. △

Suppose we are working in a category like Set where arrows are functions. An arrow $i\colon 1 \to X$ picks out an initial element from X (element in the ordinary sense); for convenience, overload notation and call this element i too. Then $f\colon X \to X$ generates a sequence of elements i, $f(i)$, $f^2(i)$, $f^3(i)$, ...

An f-generated sequence of objects in X could eventually start repeating. And our definition also allows X to contain stray elements not in that sequence. Our task therefore is to characterize (at least for categories like Set) the limiting case of a sequence object (X, i, f) where X includes just a non-repeating sequence of objects $f^n(i)$, without any stray extras. That should give us an ω-sequence which looks like the natural numbers.

To this end, we need another definition:

Definition 100. If C is a category with a terminal object, then the derived category C_{Seq} has the following data: an object is any of C's sequence objects (X, i, f), and an arrow $u\colon (X, i, f) \to (Y, j, g)$ is any C-arrow $u\colon X \to Y$ which makes this diagram commute in C:

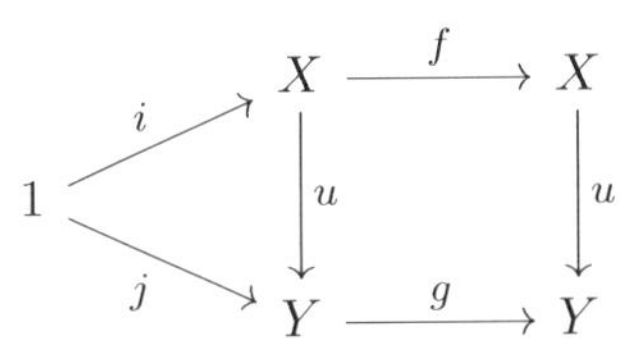

C_{Seq}'s identity arrow on (X, i, f) is C's identity arrow on X, and arrows compose in C_{Seq} in the same way that they compose in C. △

The idea is that, in a category like Set, a function u that maps the elements i, $f(i)$, $f^2(i)$, $f^3(i)$, ... to the elements j, $g(j)$, $g^2(j)$, $g^3(j)$ ... should match members of the f-sequence with same-placed members of the g-sequence. The commuting triangle ensures that u matches up the first members. Then the commuting square ensures that $f^n(i)$ gets sent by u to $g^n(j)$ for each n in turn.

An important corollary: it follows that if $g^m(j) \neq g^n(j)$, then $f^m(i) \neq f^n(i)$. Hence (C) the sequence produced by (X, i, f) can't be *more* constrained by equations of the form $f^m(i) = f^n(i)$ than (Y, j, g) is constrained by similar equations between *its* elements.

(b) We can now give the following definition:

Definition 101. If C is a category with a terminal object, then a *natural numbers object* (NNO) in C is an initial object of the derived category C_{Seq}.

Equivalently, a natural numbers object (N, z, s) in C is an object N equipped with two arrows $z\colon 1 \to N$ and $s\colon N \to N$ (think 'zero' and 'successor') such that for any sequence object (Y, j, g) there is a *unique* arrow u which makes this diagram commute:

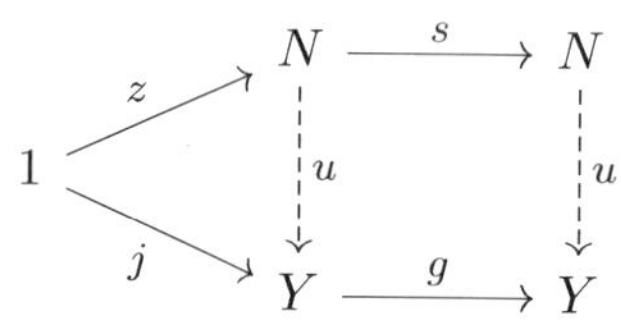

△

But why is this well-motivated? Let's think how the definition works in Set.

(1) Set is extremely rich in sequences. So if (N, z, s) is to be initial in C_{Seq}, corollary (C) above tells us that the elements of N will have to form as unconstrained a sequence $z, s(z), s^2(z), s^3(z), \ldots$ as possible, governed by no additional equations of the form $s^m(z) = s^n(z)$ (where $m \neq n$), and so never repeating.

(2) Still in Set, if N were to have a stray member in addition to $z, s(z), s^2(z), s^3(z), \ldots$, then this could be sent by different functions u to Y in more than one way while still giving a commuting diagram. Hence u wouldn't be unique, and hence (N, z, s) wouldn't be initial.

In short, in Set, the set elements $z, s(z), s^2(z), s^3(z), \ldots$ form an ω-sequence.

Reality check: Consider the standard implementation of the natural numbers in Set

$$\mathbb{N} = \{\emptyset, \{\emptyset\}, \{\emptyset, \{\emptyset\}\}, \{\emptyset, \{\emptyset\}, \{\emptyset, \{\emptyset\}\}\}, \ldots\}$$

together with the arrow $z \colon 1 \to \mathbb{N}$ which sends the object in the singleton to $\emptyset$, and the arrow $s \colon \mathbb{N} \to \mathbb{N}$ which sends a set $n \in \mathbb{N}$ to the set $n \cup \{n\}$. Then show that $(\mathbb{N}, z, s)$ constitute a NNO in Set.

Theorem 121. *If (N, z, s) and (N', z', s') are both candidate natural numbers objects in a category* C*, then $N \cong N'$.*

Proof. Immediate from the fact that NNOs are initial objects of C_{Seq}: there will be a unique isomorphism $N \xrightarrow{\sim} N'$ commuting with the NNOs' arrows in the obvious way. □

(c) I said that a NNO (N, z, s) must provide an ω-sequence at least in a rich enough category like Set. Suppose however that C is a very impoverished category with only trivial sequences. For an extreme case, suppose that for any C-object Y the only arrow $Y \to Y$ is the identity arrow. Then we can put $N = 1$ and $z = s = 1_1$, and this (N, z, s) will satisfy the condition in our definition: yet in *this* case N won't look at all like a collection of natural numbers! What to do?

As we'll see in the next section, if (N, z, s) is to behave as intended, we need at least this extra condition: the successor of zero has to be distinct from zero, i.e. in categorial terms we need $s \circ z \neq z$. We could have built this requirement into our definition as a further condition on what counts as a true NNO. Instead we'll say

Definition 102. A NNO (N, z, s) is *non-trivial* iff $s \circ z \neq z$.

But now we have an easy theorem.

Theorem 122. *If (N, z, s) is a NNO in a non-degenerate category with a subobject classifier, then it is non-trivial.*

Proof. By hypothesis, there will be arrows $\bot \colon 1 \to \Omega$ and $\neg \colon \Omega \to \Omega$ as defined in §23.3. Hence there will be a sequence $(\Omega, \bot, \neg)$. And therefore, by the definition of a NNO, there must be a unique u making the following commute:

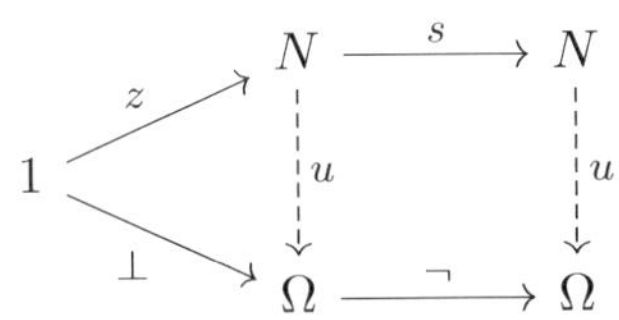

Following the outer paths, we see $u \circ s \circ z = \neg \circ \bot = \top$.

So the supposition $s \circ z = z$ implies $u \circ z = \top$. However, the left triangle tells us that $u \circ z = \bot$. Hence that supposition implies $\top = \bot$. But by Theorem 111 that contradicts our assumption about non-degeneracy. □

25.2 Proving that a NNO has an infinite object

Here are two definitions inspired by familiar set-theoretic analogues:

Definition 103. (1) An object N is *point-infinite* if some (or all) of its point elements can be put in one-one correspondence with the natural numbers.

(2) N is *Dedekind-infinite* if there is a map $f \colon N \to N$ which is point-injective but not point-surjective. △

And we are now going to prove

Theorem 123. *In a nice enough category with a NNO (N, z, s), N is both (1) point-infinite and (2) Dedekind-infinite.*

A neat route to establishing this starts from the following lemma:

Theorem 124. *A nice enough category with a NNO (N, z, s) has an arrow $p \colon N \to N$ such that (i) $p \circ z = z$ and (ii) $p \circ s = 1_N$.*

Intuitively, if s is thought of as a successor function, p is the corresponding predecessor function (counting zero as its own predecessor). And this is elementary:

Proof: Theorem 124 implies Theorem 123. We'll start with part (2) and show that $s \colon N \to N$ is (a) point-injective but (b) not point-surjective.

For (a), we just remark that if $\vec{x}, \vec{y}$ are point elements of N, then if $s \circ \vec{x} = s \circ \vec{y}$, then $p \circ s \circ \vec{x} = p \circ s \circ \vec{y}$, hence $\vec{x} = \vec{y}$.

For (b), suppose $s \circ \vec{x} = z$. Then $\vec{x} = p \circ s \circ \vec{x} = p \circ z = z$. So our supposition is equivalent to $s \circ z = z$, which is ruled out by non-degeneracy. So s is not point-surjective.

Now for part (1). Let '$\vec{n}$', for $n \geq 0$, denote the arrow $s^n z\colon 1 \to N$. Assume that $|j - k| = n + 1$ with $n \geq 0$ and suppose $\vec{j} = \vec{k}$, i.e. $s^j z = s^k z$. Apply p to each side to reduce both j and k by 1. Keep on going. And we'll arrive at $s \circ \vec{n} = z$. But we've ruled that out in arguing for (b). Hence if $j \neq k$, $\vec{j} \neq \vec{k}$, showing that N is point-infinite. □

So it remains to establish the predecessor lemma:

Proof of Theorem 124. Consider the following diagram, involving a product $N \times N$ and its two projection arrows $\pi_j\colon N \times N \to N$. By definition of the product, there is a unique mediating arrow $v = \langle\!\langle s \circ \pi_1, \pi_1 \rangle\!\rangle$ making this diagram commute (see §11.5):

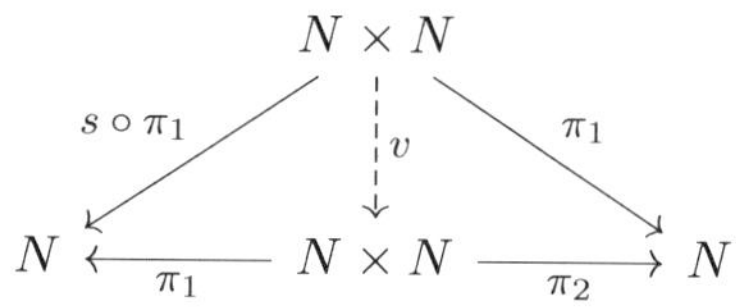

What's interesting about this? Suppose we are in Set, for example, and the members of $N \times N$ are good old-fashioned ordered pairs. Then v sends an initial pair $\langle 0, 0 \rangle$ successively to $\langle 1, 0 \rangle$, $\langle 2, 1 \rangle$, $\langle 3, 2 \rangle$, ... In other words, v will have exactly the extension of the predecessor function. So we should be able to work with this idea.

OK, take the sequence formed from $N \times N$, the arrow $\langle\!\langle z, z \rangle\!\rangle\colon 1 \to N \times N$ (which intuitively sends the initial object to a pair of zeros), and the arrow $v\colon N \times N \to N \times N$ as just defined. Then by the definition of our NNO there is a unique u such that the following commutes:

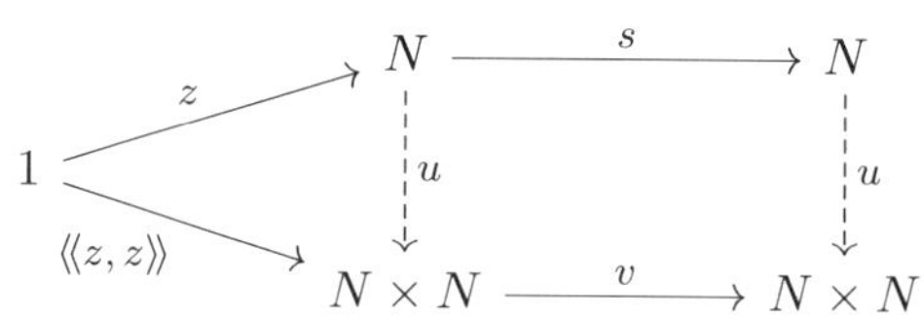

By our previous remarks, in Set, u will send a number n to a pair whose second component is n's predecessor. Hence, back to arrow talk, the predecessor arrow $p\colon N \to N$ we want should be given by putting $p = \pi_2 \circ u$.

We just need to check! So first, note that $p \circ z = \pi_2 \circ u \circ z = \pi_2 \circ \langle\!\langle z, z \rangle\!\rangle = z$. Second, we need to show $p \circ s = 1_N$. Well, note that from the original product diagram $s \circ \pi_1 = \pi_1 \circ v$; hence $s \circ \pi_1 \circ u = \pi_1 \circ v \circ u = \pi_1 \circ u \circ s$. But that means that the following commutes:

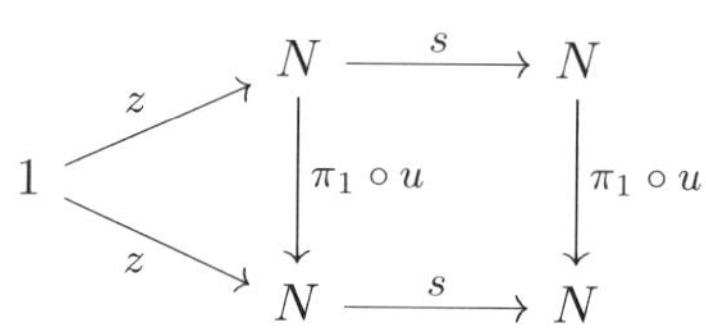

But if we replace the down arrow $\pi_1 \circ u$ by 1_N the diagram also trivially commutes: and by the definition of the NNO, the down arrow in such a diagram is unique. Therefore $\pi_1 \circ u = 1_N$.

We now have (by the definition $p = \pi_2 \circ u$, then by appeal to the last-but-one diagram, then by the preceding product diagram)

$$p \circ s = \pi_2 \circ u \circ s = \pi_2 \circ v \circ u = \pi_1 \circ u = 1_N.$$

So we are done! □

25.3 The Dedekind-Peano postulates

(a) Let's recall the Dedekind-Peano postulates for the natural numbers, familiar from elementary mathematics.

These tell us that the natural numbers include a distinguished zero object 0 and come equipped with a successor function s, and are such that:

(1) 0 is a number;

(2) If n is a number, so is its successor sn;

(3) 0 is not a successor of any number;

(4) Two numbers n, m with the same successor are equal;

(5) For any property P of natural numbers, if 0 has P, and if sn has P whenever n does, then P holds for all natural numbers.

Now, here – in the induction principle (5) – we should understand 'property' in a very generous sense, according to which *any* arbitrary selection of numbers corresponds to a property (at least the property of being one of the selection!). That's why, using the familiar set idiom, we can take (5) as equivalent to

(5′) For any subset A of the set of natural numbers $\mathbb{N}$, if $0 \in A$, and if $n \in A \Rightarrow sn \in A$, then $A = \mathbb{N}$.

(b) Now, thinking in categorial terms, we can show that a version of these postulates applies to any nice enough category with a NNO. Indeed, we've already seen that categorial versions of axioms (1) to (3). A version of (4) also holds because $s \circ n = s \circ m$ implies $p \circ s \circ n = p \circ s \circ m$ implies $n = m$.

It remains to show that we also can derive a categorial version of the induction axiom. First, then, we need to ask: what *is* the categorial version of (5′)?

(i) Taking some set A of natural numbers becomes taking a subobject of N: recycling 'A', denote this subobject $(A, a\colon A \rightarrowtail N)$.

(ii) The assumption that 0 is an element of the set A becomes the idea that there is some 'element' of the categorial counterpart A which gets sent by a to the zero $z\colon 1 \to N$: i.e. there is some arrow $z'\colon 1 \to A$ such that $a \circ z' = z$.

(iii) The assumption that $n \in A \Rightarrow sn \in A$ becomes the idea that there is some map on the category's object A which marches in step with the operation of successor applied to N: i.e. there is some $s' \colon A \to A$ such that $s \circ a = a \circ s'$.

(iv) The conclusion that $A = \mathbb{N}$ becomes: $A \cong N$.

And with that 'categorification', the claim that induction holds for natural numbers becomes this:

Theorem 125. *Assume we are in a nice enough category with a NNO (N, z, s). Then for any subobject (A, a) of N, and arrows $z' \colon 1 \to A$, $s' \colon S \to S$ such that this diagram commutes*

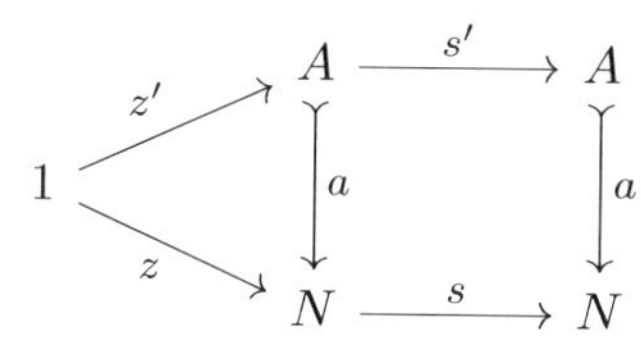

it follows that $A \cong N$.

Proof. Consider the left-hand diagram:

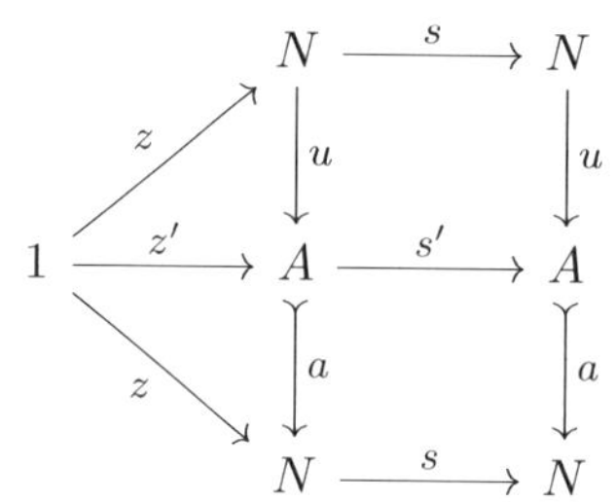

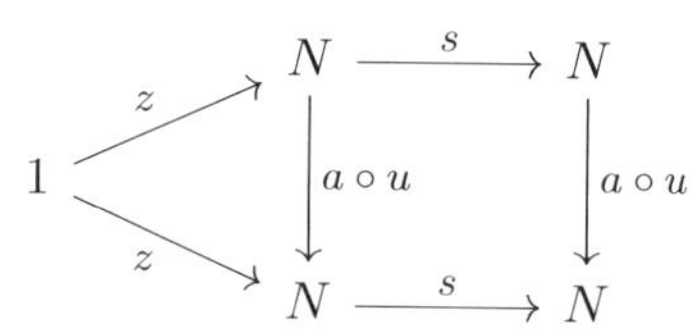

The bottom part commutes by assumption: that there is some arrow u which makes the top part commute follows from the assumption that (N, z, s) is a NNO. Hence the arrow $a \circ u$ makes the right-hand diagram commute. But 1_N would also make that diagram commute. And by the uniqueness of the completing down arrows in NNO diagrams, $a \circ u = 1_N$.

Hence a is a left-inverse and so epic. But it is monic by assumption. And in a category with a subobject classifier, epic monics are isomorphisms (by Theorem 116). Hence $A \cong N$. □

25.4 Recursion

(a) As another step towards showing that we can get a substantial amount of arithmetic working in any nice enough category with a NNO, let's now note that we can implement definitions of functions by recursion.

Suppose we fix an element $j \colon 1 \to N$, and are given an arrow $g \colon N \to N$. Then the claim that there is a *unique* arrow f making this commute

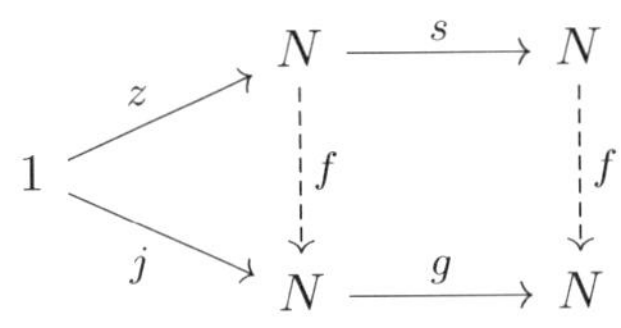

is a categorial version of the ordinary mathematical claim that – recycling the notation – a pair of equations $f(0) = j$ and $f(sn) = gf(n)$ together define a unique function f by primitive recursion.

In this way, then, we should be able to define (unary) primitive recursive functions inside our category.

(b) So far so good. But there's still some work to be done. For consider the pattern of recursive definition for a *two*-place function $f\colon N, N \to N$ in terms of a couple of given one-place functions $g, h\colon N \to N$:

(R1) $f(m, 0) = g(m)$
(R2) $f(m, sn) = h(f(m, n))$.

For example, if $g(m) = m$ and h is the successor function s again, then our equations give us a recursive definition of addition.

Call this a definition by parameterized recursion, since there is a parameter m which we hold fixed as we run the recursion on n. And our informal equations do well-define a determinate binary function f, given any determinate monadic functions g and h.

Now, to have a version of this kind of definition by parameterized recursion in a categorial framework, we will evidently need to replace the two-place function with an arrow f from a product. But suppose our category satisfies

(P) Given any arrows $g\colon N \to N$ and $h\colon N \to N$, there is a unique arrow $f\colon N \times N \to N$ in E which makes this diagram commute:

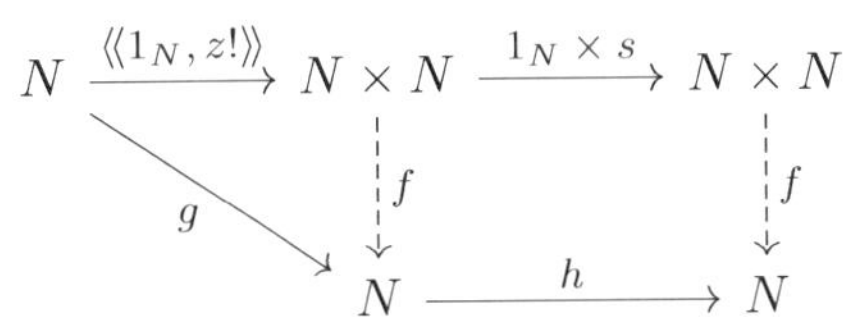

where $z!$ is the composite map $N \xrightarrow{\;!\;} 1 \xrightarrow{\;z\;} N$.

Requiring that the triangle on the left commutes is the categorial equivalent of saying that (R1) holds (since the arrow $\langle\!\langle 1_N, z! \rangle\!\rangle$ sends m to the pair $m, 0$). And requiring the square to commute is the equivalent of saying that (R2) holds. Hence in a category satisfying condition (P), then in effect parameterized recursion well-defines functions.

(c) Let's pick up the familiar example of the recursive definition of addition. Assuming (P), there will be a unique arrow – call it *add* – such that this diagram commutes:

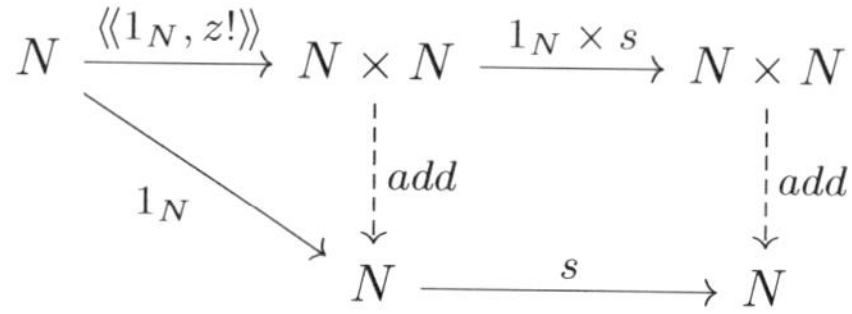

Using '$\vec{n}$' as before to denote the arrow $s^n z\colon 1 \to N$, it is then easy to show (as we want) that $add \circ \langle\langle \vec{m}, \vec{n} \rangle\rangle = \overrightarrow{m+n}$. But do check that!

(d) So when does (P) hold? Well, here's a more general result:

Theorem 126. *In a nice enough category with a natural number object (N, z, s), given any objects A, C, and arrows $g\colon A \to C, h\colon C \to C$, then there is a unique u which makes the following diagram commute:*

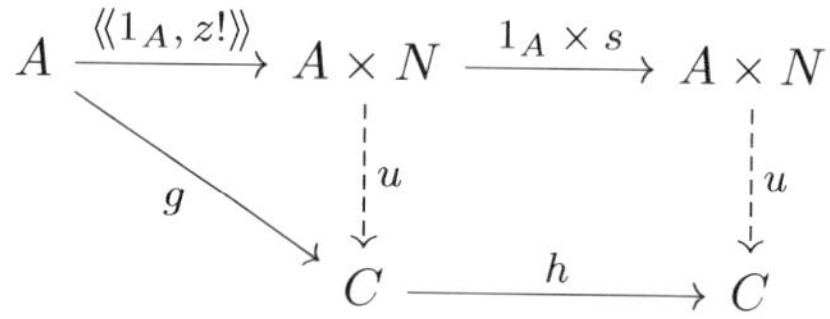

where $z!$ is now the composite map $A \xrightarrow{\;!\;} 1 \xrightarrow{\;z\;} N$.

(P) of course is the special case where $A = C = N$.

Proof sketch. The basic idea is to note that in a category with exponentials, arrows $N \times A \to C$ can be 'curried', i.e. in effect replaced by arrows $N \to C^A$. So the trick is to get ourselves from a diagram of the shape given in the statement of the theorem to a diagram of the following shape, where g' and h' depend on g and h:

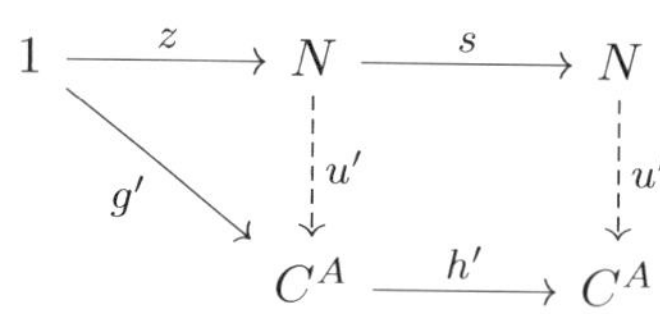

Then we use the assumption that our category has a NNO to prove that this derived diagram has a unique $u'\colon N \to C^A$ making it commute, and then we can work backwards to construct an arrow $u\colon A \times N \to C$ in terms of u' which makes our original diagram commute.

So how can we define g' and h'? Well, consider these two commutative diagrams:

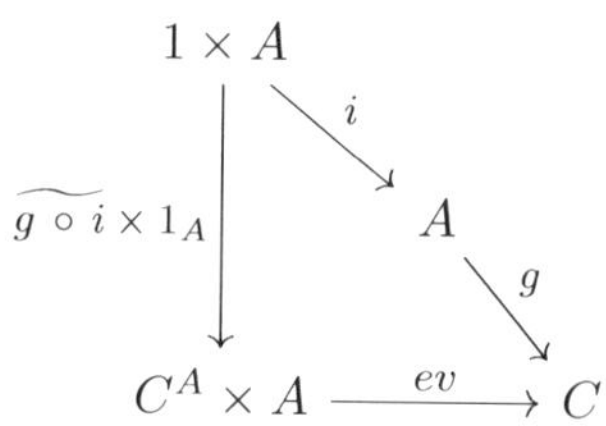

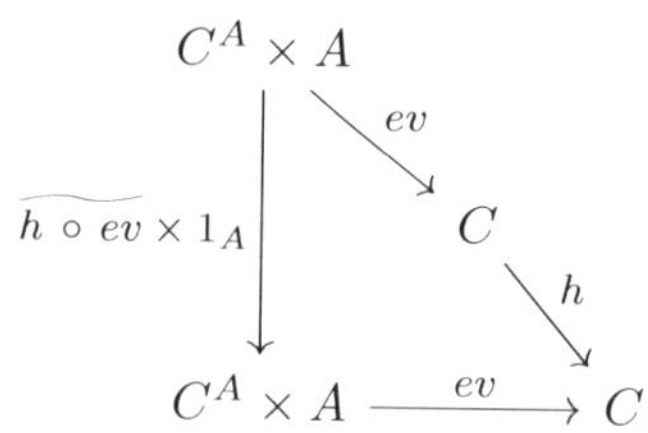

where i is the obvious isomorphism, and wavy overlining indicates an exponential transpose. Then, we get arrows with the right sources and targets if we put

$$g' = \widetilde{g \circ i} \colon 1 \to C^A, \quad h' = \widetilde{h \circ ev} \colon C^A \to C^A.$$

And how does the argument go from here? I'll meanly leave it as a rather challenging exercise to fill in the details! □

25.5 And integers too?

There's a familiar set-theoretic construction which we can use to model the integers, once we have the natural numbers and the addition function to hand. We take pairs of numbers $\langle m, n \rangle \in \mathbb{N} \times \mathbb{N}$, and define an equivalence relation on pairs by putting $\langle m_1, n_1 \rangle \sim_I \langle m_2, n_1 \rangle$ iff $m_1 + n_2 = m_2 + n_2$ (so the pairs agree on the difference between their members). Then we quotient the set of pairs $\mathbb{N} \times \mathbb{N}$ by this equivalence relation $\sim_I$ to get a set of integers.

Let's now quickly note that we can run a parallel construction in a nice enough category with a natural numbers object and co-equalizers.

Start by taking $(N \times N) \times (N \times N)$, which we can think of as collecting pairs of pairs of numbers. And define the two arrows $f, g \colon (N \times N) \times (N \times N) \to N$, by putting $f = add \circ (\pi_1 \times \pi_2)$ and $g = add \circ (\pi_2 \times \pi_1)$.[1]

Let (E, e) be an equalizer for f and g, where $e \colon E \to (N \times N) \times (N \times N)$. And now consider the pair of arrows $p, q \colon E \to N \times N$, where $p = \pi_1 \circ e$ and $q = \pi_2 \circ e$.[2]

Now take the co-equalizer (I, i) of p and q. By the discussion of §16.6, this gives us a scheme for quotienting $N \times N$ by the equivalence projection of p and q. But, at least when we can think of N as a set of natural numbers, the equivalence projection of p and q is the equivalence relation $\sim_I$ (why?). So I will serve as a corresponding set of integers.

[1]So, intuitively, given the pair of numbers $\langle m_1, n_1 \rangle$, $\langle m_2, n_2 \rangle$, f returns $m_1 + n_2$ and g returns $m_2 + n_1$. More precisely, by Theorem 51, $f \circ \langle\!\langle \langle\!\langle \overrightarrow{m_1}, \overrightarrow{n_1} \rangle\!\rangle, \langle\!\langle \overrightarrow{m_2}, \overrightarrow{n_2} \rangle\!\rangle \rangle\!\rangle = \overrightarrow{m_1 + n_2}$, while applying g instead of f returns $\overrightarrow{m_2 + n_1}$.

[2]So, intuitively, given an element of E, p and q will return pairs $\langle m, n \rangle$, $\langle m', n' \rangle$, where $m + n' = m' + n$.

Interlude

We have seen in Part I of these notes that if a category is (1) finitely complete – has all finite limits – then we can e.g. construct products of any objects in the category and e.g. construct inverse images too. If our category is also (2) finitely cocomplete – has all finite colimits – then we can e.g. form all quotients. Add (3) all exponentials and our category will, for any objects A and B, also supply a corresponding object B^A which behaves like a function space collecting the arrows from A to B. Most recently, we have seen that if a category (4) has a subobject classifier, it will have arrows that work like analogues of characteristic functions: and this combined with (1), (2) and (3) enables a range of further constructions – for example, we get power objects. As we'll see later, we also get a nicely structured algebra of subobjects.

There is a standard term for a category with (1) to (4): it is an *elementary topos*. And any topos which also has a natural numbers object should provide a generous arena in which we can now implement a lot of 'ordinary' mathematical constructions – perhaps in ways which interestingly differ from the usual set-theoretic construction.

I'll say a little more about such toposes, then, in Part III, though necessarily relatively briefly. However, to explore this fascinating area further than I can do – or to pursue other aspects of category theory – you first need to significantly upgrade your general categorial tool kit. Part II of these notes makes a start on developing some of the key ideas.

Thus far, we have spent a good deal of time looking *inside* categories, exploring how various abstract constructions recur in different categories. Indeed, a main point of packaging various familiar mathematical objects into categories is – as Emily Riehl (2017, p. xi) nicely puts it –

> to shift one's perspective from the particularities of each mathematical sub-discipline to potential commonalities between them.

But now we need to develop more apparatus for talking about relations *between* categories, tracking those commonalities more explicitly. In particular, we'll want to think about maps – functors, as we say – that can take a construction in one category to the same sort of construction in another category.

Some will in fact insist that it is only when we introduce functors that we start making headway with category theory proper (the chapters up to now being an extended prologue to the real deal). Peter T. Johnstone, in a note introducing his famed Cambridge course, writes:

> Category theory begins with the observation ... that the collection of all mathematical structures of a given type, together with all the maps between them, is itself an instance of a nontrivial structure which can be studied in its own right. In keeping with this idea, the real objects of study are not so much categories themselves as the maps between them – functors, natural transformations and (perhaps most important of all) adjunctions.

Johnstone starts discussing functors at the beginning of his second lecture. Many familiar textbooks are equally quick off the mark.[3]

Have we really not yet encountered the "real objects of study" for category theory? Be that as it may! What we *have* encountered has, I hope, already been more than interesting and illuminating enough. And I would say that – to take just one example – first meeting limits and colimits in the way in which we've so far introduced them is enormously helpful in appreciating the *point* of the fancier versions that are redefined in terms of functors. More generally, a good feel for what can happen at the base level of the hierarchy of abstraction we sketched right back in Chapter 1 makes it much easier to understand what is going on at the upper levels. As I also agreed at the very outset, however, there is no one best route for exploring through the thicket of interconnected concepts and constructions of category theory. I claim only that the route we are following has its own virtues.

But yes, it is time to ascend a level or two ...

[3]For example, Awodey (2010) defines functors on his p. 8 and Leinster (2014) on his p. 17. Riehl (2017) defines them on her p. 13 (having already talked a lot about functors in her Preface). The principal author who does things my way, exploring a lot of categorial ideas (and even, in his case, say a lot about toposes), before introducing functors as late as his p. 194, is Goldblatt (1984). Or perhaps I should rather say: I do things Goldblatt's way – for he has been a major influence. A more recent fellow-traveller who introduces the ideas of our Parts I and II in pretty much my preferred order is Yanofsky (2024).

26 Functors introduced

The quote from Peter T. Johnstone in the preceding Interlude tells us what we need to be exploring next: functors, natural transformations and adjunctions. We begin in this chapter.

26.1 Functors defined

(a) A category has two kinds of data, its objects and its arrows. A functor F mapping the category C to the category D will therefore need to have two components which we can denote F_{ob} and F_{arw}, where

> F_{ob} sends each C-object to a corresponding D-object;
>
> F_{arw} sends each C-arrow to a corresponding D-arrow.

And a first condition on these components is that their actions need to cohere. I mean that if F_{arw} sends an arrow f living in C to an arrow $F_{arw}(f)$ living in D, then F_{ob} needs to send the source and target of f to the source and target of $F_{arw}(f)$, so we have:

$$A \xrightarrow{\ f\ } B \qquad \overset{F}{\Longrightarrow} \qquad F_{ob}(A) \xrightarrow{\ F_{arw}(f)\ } F_{ob}(B)$$

Moreover, if a functor F is to respect at least C's most basic categorial structure, its component mappings must obey two further simple conditions. First,

> F must map identity arrows to identity arrows. So F_{arw} sends 1_A, the identity arrow on A, to the identity arrow on $F_{ob}(A)$, namely $1_{F_{ob}(A)}$.

Further, we want F to respect the composition of arrows so that it sends commuting diagrams to commuting diagrams. For example, consider:

$$\begin{array}{ccc} & B & \\ {}^{f}\nearrow & & \searrow^{g} \\ A & \xrightarrow[g \circ f]{} & C \end{array} \qquad \overset{F}{\Longrightarrow} \qquad \begin{array}{ccc} & F(B) & \\ {}^{F(f)}\nearrow & & \searrow^{F(g)} \\ F(A) & \xrightarrow[F(g \circ f)]{} & F(C) \end{array}$$

Here we've decluttered by dropping the subscripts making it explicit which component of F is being applied to objects and which to arrows. Then if the diagram on the left commutes, we want its F-image on the right to commute too. In other words, we require:

For any C-arrows f, g that compose, $F(g \circ_{\mathsf{C}} f) = F(g) \circ_{\mathsf{D}} F(f)$.

(b) Let's put that in a summary definition – and not only will we now drop explicit subscripts '*ob*' and '*arw*' (context always makes it clear which component of a functor is in play) but we will often drop brackets too. So we will typically write $F_{ob}(A)$ more simply as FA. And we will write $F_{arw}(f)\colon F_{ob}(A) \to F_{ob}(B)$ more simply as $Ff\colon FA \to FB$. Also we can let context fix which composition operation is in play. Then, put briskly, we have:

Definition 104. A *functor* $F\colon \mathsf{C} \to \mathsf{D}$ between categories C and D comprises a map from C-objects to D-objects and a map from C-arrows to D-arrows such that

Components cohere: if $f\colon A \to B$ is a C-arrow, then F sends it to $Ff\colon FA \to FB$ in D;

Preserving identity arrows: for any C-object A, $F1_A = 1_{FA}$;

Respecting composition: for any C-arrows f, g such that their composition $g \circ f$ exists, $F(g \circ f) = Fg \circ Ff$. △

These conditions on F are often called *functoriality*. It will emerge in §26.6 why they are said, more specifically, to be the conditions for a *covariant* functor. And by the way, a functor $F\colon \mathsf{C} \to \mathsf{C}$ from a given category to itself is said to be an *endofunctor*.[1]

(c) In presenting a functor $F\colon \mathsf{C} \to \mathsf{D}$, we could display its operation as below, following Riehl (2017, p. 19):

$$\begin{array}{ccc} \mathsf{C} & \xrightarrow{F} & \mathsf{D} \\ X & \longmapsto & FX \\ \downarrow{\scriptstyle f} & \longmapsto & \downarrow{\scriptstyle Ff} \\ Y & \longmapsto & FY \end{array}$$

But on balance I prefer a more compact linear display of the following shape:

$$F\colon \quad \begin{array}{rcl} X & \longmapsto & FX \\ f\colon X \to Y & \longmapsto & Ff\colon FX \to FY. \end{array}$$

(since it will usually be entirely clear from the context what the source and target categories of our functor are, we needn't repeat that data).

[1] A notational remark. It is quite a common convention to use the same style of lettering both to denote functors between categories and to denote objects in categories. I'm following this convention but will try to keep things clear by defaulting to the likes of F, G, H for functors, while deploying early-alphabet or late-alphabet letters $A, B, C, \ldots, X, Y, Z$ for objects.

26.2 Some forgetful functors

(a) Continue to assume, as we did at the very outset, that inclusive categories like Mon, Grp, Top and the rest live happily together in some suitably capacious arena of sets understood in not-too-unconventional a way, forming the category Set.

So we can think, for example, of a monoid belonging to Mon as a set of objects suitably equipped with a binary operation and a distinguished element. And then our first example of a functor nicely illustrates a broad class of cases:

(F1) There is a functor F : Mon $\to$ Set with the following data:

(1) F_{ob} which sends the monoid $(M, *, e)$ to its underlying set M.

(2) F_{arw} which sends $f\colon (M, *, e) \to (N, \star, d)$, a structure-respecting homomorphism mapping elements of M to elements of N, to the same map thought of simply as a set-function $f\colon M \to N$.

Or in our more concise notation:

$$F\colon \quad \begin{array}{rcl} (M,*,e) & \longmapsto & M \\ f\colon (M,*,e) \to (N,\star,d) & \longmapsto & f\colon M \to N \end{array}$$

So all F does is 'forget' about the structure carried by the collection of objects in a monoid. It's evidently a functor, a *forgetful functor* for short.

There are equally forgetful functors from other categories to Set, similarly sending sets-with-structure to the bare underlying sets (and often, 'U' for 'underlying' is used to name such a forgetful functor). For example, the functor U: Top $\to$ Set sends topological spaces to their underlying sets of points and sends continuous maps to themselves as set functions, forgetting about the topological structure.

Such amnesiac functors are not in themselves very exciting! It will turn out, however, that they can be the boring members of so-called adjoint pairs of functors, where they are married to very much more interesting companions (that's a topic for later, Chapter 41). So let's continue with the forgetful theme for a moment:

(F2) Recall: if C is a category, and X is a C-object, the slice category C$/X$'s objects are all the (A, f), for some C-object A and C-arrow $f\colon A \to X$; and C$/X$'s arrows between (A, f) and (B, g) are, economically defined, the C-arrows $j\colon A \to B$ such that $g \circ j = f$.

Then there is another kind of forgetful functor, F: C$/X \to$ C, which sends a C$/X$-object (A, f) back to A, and sends a C$/X$-arrow $j\colon (A, f) \to (B, g)$ back to the original arrow $j\colon A \to B$. Or in short

$$F\colon \quad \begin{array}{rcl} (A,f) & \longmapsto & A \\ j\colon (A,f) \to (B,g) & \longmapsto & j\colon A \to B \end{array}$$

For example, take the slice category FinSet$/I_n$ which we met in §7.3, which is the category of finite sets whose members are coloured from a palette of n colours. The forgetful functor $F\colon \mathsf{FinSet}/I_n \to \mathsf{FinSet}$ simply forgets about the colourings of a set S provided by functions $f\colon S \to I_n$.

(F3) There are somewhat less forgetful functors, such as the functor from Ring to Grp that sends a ring to the additive group it contains, ignoring the rest of the ring structure. Or take the functor from Grp to Mon, that remembers about the associative multiplicative structure and units but forgets about inverses.

(F4) There is a functor $F\colon \mathsf{Set} \to \mathsf{Rel}$ which sends sets and triples (domain, graph, codomain) thought of as, respectively, objects and arrows belonging to Set to the same items thought of as objects and arrows in Rel. In short, F forgets that the relevant graphs are functional.

(b) The forgetful functors mentioned so far send distinct objects to distinct objects, and likewise for arrows. But some functors are even more forgetful:

(F5) Suppose we take a category C, preserve all the objects, but replace parallel arrows by a single representative. This way, we get a preorder category D where there is exactly one arrow from A to B in D whenever there is at least one arrow from A to B in C. Then there is evidently a 'thinning' functor $F\colon \mathsf{C} \to \mathsf{D}$ which maps objects to themselves and sends every arrow in C with source A and target B to the unique arrow with that source and target in D.

(F6) We can thin down the objects too. An extreme case: there is a *constant functor* $\Delta_X\colon \mathsf{J} \to \mathsf{C}$ which picks out a C-object X with its identity arrow 1_X, and sends every J-object to the constant object X, and every J-arrow to the constant arrow 1_X. In short:

$$\begin{array}{lrcl} \Delta_X\colon & A & \longmapsto & X \\ & j\colon A \to B & \longmapsto & 1_X\colon X \to X. \end{array}$$

(F7) What happens to Δ_X when the source category J is the one-object category 1? Then for any chosen object X in C there is a functor – overloading notation, we can now call it simply $X\colon 1 \to \mathsf{C}$ – which sends the sole object $\star$ of 1 to the object X, and sends the sole arrow $1_\star$ to 1_X.

26.3 More examples

(a) In contrast to forgetful functors, there are – as you'd expect – functors which preserve everything. Trivially,

(F8) For any category C there is an identity functor $1_{\mathsf{C}}\colon \mathsf{C} \to \mathsf{C}$ which sends C's objects and arrows to themselves.

(F9) Suppose S is a subcategory of C in the sense of §7.1. Then there is an inclusion functor $F\colon \mathsf{S} \to \mathsf{C}$ which sends objects and arrows in S to the same items in C.

Then there are functors which double up everything. Recalling §7.1 on product categories (and using 'C^2' as shorthand for '$\mathsf{C} \times \mathsf{C}$'):[2]

(F10) For any category C there is a *binary diagonal functor* $\Delta \colon \mathsf{C} \to \mathsf{C}^2$:

$$\Delta\colon \quad \begin{array}{rcl} X & \longmapsto & \langle X, X\rangle \\ f\colon X \to Y & \longmapsto & \langle f, f\rangle\langle X, X\rangle \to \langle Y, Y\rangle. \end{array}$$

And here is an endofunctor that is in a loose sense the opposite of forgetful in that it takes a simpler widget to a more complex widget:

(F11) There is a list functor $List\colon \mathsf{Set} \to \mathsf{Set}$, where $List_{ob}$ sends a set X to $List(X)$, the set of all finite lists or sequences of elements of X, including the empty one. And $List_{arw}$ sends a function $f\colon X \to Y$ to the function $List(f)\colon List(X) \to List(Y)$ which sends the list $x_0{}^\frown x_1{}^\frown x_2{}^\frown \ldots {}^\frown x_n$ to $fx_0{}^\frown fx_1{}^\frown fx_2{}^\frown \ldots {}^\frown fx_n$ (where $^\frown$ symbolizes concatenation).

It is trivial to check that this is a functor (so do so!).

(b) Next, recall how we can cook up categories from monoids and posets in particularly simple ways: functors between such derived categories turn out to be familiar mappings.

(F12) Take the monoids $(M, *, e)$ and $(N, \star, d)$ and consider the corresponding categories M and N in the sense of §5.7.

M has a single object $\bullet_{\mathsf{M}}$, and its arrows are simply elements of M, where the composition in M of the arrows m_1 and m_2 is $m_1 * m_2$, and the identity arrow is the identity element of the monoid, e.

Likewise, of course, N has a single object $\bullet_{\mathsf{N}}$, and arrows are elements of N, where the composition of the arrows n_1 and n_2 is $n_1 \star n_2$, and the identity arrow is the identity element of the monoid, d.

So now we see that a functor $F : \mathsf{M} \to \mathsf{N}$ will need to do the following:

i. F must send $\bullet_{\mathsf{M}}$ to $\bullet_{\mathsf{N}}$.

ii. F must send the identity arrow e to the identity arrow d.

iii. F must send m_1 composed with m_2 (i.e. $m_1 * m_2$) to Fm_1 composed with Fm_2 (i.e. $Fm_1 \star Fm_2$).

Apart from the trivial first condition, that just requires F to be a monoid homomorphism. So any homomorphism between two monoids induces a corresponding functor between the corresponding monoids-as-categories.

(F13) Take the posets $(S, \preccurlyeq)$ and $(T, \sqsubseteq)$ which can be treated as corresponding categories S and T.

So the objects of S are the members of S again, and the arrows of

[2] I'm falling in with standard usage in co-opting 'Δ' for notating both constant and diagonal functors. See, for example, Agore (2023, p. 25 vs p. 69), Borceux (1994, p. 9 vs p. 50), (Leinster 2014, p. 142 vs p. 73), etc. Later, we'll make a weak connection between the two uses.

S are pairs $\langle a, b\rangle$ such that $a \preccurlyeq b$, with composition for S defined by $\langle b, c\rangle \circ \langle a, b\rangle = \langle a, c\rangle$. Similarly of course for T.

It is easy to check that a monotone function $f\colon S \to T$ (i.e. a function such that $a \preccurlyeq b$ implies $f(a) \sqsubseteq f(b)$) induces a functor $F\colon \mathsf{S} \to \mathsf{T}$ which sends an S-object a to the T-object $f(a)$, and sends an S-arrow, i.e. a pair $\langle a, b\rangle$ where $a \preccurlyeq b$, to the T-arrow $\langle f(a), f(b)\rangle$.

(c) Finally in this group of examples, here's a rather more interesting one:

(F14) Take the group $G = (G, *, e)$ and consider it as a category G – see §8.7. And suppose $F\colon \mathsf{G} \to \mathsf{Set}$ is a functor.

Then F must send G's unique object $\bullet$ to some set X. And F must send a G-arrow $m\colon \bullet \to \bullet$ (that's some member m of G) to a function $F(m)\colon X \to X$. Functoriality requires that $F(e) = 1_X$ and $F(m * m') = F(m) \circ F(m')$. But those are the conditions for F to constitute a group action of G on X.

26.4 Functors, products, exponentials

Revision! First recall again Defn. 23 where we defined a product category C^2, and also revisit the discussion in §13.3 where we defined the component-wise product $f \times g\colon X \times Y \to X' \times Y'$ of two C arrows $f\colon X \to X'$ and $g\colon Y \to Y'$, and derived some results about such products.

(a) Then first note that we can define a pair of endofunctors involving products:

(F15) Assume C has all products, and C is any object in the category. Then there is a functor $- \times C\colon \mathsf{C} \to \mathsf{C}$ which acts as follows:

$$\begin{array}{lrcl} - \times C\colon & X & \longmapsto & X \times C \\ & f\colon X \to Y & \longmapsto & f \times 1_C\colon X \times C \to Y \times C. \end{array}$$

Similarly there is a functor

$$\begin{array}{lrcl} C \times -\colon & X & \longmapsto & C \times X \\ & f\colon X \to Y & \longmapsto & 1_C \times f\colon C \times X \to C \times Y. \end{array}$$

To confirm the functoriality of $- \times C$, we need to show that $(- \times C)(g \circ f)$ equals $(- \times C)g \circ (- \times C)f$. But by definition this is just the claim $g \circ f \times 1_C = (g \times 1_C) \circ (f \times 1_C)$, which follows from Theorem 55. Similarly for the twin functor.

(b) Here's a challenge, now that you are getting the hang of things! Show that the following also defines a functor:

(F16) Assume that C has all products. Then there is a functor $\otimes\colon \mathsf{C}^2 \to \mathsf{C}$ which acts as follows:

$$\begin{array}{lrcl} \otimes\colon & \langle X, Y\rangle & \longmapsto & X \times Y \\ & \langle f\colon X \to X', g\colon Y \to Y'\rangle & \longmapsto & f \times g\colon X \times X' \to Y \times Y'. \end{array}$$

OK: we need to check that, as defined, the object and arrow components of $\otimes$ do play nicely enough together to give us a kosher functor. First then, we need to show that $\otimes$ preserves identity arrows. In other words, $\otimes$ applied to the identity arrow on $\langle X, Y\rangle$ in C^2, i.e. $\langle 1_X, 1_Y\rangle$, gives the identity arrow on $X \times Y$ in C. But that's Theorem 54.

Second, we need to show that $\otimes$ respects composition of arrows. But, as required, we have

$$\begin{aligned} \otimes(\langle f, g\rangle \circ \langle f', g'\rangle) &= \otimes(\langle f \circ f', g \circ g'\rangle) && \text{(by definition of } \circ_{\mathsf{C}^2}\text{)} \\ &= (f \circ f') \times (g \circ g') && \text{(by definition of } \otimes\text{)} \\ &= (f \times g) \circ (f' \times g') && \text{(by Theorem 55)} \\ &= \otimes\langle f, g\rangle \circ \otimes\langle f', g'\rangle && \text{(by definition of } \otimes\text{).} \end{aligned}$$

(c) A brief comment before proceeding. Suppose we use an infix notation, writing '$\otimes(\langle X, Y\rangle)$' as '$X \otimes Y$'. Then we will have (i) both $(X_1 \otimes X_2) \otimes X_3 \cong X_1 \otimes (X_2 \otimes X_3)$ and (ii) $X \otimes 1 \cong X \cong 1 \otimes X$ (plus some similar isomorphisms); so, as we put it before in §13.2(b), this gives us some monoid-like or *monoidal* behaviour.

Now we can generalize, and say that C is a *monoidal category* if there is a functor $\otimes\colon \mathsf{C}^2 \to \mathsf{C}$ for which (i) and (ii) hold (plus some similar isomorphisms which we won't spell out). Such a functor $\otimes$ may be defined not in terms of ordinary products but in terms of tensor products or coproducts, for example. And it turns out that monoidal categories in this general sense are an interesting species which crop up in many applications. But having so briefly gestured towards them, simply because you'll quite often see mention of such categories, we can't explore further here.[3]

(d) Taking a product with a fixed object is functorial; so too is exponentiation by a fixed object. In other words – assuming we are working in a category C with exponentials and products – we can again take a fixed object C and we will expect there to be a functor we can notate as $(-)^C\colon \mathsf{C} \to \mathsf{C}$ whose object-component sends any C-object X to X^C, where that is the object component of the exponential (X^C, ev).

But how will the other component work, the component that sends an arrow $f\colon X \to Y$ to a suitable arrow $f^C\colon X^C \to Y^C$ and satisfies the functoriality conditions?

Well, take the two exponentials (X^C, ev_X) and (Y^C, ev_Y) – I'll subscript 'ev' arrows like this when we have more than one in play and we need to keep them

[3]That fount of wisdom Wikipedia, in its article on monoidal categories, tells us that "Monoidal categories have numerous applications outside of category theory proper. They are used to define models for the multiplicative fragment of intuitionistic linear logic. They also form the mathematical foundation for the topological order in condensed matter physics. Braided monoidal categories have applications in quantum information, quantum field theory, and string theory." Evidently, we can't here set up the required backgrounds for explaining such applications! But it isn't straightforward either to continue the story staying within pure category theory – see, for example, Perrone (2023, Ch. 6) or Yanofsky (2024, Ch. 5) where in both cases there is a marked jump in motivational obscurity compared with what has gone before. So we'll leave the topic for enthusiasts to tackle in due course.

distinct. Then the following diagram commutes, where $\widetilde{f \circ ev_X}$ is the unique exponential transpose of $f \circ ev_X$, as defined in Defn. 72:

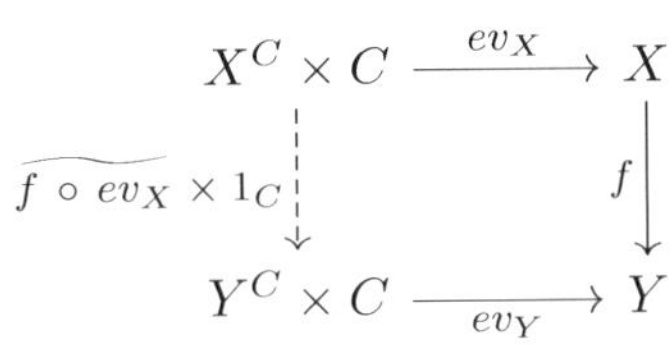

This is a straight application of the definition, given that (Y^C, ev_Y) is an exponential. So in this way, for fixed C, alongside the association between the objects X and X^C, there is a natural association between the arrows $f\colon X \to Y$ and $\widetilde{f \circ ev_X}\colon X^C \to Y^C$. We might rather hope that these associations combine to give us a functor. And they do:

(F17) Assume C has all exponentials, and that C is a C-object. Then there is a corresponding exponentiation functor $(-)^C\colon \mathsf{C} \to \mathsf{C}$ where

$$(-)^C\colon \quad \begin{array}{rcl} X & \longmapsto & X^C \\ f\colon X \to Y & \longmapsto & \widetilde{f \circ ev}\colon X^C \to Y^C. \end{array}$$

Proof. To confirm functoriality, we need to show that $(-)^C$ does preserve identities and respect composition.

The first of these is easy. $(1_X)^C$ is by definition $\widetilde{1_X \circ ev}\colon X^C \to X^C$, so this commutes:

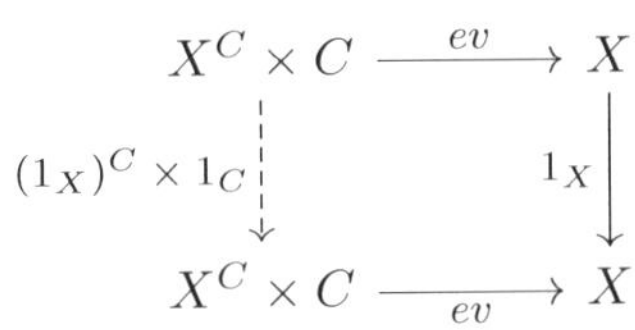

But evidently, an arrow $1_{X^C} \times 1_C$ on the left would also make the diagram commute. Hence by the uniqueness requirement that there is a single filling for $- \times 1_C$ which makes the square commute, $(1_X)^C = 1_{X^C}$, as required.

Second, we need to show that given arrows $f\colon X \to Y$ and $g\colon Y \to Z$, then $(g \circ f)^C = g^C \circ f^C$. Consider the following diagram where the top square, bottom square, and (outer, bent) rectangle commute:

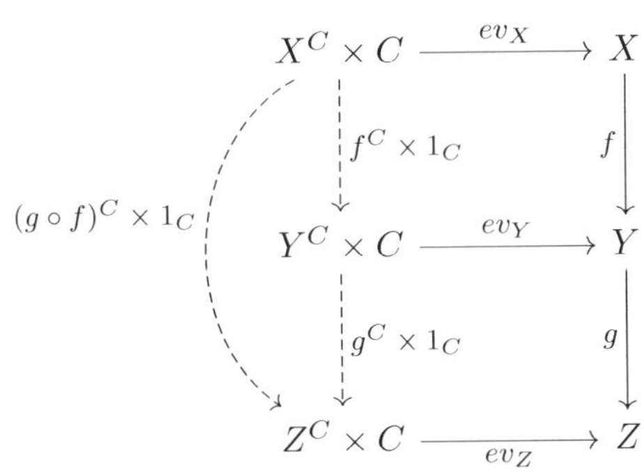

But note that, by Theorem 55, $(g^C \times 1_C) \circ (f^C \times 1_C) = (g^C \circ f^C) \times 1_C$. Hence $(g^C \circ f^C) \times 1_C$ is another arrow that makes a commuting outer rectangle. Hence, again by the requirement that there is a unique filling for $- \times 1_C$ which makes that outer rectangle commute, $(g \circ f)^C = g^C \circ f^C$. □

26.5 A functor from Set to Mon

(a) Next, we are going to define a functor going in the reverse direction to the forgetful functor in (F1), i.e. define a functor $F : \mathsf{Set} \to \mathsf{Mon}$.

There are of course utterly trivial ways of doing this. For example, pick some monoid M living in Mon. Then there is a constant functor we could call $!_M : \mathsf{Set} \to \mathsf{Mon}$ which sends every set X to M and sends every set-function $f : X \to Y$ to the identity homomorphism $1_M : M \to M$.

But it is instructive to try to come up with something rather less boring.

(b) So consider how we might methodically send sets to monoids, but this time *making as few assumptions as we possibly can* about which monoid a given set gets mapped to.

Start with a set X we are going to send to a corresponding monoid. Since we are making no more assumptions than we need to, we'll have to take the objects in X as providing us with an initial supply of objects for building our monoid, the monoid's *generators*.

We now need to equip our incipient monoid with a two-place associative function $*$. But by hypothesis we are assuming as little as we can about $*$ too, so we don't know that applying it keeps us inside the original set of generators X. So X will need to be expanded to a set M that contains not only the original members of X, e.g. $x, y, z, \ldots$, but also all the possible 'products', i.e. everything like $x * x$, $x * y$, $y * x$, $y * z$, $x * y * x$, $x * y * x * z$, $x * x * y * y * z \ldots$, etc. We know, however, that since $*$ is associative, we needn't distinguish between e.g. $x * (y * z)$ and $(x * y) * z$.

But even taking all those products is not enough, for (in our assumption-free state) we don't know whether any of the resulting elements of M will act as an identity for the $*$-function. To get a monoid, then, we need to throw into M some unit 1.

Since we are making as few assumptions as we can, we also can't assume that any of the products in M are equal (at least once we have multiplied out occurrences of the identity element), or that there are any other objects in M other than those generated from the unit and members of X.

Now, here's a neat way to model the resulting monoid generated from the set X in this assumption-free way. Represent a monoid element (such as $x{*}x{*}y{*}y{*}z$) as a *finite list of members of* X (such as $x^\frown x^\frown y^\frown y^\frown z$), so M gets represented by $List(X)$, the set of finite lists of members of X, and we model the $*$-function by simple concatenation. The identity element will then be modelled by the null list $\emptyset$. The resulting monoid $(List(X), ^\frown, \emptyset)$ is often simply called *the free monoid* on X – though perhaps it would be rather better to say that it is a standard exemplar of a monoid freely constructed from X. Which all goes to motivate the following construction:

(F18) There is a functor F : Set $\to$ Mon with the following data:

i. F_{ob} sends the set X to the free monoid $(List(X), ^\frown, \emptyset)$.

ii. F_{arw} sends the arrow $f\colon X \to Y$ to $List(f)$ as in (F11) again, where this is treated as an arrow from $(List(X), ^\frown, \emptyset)$ to $(List(Y), ^\frown, \emptyset)$.

It is again trivial to now check that F – informally the 'free functor' from Set to Mon – is a functor.

(c) We can generalize. There are similar functors that send sets to other *freely generated* structures on the set.

For example there is a functor from Set to Ab which sends a set X to the freely generated abelian group on X (which is in fact the direct sum of X-many copies of $(\mathbb{Z}, +, 0)$ – the integers $\mathbb{Z}$ with addition forming the paradigm free abelian group on a single generator).

But we need not concern ourselves with the further details of such cases.

26.6 Contravariance

(a) We now introduce an important new idea. Functors of the kind we've met so far send arrows $f\colon A \to B$ to arrows $Ff\colon FA \to FB$. Call these *covariant* functors, and compare:

Definition 105. A *contravariant functor* $F\colon$ C $\to$ D between categories C and D comprises a map from C-objects to D-objects and a map from C-arrows to D-arrows such that

Components cohere: if $f\colon A \to B$ is a C-arrow, then F sends it to $Ff\colon FB \to FA$ in D;[4]

Preserving identity arrows: for any C-object A, $F1_A = 1_{FA}$;

Respecting composition: for any C-arrows f, g such that their composition $g \circ f$ exists, $F(g \circ f) = Ff \circ Fg$. (NB the order of the compositions!) △

And why do we want to introduce this second style of functor, alongside the original covariant ones?

Well, consider going from a category C to its opposite C^{op}. We map any object C to itself: but on arrows, we reverse direction, mapping $f\colon A \to B$ to $f\colon B \to A$. So we could think of this as the operation of a contravariant functor. However, you might regard this as a rather artificial example. So let's have a couple of elementary examples of contravariant functors which arise 'in nature', so to speak:

[4]Write those as $f\colon A \to B$ and $Ff\colon FA \leftarrow FB$ so the arrows do go in contrary directions, and you can see why 'contravariance' is an apt description!

(F19) The covariant powerset functor $P\colon \mathsf{Set} \to \mathsf{Set}$ maps a set X to its powerset $\mathcal{P}X$ and maps a set-function $f\colon X \to Y$ to the function $Pf\colon \mathcal{P}X \to \mathcal{P}Y$ which sends $U \in \mathcal{P}X$ to its f-image $f[U] = \{f(x) \mid x \in U\} \in \mathcal{P}Y$. (Check that this *is* a functor!)

Now note that this functor has a contravariant twin $\overline{P}\colon \mathsf{Set} \to \mathsf{Set}$ which again maps a set to its powerset, and this time maps a set-function $f\colon X \to Y$ to the function $\overline{P}f\colon \mathcal{P}Y \to \mathcal{P}X$ which sends $U \in \mathcal{P}Y$ to its inverse image $f^{-1}[U] \in \mathcal{P}X$ (where $f^{-1}[U] = \{x \mid f(x) \in U\}$). (Check that this works too!)

(F20) Take FVect, the category whose objects are the finite-dimensional vector spaces over the reals, and whose arrows are linear maps between spaces.

Now recall, the dual space of a given finite-dimensional vector space V over the reals is V^*, the set of all linear functions $f\colon V \to \mathbb{R}$ (where this set is equipped with vectorial structure in the obvious way). V^* has the same dimension as V (so, a fortiori, is also finite dimensional and belongs to FVect). We'll construct a dualizing functor $D\colon \mathsf{FVect} \to \mathsf{FVect}$, where D_{ob} sends a vector-space to its dual.

So how is our functor's component D_{arw} going to act on arrows in the category FVect? Take the spaces V, W and consider any linear map $f\colon V \to W$. Then, on the dual spaces, there will be a corresponding map $(- \circ f) : W^* \to V^*$ which sends a function $g\colon W \to \mathbb{R}$ to $g \circ f\colon V \to \mathbb{R}$. This suggests what we want the action of the component D_{arw} to be: it will send a linear map f to the functional $(- \circ f)$.

It is readily checked that these components D_{ob} and D_{arw} do give us a contravariant functor.

(b) We noted that going from a category to the opposite category could be regarded as the operation of a contravariant functor. But there's more to be said about functors and opposite categories. First:

Theorem 127. *The data for a covariant functor $F\colon \mathsf{C} \to \mathsf{D}$ provides us with a covariant functor $F^{op}\colon \mathsf{C}^{op} \to \mathsf{D}^{op}$.*

Proof. Recall, the objects of C^{op} are exactly the same as the objects of C. We can therefore sensibly define the object-mapping component of F^{op} as acting on C^{op} objects exactly as the object-mapping component of F acts on C-objects. And then, allowing for the fact that taking opposites reverses arrows, we can define the arrow-mapping component of F^{op} as acting on the C^{op}-arrow $f\colon A \to B$ exactly as the arrow-mapping component of F acts on the C-arrow $f\colon B \to A$. F^{op} will then obey the axioms for being a functor because F does. □

(c) Next, a *much* more important general point about contravariant functors. Given our definition,

Theorem 128. *$F\colon \mathsf{C} \to \mathsf{D}$ is a* contravariant *functor from C to D if and only if $F\colon \mathsf{C}^{op} \to \mathsf{D}$ with the same data is a functor in the original* covariant *sense.*

Proof. Suppose $F\colon \mathsf{C} \to \mathsf{D}$ is a contravariant functor. Then we know that the object-mapping F_{ob} sends the C^{op}-object A to some D-object $F(A)$, because the C^{op}-object A is none other than the C-object A. Further, the arrow-mapping F_{arw} sends the C^{op}-arrow $f\colon A \to B$ to the D-arrow $F(f)\colon\ FA \to FB$, since the C^{op}-arrow $f\colon A \to B$ is none other than the C-arrow $f\colon B \to A$.

Moreover, F_{arw} preserves identity arrows on C^{op}-objects, and respects (covariant) composition: for any C^{op}-arrows f, g such that their composition $g \circ f$ exists, $F(g \circ f) = Fg \circ Ff$. Being more explicit about the composition operator for once, the point is that if $g \circ_{\mathsf{C}^{op}} f$ in C^{op} exists, then by definition it is $f \circ_{\mathsf{C}} g$ in C. And we know that the contravariant F sends $f \circ_{\mathsf{C}} g$ to $Fg \circ_{\mathsf{D}} Ff$.

All that simply adds up to our contravariant F having the data and satisfying the conditions to be a standard, covariant, functor from C^{op} to D. □

So whenever we are tempted to talk of a contravariant functor from the category C, we *could* talk instead of the same data in its guise as a covariant functor from C^{op}. But I don't think that's always the best way to keep things clear, and will often take contravariant functors as nature intended.

However, I do adopt one common convention: from now on, unqualified talk of a functor should be read as referring by default to a covariant one.

26.7 Composing functors

(a) Functors are maps; maps can compose; so functors can compose. We can spell this out in the form of theorem for future use. Pause to check it:

Theorem 129. *Suppose there exist (covariant) functors $F\colon \mathsf{C} \to \mathsf{D}$, $G\colon \mathsf{D} \to \mathsf{E}$. Then there is also a composite (covariant) functor $G \circ F\colon \mathsf{C} \to \mathsf{E}$ with the following data:*[5]

(i) *A mapping $(G \circ F)_{ob}$ which sends a C-object A to the E-object $G(FA)$.*
(ii) *A mapping $(G \circ F)_{arw}$ which sends a C-arrow $f\colon A \to B$ to the E-arrow $G(Ff)\colon\ G(FA) \to G(FB)$.*

Further, such composition of functors is associative. □

And to further reduce clutter, we will later allow ourselves to write simply 'GF' rather than '$G \circ F$' and 'GFA' rather than '$G(FA)$', etc.[6]

[5] We are assuming that maps of the kind which constitute the components of functors compose-as-maps in the usual way that maps do! Then what we want to check is that functors that are built out of composable maps will themselves compose in such a way as to result in another functor.

[6] Another notational remark. As I said before, it is quite standard to use the same style of letters to represent both functors and the objects on which functors might operate. And we don't usually notationally differentiate the components of a functor acting on objects and on arrows. It is very common not to use brackets when notating functor application. And I'm now noting that is also entirely standard to use a minimalist notation which doesn't use an explicit sign like '$\circ$' in notating the composition of functors. So that means that on one page 'FC' (rather than the more explicit '$F(C)$') might represent the application of the functor F

(b) Let's finish the chapter with another mini-theorem (again, pause to prove it):

Theorem 130. *If two contravariant functors compose, the result is a covariant functor.* □

Reality check: explain how it can be that contravariant functors don't compose to give a contravariant functor while covariant functors compose to give a covariant functor, even though contravariant functors can be treated as covariant functors (from the opposite category).

---

to an object C, while on the next page 'FG' (rather than '$F \circ G$') represents the composition of the functor F with the functor G.

You might very reasonably think this overloading of notation could potentially get confusing. True, context will always disambiguate if you pay enough attention. But still, why not make things easier for ourselves by being more explicit? For example, how about using a different font for functors and symbolically marking off functor application from composition of functors? Then we can clearly distinguish '$\mathcal{F}(G)$' vs '$\mathcal{F} \circ \mathcal{G}$'?

Good question. I can only report that experimentation didn't lead to happy results. Mixing symbol salads from multiple fonts can – to my eyes – make for a rebarbative reading experience, and as a consequence doesn't sufficiently aid comprehension. And, in the end, overdoing brackets or composition signs can be the opposite of helpful. So I'm not just mindlessly following convention here. However, I will try to remember to add enough contextual commentary to ensure that my generally minimalist notational practice won't cause confusion or lead you astray.

27 What functors can do

A functor $F\colon \mathsf{C} \to \mathsf{D}$ sends each C-object A to its image FA and sends each C-arrow $f\colon A \to B$ to its image $Ff\colon FA \to FB$. These resulting images assemble into an overall image or representation of the category C living in the category D. But how good a representation do we get in the general case? What features of C get carried over by a functor?

27.1 Images assembled by a functor needn't be categories

First, an important general observation worth highlighting as a theorem.

Recall, the image of a group G under a group homomorphism $f\colon G \to H$ will itself be a group, a subgroup of H (see Theorem 6). Similarly for other algebraic homomorphisms. By contrast, however,

Theorem 131. *The image of the category* C *assembled by a functor* $F\colon \mathsf{C} \to \mathsf{D}$ *need* not *be a subcategory of* D.

Proof. A toy example establishes the point. Let C be the four-object category we can diagram (omitting identity arrows) as

$$A \longrightarrow B \qquad\qquad C \longrightarrow D$$

and let D be the three-object category (again omitting identity arrows)

$$X \longrightarrow Y \longrightarrow Z \quad \text{(with an arrow } X \to Z \text{ curving over)}$$

Suppose F_{ob} sends A to X, both B and C to Y, and D to Z; and let F_{arw} send identity arrows to identity arrows, and send the arrows $A \to B$ and $C \to D$ respectively to $X \to Y$ and $Y \to Z$.

Trivially, F with those components is functorial. But the image of C under F is not a category (and so not a subcategory of D), since that image contains the arrows $X \to Y$ and $Y \to Z$ but not their composition. □

27.2 Preserving and reflecting

(a) We next introduce a pair of standard notions for describing the actions of functors:

Definition 106. Suppose $F\colon \mathsf{C} \to \mathsf{D}$ and P is some property of objects or arrows (or of some combination). Then

(1) F *preserves* P iff, for any relevant C-data D, if D has property P, so does the result of applying F to D, $F(D)$ for short.
(2) F *reflects* P iff, for any C-data D, if $F(D)$ has property P, so does D.

We will also say that F preserves (reflects) Xs if F preserves (reflects) the property of being an X. △

For example, to say that a functor $F\colon \mathsf{C} \to \mathsf{D}$ preserves commutative diagrams is to say that if some objects/arrows have the property of forming a commutative diagram in C, then the result of applying some functor $F\colon \mathsf{C} \to \mathsf{D}$ to those objects and arrows always gives us another commutative diagram in D. And we in fact have an easy theorem (why?):

Theorem 132. *Functors always preserve commutative diagrams.* □

(b) So what properties of arrows in particular do or don't get preserved or reflected by functors? Here's a composite theorem:

Theorem 133. *Functors do not necessarily preserve or reflect monomorphisms and epimorphisms.*

Functors do preserve right inverses, left inverses, and isomorphisms. But functors do not necessarily reflect those.

And to help fix ideas, here's a challenge: prove these claims before reading on.

Proof: functors needn't preserve epics. In §8.4, Ex. (3) we saw that the inclusion map $i_M\colon (\mathbb{N}, +, 0) \to (\mathbb{Z}, +, 0)$ in Mon is epic. But plainly the inclusion map $i_S\colon \mathbb{N} \to \mathbb{Z}$ in Set is not epic (as it isn't surjective). Therefore the forgetful functor $F\colon \mathsf{Mon} \to \mathsf{Set}$ maps an epic map (i_M) to a non-epic one (i_S), so does not preserve epics. □

Proof: functors needn't preserve monics. We could appeal to duality: since the forgetful $F\colon \mathsf{Mon} \to \mathsf{Set}$ doesn't preserve epics, $F^{op}\colon \mathsf{Mon}^{op} \to \mathsf{Set}^{op}$ won't preserve monics.

Or here's a toy example to establish the claim. Recall 2, the two-object category which we can diagram on the left (omitting identity arrows), and let C be the cofork-shaped category on the right, where $f \neq g$ but $k \circ f = k \circ g$ (again omitting identity arrows).

$$V \xrightarrow{\;j\;} W \qquad\qquad X \underset{g}{\overset{f}{\rightrightarrows}} Y \xrightarrow{\;k\;} Z$$

Trivially, the non-identity arrow j on the left is monic in 2. Equally trivially, the arrow k on the right is not monic in C. And so the inclusion functor $F\colon 2 \to \mathsf{C}$ which sends V to Y, sends W to Z, and sends j to k, doesn't preserve monics. □

Are there less contrived examples? Well, we'll find that lots of nice functors preserve limits like pullbacks, and Theorem 134 will tell us that such functors *do* preserve monos. But for a negative example 'in nature', so to speak, it can be shown that the functor that sends a group to its 'abelianization' doesn't preserve monomorphisms.

Proof: functors needn't reflect monics or epics. For an easy example of a functor which need not reflect monics or epics, consider a collapse functor which maps C to the one-object category $\mathbf{1}$, thereby sending arrows of all sorts to the trivially monic and epic identity arrow on the sole object of $\mathbf{1}$. □

Proof: functors preserve inverses and isomorphisms. Assume $f\colon A \to B$ is a right inverse in the category C. Then there exists an arrow g such that $g \circ f = 1_A$. If $F\colon \mathsf{C} \to \mathsf{D}$ is a functor, then $F(g \circ f) = F(1_A)$, and by functoriality that implies $F(g) \circ F(f) = 1_{FA}$. So $F(f)$ is a right inverse in the category D. Therefore F preserves right inverses.

Similarly, left inverses are preserved. And putting the two results together shows that isomorphisms are preserved. □

Proof: functors need not reflect right inverses, left inverses, and isomorphisms. Consider again the collapse functor sending C to $\mathbf{1}$. The only arrow in $\mathbf{1}$, the identity arrow, is trivially an isomorphism (and so a left and right inverse). The C-arrows the collapse functor sends to it will generally not be inverses or isomorphisms. □

So functors needn't reflect isomorphisms, but there is a special term for those which do:

Definition 107. A functor F is *conservative* iff it reflects all isomorphisms. △

(c) Finally, a result about how preservation properties can be interrelated:

Theorem 134. *If a functor preserves pullbacks then it preserves monomorphisms. Dually, if it preserves pushouts it preserves epimorphisms.*

Proof. We use Theorem 92. This tells us that if $f\colon X \to Y$ in C is monic then it is part of a pullback square, as on the left:

$$\begin{array}{ccc} X & \xrightarrow{1_X} & X \\ {\scriptstyle 1_X}\downarrow & & \downarrow{\scriptstyle f} \\ X & \xrightarrow{f} & Y \end{array} \quad \Rightarrow \quad \begin{array}{ccc} FX & \xrightarrow{1_{FX}} & FX \\ {\scriptstyle 1_{FX}}\downarrow & & \downarrow{\scriptstyle Ff} \\ FX & \xrightarrow{Ff} & FY \end{array}$$

Now suppose $F\colon \mathsf{C} \to \mathsf{D}$ sends pullback squares to pullback squares. Then the square on the right is also a pullback square. Therefore, by Theorem 92 again, Ff is monic too. Duality gives the other half of the Theorem. □

27.3 Faithful, full, and essentially surjective functors

(a) We will now define functorial analogues of the notions of injective and surjective functions.

First, as far as their behaviour on *arrows* is concerned, the useful notions for functors turn out to be these:

Definition 108. A functor $F\colon \mathsf{C} \to \mathsf{D}$ is *faithful* iff, given any C-objects A, B and any pair of parallel arrows $f, g\colon A \to B$, then $F(f) = F(g)$ implies $f = g$.

F is *full* (that's the standard term) iff given any C-objects A, B, then for any D-arrow $g\colon FA \to FB$ there is an arrow $f\colon A \to B$ such that $g = Ff$.

F is *fully faithful*, some say, iff it is full and faithful. △

Note, a faithful functor needn't be, overall, injective on arrows. For suppose C is in effect two copies of D, and F sends each copy faithfully to D: then F sends two copies of an arrow to the same image arrow. However, for each particular pair of objects A and B, a faithful functor is injective from the arrows $A \to B$ to the arrows $FA \to FB$. Likewise, a full functor needn't be, overall, surjective on arrows: but it is locally surjective from the arrows $A \to B$ to the arrows $FA \to FB$.

Later, we will introduce the notation $\mathsf{C}(A, B)$ for the collection of C-arrows from A to B. Using this notation, we can then say that a functor $F\colon \mathsf{C} \to \mathsf{D}$ is faithful iff, for any C-objects A, B, it is injective from $\mathsf{C}(A, B)$ to $\mathsf{D}(FA, FB)$, and is full iff it is surjective from $\mathsf{C}(A, B)$ to $\mathsf{D}(FA, FB)$.

(b) Second, regarding how functors treat *objects*, the notions worth highlighting are these:

Definition 109. A functor $F\colon \mathsf{C} \to \mathsf{D}$ is *essentially injective on objects* iff $FC \cong FC'$ implies $C \cong C'$, for any C-objects C and C'.

More importantly: a functor $F\colon \mathsf{C} \to \mathsf{D}$ is *essentially surjective on objects* (e.s.o.) iff for any D-object D, there is a C-object C such that $FC \cong D$. △

Plain injectivity on objects (requiring that $FC = FC'$ implies $C = C'$) is less interesting, given that we usually only care, categorially speaking, about the identity of objects up to isomorphism. Likewise plain surjectivity on objects (requiring, for every object D in D, an object C such that $FC = D$) is less interesting, given that we usually won't care whether D has extra non-identical-but-isomorphic copies of objects.

In fact, caring about whether objects are actually identical as opposed to isomorphic is often jokingly said to be 'evil' as far as category theory is concerned.

(c) Some examples:

(1) The forgetful functor $F : \mathsf{Mon} \to \mathsf{Set}$ is faithful, as F sends a set-function which happens to be a monoid homomorphism to itself, so different arrows in Mon get sent to different arrows in Set. But the functor is not full: there will be many arrows in Set that don't correspond to a monoid homomorphism.

(2) The forgetful functor $F\colon \mathsf{Ab} \to \mathsf{Grp}$ is faithful. And this one is full, because any objects C, C' in Ab (in other words, any two abelian groups) also live in Grp, and a group homomorphism between them as an arrow in Grp is also an abelian group homomorphism in Ab. But lots of groups aren't abelian, so F is not essentially surjective on objects.

(3) Take the category 2^* with exactly two isomorphic objects $\bullet$ and $\star$, whose four arrows are the identity arrows and inverse arrows each way between the two objects. Then the only possible functor $F\colon 2^* \to 1$ is trivially faithful (since there are no parallel arrows in 2^*) and trivially full. But note that it is not injective on objects.

(4) The 'thinning' functor (F5) from §26.2, $F\colon \mathsf{C} \to \mathsf{D}$, is full but not faithful unless C is already a preorder category. But it is e.s.o.

(5) Suppose M and N are the categories that correspond to the monoids $(\mathrm{M}, *, e)$ and $(\mathrm{N}, \star, d)$. And let f be a monoid homomorphism between those monoids which is surjective but not injective. Then the functor $F\colon \mathsf{M} \to \mathsf{N}$ corresponding to f is full but not faithful.

(6) You might be tempted to say that the 'total collapse' functor $\Delta_1\colon \mathsf{Set} \to 1$ (which sends every set to the sole object of 1, and every set-function to the identity arrow of 1) is full but not faithful. But it isn't full. Take A, B in Set to be a singleton and the empty set respectively. There is a trivial identity map in 1, namely $1\colon \Delta_1 A \to \Delta_1 B$; but there is no arrow in Set from A to B.

(7) An inclusion functor $F\colon \mathsf{S} \to \mathsf{C}$ is faithful; if S is a full subcategory of C, then the inclusion map is fully faithful, but usually not e.s.o.

(8) Consider again the free functor (F18) from §26.5. It sends different set functions $f, g\colon X \to Y$ to different functions $Ff, Fg\colon \mathit{List}(X) \to \mathit{List}(Y)$ (if f and g give different values when applied to the object x, then Ff and Fg will give different values applied to the corresponding list whose sole object is x). So F is faithful.

Now consider a singleton set 1. This gets sent by F to the free monoid with a single generator – which is tantamount to N, the monoid $(\mathbb{N}, +, 0)$. The sole set-function from 1 to itself, the identity function, gets sent by F to the identity monoid homomorphism on N. But there are other monoid homomorphisms from N to itself, e.g. $n \mapsto 2n$. So F is not full.

(d) How do faithful, full, or fully faithful functors behave?

Being faithful means being locally injective on arrows, and compositions of injective functions are injective; being full means being locally surjective, and compositions of surjective functions are surjective. Hence

Theorem 135. *The composition of faithful functors is faithful and the composition of full functors is full.* □

Next, we note

Theorem 136. *A faithful functor $F\colon \mathsf{C} \to \mathsf{D}$ reflects monomorphisms and epimorphisms.*

Proof. Suppose Ff is monic, and suppose $f \circ g = f \circ h$. Then $F(f \circ g) = F(f \circ h)$, so by functoriality $Ff \circ Fg = Ff \circ Fh$, and since Ff is monic, $Fg = Fh$. Since F is faithful, $g = h$. Hence f is monic. Dually for epics. □

Theorem 137. *If a functor is fully faithful it reflects right inverses and left inverses, and hence is conservative.*

Proof. Suppose $F\colon \mathsf{C} \to \mathsf{D}$ is a fully faithful functor, and let Ff be a right inverse, with f an arrow in C with source A. Since F is full, Ff must be the right inverse of Fg for some arrow g in C. So $Fg \circ Ff = 1_{FA}$, whence $F(g \circ f) = 1_{FA} = F(1_A)$. Since F is faithful, it follows that $g \circ f = 1_A$, and f is a right inverse.

Dually, F reflects left inverses, and combining the two results shows that F reflects isomorphisms, i.e. is conservative. □

Note, however, that the reverse of the last result is not true. A functor can reflect isomorphisms without being fully faithful. An example is the forgetful functor $F\colon \mathsf{Mon} \to \mathsf{Set}$. This is faithful but not full. But it is conservative because if the set function Ff is an isomorphism, so is the monoid homomorphism f, because a monoid homomorphism is an isomorphism if and only if its underlying function is one too.

Finally, we have seen that fully faithful functors need neither be injective on objects nor essentially surjective on objects (and hence not plain surjective). However,

Theorem 138. *A fully faithful functor* is *essentially injective on objects.*

Proof. If $F\colon \mathsf{C} \to \mathsf{D}$ is full then if $FC \cong FC'$, i.e. if there is an isomorphism $g\colon FC \xrightarrow{\sim} FC'$, then there is some $f\colon C \to C'$ such that $g = Ff$. But then, by the previous theorem, if F is faithful as well, it reflects isomorphisms, so f is an isomorphism, witnessing that $C \cong C'$. □

27.4 An example from topology

(a) Let's descend from these airy generalities and apply the simple result that functors preserve left inverses. Our example comes from topology, but is easy enough to get a glimmer of what's going on even if you know almost nothing about the setting. You only need the idea of the fundamental group of a topological space (at a point), roughly as follows.

Given a space and a chosen base point in it, consider all the directed paths that start at this base point then wander around and eventually loop back to their starting point. Such directed loops can be "added" together in a natural way: you traverse the "sum" of two loops by going round the first loop, then round the second. Every loop has an "inverse" (you go round the same path in the opposite direction). Two loops are considered homotopically equivalent if one can be continuously deformed into the other. We can take, then, the set of all such equivalence classes of loops – so-called homotopy equivalence classes – and define "addition" for these classes in the obvious derived way. This set, when equipped with addition, evidently forms a group: it is the *fundamental group* for that particular space, with the given basepoint. (Though for many spaces, the nature of the group is independent of the basepoint.)

Suppose, therefore, that Top_* is the category of pointed topological spaces: an object in the category is a topological space X equipped with a distinguished base point x_0, and the arrows in the category are continuous maps that preserve basepoints.

Then here's our new example of a functor:

(F3) There is a functor $\mathsf{Top}_* \to \mathsf{Grp}$, conventionally called π_1, with the following data:

i. π_1 sends a pointed topological space (X, x_0) – i.e. X with base point x_0 – to the fundamental group $\pi_1(X, x_0)$ of X at x_0.

ii. π_1 sends a basepoint-preserving continuous map $f\colon (X, x_0) \to (Y, y_0)$ to a corresponding group homomorphism $f_*\colon \pi_1(X, x_0) \to \pi_1(Y, y_0)$.

To explain: f will send a loop based at x_0 to a loop based at y_0. And since f is continuous, it can be used to send a continuous deformation of a loop in (X, x_0) to a continuous deformation of a loop in (Y, y_0). And that induces a corresponding association f_* between the homotopy equivalence classes of (X, x_0) and (Y, y_0), and this will respect the group structure of adding and reversing loops. Moreover, a composition of continuous maps f, g will give rise to a composition of corresponding homomorphisms f_*, g_*. OK – that's a bit hand-waving: but it should be enough to persuade you that π_1 is functorial.

(b) Here, then, is a nice application. We'll prove Brouwer's famed Fixed Point Theorem:

Theorem 139. *Any continuous map of the closed unit disc to itself has a fixed point.*

Proof Suppose, for reductio, that there is a continuous map f on the two-dimensional disc D (considered as a topological space) *without* a fixed point, i.e. such that for any point x in D, we always have $f(x) \neq x$.

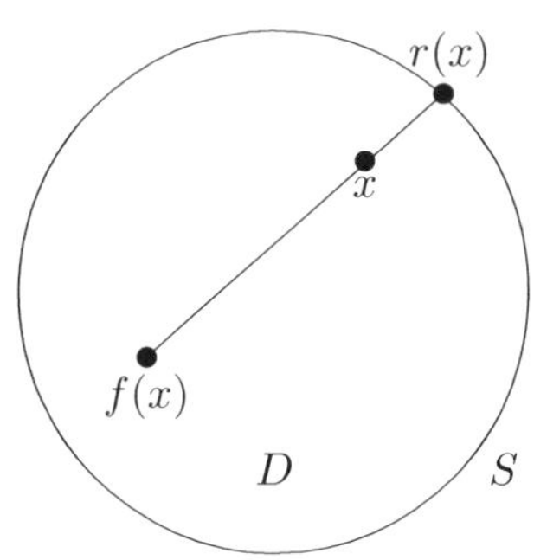

Let the boundary of the disc be the circle S (again considered as a topological space). Then we can define a map that sends the point x in D to the point $r(x)$ on S at which the straight line starting from the point $f(x)$ and passing through the distinct point x intersects the boundary of the disc.

This map sends a point on the boundary to itself. Pick a boundary point $*$ to be the base point of the pointed space D_* and also of the pointed space S_*, then our map induces a map $r\colon D_* \to S_*$. Moreover, this map is evidently continuous (intuitively: nudge a point x and since f is continuous that only nudges $f(x)$, and hence the ray from $f(x)$ through x is only nudged, and the point of intersection with the boundary is only nudged). And r is a left inverse of the inclusion map $i\colon S_* \to D_*$ in Top_*, since $r \circ i = 1$.

Functors preserve left inverses by Theorem 133, so $\pi_1(r)$ will be a left inverse of $\pi_1(i)$, which means that $\pi_1(i)\colon \pi_1(S_*) \to \pi_1(D_*)$ is an injection by Theorem 17.

But that's impossible. $\pi_1(S_*)$, the fundamental group of S_*, is equivalent to the group of integers under addition (think of looping round a circle, one way or another, n times – each positive or negative integer corresponds to a different path). While $\pi_1(D_*)$, the fundamental group of D_*, is a one element group (for every loop in the disk D_* can be smoothly shrunk to a point). And there is no injection between the integers and a one-element set! □

(c) What, if anything, do we gain by putting the proof in category-theoretic terms? It might be said: the proof crucially depends on facts of algebraic topology – continuous maps preserve homotopic equivalences in a way that makes π_1 a functor, and the fundamental groups of S_* and D_* are respectively the group of integers and the trivial group. And we run the whole proof without actually mentioning categories at all.

Of course. Still, what we've done is to very clearly demarcate those bits of the proof that depend on topic-specific facts of topology and those bits which depend on general proof-ideas about functoriality and about kinds of maps (inverses, injections), ideas which are thoroughly *portable* to other contexts. And *that* arguably counts as a real gain in understanding.

27.5 An afterword on the idea of concrete categories

A short aside for the record, before moving on, this time applying the notion of a faithful functor.

(a) At the end of §5.6, I mentioned that categories like Mon and Preord whose objects are sets-equipped-with-some-structure and whose arrows are structure-respecting-set-functions are often informally called concrete categories. As we also saw right at the outset, lots of categories are *not* concrete in this intuitive sense – for example, neither a monoid-as-category nor a preordered-collection-as-category need qualify.

Well, now that we have the notion of a faithful functor in play, I guess I should mention a conventional formal definition that you might well meet elsewhere:

Definition 110. A *concrete category* is a pair (C, U) such that C is a category and $U\colon \mathsf{C} \to \mathsf{Set}$ is a faithful functor.

A category C is *concretizable* if there exists a faithful functor $U\colon \mathsf{C} \to \mathsf{Set}$. △

A paradigm case is provided by, say, Mon equipped with the forgetful functor U which sends a monoid to its underlying set and sends a monoid homomorphism to its underlying set-function.

(b) So far, it might seem, so good. However, it is easy to see that this new definition does *not* capture the original intuitive notion of a concrete category as a category of more-or-less structured sets.

For example, suppose we take some objects P (non-sets, and not too many!) preordered by the relation $\preccurlyeq$. Then, as noted in §5.4, there is a corresponding category P whose objects are P again, and which has a single arrow from a particular object q to an object r if and only if $q \preccurlyeq r$. Now, we can presumably index these objects by associating them one-by-one with pure sets. Let p' be the index of the object p. We can then define a functor $F\colon \mathsf{P} \to \mathsf{Set}$ by stipulating

1. F_{ob} sends a P-object q to the set $P_{\leqslant q}$ of all p' such that $p \leqslant q$,
2. F_{arw} sends a P-arrow $q \to r$ to the inclusion function $i\colon P_{\leqslant q} \hookrightarrow P_{\leqslant r}$.

It is trivial to check that this *is* a functor, and that it is faithful.

But this makes P equipped with F officially count as a concrete category according to our new-fangled definition: yet the likes of P are exactly the sort of example that we originally *contrasted* with concrete categories in our intuitive sense. And it gets worse. We can similarly show that *any* category whose objects and arrows can be indexed by sets can be made concrete in our new sense.[1]

(c) What to do? Should we aim to find a revised formal definition which sticks closer to that original intuitive idea of concrete categories as being actually built out of suitably equipped sets?

Perhaps not. For category theory is centrally concerned with the *structural* properties of structures, not what they are 'made of'. So perhaps our Defn. 110 does after all home in on the categorially significant idea hereabouts. And while all categories that aren't too big compares with Set do count as concretizable on *this* definition, it turns out that our new officially defined notion still draws an interesting distinction among the very large categories. For example, while Set is trivially concrete, it can be proved that the category hTop is not concretizable, where that is the category of topological spaces whose arrows are whole classes of maps which can be continuously deformed into each other. But following *this* theme any further would take us much too far from the elementary focus of these notes. So we must let the topic rest here.[2]

[1] See too Awodey (2010, p. 14, Remark 1.7).

[2] The highly non-trivial result about hTop is due to Freyd (1970). I note that the elementary texts by e.g. McLarty (1992), Goldblatt (1984), Simmons (2011), and Leinster (2014) perhaps sensibly avoid using the notion of concreteness altogether!

28 Functors, diagrams, and limits

As we have seen, a functor $F\colon \mathsf{J} \to \mathsf{C}$ will, in virtue of its functoriality, preserve/reflect some minimal aspects of the categorial structure of J as it sends objects and arrows into C. And if the functor has properties like being full or faithful, it will preserve/reflect more.

So now here's an obvious question: how do things stand with respect to preserving/reflecting products, equalizers, quotients, etc.? More generally, we want to know about preserving/reflecting limits and colimits. The theorems in this chapter give the headline news.

28.1 Diagrams redefined as functors

We start by showing that, now that we have the notion of a functor in play, we can redefine the notion of a diagram and the notion of a limit over a diagram (or colimit under a diagram) in a particularly neat way.

(a) First, diagrams. The idea is this:

(i) A functor $D\colon \mathsf{J} \to \mathsf{C}$ will send the objects and arrows of J to some corresponding objects and arrows sitting inside C. So we can think of the functor D as generating in C a *diagram* in the sense introduced rather loosely in §6.1 and then refined a little in §19.2.

(ii) This induced diagram might not be a faithful representation of J – because distinct arrows of J might get diagrammed by a single C arrow, and likewise for objects.

(iii) Still, we could say that the induced diagram in C retains at least something of the overall shape of the original category J.

And this is enough to motivate some absolutely standard terminology:

Definition 111. Given categories C and J, a functor $D\colon \mathsf{J} \to \mathsf{C}$ is said to be a *diagram of shape* J *in* C. △

Yes, the functor D with source J which in an intuitive sense generates a diagram that retains something of the shape of J is said itself to *be* a diagram. And this diagram-as-functor is said, without qualification, to *have* shape J – even if the functor isn't faithful or essentially injective on objects and merges J-data. We can learn to live with this useful idiom!

(b) To go along with this redefinition of diagrams as functors, we then have predictable redefinitions of cones and limit cones (we'll leave co-cones and colimits to look after themselves). We simply rework the definitions we met earlier in §19.1, 19.2.

So, look again at Defn. 78. This told us that a cone with vertex C over a diagram (old sense!) with objects D_j is a suite of arrows $c_j\colon C \to D_j$ where, for any arrow $d\colon D_k \to D_l$ in the diagram, the triangle formed with the arrows c_k and c_l commutes, so $c_l = d \circ c_k$. And look again at Defn. 79: this told us that, among the cones over a given diagram (old sense!), (L, λ_j) is a limit cone if, for any other cone (C, c_j), there is a unique arrow $u\colon C \to L$ such that, for every c_j, we have $c_j = \lambda_j \circ u$.

These old definitions naturally lead to the following two-part definition for cones and limit cones over diagrams in our new sense:

Definition 112. Suppose we are given a category C, together with a diagram-as-functor $D\colon \mathsf{J} \to \mathsf{C}$. By definition, D sends a J-arrow $j\colon J \to K$ to the C-arrow $Dj\colon DJ \to DK$. Then:

(1) *A cone over* D in C has an object C as vertex and an arrow $c_J\colon C \to DJ$ for each J-object J, subject to the following condition: for any J-arrow $d\colon K \to L$, we have $c_L = Dd \circ c_K$ in C. We again use (C, c_J) to denote such a cone.

(2) *A limit cone over* D is a cone (L, λ_J) such that for every cone (C, c_J) over D, there is a unique arrow $u\colon C \to L$ such that, for all J-objects J, $c_J = \lambda_J \circ u$. △

(c) How does our original way of talking of diagrams and limits relate to our new idiom?

Two quick general points:

(1) Evidently, not every diagram-in-C in the original sense of §6.1 corresponds exactly to a diagram-as-functor. There's a trivial reason. A diagram of shape J in C will always need to carry over the required identity arrows on all the objects in J to identity arrows on all their images. But a diagram-in-a-category as we first defined it doesn't need to have identity arrows on all (or any) of its objects.

(2) Still, the lack of an exact one-to-one correspondence between diagrams in the two senses makes no real difference when thinking about limits.

A limit (new sense) over the diagram $D\colon \mathsf{J} \to \mathsf{C}$ will of course be a limit (old sense) over the D-image of J living in C.

Conversely, suppose (L, λ_j) is a limit cone over some diagram D (in the old sense of diagram). Then by Theorem 85, (L, λ_j) is also a limit over the reflexive, transitive closure of the old-style diagram D because *every* cone over D is equally a cone over its closure. But we noted that we can think of this closure as a subcategory of C : call it J. So now take the inclusion functor $D_I \colon \mathsf{J} \to \mathsf{C}$. Then, by our new definition, (L, λ_j) becomes a limit cone over the diagram-as-functor $D_I \colon \mathsf{J} \to \mathsf{C}$.

In short, limits in the old and new senses come to the same; from now on, we can follow the widely-adopted line of taking the revised definitions of diagrams and their limits as our preferred one.[1]

(d) Our initial key examples of limits earlier were terminal objects, products, equalizers, and pullbacks. And in our original discussion in §19.2, we saw these to be respectively limits over the following miniature diagrams (in the old sense):

(1) the empty diagram,

(2) a two-object, zero-arrow, diagram,

(3) a two-object two-parallel-arrow, diagram, and

(4) a three-object, two-arrow, corner diagram.

Now thinking of the diagrams and limits in the new way, terminal objects, products, equalizers, and pullbacks become limit cones over the following kinds of diagrams-as-functors (with the sources of the respective functors being now some miniature categories):

(1′) a diagram of the shape of the empty category (we allowed the limiting case of an empty category in §5.1, and here's a context where we see why this is a convenient policy),

(2′) a diagram of the shape of the discrete two-object category $\overline{2}$: 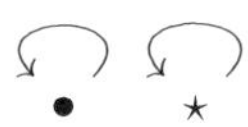

(3′) a diagram of the shape of the category 

(4′) a diagram of the shape of the category 

(e) Finally, some brisk terminology for future use (compare Defn. 87, and also compare Defn. 88 and the comment there on the use of 'small'):

Definition 113. A category C *has all limits of shape* J iff, for every functor $D \colon \mathsf{J} \to \mathsf{C}$, the category has a limit cone over D.

A category C *has all finite limits* iff, for any J with a finite number of objects and arrows, and every functor $D \colon \mathsf{J} \to \mathsf{C}$, the category has a limit cone over D. Such a category is said to be *finitely complete*.

A category C *has all small limits* iff, for any J which has no more than a set's worth of objects and of arrows,[2] and every functor $D \colon \mathsf{J} \to \mathsf{C}$, the category has a limit cone over D. Such a category is said to be *complete*. △

[1] See e.g. Borceux (1994, p. 56), Leinster (2014, p. 118) and Riehl (2017, §3.1) who take the definition of diagrams and limits in terms of functors as not merely preferred but basic.

[2] In other words, the objects form a set, or at least can be indexed by a set: likewise for the arrows. Compare §21.5.

28.2 Preserving limits

(a) Let's start with another definition, extending the notion of preservation we met in §27.1: we say a functor preserves limits if it sends limits of a given shape to limits of the same shape (and it preserves colimits if it sends colimits to colimits; but I won't keep mentioning the dual case). More carefully,

Definition 114. A functor $F\colon \mathsf{C} \to \mathsf{D}$ *preserves the limit* (L, λ_J) over $D\colon \mathsf{J} \to \mathsf{C}$ if and only if $(FL, F\lambda_J)$ is a limit over $F \circ D\colon \mathsf{J} \to \mathsf{D}$. △

But limits, we know, are only unique up to isomorphism. So let's quickly check that a functor will treat all limits over a given diagram the same way:

Theorem 140. *If* $F\colon \mathsf{C} \to \mathsf{D}$ *preserves a limit over the diagram* $D\colon \mathsf{J} \to \mathsf{C}$*, it preserves* all *limits over that diagram* D*.*

Proof. Suppose (L, λ_J) is a limit cone over $D\colon \mathsf{J} \to \mathsf{C}$. Then, by the argument of Theorem 82, if (L', λ'_J) is another such cone, there is an isomorphism $f : L' \to L$ in C such that $\lambda'_J = \lambda_J \circ f$.

Suppose now that F preserves (L, λ_J) so $(FL, F\lambda_J)$ is a limit cone over $F \circ D$. Then F will send (L', λ'_J) to $(FL', F\lambda'_J)$, i.e. $(FL', F\lambda_J \circ Ff)$. But then this cone factors through $(FL, F\lambda_J)$ via the arrow $Ff\colon FL' \to FL$, and Ff is an isomorphism (remember, functors preserve isomorphisms). Hence, by the argument of Theorem 83, $(FL', F\lambda'_J)$ is also a limit over $F \circ D$. In other words, F preserves (L', λ'_J) too. □

(b) We have been talking about preserving limits over one particular given diagram D. Next, let's say

Definition 115. A functor $F\colon \mathsf{C} \to \mathsf{D}$ *preserves all limits of shape* J *in* C iff, for *any* diagram $D\colon \mathsf{J} \to \mathsf{C}$, F preserves the limits over D. △

We should immediately note, however, that preserving all limits of one shape can go along with preserving none of some other shape. A toy example establishes the point. Take the two posets $(\{0, 1, 2, 3, 4, 5\}, \leqslant)$ and $(\mathbb{N}, \leqslant)$ thought of as categories. There is a trivial inclusion functor I from the first category to the second. This functor preserves all limits of the shape of the discrete category $\underline{2}$, i.e. all products (recall that the product of two elements in a poset, when it exists, is their greatest lower bound). But the inclusion functor doesn't preserve limits of the shape of the null category, i.e. doesn't map a terminal object to a terminal object (5 is terminal in the first category, but $5 = I(5)$ is not terminal in the second one).

However, some functors do preserve limits more generally. In particular, let's say:

Definition 116. A functor $F\colon \mathsf{C} \to \mathsf{D}$ *preserves all finite limits* iff it preserves limits of shape J whenever J is finite (i.e. has only finitely many objects and arrows).

$F\colon \mathsf{C} \to \mathsf{D}$ *preserves all small limits* iff it preserves limits of shape J whenever J's objects are not too many to form a set, and likewise for its arrows. △

28.3 A limit preservation theorem

(a) Given our earlier discussion of the existence of limits in Chapter 21 we can easily establish an important general result:

Theorem 141. *If* C *is finitely complete, and a functor* $F\colon \mathsf{C} \to \mathsf{D}$ *preserves terminal objects, binary products and equalizers, then* F *preserves all finite limits.*

Proof. Assume C is finitely complete so that for any functor $D\colon \mathsf{J} \to \mathsf{C}$ with J finite, then C has a limit cone (L, λ_J) over D. We want to show F preserves any such (L, λ_J), i.e. $(FL, F\lambda_J)$ is a limit over $F \circ D$.

By the argument in the proof of Theorem 97, there will also be a limit cone (L', λ'_J) over the diagram D constructed from equalizers and finite products. And by the general uniqueness-up-to-isomorphism result for limit cones, (L', λ'_J) is isomorphic to (L, λ_J).

Now, by assumption, F preserves terminal objects, binary products and equalizers, so F will send the construction (L', λ'_J) to a construction $(FL', F\lambda'_J)$ which will similarly be constructed from terminal objects, binary products and equalizers in a way making $(FL', F\lambda'_J)$ a limit cone over $F \circ D\colon \mathsf{J} \to \mathsf{D}$.

But F preserves isomorphisms, so F also sends the isomorphism in C between (L, λ_J) and (L', λ'_J) to an isomorphism in D between $(FL, F\lambda_J)$ and $(FL', F\lambda'_J)$. But whatever is isomorphic to a limit cone over $F \circ D$ is itself a limit cone over $F \circ D$. Hence $(FL, F\lambda_J)$ is indeed such a limit cone, showing that F preserves that limit. □

(b) Here's a simple application of that general result:

Theorem 142. *The forgetful functor* $F\colon \mathsf{Mon} \to \mathsf{Set}$ *preserves all finite limits.*

Of course, a functor that forgets that relevant arrows are structure-preserving should still leave the arrows that form limit cones in place. But let's check:

Proof. First, the forgetful functor $F\colon \mathsf{Mon} \to \mathsf{Set}$ evidently sends a terminal object in Mon, a one-object monoid, to its underlying singleton set, which is terminal in Set. So F preserves terminal objects (limits of the null shape).

Second, the same functor sends a product $(M, *, e) \times (N, \star, d)$ in Mon to its underlying set of pairs of objects from M and N, which is a product in Set. So the forgetful F also preserves products.

Third, we saw in §16.2, Ex. (2), the equalizer of two parallel monoid homomorphisms $(M, *, e) \underset{g}{\overset{f}{\rightrightarrows}} (N, \star, d)$ is E equipped with the inclusion map $E \to M$, where E is the set on which f and g agree. This means that the forgetful functor takes the equalizer of f and g as monoid homomorphisms to their equalizer as set functions. So F preserves equalizers.

So: we know from Theorem 100 that Mon is finitely complete, and have now seen that the forgetful functor $F\colon \mathsf{Mon} \to \mathsf{Set}$ preserves terminal objects, binary products and equalizers. Hence F preserves all the finite limits in Mon. □

(c) We've been concentrating on finite limits because in many contexts this is the interesting case. But we can go further, as we already briefly noted in §21.5. We just note again that our proof of Theorem 97 still goes through even when dealing with non-finite diagrams – and if we assume everything is set-sized, then the argument could still be dressed up as set-theoretically respectable. So, we will be able to beef up Theorem 142 to give

Theorem 143. *The forgetful functor* $F\colon \mathsf{Mon} \to \mathsf{Set}$ *preserves all small limits.*

But we needn't go into more details at this stage.

(d) Rather, let's quickly note

Theorem 144. *The forgetful functor* $F\colon \mathsf{Mon} \to \mathsf{Set}$ *does* not *preserve all co-limits.*

Proof. Take the simplest kind of colimit – initial objects (i.e. colimits under diagrams-as-functors with the 'shape' of the empty category). Then note that a one-object monoid is initial in Mon; but its underlying singleton set is not initial in Set. □

We might usefully also note that the forgetful F does not preserve coproducts either – essentially because coproducts in Mon can be larger than coproducts in Set. Recall our discussion in §11.7 of coproducts in Grp : similarly, $F(M + N)$, the underlying set of a coproduct of monoids M and N, is (isomorphic to) the set of finite sequences of alternating non-identity elements from M and N. Contrast $FM + FN$, which is simply the disjoint union of the underlying sets.

Our example of the forgetful functor $F\colon \mathsf{Mon} \to \mathsf{Set}$ generalizes, by the way. A forgetful functor from a category of structured sets to Set typically preserves finite limits but does not preserve all colimits. Though the forgetful functor $F\colon \mathsf{Top} \to \mathsf{Set}$ preserves not only limits but colimits too.

28.4 Reflecting limits

(a) Here's a companion definition to set alongside the definition of preserving limits, together with a couple of general theorems:

Definition 117. A functor $F\colon \mathsf{C} \to \mathsf{D}$ *reflects limits of shape* J iff, given a cone (C, c_J) over a diagram $D\colon \mathsf{J} \to \mathsf{C}$, if (FC, Fc_J) is a limit cone over $F \circ D\colon \mathsf{J} \to \mathsf{D}$, then (C, c_J) is itself a limit cone over D.

Reflecting colimits is defined dually. △

Theorem 145. *Suppose* $F\colon \mathsf{C} \to \mathsf{D}$ *is fully faithful. Then* F *reflects limits.*

Proof. Suppose (C, c_J) is a cone over a diagram $D\colon \mathsf{J} \to \mathsf{C}$, and (FC, Fc_J) is a *limit* cone over $F \circ D\colon \mathsf{J} \to \mathsf{D}$. We need to show that (C, c_J) must already be a limit cone too.

So take any other cone (B, b_J) over D. F sends this to a cone (FB, Fb_J) which must uniquely factor through the limit cone (FC, Fc_J) via some $u\colon FB \to FC$ which makes $Fb_J = Fc_J \circ u$ for each J-object J. Since F is full and faithful, $u = Fv$ for some unique $v\colon B \to C$ such that $b_J = c_J \circ v$ for each J. So (B, b_J) factors uniquely through (C, c_J). Which shows that (C, c_J) is a limit cone. □

Theorem 146. *Suppose $F\colon \mathsf{C} \to \mathsf{D}$ preserves finite limits. Then if C is finitely complete and F reflects isomorphisms, then F also reflects finite limits.*

Proof. Since C is complete there exists a limit cone (L, λ_J) over any diagram $D\colon \mathsf{J} \to \mathsf{C}$ (where J is finite), and so – since F preserves limits – $(FL, F\lambda_J)$ is a limit cone over $F \circ D\colon \mathsf{J} \to \mathsf{D}$.

Now suppose that there is a cone (C, c_J) over D such that (FC, Fc_J) is another limit cone over $F \circ D$. We want to show that (C, c_J) is also a limit cone.

(C, c_J) must uniquely factor through (L, λ_J) via a map $f\colon C \to L$. Which means that (FC, Fc_J) factors through $(FL, F\lambda_J)$ via Ff. However, since these are by hypothesis both limit cones over $F \circ D$, Ff must be an isomorphism. Hence, since F reflects isomorphisms, f must be an isomorphism. (C, c_J) must therefore be a limit cone by Theorem 83. □

(b) Since Mon is finitely complete, and the forgetful functor $F\colon \mathsf{Mon} \to \mathsf{Set}$ preserves limits and reflects isomorphisms, the last theorem shows that

(1) The forgetful functor $F\colon \mathsf{Mon} \to \mathsf{Set}$ reflects all finite limits. Similarly for some other forgetful functors from familiar categories of structured sets to Set.

However, be careful! For we also have ...

(2) The forgetful functor $F\colon \mathsf{Top} \to \mathsf{Set}$ which sends a topological space to its underlying set *preserves* all limits (as noted at the end of the previous section) but does not *reflect* all limits.

Here's a case involving binary products. Suppose X and Y are a couple of spaces with a coarse topology, and let Z be the space $FX \times FY$ equipped with a finer topology. Then, with projection arrows, $X \leftarrow Z \to Y$ is a wedge to X, Y but not a limit wedge in Top: but $FX \leftarrow FX \times FY \to FY$ is a limit wedge in Set.

Given the previous theorem, since Top is finitely complete (as remarked in Theorem 100), we can conclude that $F\colon \mathsf{Top} \to \mathsf{Set}$ doesn't reflect isomorphisms. Which is also something we can show directly.[3]

[3]Consider the continuous bijection from the half-open interval $[0, 1)$ to S^1. Think of this bijection as a topological map f; then f is not a homeomorphism in Top. However, treating the bijection as a set-function, i.e. as Ff, it *is* an isomorphism in Set.

29 Functors and comma categories

In the previous chapter, we were able to backtrack and illuminatingly rework our old ideas of diagrams and of limits over/under diagrams. That's just one example – though a centrally important one – of how we can use functors in redefining some familiar ideas.

In this chapter, we introduce the concept of a comma category. The resulting construction will be rather significant later, but working through the definition now will give us a nice exercise to help fix some ideas about functors. And we can immediately apply it to rework the ideas of slice categories and arrow categories that we met in Part I. In addition, we can get a rather nice result about free monoids.

29.1 Comma categories defined

(a) Suppose that we start with three categories $\mathsf{A}, \mathsf{B}, \mathsf{C}$ and a pair of functors involving them, $S\colon \mathsf{A} \to \mathsf{C}$ and $T\colon \mathsf{B} \to \mathsf{C}$. We are going to use these two functors to build a new category.

Now, our two functors S and T give us a way of indirectly connecting an object A in A to an object B in B: we look at their respective images SA and TB and then consider C-arrows $f\colon SA \to TB$ between them. (So the functor S provides the Source for the connecting arrow f and the functor T the Target – hence our chosen labels for the functors!)

We are going to build our new category out of such indirect connections. So let's stipulate that this category's objects are triples (A, B, f) comprising an A-object A and a B-object B together with a C-arrow $f\colon SA \to TB$.

What then could be the arrows in our new category (assuming these arrows are also assembled from ingredients available in $\mathsf{A}, \mathsf{B}, \mathsf{C}$)? Suppose we have two triples (A, B, f), (A', B', f'): an arrow between them should presumably somehow involve an A-arrow $a\colon A \to A'$ from A and a B-arrow $b\colon B \to B'$. But note that these two arrows are sent by S and T respectively to the arrows $Sa\colon SA \to SA'$ and $Tb\colon TB \to TB'$ in C; and we will presumably want these two C-arrows to interact appropriately with the given C-arrows f and f'.

These thoughts suggest the following definition:

Definition 118 (?). Given functors $S\colon \mathsf{A} \to \mathsf{C}$ and $T\colon \mathsf{B} \to \mathsf{C}$, then the *comma category* $(S \downarrow T)$ is the category with the following data:

(1) The objects of $(S \downarrow T)$ are triples (A, B, f) where A is an A-object, B is a B-object, and $f\colon SA \to TB$ is an arrow in C.

(2) An arrow of $(S \downarrow T)$ from (A, B, f) to (A', B', f') is a pair (a, b), where $a\colon A \to A'$ is an A-arrow, $b\colon B \to B'$ is an B-arrow, such that the following diagram commutes:

$$\begin{array}{ccc} SA & \xrightarrow{\;f\;} & TB \\ \downarrow{\scriptstyle Sa} & & \downarrow{\scriptstyle Tb} \\ SA' & \xrightarrow{\;f'\;} & TB' \end{array}$$

(3) The identity arrow on the object (A, B, f) is the pair $(1_A, 1_B)$.

(4) Composition in $(S \downarrow T)$ is induced by the composition laws of A and B thus: $(a', b') \circ (a, b) = (a' \circ_{\mathsf{A}} a,\, b' \circ_{\mathsf{B}} b)$. △

So at any rate runs the usual definition.[1]

The label 'comma category', by the way, comes from a slightly unhappy earlier notation '(S, T)' – that notation has long been abandoned but the name has stuck.

(b) However, as with our initial attempt at characterizing a slice category back in §7.3, there is a snag. Suppose there are two C-arrows $f_1, f_2\colon SA \to TB$; then we have two distinct $(S \downarrow T)$-objects (A, B, f_1) and (A, B, f_2). But our proposed new definition would assign these distinct objects the same identity arrow in $(S \downarrow T)$, namely the pair $(1_A, 1_B)$. However, distinct objects can't share an identity arrow.

Hence our account of $(S \downarrow T)$ doesn't quite work as it stands as a kosher definition of a category. What to do? In some way, we want to recast an arrow from the object (A, B, f) to the object (A', B', f') into a triple $(a, b, ?)$ where the mystery ingredient tells us something about the third components of the object-triples involved. Well, why not simply choose the commutative squares that are in the story anyway? And then the identity arrow on (A, B, f_1) will involve a different third component from the identity arrow on (A, B, f_2). Hooray! So let's officially tinker with our suggested definition to get

Definition 118. As before, except in (2) we define the arrow as the pair (a, b) together with the commuting diagram, and adjust (3) and (4) to match. △

However, it would be a very annoying complication to stick religiously to the official story. From now on, we'll cheat a tiny bit, just as we did with slice categories. So when we talk of comma categories, we'll continue to talk of an arrow from (A, B, f) and (A', B', f') as if it is simply a suitable pair of arrows which (when hit with S and T) gives rise to a commuting square. Since that story gives us the square anyway, no information at all is lost if we don't beef up the pair into a triple including that square. Allowing ourselves that pinch of salt, $(S \downarrow T)$ can count as a category.

[1]See for example Adámek et al. (2009, p. 46), Barr and Wells (1985, p. 13), Riehl (2017, p. 22), Simmons (2011, p. 86).

29.2 Three types of comma category

But why on earth we should be bothering with this construction?

Well, for a start, the notion of a comma category nicely generalizes a number of simpler constructions. And we have in fact already met two comma categories in thin disguise.

(a) First take the minimal case where $\mathsf{A} = \mathsf{B} = \mathsf{C}$, and where both S and T are the identity functor on that category, 1_{C} (as introduced in Theorem 129). Then,

(1) The objects in this category $(1_{\mathsf{C}} \downarrow 1_{\mathsf{C}})$ are triples $(X, Y, X \xrightarrow{f} Y)$ for X, Y both C-objects.

(2) An arrow in $(1_{\mathsf{C}} \downarrow 1_{\mathsf{C}})$ from $(X, Y, X \xrightarrow{f} Y)$ to $(W, Z, Y \xrightarrow{g} Z)$ is a pair of C-arrows $j\colon X \to W$, $k\colon Y \to X$ such that this square commutes:

$$\begin{array}{ccc} X & \xrightarrow{\ f\ } & Y \\ \downarrow{\scriptstyle j} & & \downarrow{\scriptstyle k} \\ W & \xrightarrow{\ g\ } & Z \end{array}$$

So the only difference between $(1_{\mathsf{C}} \downarrow 1_{\mathsf{C}})$ and the arrow category $\mathsf{C}^{\rightarrow}$ as defined in §7.4 is that we have now 'decorated' the objects of $\mathsf{C}^{\rightarrow}$, which were plain C-arrows $f\colon X \to Y$, with explicit assignments of their sources and targets as C-arrows, to give triples $(X, Y, X \xrightarrow{f} Y)$. Hence $(1_{\mathsf{C}} \downarrow 1_{\mathsf{C}})$ and $\mathsf{C}^{\rightarrow}$, although not strictly identical, evidently amount to the same in some good sense (more about that last thought in a moment).

(b) Let's secondly take another special case, this time one where $\mathsf{A} = \mathsf{C}$, while $\mathsf{B} = \mathsf{1}$ (the category with a single object $\star$ and the single identity arrow $1_\star$). So in place of the functor $S\colon \mathsf{A} \to \mathsf{C}$ we can have the identity functor 1_{C}. And in place of the functor $T\colon \mathsf{B} \to \mathsf{C}$ there will be some functor $X\colon \mathsf{1} \to \mathsf{C}$, that sends $\star$ to an individual C-object which we'll also call X and sends $1_\star$ to 1_X – see §26.2, Ex. (F7). Our definition will now grind out the following specification for the relevant comma category:

(1) The objects of $(1_{\mathsf{C}} \downarrow X)$ will be triples $(A, \star, f)$, where A is any C-object and f any C-arrow from A to X.

(2) And a $(1_{\mathsf{C}} \downarrow X)$-arrow between $(A, \star, A \xrightarrow{f} X)$ and $(B, \star, B \xrightarrow{g} X)$ will be a pair $(j, 1_\star)$, with $j\colon A \to B$ an arrow such that this square commutes:

$$\begin{array}{ccc} A & \xrightarrow{\ f\ } & X \\ \downarrow{\scriptstyle j} & & \downarrow{\scriptstyle 1_X} \\ B & \xrightarrow{\ g\ } & X \end{array}$$

But of course this square is trivially equivalent to the triangle

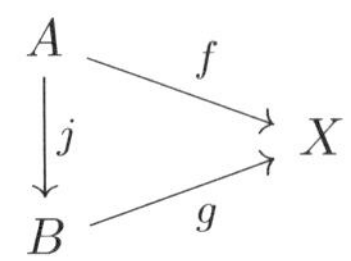

And *that* should look rather familiar! We've ended up with something tantamount to a slice category C/X. The only differences are that (i) instead of the original slice category's objects, i.e. pairs like (A, f), in the category $(1_\mathsf{C} \downarrow X)$ we now have corresponding 'decorated' pairs $(A, \star, f)$. And (ii) instead of the slice category's arrows like $j\colon A \to B$ satisfying a certain condition, in the category $(1_\mathsf{C} \downarrow X)$ we have corresponding 'decorated' arrows $(j, 1_\star)$, with j still satisfying the same condition.

Hence we can again say: the comma category $(1_\mathsf{C} \downarrow X)$ and the slice category C/X in some good sense amount to the same. Exactly similarly, of course, the comma category $(X \downarrow 1_\mathsf{C})$ and the co-slice category X/C amount to the same.

(c) Let's add a third illustrative case for future use. This time A and C may be distinct categories, and we make no special assumptions about the functor $S\colon \mathsf{A} \to \mathsf{C}$. However, we again put $\mathsf{B} = \mathsf{1}$ and take the second functor used in constructing our comma category to be some $X\colon \mathsf{1} \to \mathsf{C}$ which sends the unique object $\star$ of $\mathsf{1}$ to an individual C-object X and sends $1_\star$ to 1_X.

Turning the handle once more, our definition of a comma category now grinds out this:

(1) The objects of $(S \downarrow X)$ will be triples $(A, \star, f)$, where A is any A-object and f is any C-arrow from $SA \to X$.

(2) And a $(S \downarrow X)$-arrow between $(A, \star, SA \xrightarrow{f} X)$ and $(A', \star, SA' \xrightarrow{f'} X)$ will be a pair $(j, 1_\star)$, with $j\colon A \to A'$ an arrow in A such that this square commutes in C:

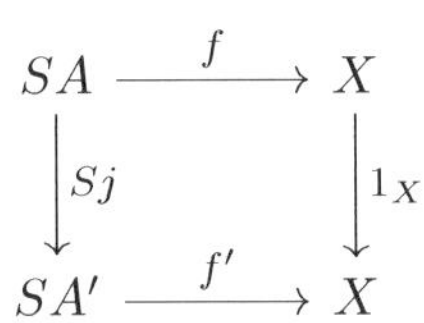

But as with our last example, the $\star$ component of objects and the $1_\star$ component of arrows are just coming along for the ride, doing no real work. And our commuting square is equivalent to a triangle. So we might as well say, more snappily,

(1′) The objects of $(S \downarrow X)$ are pairs (A, f), where A is any A-object and f is any C-arrow from $SA \to X$.

(2′) And a $(S \downarrow X)$-arrow between $(A, SA \xrightarrow{f} X)$ and $(A', SA' \xrightarrow{f'} X)$ will be an A-arrow $j\colon A \to A'$ such that this triangle commutes in C:

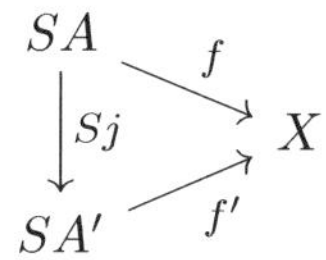

Likewise, for a companion dual definition, now assume instead that $\mathsf{A} = 1$ and take our two functors to be some $X\colon 1 \to \mathsf{C}$ and $T\colon \mathsf{B} \to \mathsf{C}$; then we can say – jumping straight to the snappy version –

(1″) The objects of $(X \downarrow T)$ will be pairs (B, f), where B is any B-object and f is any C-arrow from X to TB.

(2″) And a $(X \downarrow T)$-arrow between $(B, X \xrightarrow{f} TB)$ and $(B', X \xrightarrow{f'} TB')$ will be an B-arrow $j\colon B \to B'$ such this triangle commutes in C:

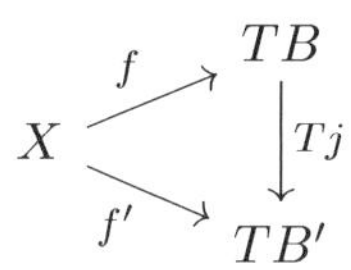

(d) A minor terminological point about our third kind of case of a comma category.

Officially, a comma category $(S \downarrow T)$ is defined in terms of two functors, S and T. When we just defined $(S \downarrow X)$, the 'X' still denotes a *functor*, a functor $X\colon 1 \to \mathsf{C}$ that picks out a particular object X. But of course, it would make no difference in this sort of case if we took the 'X' in notations like $(S \downarrow X)$ to officially denote an *object*, the one which gives rise to the corresponding functor X. I mention this as you'll often find the notation being interpreted that way, and find corresponding definitions of the likes of $(S \downarrow X)$ that consequently only need to explicitly mention one functor, in this case S.

(e) Now, three times in this section I've done a bit of hand-waving! I said that the categories $(1_\mathsf{C} \downarrow 1_\mathsf{C})$ and $\mathsf{C}^\to$, although not strictly identical, 'evidently amount to the same'. Similarly, I said the comma category $(1_\mathsf{C} \downarrow X)$ and the slice category C/X 'in some good sense amount to the same'. And just now, I claimed that instead of defining the objects of $(S \downarrow X)$ as triples $(A, \star, f)$, 'we might as well' treat them as pairs (A, f). Now, those claims hopefully looked intuitively sensible; but still, we ought to set down an official account of when categories do 'come to the same'.

You can guess what the shape of the account will be: categories will 'come to the same', at least in the strongest sense, when there is a functor that is an isomorphism between them. And you should already be able to guess too what it takes for a functor to be an isomorphism in the required sense. However, I won't spell this out right now: we will return to this theme soon, in Chapter 34.

29.3 An application: free monoids again

Recall from §26.5 that the free monoid on X can be identified with $L_X = (List(X), ^\frown, \emptyset)$, where $List(X)$ is the set of finite lists of members of X, the monoid operation is simply concatenation, and the monoid's unit is the empty list. We can now make a neat connection between L_X and a certain comma category.

Start with the categories 1, Mon, and Set; take the functor $X\colon 1 \to \mathsf{Set}$ that picks out the particular set X, and the forgetful functor $F\colon \mathsf{Mon} \to \mathsf{Set}$. Applying the definition from the last section, we get a comma category $(X \downarrow F)$:

(1) The objects of $(X \downarrow F)$ are pairs (M, f) of a monoid M and a function f in Set from X to FM, i.e. a function f from X to $\underline{M}$, the underlying set of the monoid.

(2) An $(X \downarrow F)$-arrow from the pair (M, f) to the pair (N, g) is a monoid homomorphism $j\colon M \to N$, such that $g = Fj \circ f$ in Set.

Then we have the following result:

Theorem 147. *Equip the free monoid on X, $L_X = (List(X), ^\frown, \emptyset)$, with the function $l\colon X \to List(X)$ that sends an element x of X to the one-element list with x as sole element. Then (L_X, l) is an initial object of $(X \downarrow F)$.*

Proof. Suppose (N, g) is an object of $(X \downarrow F)$ – so N is a monoid and $g\colon X \to \underline{N}$ is a set function. We need to show that there is a unique $(X \downarrow F)$-arrow from (L_X, l) to (N, g). In other words, there exists a unique monoid homomorphism $j\colon L_X \to N$ such that $g = Fj \circ l$.

For existence, let j send the empty list $\emptyset$ in $List(X)$ to e, the unit of N, and send a one-element list x from $List(X)$ to $g(x)$. Extend the function j to all members of $List(X)$ by putting $j(x_1 ^\frown x_2 ^\frown \ldots ^\frown x_n) = j(x_1) \star j(x_2) \star \ldots \star j(x_n)$. Then j is easily seen to be a monoid homomorphism; and by construction, $g = Fj \circ l$.

For uniqueness, suppose k is another monoid homomorphism $k\colon L_X \to N$ such that $g = Fk \circ l$. Then, being a homomorphism, k needs to send the empty list to the unit of N. And because $g = Fk \circ l$, k has to agree with j on single elements of X, both sending an element x to $g(x)$. Hence

$$\begin{aligned} k(x_1 ^\frown x_2 ^\frown \ldots ^\frown x_n) &= k(x_1) \cdot k(x_2) \cdot \ldots \cdot k(x_n) \\ &= j(x_1) \cdot j(x_2) \cdot \ldots \cdot j(x_n) \\ &= j(x_1 ^\frown x_2 ^\frown \ldots ^\frown x_n). \end{aligned}$$

Whence j and k must agree on all members of $List(X)$. □

We originally defined the idea of a free monoid on some given objects by a concrete construction. Now we see that we can define it, up to isomorphism of course, by a unique mapping property, as part of a colimit, the initial object in a certain category. That's a rather neat result!

30 Hom-sets (and some matters of size)

At this point, I do need to return to our original definition of a category and to compare it with another style of definition which you are very likely to encounter elsewhere. This alternative type of definition pivots on the key notion of a *hom-set* which we are now going to need, and builds in the idea that there should be only set-many arrows between any two objects in a category. I'll explain.

So here we will need to touch lightly on foundational issues to do with the 'size' of categories. But I'm with Emily Riehl (2017, p. 6): mostly, we will "sweep these foundational issues under the rug, not because these issues are not serious or interesting, but because they distract from the task at hand."

30.1 Defining categories again

(a) We have been working with a definition – Defn. 14 – of the following type:[1]

Type I Defn. A category C comprises two kinds of things:

(1) Some C-objects,
(2) Some C-arrows.

These objects and arrows are governed by such-and-such axioms. △

Three comments. First, this definition says nothing about what kinds of entities the objects and arrows of a category might be. This seems entirely consonant with the categorial philosophy that we should really focus on the ways that the relevant objects and arrows inter-relate, not on their intrinsic make-up.

Second, our Type I definition also puts no limit on the number of objects a category might have, nor on the number of arrows there might be between any two objects in the category.

Third, this definition says that what we need for a category are some objects and some arrows satisfying certain conditions. It makes no requirement that there exist *collections*, *classes* or *sets* of these objects or arrows, at least in any sense which conceives of these as being further entities in their own right over and above their members.

[1] As previously noted in §5.1, fn. 1, you will find definitions of categories of Type I in e.g. Awodey (2010, p. 4), Lawvere and Schanuel (2009, p. 21) and McLarty (1992, p. 13).

(b) Here's a (minor?) variant of the Type I definition:[2]

Type I* Defn. A category C comprises two (non-overlapping) collections:

(1) A collection of C-objects,
(2) A collection of C-arrows.

These objects and arrows are governed by such-and-such axioms. △

Again, this version puts no constraints on the nature or number of the objects and arrows in a category. The two versions will only peel apart to the extent that we take the talk of 'collections' as committing us to additional entities, over and above the objects and arrows. Proponents of Type I* definitions vary in their views here: but let's not worry about this too much.

(c) Now compare the Type I versions with an alternative style of definition that is often proposed:[3]

Type II Defn. A category C comprises:

(1) A class or collection $Ob(\mathsf{C})$ whose elements are called the objects of the category,
(2) For each pair A, B of such objects, a set $Hom_{\mathsf{C}}(A, B)$ whose elements are called the arrows from A to B.

These objects and arrows are governed by such-and-such axioms. △

The old-school notation 'Hom_{C}' reminds us that, in many typical concrete categories C, sets of arrows will be sets of homomorphisms of some kind. And the hom-sets $Hom_{\mathsf{C}}(A, B)$ and $Hom_{\mathsf{C}}(A', B')$ for distinct pairs of objects (A, B) and (A', B') are usually assumed to be disjoint.

There's a question about how seriously we are to take the talk of a 'class' or 'collection' in the first clause; but again, let's not pause over that now. For the first point I want to highlight is that our Type II definition *does* put a constraint on the number of arrows there can be between any pair of objects in a category – there can only be 'a set's worth' (cf. Defn. 113).

Let's introduce a definition:

Definition 119. A category is *small* iff it has only a set's worth of objects and a set's worth of arrows.

A category is *locally small* iff it has only a set's worth of arrows between any pair of its objects. △

Then a Type II definition builds into the very definition of a kosher category that it must be at least locally small (a nice property for certain purposes). But how restrictive is this?

[2]See e.g. Crole (1993, p. 40), Goldblatt (1984, p. 24), Pierce (1991, p. 1), Riehl (2017, p. 3), Simmons (2011, p. 2), Yanofsky (2024, p. 34), who all talk of 'collections' in this way.

[3]For definitions of Type II, in books old and new, you could see e.g. Agore (2023, p. 2), Borceux (1994, p. 4), Pareigis (1970, p. 1), Richter (2020, p. 7), Roman (2017, p. 2), Schubert (1972, p. 1), and Taylor (1999, p. 184). 'Collection' is used in the first clause instead of 'class' by e.g. Adámek et al. (2009, p. 21) and Spivak (2014, p. 93).

(d) Consider the category Set. Assuming, as before, that our universe of sets is a model of standard set theory (rather than e.g. of Quine's deviant NF which countenances a universal set), the objects of Set are too many to themselves form a set. And since each object has its proprietary identity arrow, the arrows of the category are again too many to themselves form a set. Hence Set is certainly not a small category.

However – an important new point – when we home in on the arrows, i.e. functions, between any two given objects of Set, there will always be a set containing just *them*. In the standard construction, functions from A to B are treated as graphs, and these live only a few levels up from $(A \cup B)$; and if, as we should, we treat Set-arrows from A to B as triples of domain/graph/source, these will live only a bit higher. There is therefore only ever a set's worth of such arrows. Hence the second clause of our Type II definition – although it puts a constraint on how many arrows a category has between any given pair of objects – *does* happily cover Set and other sweepingly inclusive categories like Grp, Top and the like.

So far, so good.

(e) It is, however, actually quite easy to cook up somewhat artificial examples which count as categories according to a Type I definition but are not locally small, so don't accord with a Type II definition.[4] And later, we'll find that there are quite natural ways of building 'large' category-like structures.

Now, we could say that these larger structures which don't fit the Type II story are, as it were, hyper-categories, to be discussed separately. But perhaps a better response starts from the reflection that being 'small' or 'locally small' is only defined relative to our current ambient universe of sets. So let's make the relativity explicit, and revise our earlier definition accordingly:

Definition 119*. Let U denote our current default universe of sets. Then a category is *U-small* iff its objects and its arrows can both be indexed by U-sets.

A category is *locally U-small* if, for any of its objects A and B, there is only a U-set's worth of arrows from A to B, i.e. in each case the arrows from A to B can be indexed by some U-set. △

So the situation can be this. We start off working in a universe U and define Set, Grp, Top and the like to be the categories of all the sets, groups, topological spaces, etc., which can be implemented in U. Those categories will be locally U-small. Later, we can go on to construct larger categories which can have more arrows between a pair of objects than can be indexed by U-sets. But then, to be keep things under control, we can upgrade to a more capacious ambient universe

[4] For example, recall from §5.4 that any monoid $(\mathsf{M}, *, e)$ can be treated as a Type I category M with a single object $\bullet$ and with each m from among M counting as an M-arrow $m\colon \bullet \to \bullet$. So if there are too many *objects* M in the monoid to form a set, then the associated category M will have too many *arrows* from $\bullet$ to $\bullet$ to form a set. Consider then the particular case of, say, the monoid of von Neumann ordinals under addition. There is no set of all ordinals; hence the corresponding Type I monoid-as-category will not be locally small and hence won't be a Type II category. (Fine print: if you think that a monoid properly so-called can only have a set's worth of arrows, you'll need to rephrase the argument.)

of sets U^+, and find that these larger categories which aren't locally U-small are well-behaved enough to be locally U^+-small (or outright U^+-small). And so it goes – when we find that we want to consider bigger categories than can be implemented in our current universe U, avail ourselves of a bigger set-universe U^+ to work in.

Of course, that's *very* arm-waving! But the story can be made good at a price, and then the Type II idea can fly: our categories at any point of the story can be treated as (at least locally) small relative to our current working universe.[5]

30.2 Where do hom-sets live?

(a) The Type II definition, however, does more than claim that there should only be a set's worth of arrows between any pair of objects (in other words, those arrows can at least be indexed by a set). It explicitly says that the arrows between any objects A to B will form a set, the hom-set $Hom_{\mathsf{C}}(A, B)$. OK: but, to put the question snappily, *where do these hom-sets live?*

The standard assumption is that, whatever the category C, all of its hom-sets can be found over in the category Set.

Typically, authors at some point define, for each C-object A, a corresponding map which sends each object X in the category C to the corresponding hom-set $Hom_{\mathsf{C}}(A, X)$. And then each such map is taken to be the object-component of a so-called hom-functor from C to Set. That requires each hom-set $Hom_{\mathsf{C}}(A, X)$ to be an object of Set.[6]

"Surprise, surprise! Hom-*sets* are objects in the category of all *sets*! Where else should we find them?" But hold on a moment. As noted before, Set is usually taken to be – not the world of all sets in some loose and generous sense, but more narrowly – a universe of *pure* sets, a model of the standard set theory ZFC or some extension: see again the discussion in §4.3 (with its fn. 12) and §5.6. So if a set of C-arrows like $Hom_{\mathsf{C}}(A, X)$ actually is a pure set belonging to Set then *its members, the* C*-arrows, must themselves be pure sets.*

Now, assuming that the arrows of a category C are pure sets doesn't, strictly speaking, require that C's objects are pure sets too. But that would be the natural companion assumption. So assuming that, for any locally small category

[5]What I'm alluding to here is the trick of adding to our standard ZFC set theory Grothendieck's Axiom of Universes. You'll find a story about this briefly in Borceux (1994, §1.1), or with a little more detail online at tinyurl.com/GrUniv.

But this is only one option in the marketplace of ideas for coping with 'large' categories in an ambient set theory: enthusiasts can glance at the rightly much-cited paper on 'Set Theory for Category Theory' by Mike Shulman (2008). But if you follow up that reference, you'll very quickly see why I don't want to try to unknot the tangles here in these introductory notes. Verdicts on how best eventually to deal with very large categories won't affect the topics we want to discuss at the level of this 'naive' presentation of entry-level category theory. So we'll cut ourselves some slack.

[6]Just to take our Type II authors, you'll find this move clearly made by Agore (2023, pp. 26–27), Borceux (1994, p. 9), Pareigis (1970, p. 11), Richter (2020, p. 20), Roman (2017, p. 42), Schubert (1972, p. 7), and Taylor (1999, p. 237(f)).

C, its arrows between two objects assemble into a set living in Set, pushes us towards the idea that all the ingredients of C are sets, pure if not simple!

But at the end of the day do we *really* want to be restrictive and assume that categories – or at any rate, all the categories we want to theorize about which aren't too big – all live in the universe of pure sets? We don't want to rule out conceiving of category theory as offering a framework for organizing the mathematical universe which provides an alternative to full-strength set-theoretic reductionism.[7]

Let's be clear. It is one thing to highlight a category like Grp which neatly corrals together all the groups that we finding living in some favoured generous arena of sets, as recommended in §4.3. It is quite another thing to suppose that *all* structured families of groups or, more generally, *all* categories live there. Indeed, don't we want to be able to tell stories e.g. about how faithful functors can map categories out in the wild into our favoured universe of sets?

So there is reason after all to prefer a Type I definition of categories which allows us to be more ecumenical about the nature of the gadgets of a category. We can then echo Awodey (2010, p. 5): "A category is anything that satisfies [our Type I] definition. ... I want to emphasize that the objects do not have to be sets" and the arrows too need not be implemented as sets.[8]

30.3 Hom-sets, officially?

(a) Note, however, that whether or not we adopt a Type I or Type II definition of category, we will still want to talk about hom-sets (living in Set) and hom-functors (from C to Set). In fact, these will be centre stage in coming chapters.

But how can we treat collections of C-arrows between objects A and B *as if* they are hom-sets living in Set without assuming that C-arrows are sets?

By officially regarding a hom-set $Hom_{\mathsf{C}}(A, B)$ living in Set as faithfully *representing* the associated collection of C-arrows. If C is locally small, it only has a set's worth of arrows from A to B, meaning that these arrows can be indexed one-to-one by the members of some set: so – the basic point will be – we can take a suitable indexing set in Set to be the hom-set $Hom_{\mathsf{C}}(A, B)$.[9]

[7] Contrast: "Presumably, you know what set theory is important. You may not know that set theory is *all*-important. *All* abstract mathematical concepts are set-theoretic. *All* concrete mathematical objects are specific sets." (Kunen 2012, p. 14) Really?!

[8] Interestingly, Mac Lane himself, who initially (as we noted in §4.3) takes categories to be all implemented in a universe of sets, rather changes his official position in a short Appendix written for the second edition of *Categories for the Working Mathematician*. He now writes

> We have described a category in terms of sets, as a set of objects and a set of arrows. However, categories can be described directly – and they can then be used as a possible foundation for all of mathematics, thus replacing the use in such a foundation of the usual Zermelo-Fraenkel axioms for set theory. Here is the direct description: ...

And what follows is a straight Type I definition.

[9] I learnt this way of thinking of a hom-set as indexing the relevant arrows from Crole (1993, e.g. at p. 61).

This trick allows us to have our cake and eat it: we get hom-sets $Hom_{\mathsf{C}}(A, B)$ living in Set and yet also avoid having to assume that C-arrows are sets. Great. However, having made the basic point, how much do we want to keep emphasizing it or fussing about the details? Let's not! And this is one of those occasions where it makes for ease of exposition if we go on to abuse notation: so when $f\colon A \to B$ is a C-arrow, then we will also denote its representation in our chosen $Hom_{\mathsf{C}}(A, X)$ by $f\colon A \to B$. Then, in general, it will do no harm if we continue to talk as if the members of $Hom_{\mathsf{C}}(A, X)$ are, after all, the C-arrows themselves rather than representations of them. This makes life easier, and again keeps us marching in step with conventional presentations.[10] So the price is right.

[10] A famous-to-philosophers quote: "To think with the learned, and speak with the vulgar, is the proper way to gain [i.e. gain the understanding of] others, and to preserve one's own clearness and frankness." George Berkeley (1685–1753).

31 Hom-functors

This chapter gets to the heart of things by introducing the key notion of a *hom-functor*, a type of functor that will play a starring role in some later chapters (featuring essentially, for example, in the famed Yoneda Lemma). After distinguishing two kinds of hom-functors, we can link up with the discussion in Chapter 28 by showing that one sort plays very nicely with limits.

31.1 Two kinds of hom-functors

For brevity's sake, let's introduce some (standard) snappier notation:

Definition 120. We will henceforth denote the hom-set $Hom_{\mathsf{C}}(A, B)$ simply by $\mathsf{C}(A, B)$.

We say that a category C is locally small if and only if – for any of its objects A, B – the arrows from A to B can be indexed by a hom-set $\mathsf{C}(A, B)$ living in Set. But, in the spirit of §30.2(a), for brevity's sake we will talk as if the members of the hom-set really are the relevant C-arrows themselves (not just representations of them).

(a) Suppose, then, that C is locally small, and choose a fixed C-object A. Then, as we vary X through the objects in C, we get varying hom-sets $\mathsf{C}(A, X)$. In other words, for a given A, there is a function that sends X in C to the – possibly empty! – corresponding hom-set $\mathsf{C}(A, X)$ in Set.

Can we treat this function on C-*objects* as the first component of a functor – a *hom-functor* – from C to Set?

Well, how could we fix the second component of the functor, the component needed to deal with the C-*arrows*? By definition, such a component must send an arrow $j\colon X \to Y$ in C to a corresponding Set-function from $\mathsf{C}(A, X)$ to $\mathsf{C}(A, Y)$. And the obvious candidate for the latter function is the one we can notate as $j \circ -$, i.e. the function that maps any $h\colon A \to X$ in $\mathsf{C}(A, X)$ to $j \circ h\colon A \to Y$ in $\mathsf{C}(A, Y)$.[1]

Do these two components assemble into a hom-functor we can notate $\mathsf{C}(A, -)$? Yes! We have:

[1]Note: assuming that an arrow $h\colon A \to X$ exists, the arrow $j \circ h\colon A \to Y$ must be in $\mathsf{C}(A, Y)$ – because C is a category that by hypothesis contains h and j and hence contains their composite.

Theorem 148. *If* C *is a locally small category, and* A *is one of its objects, then there is a (covariant) functor* $\mathsf{C}(A,-)$ *from* C *to* Set *which operates like this:*

$$\begin{array}{lrcl} \mathsf{C}(A,-)\colon & X & \longmapsto & \mathsf{C}(A,X) \\ & j\colon X \to Y & \longmapsto & j \circ -\colon \mathsf{C}(A,X) \to \mathsf{C}(A,Y). \end{array}$$

Proof. Temporarily write $\mathsf{C}(A,-)$ as simply F for short. Then Fj applied to a member h of $\mathsf{C}(A,X)$ yields $j \circ h$: in other words, $Fj(h) = j \circ h$.

To confirm functoriality, note first that $F1_X(h) = h$ for any $h\colon A \to X$. Hence $F1_X$ is the function that maps any member of $\mathsf{C}(A,X)$ to itself, making it the identity function for $\mathsf{C}(A,X)$, i.e. $1_{F(X)}$.

Second, note that for any h, $F(j \circ k)(h) = (j \circ k) \circ h = j \circ (k \circ h) = F(j)(k \circ h) = F(j)(F(k)(h)) = (Fj \circ Fk)(h)$ Hence $F(j \circ k) = Fj \circ Fk$. □

(b) Now, start again from the hom-set $\mathsf{C}(A,B)$ but this time keep B fixed: as we vary X through the objects in C, we get corresponding hom-sets $\mathsf{C}(X,B)$. This generates a function that sends an object X in C to an object $\mathsf{C}(X,B)$ in Set.

To upgrade *this* into a hom-functor $\mathsf{C}(-,B)$, we must again add a component to deal with C-arrows. This will need to send $j\colon X \to Y$ in C to some function from $\mathsf{C}(X,B)$ to $\mathsf{C}(Y,B)$. But if we are going to use the same sort of idea as before, in order to get functions to compose properly, things have to go the other way about. I mean $\mathsf{C}(-,B)$ will have to send an arrow $j\colon X \to Y$ to the function we can notate as $- \circ j$ that maps any arrow $h\colon Y \to B$ in $\mathsf{C}(Y,B)$ to $h \circ j\colon X \to B$ in $\mathsf{C}(X,B)$.

We can leave it as an exercise to check:

Theorem 149. *If* C *is a locally small category, and* B *is one of its objects, then there is a (contravariant) functor* $\mathsf{C}(-,B)$ *from* C *to* Set *which operates like this:*

$$\begin{array}{lrcl} \mathsf{C}(-,B)\colon & X & \longmapsto & \mathsf{C}(X,B) \\ & j\colon Y \to X & \longmapsto & - \circ j\colon \mathsf{C}(X,B) \to \mathsf{C}(Y,B). \end{array}$$ □

Or noting the point made at the end of §26.6, you might prefer to put it this way:

Theorem 149*. *If* C *is a locally small category, and* B *is one of its objects, then there is a (covariant) functor* $\mathsf{C}(-,B)$ *from* C^{op} *to* Set *which operates like this:*

$$\begin{array}{lrcl} \mathsf{C}(-,B)\colon & X \text{ (in } \mathsf{C}^{op}) & \longmapsto & \mathsf{C}(X,B) \\ & j\colon X \to Y \text{ (in } \mathsf{C}^{op}) & \longmapsto & - \circ j\colon \mathsf{C}(X,B) \to \mathsf{C}(Y,B). \end{array}$$ □

And yes, the second clause here is right: we need to combine the C-arrow h from $\mathsf{C}(X,B)$ with another C-arrow, and in its guise as a C-arrow j does go from Y to X. So $h \circ j$ is correctly composed.

(c) A quick note on notation. The use of a blank in the notation '$\mathsf{C}(A,-)$' invites a handy shorthand: instead of writing e.g. '$\mathsf{C}(A,-)_{arw}(j)$' or '$\mathsf{C}(A,-)j$' to indicate the result of taking the component of the functor that acts on arrows and applying it to the particular arrow j, we can more snappily write '$\mathsf{C}(A,j)$'. Similarly for the dual case.

31.2 Points of view

Those definitions were straightforward enough; but why care? What's so great about hom-functors?

Look at it this way. Fix a C-object A. Then given any C-object X, the covariant hom-functor $\mathsf{C}(A, -)$ functor outputs the set of all the arrows from A to X. So we can think of the hom-functor as giving all the various ways that we can probe a given object X from A: as we vary X, the hom-functor encapsulates how A 'sees' its world. Now start from any other object A'; the corresponding hom-functor $\mathsf{C}(A', -)$ likewise encapsulates A''s view of its world. Covariant hom-functors, in short, survey their categorial world from various possible points of view.

An obvious issue arises. The views of the world from distinct objects A and A' won't be the same: but the perspectives should coherently fit together. There should be a story to be told about how going from A to A' transforms the view, depending on how A and A' themselves relate. Dropping the metaphor, we want a coherent story about the relations between the hom-functors $\mathsf{C}(A, -)$ and $\mathsf{C}(A', -)$. That is going to presuppose a general story about 'transformations' between functors, and we haven't got one yet: this general story will be the business of the next two chapters. We then return to the story about transformations between hom-functors in particular in Chapter 37.

Similarly for a contravariant hom-functor like $\mathsf{C}(-, B)$. This encapsulates how B is seen by its world. For given any object X, the functor outputs the set of all the arrows from X to B, i.e. all the various ways that B is seen by X. Another natural issue arises, the dual of the one before. The ways that the world sees B and B' won't be the same; but there should be a story to be told about how the views fit together, depending on how B and B' relate. More later on this too.

31.3 Covariant hom-functors preserve limits

There is, however, an important result which we *can* prove right now:

Theorem 150. *Suppose that* C *is a locally small category. Then the covariant hom-functor* $\mathsf{C}(A, -)\colon \mathsf{C} \to \mathsf{Set}$, *for any* A *in the category* C, *preserves all limits that exist in* C.

At this point, we haven't any fancy apparatus that could give us a slick proof. So we have to go for a brute-force, just-apply-the-definitions-and-see-what-happens, demonstration (but it is illuminating to run through the details).

Proof. We'll first show that $\mathsf{C}(A, -)\colon \mathsf{C} \to \mathsf{Set}$ sends a cone over the diagram $D\colon \mathsf{J} \to \mathsf{C}$ to a cone over $\mathsf{C}(A, -) \circ D\colon \mathsf{J} \to \mathsf{Set}$.

By definition, a cone over D comprises a vertex C and arrows $c_J\colon C \to DJ$ for each J-object J where, for any $d\colon J \to K$ in J, $c_K = Dd \circ c_J$.

Now, acting on objects, $\mathsf{C}(A, -)$ sends C to $\mathsf{C}(A, C)$ and sends DJ to $\mathsf{C}(A, DJ)$. Acting on arrows, $\mathsf{C}(A, -)$ sends $c_J\colon C \to DJ$ to the set function $c_J \circ -$ (i.e. the function that takes $g\colon A \to C$ and outputs $c_J \circ g\colon A \to DJ$). And it sends

$Dd\colon DJ \to DK$ to the set-function $Dd \circ -$ (the function that takes $h\colon A \to DJ$ and outputs $Dd \circ h\colon A \to DK$). So, in summary, the functor $\mathsf{C}(A, -)$ sends the triangle on the left in C to the one on the right in Set:

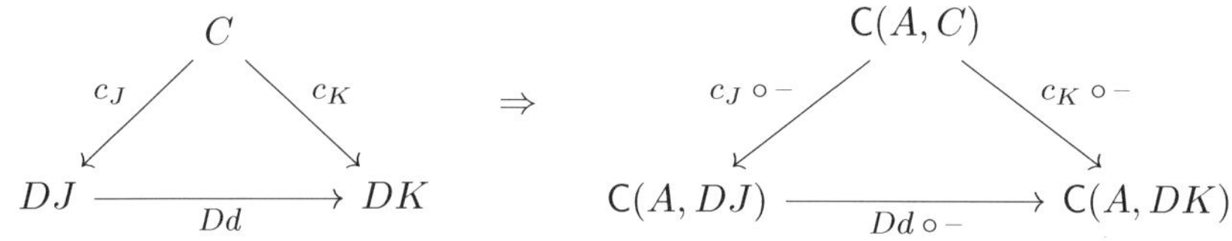

Given that $c_K = Dd \circ c_J$, we have $c_K \circ - = (Dd \circ c_J) \circ - = (Dd \circ -) \circ (c_J \circ -)$.[2] Hence, if the triangle on the left commutes, so does the triangle on the right. Likewise for other such triangles. Which means that if (C, c_J) is a cone over D, then $(\mathsf{C}(A, C), c_J \circ -)$ is a cone over $\mathsf{C}(A, -) \circ D$.

So far, so good! But we now need to show that $\mathsf{C}(A, -)$ doesn't just send cones to cones, but sends limit cones to limit cones.

Suppose, then, that (L, λ_J) is a limit cone in C over D. For any $d\colon J \to K$ in J the functor $\mathsf{C}(A, -) : \mathsf{C} \to \mathsf{Set}$ sends the left-hand commuting diagram below to the commuting triangle at the bottom of the right-hand diagram. And we now suppose that (M, m_J) is any other cone over $\mathsf{C}(A, -) \circ D$:

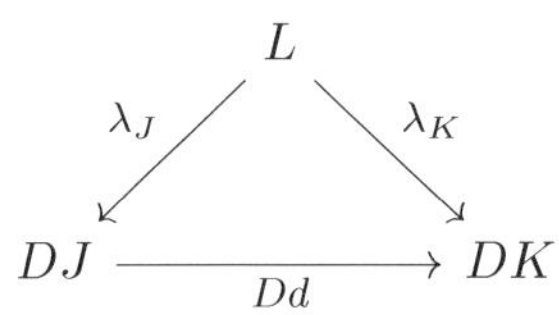

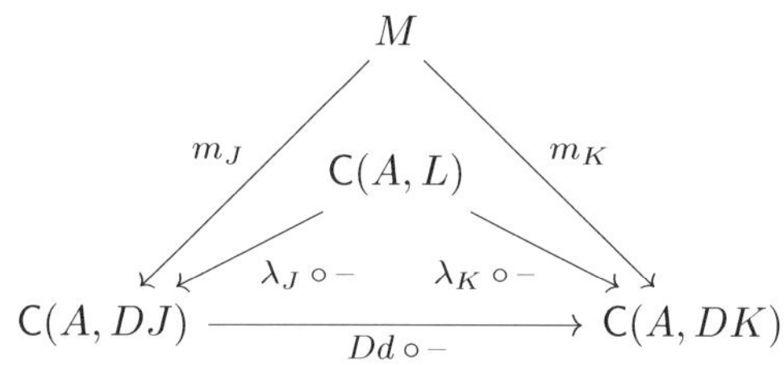

Hence, again for each $d\colon J \to K$, we have $m_K = (Dd \circ -) \circ m_J$.

Now remember that M lives in Set: so take a member x. Then $m_J(x)$ is an arrow in $\mathsf{C}(A, DJ)$; in other words, $m_J(x)\colon A \to DJ$. Likewise we have $m_K(x)\colon A \to DK$. But $m_K(x) = Dd \circ m_J(x)$. Hence for all d the outer triangles in the left-hand diagram below must commute in C, so $(A, m_J(x))$ is a cone over D. By the assumption that (L, λ_J) is a limit cone, $(A, m_J(x))$ must factor uniquely through an arrow we can call $u(x)$, again as on the left:

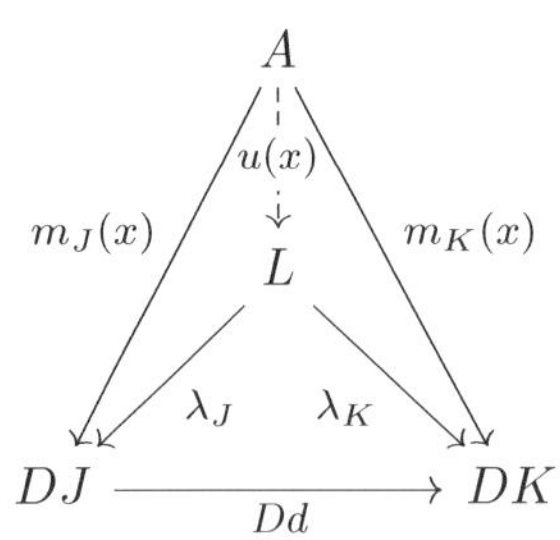

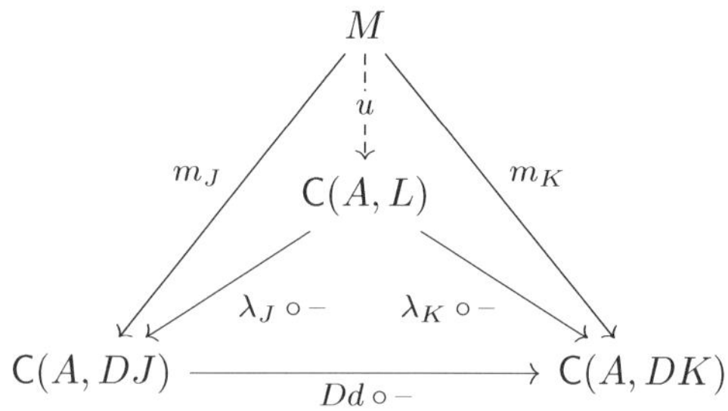

[2]For the second equality think: composing a function with c_J-followed-by-Dd is the same as composing-with-c_J and then following that by composing-with-Dd.

Hence $u(x)$ is an arrow from A to L, i.e. an element of $\mathsf{C}(A, L)$. So consider the map $u\colon M \to \mathsf{C}(A, L)$ that sends any x in M to $u(x)$. Since $m_J(x) = \lambda_J \circ u(x)$ for each x, we have $m_J = (\lambda_J \circ -) \circ u$. And since this applies for each object J, (M, m_J) factors through the image of the cone (L, λ_J) via u.

We are almost there: we have shown that any cone over the bases on the right factors through the cone with vertex $\mathsf{C}(A, L)$. It remains to prove that it factors *uniquely* via u.

Suppose that there is another map $v\colon M \to \mathsf{C}(A, L)$ such that we also have each $m_J = (\lambda_J \circ -) \circ v$. Take an element x in M: then $m_J(x) = \lambda_J \circ v(x)$. So again, $(A, m_J(x))$ factors through (L, λ_J) via $v(x)$ – which, by the uniqueness of factorization through limits, means that $v(x) = u(x)$. Since that obtains for all x in M, we have $v = u$. Hence, as we wanted to show, (M, m_j) factors uniquely through the image of (L, λ_J).

Since (M, m_J) was an arbitrary cone, we have therefore proved that the image of the limit cone (L, λ_J) is also a limit cone. □

31.4 A dual result?

What is the dual of Theorem 150? We have two dualities to play with: limits vs colimits and covariant functors vs contravariant functors.

Two initial observations. First, a covariant hom-functor need not preserve colimits such as initial objects. For example, take the hom-functor $\mathsf{Grp}(A, -)$. In Grp the initial object 0 is also the terminal object, so for any group A, $\mathsf{Grp}(A, 0)$ is a singleton, which is not initial in Set.

Second, contravariant hom-functors from a category C can't preserve either limits or colimits in C, because contravariant functors reverse arrows.

The dual result we *can* get is this:

Theorem 151. *Suppose that* C *is a locally small category. Then the contravariant hom-functor* $\mathsf{C}(-, A)\colon \mathsf{C} \to \mathsf{Set}$, *for any A in the category* C, *sends a colimit of shape* J *to a limit of that shape.* □

Yes, that's right: contravariant functors send colimits to limits. I'll leave you to prove that as a merry dualizing challenge.

32 Natural isomorphisms

Category theory is an embodiment of Klein's dictum that it is the maps that count in mathematics. If the dictum is true, then it is the functors between categories that are important, not the categories. And such is the case. Indeed, the notion of category is best excused as that which is necessary in order to have the notion of functor. But the progression does not stop here. There are maps between functors, and they are called natural transformations.
(Freyd 1965, quoted in Marquis 2008.)

Natural transformations between functors – and more specifically, natural isomorphisms – were there from the very start. The founding document of category theory is the paper by Samuel Eilenberg and Saunders Mac Lane 'General theory of natural equivalences' (Eilenberg and Mac Lane 1945). But the pivotal idea had already been introduced, three years previously, in a paper on 'Natural isomorphisms in group theory', before the categorial framework was invented in order to provide a proper setting for the account (Eilenberg and Mac Lane 1942). These natural isomorphisms and the more general natural transformations will be our main topic for the next couple of chapters.

32.1 Natural isomorphisms between functors defined

(a) Suppose we have a pair of functors $F, G\colon \mathsf{C} \to \mathsf{D}$. Then each of the functors projects the objects and arrows of C into D, giving us two images of C within D. For example, we might in small part have:

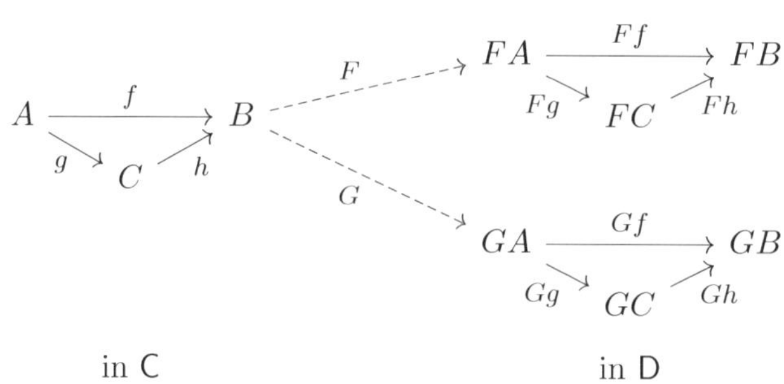

Now, in general, the images of C projected by F and G could be significantly different (why?). But suppose (i) there is a suite ψ of isomorphisms in D, $\psi_A\colon FA \xrightarrow{\sim} GA$, $\psi_B\colon FB \xrightarrow{\sim} GB$, $\psi_C\colon FC \xrightarrow{\sim} GC$, etc., one for every C-object A, B, C etc., ensuring that $FA \cong GA$, $FB \cong GB$, $FC \cong GC$, etc. And suppose (ii) that all the squares like these in D commute:

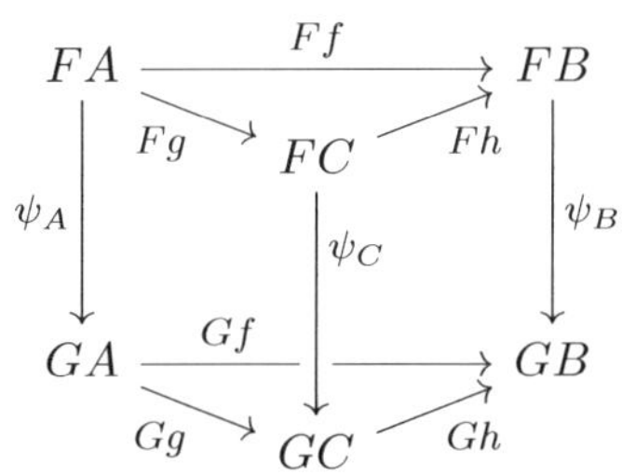

Then that suite ψ of isomorphisms, nicely interacting with arrows like Ff and Gf etc., means that the G-image of C does behave like a copy of the F-image (a faithful copy except perhaps in collapsing isomorphic objects together). So, stretching our terminology a bit, we might say that in this case the functors F and G are themselves isomorphic.[1]

(b) Which all goes to motivate the following standard definition – or rather, it's a pair of definitions, one for each flavour of functor:

Definition 121. Let C and D be categories, let $F, G\colon \mathsf{C} \to \mathsf{D}$ be covariant functors (respectively, contravariant functors), and suppose that for each C-object C there is a D-isomorphism $\psi_C\colon FC \xrightarrow{\sim} GC$. Then ψ, the family of arrows ψ_C, is said to be a *natural isomorphism* between F and G iff for every arrow $f\colon A \to B$ (respectively, $f\colon B \to A$, note the reversal!) in C the following *naturality square* commutes in D:

$$\begin{array}{ccc} FA & \xrightarrow{Ff} & FB \\ \downarrow{\scriptstyle \psi_A} & & \downarrow{\scriptstyle \psi_B} \\ GA & \xrightarrow{Gf} & GB \end{array}$$

In this case, we write $\psi\colon F \xRightarrow{\sim} G$, and the ψ_C are said to be components of ψ. If there is such a natural isomorphism, the functors F and G will be said to be naturally isomorphic, and we write $F \cong G$. △

(c) Let's have an immediate toy example. In §26.4, (F15) introduced us to the functors $(- \times C)\colon \mathsf{C} \to \mathsf{C}$ and $(C \times -)\colon \mathsf{C} \to \mathsf{C}$. These two functors evidently output products that are isomorphic to each other. So intuitively the functors

[1] Careful! Let me flag a distinction. What we are defining here is the idea of isomorphic functors – i.e. explaining a notion of isomorphism *between functors*. What we mentioned at the very end of §29.2, and will return to in Chapter 34, is the idea of functors which are themselves isomorphisms – i.e. isomorphisms *between categories*.

ought to be 'isomorphic' on any nice definition of isomorphism for functors. And they are.

Let ψ_X be the isomorphism that maps a product $X \times C$ to $C \times X$ (for any C). Then the following square commutes for any $f\colon A \to B$:

$$\begin{array}{ccc} (-\times C)_{ob}(A) & \xrightarrow{(-\times C)_{arw}(f)} & (-\times C)_{ob}(B) \\ \Big\downarrow{\scriptstyle \psi_A} & & \Big\downarrow{\scriptstyle \psi_B} \\ (C\times -)_{ob}(A) & \xrightarrow{(C\times -)_{arw}(f)} & (C\times -)_{ob}(B) \end{array}$$

because by the definition in (F15) that is none other than the square

$$\begin{array}{ccc} A\times C & \xrightarrow{f\times 1_C} & B\times C \\ \Big\downarrow{\scriptstyle \psi_A} & & \Big\downarrow{\scriptstyle \psi_B} \\ C\times A & \xrightarrow{1_C\times f} & C\times B \end{array}$$

which commutes because both routes send a pair $\langle a, c\rangle$ round to $\langle c, fa\rangle$.

Hence assembling all the components ψ_X gives us the desired natural isomorphism $\psi\colon (-\times C) \overset{\sim}{\Longrightarrow} (C\times -)$.

32.2 Some basic properties

Theorem 152. *Suppose F, G, H are covariant functors from C to D:*

(1) There is an identity natural isomorphism $1_F\colon F \overset{\sim}{\Longrightarrow} F$.

(2) Given natural isomorphisms $\psi\colon F \overset{\sim}{\Longrightarrow} G$ and $\chi\colon G \overset{\sim}{\Longrightarrow} H$, there is a composite natural isomorphism $\chi\circ\psi\colon F \overset{\sim}{\Longrightarrow} H$, and composition is associative.

(3) Any $\psi\colon F \overset{\sim}{\Longrightarrow} G$ has an inverse $\psi^{-1}\colon G \overset{\sim}{\Longrightarrow} F$ such that $\psi^{-1}\circ\psi = 1_F$ and $\psi\circ\psi^{-1} = 1_G$.

(4) If $F \cong G$ then F is faithful if and only if G is.

(5) If $F \cong G$ then F is full if and only if G is.

(6) If $F \cong G$ then F is essentially surjective on objects if and only if G is.

(1) to (3) tell us that natural isomorphisms do behave like isomorphisms. While (4) to (6) illustrate that naturally isomorphic functors share properties in predictable ways. (Pause to find the following unchallenging proofs for yourself!)

Proof: there's an identity natural isomorphism. The following diagram trivially commutes for any C-arrow $f\colon A \to B$:

$$\begin{array}{ccc} FA & \xrightarrow{Ff} & FB \\ \Big\downarrow{\scriptstyle 1_{FA}} & & \Big\downarrow{\scriptstyle 1_{FB}} \\ FA & \xrightarrow{Ff} & FB \end{array}$$

So we have a natural isomorphism $1_F\colon F \overset{\sim}{\Longrightarrow} F$, where the components $(1_F)_A$ of the isomorphism are the identity arrows 1_{FA}. □

Proof: natural isomorphisms compose. Given $\psi\colon F \overset{\sim}{\Longrightarrow} G$ and $\chi\colon G \overset{\sim}{\Longrightarrow} H$, the inner squares below commute for any C-arrow $f\colon A \to B$:

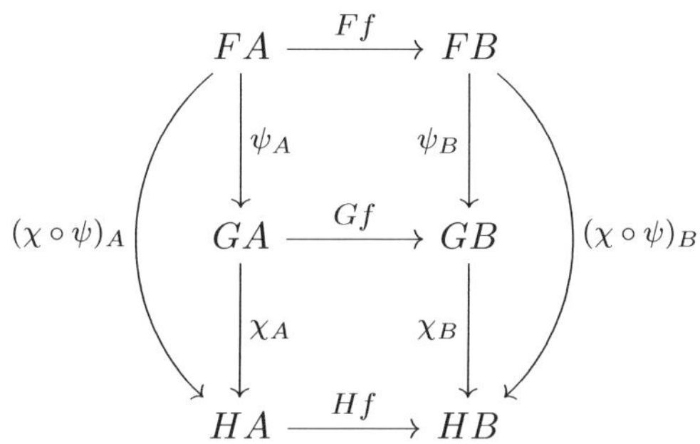

So, let's put $(\chi \circ \psi)_C = \chi_C \circ \psi_C$ for every C-object C (more isomorphisms, of course). Then, with those components, it is immediate that $(\chi \circ \psi)\colon F \overset{\sim}{\Longrightarrow} H$.

The associativity of composition for natural isomorphisms is then inherited from the associativity of composition for their components. □

Proof: natural isomorphisms have inverses. The commuting naturality squares which show that $\psi\colon F \overset{\sim}{\Longrightarrow} G$ tell us that (for any $f\colon A \to B$) we have $\psi_B \circ Ff = Gf \circ \psi_A$. The components of ψ, being isomorphisms, have inverses. So

$$(\psi_B)^{-1} \circ (\psi_B \circ Ff) \circ (\psi_A)^{-1} = (\psi_B)^{-1} \circ (Gf \circ \psi_A) \circ (\psi_A)^{-1}$$

whence $(\psi_B)^{-1} \circ Gf = Ff \circ (\psi_A)^{-1}$. In other words, all the squares

$$\begin{array}{ccc} GA & \xrightarrow{\;Gf\;} & GB \\ \big\downarrow{\scriptstyle(\psi^{-1})_A} & & \big\downarrow{\scriptstyle(\psi^{-1})_B} \\ FA & \xrightarrow{\;Ff\;} & FB \end{array}$$

commute, if we put $(\psi^{-1})_C = \psi_C^{-1}$ for all C. Hence $\psi^{-1}\colon G \overset{\sim}{\Longrightarrow} F$.

It is then immediate that $\psi^{-1} \circ \psi = 1_F$ and $\psi \circ \psi^{-1} = 1_G$. □

Next, we need only prove one direction of each of the following three results about biconditionals. So:

Proof: $F \cong G$ implies that if F is faithful, so is G. By the naturality square, for any $f\colon A \to B$, $Gf = \psi_B \circ Ff \circ \psi_A^{-1}$. Hence if $Gf = Gg$, then $\psi_B \circ Ff \circ \psi_A^{-1} = \psi_B \circ Fg \circ \psi_A^{-1}$, whence (composing with ψ_B^{-1} and ψ_A) we have $Ff = Fg$.

By definition, F is faithful iff given any pair of parallel arrows $f, g\colon A \to B$, then if $Ff = Fg$ then $f = g$. So, given $F \cong G$ and F is faithful, then if $Gf = Gg$ we have $Ff = Fg$ and so $f = g$, making G faithful. □

Proof: $F \cong G$ implies that if F is full so is G. F is full iff for any $k\colon FA \to FB$ there is an arrow $f\colon A \to B$ such that $k = Ff$. Now, trivially, this commutes:

$$\begin{array}{ccc} FA & \xrightarrow{\psi_B^{-1} \circ g \circ \psi_A} & FB \\ \downarrow{\scriptstyle \psi_A} & & \downarrow{\scriptstyle \psi_B} \\ GA & \xrightarrow{\quad g \quad} & GB \end{array}$$

So if F is full, there is an arrow f such that $Ff = \psi_B^{-1} \circ g \circ \psi_A$, and hence $g = \psi_B \circ Ff \circ \psi_A^{-1} = Gf$ (with the second equation as before). So G is full too. □

Proof: $F \cong G$ implies that if F is e.s.o., then G is too. Let D be any D-object. If $F\colon \mathsf{C} \to \mathsf{D}$ is e.s.o., then there is a C-object C such that $D \cong FC$. But if $F \cong G$, $FC \cong GC$, and then by transitivity, we also get $D \cong GC$. Which makes G e.s.o. □

Challenge: think through what the analogue of Theorem 152 is for contravariant functors.

32.3 Why 'natural'?

But why call what we've defined a *natural* isomorphism? There's a mathematical back-story which I alluded to in the preamble of the chapter and which I should now pause to explain, using one of Eilenberg and Mac Lane's own examples. (Hopefully, even if your grip on the theory of vector spaces is a bit shaky, the general drift of the example should be reasonably clear.)

(a) Consider a finite dimensional vector space V over the reals $\mathbb{R}$, and the corresponding dual space V^* of linear functions $g\colon V \to \mathbb{R}$.

It is elementary to show that V is isomorphic to V^* (there's a bijective linear map between the spaces). Proof sketch: take a basis $B = \{v_1, v_2, \ldots, v_n\}$ for the space V. Define the functions $v_i^*\colon V \to \mathbb{R}$ by putting $v_i^*(v_j) = 1$ if $i = j$ and $v_i^*(v_j) = 0$ otherwise. Then $B^* = \{v_1^*, v_2^*, \ldots, v_n^*\}$ is a basis for V^*, and the linear function $\varphi_B\colon V \to V^*$ generated by putting $\varphi_B(v_i) = v_i^*$ gives us an isomorphism. QED

Note, however, that the isomorphism we have arrived at here depends on the initial choice of basis B. And no choice of basis B is more 'natural', no more 'canonical', than any other. Hence no one of the isomorphisms $\varphi_B\colon V \to V^*$ of the kind just defined is to be especially preferred.

To get a sharply contrasting case, now consider V^{**} the double dual of V, i.e. the space of functionals $h\colon V^* \to \mathbb{R}$ (so each h takes any linear function $V \to \mathbb{R}$ as input and outputs a corresponding real in $\mathbb{R}$).

Suppose we select a basis B for V, define a derived basis B^* for V^* as we just did, and then use this new basis in turn to define a basis B^{**} for V^{**} by repeating the same little trick. Then we can construct an isomorphism from V to V^{**} by mapping the elements of B to the corresponding elements of B^{**}. However, and this is the key observation, *we don't have to go through any such palaver of initially choosing a basis.* Suppose we simply define $\psi_V : V \to V^{**}$ as acting on an element $v \in V$ to give as output the functional $\psi_V(v) : V^* \to \mathbb{R}$ which sends a function $g : V \to \mathbb{R}$ to the value $g(v)$: in short, we set $\psi_V(v)(g) = g(v)$. It is a simple result that ψ_V is an isomorphism (we can rely on the fact that V is finite-dimensional). And we get *this* isomorphism independently of any arbitrary choice of basis.

Interim summary: it is very tempting to say that the isomorphisms of the kind we described between V and V^* are not particularly 'natural': they are cooked up on the basis(!) of some arbitrary choices. By contrast there *is* a 'natural' isomorphism between V and V^{**}, generated by a procedure that doesn't involve any arbitrary choices.

Now, there are many other cases where we might similarly want to contrast intuitively 'natural' or 'canonical' maps with more arbitrarily cooked-up maps between structured objects. So a question arises: can we give a general account of what makes for naturality here? Eilenberg and Mac Lane were aiming to provide such a story.

(b) To continue with our example, the nice isomorphism $\psi_V : V \xrightarrow{\sim} V^{**}$ only depended on the fact that V is a finite dimensional vector space over the reals. Which implies that our construction will work in exactly same way for any other such vector space W, so we get in each case a corresponding isomorphism $\psi_W : W \xrightarrow{\sim} W^{**}$.

Now, we will expect such naturally constructed isomorphisms to respect the relation between a structure-preserving map f between the spaces V and W and its double-dual correlate map between V^{**} to W^{**}. Putting that more carefully, we want the following informal diagram to commute, whatever vector spaces we take and for any linear map $f : V \to W$,

$$\begin{array}{ccc} V & \xrightarrow{\quad f \quad} & W \\ \downarrow{\scriptstyle \psi_V} & & \downarrow{\scriptstyle \psi_W} \\ V^{**} & \xrightarrow{DD(f)} & W^{**} \end{array}$$

where $DD(f)$ is the double-dual correlate of f.

Now recall that back in §26.6 (F20), we saw that the (contravariant) dualizing functor D that sends a vector space V to its dual V^* will send an arrow $f : V \to W$ to $D(f) : W^* \to V^*$, where $D(f)$ takes a member of W^* such as $g : W \to \mathbb{R}$ and outputs $g \circ f$ which is a member of V^*.

But what about the double-dualizing functor DD that sends a vector space V to its double dual V^{**}? Where should it send an arrow $f : V \to W$? To $DD(f) : V^{**} \to W^{**}$, where $DD(f)$ takes a member of V^{**} such as $h : V^* \to \mathbb{R}$ and outputs $h \circ D(f)$ which is a member of W^{**}. (It is readily checked that this is functorial.)

Thus understood, our square does indeed commute. By either route, a vector v in V gets sent to the functional living in W^{**} which sends a function $k\colon W \to \mathbb{R}$ to the value $k(f(v))$. Think about it![2]

(c) So far, so good. Now let's pause to consider why there *can't* be a similarly 'natural' way of setting up isomorphisms from vector spaces V to their duals V^*. (The isomorphisms we mentioned which are based on an arbitrary choice of basis aren't natural: but we want to show that there is no other way of getting a 'natural' isomorphism.)

Suppose then that there were a construction which gave us an isomorphism $\varphi_V\colon V \xrightarrow{\sim} V^*$ which again does not depend on information about V other than that it has the structure of a finite dimensional vector space. So again we will want the construction to work the same way on other such vector spaces, and to be preserved by structure-preserving maps between the spaces. This time, therefore, we will presumably want the following diagram to commute for any structure-preserving f between vector spaces (note, however, that we have to reverse an arrow for things to make any sense, given our definition of the contravariant dualizing functor D):

$$\begin{array}{ccc} V & \xrightarrow{\quad f \quad} & W \\ \downarrow{\scriptstyle \varphi_V} & & \downarrow{\scriptstyle \varphi_W} \\ V^* & \xleftarrow[\quad D(f) \quad]{} & W^* \end{array}$$

Hence $D(f) \circ \varphi_W \circ f = \varphi_V$. But by hypothesis, the φs are isomorphisms; so in particular φ_V has an inverse. So we have $(\varphi_V^{-1} \circ D(f) \circ \varphi_W) \circ f = 1_V$. Therefore f has a left inverse. In general, however, a linear map $f\colon V \to W$ need not have a left inverse.

Hence there can't in general be isomorphisms φ_V, φ_W making that diagram commute.

(d) We started off by saying that, intuitively, there's always a 'natural', intrinsic, isomorphism between a (finite dimensional) vector space and its double dual, one that depends only on their structures as vector spaces. And we've now suggested that this intuitive idea can be reflected by saying that a certain diagram involving such isomorphisms always commutes, for any choice of vector spaces and structure-preserving maps between them.

We have also seen that we can't get analogous always-commuting diagrams for the case of isomorphisms between a vector space and its dual – which chimes with the intuition that the obvious examples are *not* 'natural' isomorphisms.

[2] OK, if you insist. The top route sends v to $f(v)$; but then $\psi_W(f(v))$ by definition outputs the functional which sends a function $k\colon W \to \mathbb{R}$ to $k(f(v))$. For the bottom route, $\psi_V(v)$ by definition outputs the functional h which sends any function $j\colon V \to \mathbb{R}$ to $j(v)$. Now we need to hit that functional with DD, and that gives as another functional with takes any $k\colon W \to \mathbb{R}$, applies Df which gives us $k \circ f\colon V \to \mathbb{R}$, and then applies h to that which sends it to the output $(k \circ f)(v) = k(f(v))$ again.

So this gives us a promising way forward: characterize 'naturality' here in terms of the availability of a family of isomorphisms which make certain diagrams commute.

(e) But now the key move. The claim that the diagram

$$\begin{array}{ccc} V & \xrightarrow{f} & W \\ \downarrow{\scriptstyle \psi_V} & & \downarrow{\scriptstyle \psi_W} \\ V^{**} & \xrightarrow{DD(f)} & W^{**} \end{array}$$

always commutes can be recast as a claim about two functors. For we have been talking about the category FVect (of finite-dimensional spaces over the reals and the structure-preserving maps between them), and about the functor DD: FVect $\to$ FVect which takes a vector space to its double dual, and maps each arrow between vector spaces to its double-dual correlate as explained. But of course there is also a trivial functor 1: FVect $\to$ FVect that maps each vector space to itself and each FVect-arrow to itself. So we can re-express the claim that the last diagram commutes as follows: for every arrow $f\colon V \to W$ in FVect, there are isomorphisms ψ_V and ψ_W in FVect such that *this* diagram commutes:

$$\begin{array}{ccc} 1(V) & \xrightarrow{1(f)} & 1(W) \\ \downarrow{\scriptstyle \psi_V} & & \downarrow{\scriptstyle \psi_W} \\ DD(V) & \xrightarrow{DD(f)} & DD(W) \end{array}$$

In other words, in the terms of the previous section, *the suite of isomorphisms* ψ_V *provide a natural isomorphism* $\psi\colon 1 \overset{\sim}{\Longrightarrow} DD$.

(f) In sum: our claim that there is an intuitively 'natural' isomorphism between two *spaces*, a vector space and its double dual, now becomes reflected in the claim that there is an isomorphism in our official sense between two *functors*, the identity and the double-dual functors from the category FVect to itself. Hence the aptness of calling the latter isomorphism between functors a *natural* isomorphism.

32.4 More examples of natural isomorphisms

(a) We now have one important case to hand. Many more are to be had. For example, Riehl (2017, p. 26) gives a nice example of a classic representation theorem relating – as we can now say – the category of compact Hausdorff spaces and the category of Banach spaces and continuous maps, a theorem which amounts to a claim of natural isomorphism between relevant functors. But it would take us too far afield to chase down the details. So here, let's stick to a few *much* more elementary examples.

(1) Given a group $G = (G, *, e)$ we can define its mirror-image or opposite $G^{op} = (G, *^{op}, e)$, where $a *^{op} b = b * a$.

We can also define a functor Op: Grp $\to$ Grp which sends a group G to its opposite G^{op}, and sends an arrow f in the category, i.e. a group homomorphism $f: G \to H$, to $f^{op}: G^{op} \to H^{op}$ where $f^{op}(a) = f(a)$ for all a in G. f^{op} so defined is indeed a group homomorphism, since

$$f^{op}(a *^{op} b) = f(b * a) = f(b) * f(a) = f^{op}(a) *^{op} f^{op}(b)$$

Claim: there is a natural isomorphism $\psi: 1 \overset{\sim}{\Longrightarrow} Op$ (where 1 is the trivial identity functor in Grp). Which is as it should be, since the functors produce isomomorphic results.

Proof. We need to find a family of isomorphisms $\psi_G, \psi_H, \ldots$ in Grp such that the following diagram commutes for any homomorphism $f: G \to H$:

$$\begin{array}{ccc} G & \xrightarrow{\quad f \quad} & H \\ \downarrow{\scriptstyle \psi_G} & & \downarrow{\scriptstyle \psi_H} \\ G^{op} & \xrightarrow{\quad f^{op} \quad} & H^{op} \end{array}$$

Now, since going *between* opposite groups involves reversing the order of multiplication and taking inverses *inside* a group in effect does the same, let's put $\psi_G(a) = a^{-1}$ for any G-element a, and likewise for ψ_H, etc. It is easy to check that each ψ_G is a group isomorphism, and that they assemble to give a natural isomorphism. □

(2) Recall from §26.5 the functor $List$: Set $\to$ Set which sends a set X to the set of finite lists of members of X. There is a natural isomorphism ρ: $List \overset{\sim}{\Longrightarrow} List$, whose component ρ_X: $List(X) \to List(X)$ acts on a list of X-elements to reverse their order.

(3) Now for a slightly more interesting example, this time involving contravariant functors from Set to Set. First, recall from §26.6 the contravariant powerset functor $\overline{P}$: Set $\to$ Set which maps a set X to its powerset $\mathcal{P}X$, and maps a set-function $f: Y \to X$ to the function $Inv(f): \mathcal{P}X \to \mathcal{P}Y$ which sends $U \subseteq X$ to its inverse image $f^{-1}[U] \subseteq Y$.

And let now C be the hom-functor Set$(-, 2)$, where 2 is some nice two-element set which we can think of as $\{true, false\}$. So C sends a set X to Set$(X, 2)$, i.e. the set of functions from X to 2: and C sends an arrow $f: Y \to X$ to the function $- \circ f$: Set$(X, 2) \to$ Set$(Y, 2)$ (i.e. the function which sends an arrow $g: X \to 2$ to the arrow $g \circ f: Y \to 2$).

Claim: $\overline{P} \cong C$.

Proof. We need to find a family of isomorphisms $\psi_X, \psi_Y, \ldots$ in Set such that the following diagram always commutes, for any $f: Y \to X$:

$$\begin{array}{ccccccc}
\overline{P}X & \xrightarrow{\overline{P}f} & \overline{P}Y & & \mathcal{P}X & \xrightarrow{Inv(f)} & \mathcal{P}Y \\
\downarrow{\scriptstyle \psi_X} & & \downarrow{\scriptstyle \psi_Y} & \text{i.e.} & \downarrow{\scriptstyle \psi_X} & & \downarrow{\scriptstyle \psi_Y} \\
CX & \xrightarrow{Cf} & CY & & \mathsf{Set}(X, 2) & \xrightarrow{-\circ f} & \mathsf{Set}(Y, 2)
\end{array}$$

Well, for any X, let ψ_X send the set $U \in \mathcal{P}X$ to its characteristic function – i.e to the function which sends an element of X to *true* iff it is in U. ψ_X is evidently bijective and hence an isomorphism in Set. And it is easy to see that our diagram will always commute. Both routes sends a set $U \in \mathcal{P}X$ to the function which sends y in Y to *true* iff $fy \in U$.

Here the natural isomorphism reflects of course the non-arbitrary association of subsets with characteristic functions (non-arbitrary, at any rate, once we have decided how to represent *true* and *false*.) □

(4) Recall from §17.1 that there is a naturally arising bijection between two-place set functions from A and B to C and one-place functions from A to functions-from-B-to-C. That is to say, there is a natural way of constructing an isomorphism between the hom-sets $\mathsf{Set}(A \times B, C)$ and $\mathsf{Set}(A, C^B)$. And the point applies more generally to the hom-sets of a locally small category with exponentials: $\mathsf{C}(A \times B, C)$ will be isomorphic to $\mathsf{C}(A, C^B)$.

So consider the two contravariant functors $\mathsf{C}(- \times B, C)$ and $\mathsf{C}(-, C^B)$ from C to Set. The object and arrow components of the first of these functors work as follows:

$$\begin{array}{rcl}
X & \longmapsto & \mathsf{C}(X \times B, C) \\
j\colon X \to Y & \longmapsto & - \circ (j \times 1_B)\colon \mathsf{C}(Y \times B, C) \to \mathsf{C}(X \times B, C).
\end{array}$$

In other words, $\mathsf{C}(- \times B, C)$ applied to $j\colon X \to Y$ gives the function which sends an arrow $f\colon Y \times B \to C$ to the arrow $f \circ (j \times 1_B)\colon X \times B \to C$. (Check that this does define a functor!) While our second functor is going to be a hom-functor of the familiar kind, which works as follows:

$$\begin{array}{rcl}
X & \longmapsto & \mathsf{C}(X, C^B) \\
j\colon X \to Y & \longmapsto & - \circ j\colon \mathsf{C}(Y, C^B) \to \mathsf{C}(X, C^B).
\end{array}$$

Now, our two functors systematically send any object X to isomorphic outputs, and 'do the simple, obvious, thing' to arrows $j\colon X \to Y$. So we might expect them to be naturally isomorphic functors. Which they are.

Proof. We need to find a suite of isomorphisms $\psi_X, \psi_Y, \ldots$ such that for every $j\colon X \to Y$ in C, the following diagram commutes in Set, with the direction of arrows dictated by the contravariance of the functors:

$$\begin{array}{ccc}
\mathsf{C}(X \times B, C) & \xleftarrow{-\circ (j \times 1_B)} & \mathsf{C}(Y \times B, C) \\
\downarrow{\scriptstyle \psi_X} & & \downarrow{\scriptstyle \psi_Y} \\
\mathsf{C}(X, C^B) & \xleftarrow{-\circ j} & \mathsf{C}(Y, C^B)
\end{array}$$

But the isomorphism we know about between $\mathsf{C}(X \times B, C)$ and $\mathsf{C}(X, C^B)$ is the one given in the chapter about exponentials by Theorem 75: so let's try putting ψ_X, for a given X, to be the function which sends an arrow $f\colon X \times B \to C$ to its exponential transpose $\widetilde{f}\colon X \to C^B$.

Then our square will commute if for any $g\colon Y \times B \to C$,

$$\widetilde{g} \circ j = \widetilde{g \circ (j \times 1_B)}.$$

But that's true, for consider the following diagram:

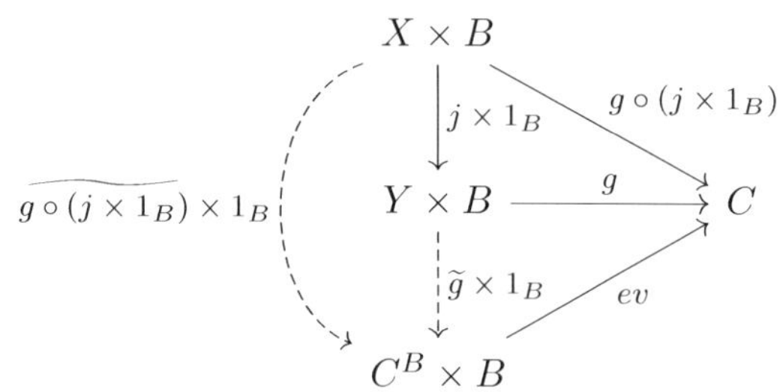

The top-right triangle trivially commutes. The bottom-right triangle commutes by the definition of the existential transpose. So the composite vertical arrow $(\widetilde{g} \circ j) \times 1_B$ gives us a commuting large triangle together with the arrows $g \circ (j \times 1_B)$ and ev. But by definition, $\widetilde{g \circ (j \times 1_B)}$ is the unique arrow whose product with 1_B makes that triangle commute. So, as we needed to show, it must be the case that $\widetilde{g} \circ j = \widetilde{g \circ (j \times 1_B)}$. □

(5) We'll make use of that last particular result later, in §37.6. But our proof also illustrates a general fact. We can have functors that systematically 'do the same thing, up to isomorphism', and hence *ought* to be naturally isomorphic; but often, showing that they *are* isomorphic can be somewhat tediously fiddly.

For another example, working in a locally small category with exponentials, take first the hom-functor $\mathsf{C}(A \times B, -)$. And compare this with the functor we can notate

$$\mathsf{C}(A, -^B)\colon \quad \begin{array}{rcl} X & \longmapsto & \mathsf{C}(A, X^B) \\ f\colon X \to Y & \longmapsto & \widetilde{f \circ ev} \circ -\colon \mathsf{C}(A, X^B) \to \mathsf{C}(A, Y^B). \end{array}$$

Two challenges. First, why is the given operation on arrows the natural way to define the arrow component of a functor which operates on objects as described? (Hint: compare §26.4, (F17).) Second, noting that the two functors systematically send given objects to isomorphic outputs, we might conjecture that they are naturally isomorphic: prove that they are. (Hint: the natural isomorphism you need isn't new, but you'll want to prove something of the form $\widetilde{f \circ ev} \circ \widetilde{g} = \widetilde{f \circ g}$, where $g\colon A \times B \to X$.)

(6) Theorem 110 tells us that equivalence classes of subobjects of X in C line up one-to-one with arrows from X to Ω. Let's use $Sub(X)$ to denote the set

of equivalence classes of monic arrows from X in C – and we'll cheerfully assume for now that there *is* such a set. Then what we've just said is that there is an isomorphism between this set and the hom-set $\mathsf{C}(X, \Omega)$.

Moreover this isomorphism arises in a natural way, without relying on arbitrary choices (at least, once we've fixed on our Ω). Let's show how to construe this naturally-arising isomorphism in fact results from a natural isomorphism between relevant associated functors.

One of the relevant functors must be the contravariant hom-functor $(\mathsf{C})(-, \Omega)\colon \mathsf{C} \to \mathsf{Set}$, of a type familiar from Theorem 149.

The other functor we need will then have to be a matching contravariant functor $Sub(-)\colon \mathsf{C} \to \mathsf{Set}$ which sends a C-object X to $Sub(X)$ and sends a C-arrow $f\colon Y \to X$ to some appropriate Set arrow $f^*\colon Sub(X) \to Sub(Y)$. So what's f^*?

Let's use the notation $[s]$ for the class of monics equivalent to the particular subobject-as-monic s. Then f^* needs to act on an element $[s]$ of $Sub(X)$, i.e. a class of monics equivalent to a particular $s\colon S \to X$, and return an element $[r]$ of $Sub(Y)$, i.e. a class of monics equivalent to a particular $r\colon R \to Y$. OK: how do we use $f\colon Y \to X$ to get from a monic with target X to some monic with target Y? The obvious thing to try is pulling back the first monic along f (what else?).

So the plan will be to get f^* to send $[s]$ to $[r]$ where r is the result of pulling back s along f. It is easily checked that, so defined, $Sub(-)$ really is a contravariant functor

It remains to show that there is a natural isomorphism $\psi\colon Sub(-) \overset{\sim}{\Longrightarrow} \mathsf{C}(-, \Omega)$. We just need there to be a suite of isomorphisms $\psi_X, \psi_Y, \ldots$ such that for any C-arrow $f\colon Y \to X$, the following commutes in Set:

$$\begin{array}{ccc} Sub(X) & \xrightarrow{\ f^*\ } & Sub(Y) \\ \big\downarrow{\scriptstyle \psi_X} & & \big\downarrow{\scriptstyle \psi_Y} \\ \mathsf{C}(X, \Omega) & \xrightarrow{\ -\circ f\ } & \mathsf{C}(Y, \Omega) \end{array}$$

So define ψ_X to be the bijection which sends the class of subobjects of X equivalent to s to the characteristic arrow $\chi_s\colon X \to \Omega$, and define ψ_Y similarly. Then our square commutes as we want (it's a nice reality-check to confirm this).

(7) Lastly, a very simple example illustrating an important general point.

Take again the category FVect whose objects are the finite dimension vector spaces over the reals, and whose arrows are linear maps between spaces. For any V, let $\sigma^s_V\colon V \to V$ be the linear function which maps any vector v to its scalar product with the real number $s \neq 0$, i.e. to sv, and let $f\colon V \to V$ be any linear map from a space V to itself. Let $1\colon \mathsf{FVect} \to \mathsf{FVect}$ be the identity functor that sends a vector space to itself, and any arrow between spaces to itself. Then by the definition of linearity, the square on the left of course nicely commutes:

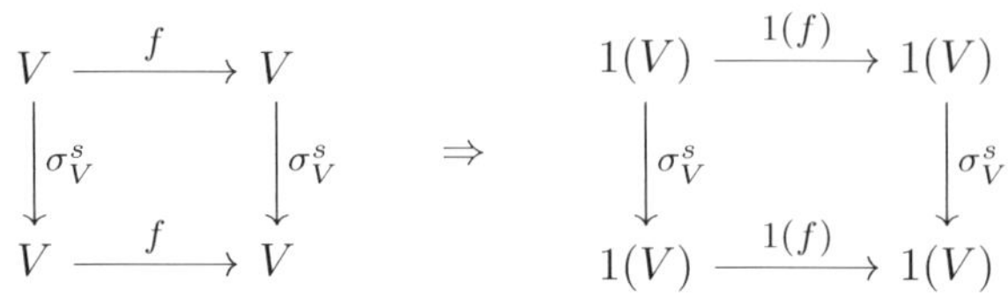

Hence, by definition of the functor 1, the square on the right (in fact, the very same square!) commutes.

But each σ^s_V is a bijection for $s \neq 0$, so is an isomorphism. Which means that σ^s (assembled from the σ^s_V) is a natural isomorphism from the functor 1 to itself.

And note we get a different such isomorphism for each choice of the scaling factor s. That's not at all surprising – it simply reflects the fact that vector spaces that differ by a scaling factor can be thought of as in effect being the same, and in a natural way.

This trite example, though, illustrates an important general moral: that there can be multiple natural isomorphisms between given functors – even infinitely many.

32.5 Another basic property of isomorphic functors

Parts (4) to (6) of Theorem 152 tell us that if the functors F and G are naturally isomorphic, then one of them is full (faithful, essentially surjective) if and only the other is.

Here, for future reference, is another crucial (and predictable!) respect in which naturally isomorphic functors behave in the same way:

Theorem 153. *Suppose the functors $F, G\colon \mathsf{C} \to \mathsf{D}$ are naturally isomorphic. Then if F preserves a given limit so does G.*

Proof. We mostly apply definitions.[3] So let (L, λ_J) be a limit cone for $D\colon \mathsf{J} \to \mathsf{C}$. Then this diagram commutes in C for any $f\colon J \to K$ in J:

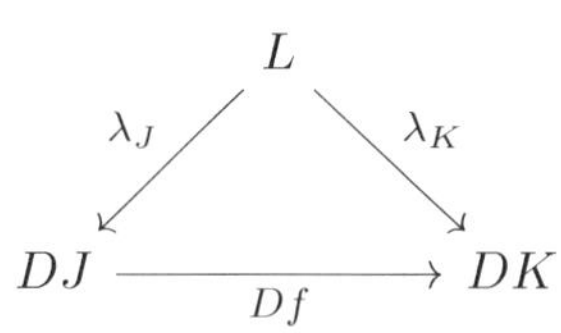

By definition, F and G both map this triangle into D, giving the two commuting triangles in the next diagram. And the assumed natural isomorphism $\alpha\colon F \xRightarrow{\sim} G$ by definition then gives us *three* naturality squares, making a commuting prism in D:

[3]The following argument would look much simpler if I could wave my hands at diagrams drawn with different coloured chalks and growing in real time on a blackboard!

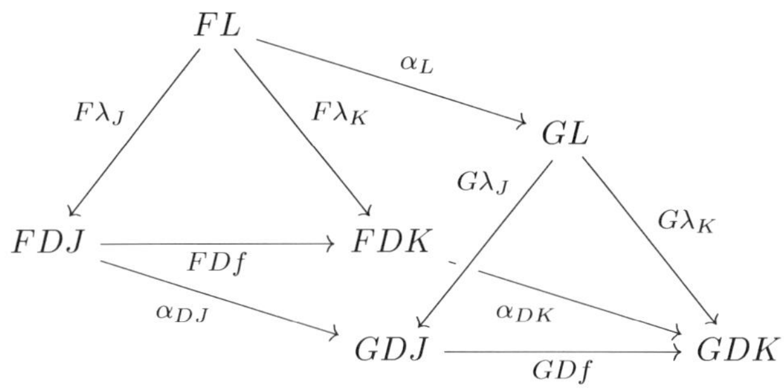

Now consider any cone (C, c_J) over the functor $G \circ D\colon \mathsf{J} \to \mathsf{D}$. Being part of a cone, each new triangle like the one below commutes by definition:

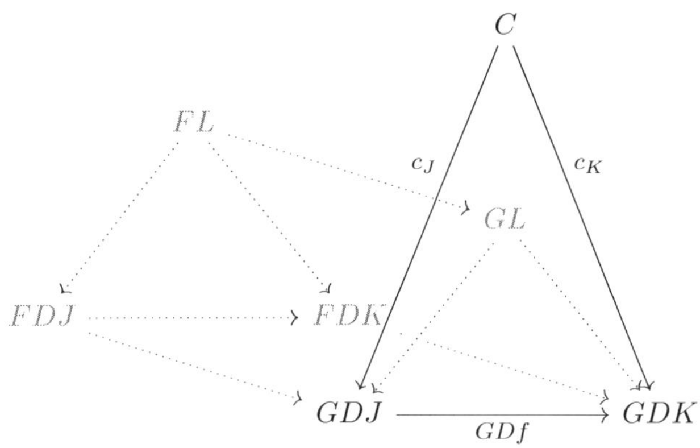

Further, using the commuting base square of the relevant prisms, we can extend each leg c_J of the cone by composition with the corresponding α_{DJ}^{-1} to get a cone $(C, \alpha_{DJ}^{-1} \circ c_J)$ over FD, making further triangles like this:

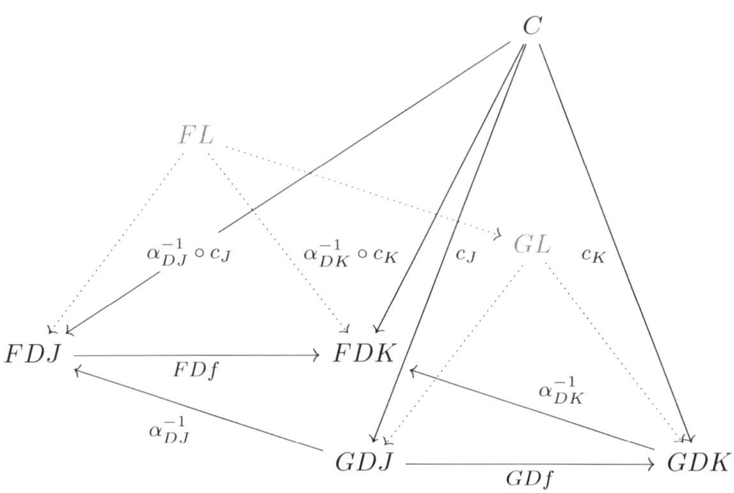

OK: now *suppose for the sake of argument that F preserves the limit (L, λ_J)*. Then by definition $(FL, F\lambda_J)$ is a limit cone over FD. Which means that our cone $(C, \alpha_{DJ}^{-1} \circ c_J)$ over FD must factor through this limit cone via a unique $u\colon C \to FL$, so e.g. $\alpha_{DJ}^{-1} \circ c_J = F\lambda_j \circ u$. So all this, for example, commutes:

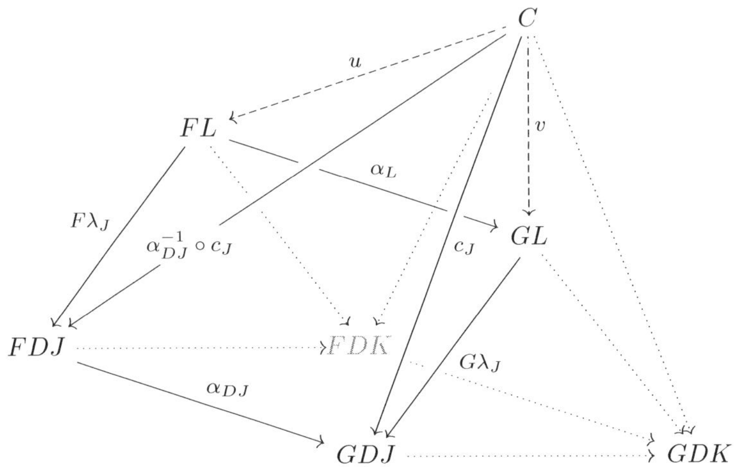

But it is easy to check – chasing arrows round the diagram, using the sloping sides of the prism – that it is also the case that the cone (C, c_J) over GD factors through $(GL, G\lambda_J)$ via $v = \alpha_L \circ u$. For example, we have

$$c_J = \alpha_{DJ} \circ \alpha_{DJ}^{-1} \circ c_J = \alpha_{DJ} \circ F\lambda_j \circ u = G\lambda_j \circ \alpha_L \circ u = G\lambda_j \circ v$$

And further, (C, c_J) can't factor through a distinct v': or else there would be a distinct $u' = \alpha_L^{-1} \circ v'$ which makes everything commute, which is impossible by the uniqueness of u.

Hence, in sum, any (C, c_J) factors through $(GL, G\lambda_J)$ via a unique v, and therefore $(GL, G\lambda_J)$ is a limit cone. Therefore, as we wanted to show, G *also preserves the limit* (L, λ_J). Phew! □

32.6 Natural and unnatural isomorphisms between objects

To finish the chapter, we return to some more general considerations about isomorphic functors and isomorphic objects.

(a) Suppose we have functors $F, G\colon \mathsf{C} \to \mathsf{D}$; and let $A, B, C, \ldots$ be objects in C. Then there will be objects $FA, FB, FC \ldots$ and $GA, GB, GC \ldots$ in D. And in some cases these will be pairwise isomorphic, so that we have $FA \cong GA$, $FB \cong GB$, $FC \cong GC \ldots$.

One way this can happen, as we have been emphasizing, is if there is a natural isomorphism between the functors F and G. But it is also important to stress that it can happen in other, 'unnatural', ways. Let's have a couple of examples, first a toy example to hammer home the point of principle, then a standard illustrative case which is worth thinking through:

(1) Suppose C is a category with exactly one object A, and two arrows, the identity arrow 1_A, and a distinct arrow $f \colon A \to A$, where $f \circ f = f$. And now consider two functors, the identity functor $1_{\mathsf{C}} \colon \mathsf{C} \to \mathsf{C}$, and the functor $F \colon \mathsf{C} \to \mathsf{C}$ which sends the only object to itself, and sends both arrows to the identity arrow.

Then, quite trivially, we have $1_{\mathsf{C}}(A) \cong F(A)$ for the one and only object in C. But there isn't a natural isomorphism between the functors, because by hypothesis $1_A \neq f$, and hence the square

$$\begin{array}{ccc} F(A) & \xrightarrow{F(f)} & F(A) \\ \Big\downarrow{\scriptstyle 1_A} & & \Big\downarrow{\scriptstyle 1_A} \\ 1_{\mathsf{C}}(A) & \xrightarrow{1_{\mathsf{C}}(f)} & 1_{\mathsf{C}}(A) \end{array} \quad \text{i.e.} \quad \begin{array}{ccc} A & \xrightarrow{1_A} & A \\ \Big\downarrow{\scriptstyle 1_A} & & \Big\downarrow{\scriptstyle 1_A} \\ A & \xrightarrow{f} & A \end{array},$$

cannot commute.

(2) For a much more interesting example, we'll work in the category of finite sets and *bijections* between them which I'll locally call simply F for short.

There is a functor $Sym \colon \mathsf{F} \to \mathsf{F}$ which (i) sends a set A in F to the set of permutations on A (treating permutation functions as sets, this is a finite set), and (ii) sends a bijection $f \colon A \to B$ in F to the bijection that sends the permutation p on A to the permutation $f \circ p \circ f^{-1}$ on B. Note: if A has n members, there are $n!$ members of the set of permutations on A.

There is also a functor $Ord \colon \mathsf{F} \to \mathsf{F}$ which (i) sends a set A in F to the set of total linear orderings on A (you can identify an order-relation with a set, so we can think of this too as a finite set), and (ii) sends a bijection $f \colon A \to B$ in F to the bijection $Ord(f)$ which sends a total order on A to the total order on B where $x <_A y$ iff $f(x) <_B f(y)$. Again, if A has n members, there are also $n!$ members of the set of linear orderings on A.

Now, for any object A of F, $Sym(A) \cong Ord(A)$ (since they are equinumerous finite sets). But there cannot be a natural isomorphism ψ between the functors Sym and Ord. For suppose otherwise, and consider the functors acting on a bijection $f \colon A \to A$. Then the following naturality square would have to commute:

$$\begin{array}{ccc} Sym(A) & \xrightarrow{Sym(f)} & Sym(A) \\ \Big\downarrow{\scriptstyle \psi_A} & & \Big\downarrow{\scriptstyle \psi_A} \\ Ord(A) & \xrightarrow{Ord(f)} & Ord(A) \end{array}$$

Consider then what happens to the identity permutation i in $Sym(A)$: it gets sent by $Sym(f)$ to $f \circ i \circ f^{-1} = i$. So the naturality square would tell us that $\psi_A(i) = Ord(f)(\psi_A(i))$. But that in general won't be so – suppose f swaps around elements, so $Ord(f)$ is not the 'do nothing' identity map.

Think of it this way: yes, $Sym(A)$ and $Ord(A)$ are isomorphic; but there is no privileged, especially natural, way of setting up an isomorphism between them. In a summary slogan: pointwise isomorphism doesn't entail natural isomorphism.

(b) We are, however, going to be particularly interested in cases where an isomorphism between objects $FA \cong GA$ is the result of a natural isomorphism between functors. And we can introduce the following key definition:

Definition 122. Two objects in a category D are said to be *naturally isomorphic (in A)* if they are the images FA and GA of the same object A under a couple of naturally isomorphic functors $F, G\colon \mathsf{C} \to \mathsf{D}$. △

Let's have some quick examples:

(1) A toy case. The products $A \times C$ and $C \times A$ are of course isomorphic, and intuitively the isomorphism here arises in an entirely natural way in a category C with all products (we don't have to make arbitrary choices to set it up). And so we will want to show that $A \times C$ and $C \times A$ are naturally isomorphic in A, in the sense defined.

But that follows immediately from the result in §32.1 that the functors $- \times C$ and $C \times -$ are naturally isomorphic.

(2) Another toy case. We've noted that the elements (ordinary sense) of a set X correspond one-to-one to the elements (categorial sense) $1 \to X$, where 1 is your favourite singleton. So in Set we have $X \cong \mathsf{Set}(1, X)$, and that's intuitively a naturally arising bijection.

And those isomorphic objects are indeed naturally isomorphic (in X) in our official sense. For it is obvious that there is a natural isomorphism α between the identity functor 1_{Set} and the hom-functor $\mathsf{Set}(1, -)$, whose components $\alpha_X\colon X \to \mathsf{Set}(1, X)$ send $x \in X$ to the corresponding function $\vec{x}\colon 1 \to X$.

(3) More excitingly, we have seen that $V \cong DDV$ naturally in V in FVect: that is the message of §32.3.

(4) And for a fourth case, given a category C with exponentials, $\mathsf{C}(A \times B, C) \cong \mathsf{C}(A, C^B)$ both naturally in A and naturally in C: that is the combined message of §32.4 (4) and (5). Challenge: use similar arguments to show that the isomorphism is natural in B too.

We will see more examples of naturally isomorphic objects in due course. For the moment, let's note a mini-theorem:

Theorem 154. *Given functors $F, G, H\colon \mathsf{C} \to \mathsf{D}$ and an object A in C, then if $FA \cong GA$ and $GA \cong HA$, both naturally in A, then $FA \cong HA$ naturally in A.*

Why so? Just recall how natural isomorphisms compose – see Theorem 32.2.

32.7 An 'Eilenberg/Mac Lane Thesis'?

Can we generalize from our various examples, and say that whenever we have a 'natural' or 'canonical' isomorphism between widgets and wombats (i.e. one that doesn't depend on arbitrary choices of co-ordinates, or the like), this can be seen as resulting from a natural isomorphism between suitable associated functors in the way we've defined? Let's call the claim that we *can* generalize like this the 'Eilenberg/Mac Lane Thesis'.

I choose the label to be reminiscent of the Church/Turing Thesis that we all know and love, which asserts that every algorithmically computable function (in an informally characterized sense) is in fact recursive/Turing computable/lambda computable. A certain intuitive concept, this Thesis claims, actually picks out the same functions as certain (provably equivalent) sharply defined concepts.

What kind of evidence do we have for the Church/Turing Thesis? Two sorts: (1) 'quasi-empirical', i.e. no unarguable clear exceptions have ever been found, and (2) more conceptual, as in for example Turing's own efforts to show that when we reflect hard on what we mean by algorithmic computation we get down to the sort of operations that a Turing machine can emulate, so a computable function just ought to be Turing computable. And the support is so overwhelming that it is now routine to appeal to the Church/Turing Thesis as a labour-saving device: if we can give an outline sketch of an argument that a certain function is algorithmically computable, we are allowed to assume that it is recursive/Turing computable/lambda computable without doing the hard work of e.g. defining a Turing maching to compute it.

We now seem to have on the table another Thesis of the same general type: an informal intuitive concept of a canonical or natural isomorphism, the Eilenberg/Mac Lane Thesis claims, in effect picks out the same isomorphisms as a certain sharply defined categorial concept.

Evidence? We would expect again two sorts. (1*) 'quasi-empirical', a lack of clear exceptions, and maybe (2*) conceptual, an explanation of why the Thesis just ought to be true.

It is, however, not clear exactly how things stand evidentially here, and the usual textbook discussions of natural isomorphisms usually don't pause to do much more than give a few examples.

More needs to be said. We therefore can't suppose that the new Eilenberg/Mac Lane Thesis is so secure that we can happily appeal to it in the same labour-saving way as the old Church/Turing Thesis. In other words, even if (i) intuitively an isomorphism between objects seems to be set up in a very 'natural' way, without appeal to arbitrary choices, and (ii) we can readily massage the claim of an isomorphism into a claim about at least pointwise isomorphism of relevant functors, we need to pause to work through a proof if we are to conclude that (iii) there is a natural isomorphism here in the official categorial sense. Annoying: for as we have already seen, such proofs can be a bit tedious.

33 Natural transformations

We typically think of isomorphisms categorially as special cases of some wider class of morphisms – they are the morphisms that have two-sided inverses. OK: so we'll want to think of natural isomorphisms between functors as special cases of .... Well, what? This chapter explores.

33.1 Natural transformations defined

(a) The generalized notion of morphisms between functors that we need is perhaps obvious enough. As before, we'll give a pair of definitions, one for each flavour of functor:

Definition 123. Let C and D be categories, let $F, G\colon \mathsf{C} \to \mathsf{D}$ be covariant functors (respectively, contravariant functors), and suppose that for each C-object C there is a D-arrow $\alpha_C\colon FC \to GC$. Then α, the family of arrows α_C, is a *natural transformation* between F and G iff for every $f\colon A \to B$ (respectively $f\colon B \to A$, note the reversal!) in C the following naturality square commutes in D:

$$\begin{array}{ccc} FA & \xrightarrow{Ff} & FB \\ \downarrow{\scriptstyle \alpha_A} & & \downarrow{\scriptstyle \alpha_B} \\ GA & \xrightarrow{Gf} & GB \end{array}$$

In this case, we write $\alpha : F \Rightarrow G$, and the α_C are said to be components of α. △

So, to compare: natural isomorphisms are the special case of natural transformations where the components are themselves all isomorphisms. And looking at the diagrams in §32.1, we see that a natural transformation between functors $F, G\colon \mathsf{C} \to \mathsf{D}$ again sends an F-image of (some or all of) C to its G-image in a way that respects some of the internal structure of the original – perhaps now collapsing even non-isomorphic objects together, but at least preserving composition of arrows.

(b) On notation: different styles of arrows can be found in use for talking about natural transformations, but Greek letters are almost universally used for their names.

A common diagrammatic convention will prove useful. When we have two functors $F, G\colon \mathsf{C} \to \mathsf{D}$, together with a natural transformation $\alpha\colon F \Rightarrow G$, we can neatly represent the whole situation thus:

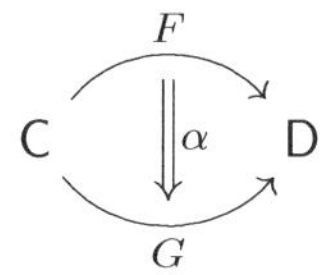

(c) Suppose now that we have three functors $F, G, H\colon \mathsf{C} \to \mathsf{D}$, together with two natural transformations $\alpha\colon F \Rightarrow G$ and $\beta\colon G \Rightarrow H$. Then we can simply generalize what we said about composing natural isomorphisms in §32.2 to apply to all natural transformations. In other words, we can compose $\alpha\colon F \Rightarrow G$ and $\beta\colon G \Rightarrow H$ to give a natural transformation $\beta \circ \alpha\colon F \Rightarrow H$ defined componentwise by putting $(\beta \circ \alpha)_C = \beta_C \circ \alpha_C$ for all objects C in C. Composing transformations 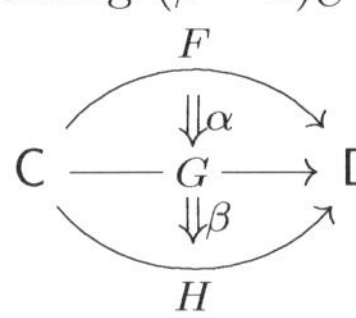

in this way to get 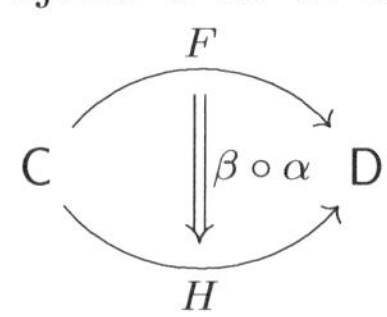

is rather predictably called *vertical composition*, and as with natural isomorphisms, vertical composition is associative. We'll meet a companion notion of horizontal composition shortly.

So, summarizing, we have (compare Theorem 152, parts (1) and (2)):

Theorem 155. *Suppose F, G, H are covariant functors from C to D:*

(1) There is an identity natural transformation (isomorphism!) $1_F\colon F \Rightarrow F$.

(2) Given natural transformations $\alpha\colon F \Rightarrow G$ and $\beta\colon G \Rightarrow H$, there is a composite natural transformation $\beta \circ \alpha\colon F \Rightarrow H$, and composition is associative.

(d) Since it makes sense to compose natural transformations, we can talk of a natural transformation $\alpha\colon F \Rightarrow G$ as being an isomorphism in the sense of having a two-sided inverse (so there is some $\beta\colon G \Rightarrow F$ such that $\beta \circ \alpha = 1_F$ and $\alpha \circ \beta = 1_G$). We then have a predictable result:

Theorem 156. *A natural transformation is an isomorphism (in the sense of having a two-sided inverse) if and only if it is a natural isomorphism.* □

Proof of 'if'. The proof of part (3) of Theorem 152 showed that a natural transformation that is an isomorphism has a two-sided inverse. □

Proof of 'only if'. Suppose the natural transformation $\alpha\colon F \Rightarrow G$ has an inverse α^{-1}, so $\alpha^{-1} \circ \alpha = 1_F$, and $\alpha \circ \alpha^{-1} = 1_G$. But composition of natural transformations is defined component-wise, so this requires for each component that $\alpha_C^{-1} \circ \alpha_C = 1_{FC}$, $\alpha_C \circ \alpha_C^{-1} = 1_{GC}$. Therefore each component of α has an inverse, so is an isomorphism, and hence α is a natural isomorphism. □

33.2 Some examples

(a) Let's have a couple of initial toy examples of natural transformations that aren't isomorphisms:

(1) Suppose D has a terminal object 1, and let $F\colon \mathsf{C} \to \mathsf{D}$ be any functor. Then there is also a parallel constant functor $\Delta_1\colon \mathsf{C} \to \mathsf{D}$ which sends every C-object to D's terminal object 1, and every C-arrow to the identity arrow on the terminal object.

Claim: there is a natural transformation $\alpha\colon F \Rightarrow \Delta_1$.

Proof. We need a suite of D-arrows $\alpha_C\colon FC \to \Delta_1 C$ (one for each C in C) which makes the following commute for any $f\colon A \to B$ in C, remembering that any $\Delta_1 C = 1$:

$$\begin{array}{ccc} FA & \xrightarrow{Ff} & FB \\ \downarrow{\scriptstyle \alpha_A} & & \downarrow{\scriptstyle \alpha_B} \\ 1 & \xrightarrow{1_1} & 1 \end{array}$$

But there is only one candidate for each of α_A, α_B, etc. – i.e. the unique arrow from its source to the terminal object. The diagram must then commute because all arrows (including composites) from FA to 1 are equal. □

And now here's a companion claim: there is no natural transformation $\alpha\colon \Delta_1 \Rightarrow F$. Essentially, that's because there is no canonical way of choosing where a candidate transformation should send the Δ_1-image of an object in C. Let's confirm that:

Proof. If there were a natural transformation, there would need to be a suite of arrows $\gamma_C\colon \Delta_1 C \to FC$, i.e. $\gamma_C\colon 1 \to FC$, one for every C-object C, which makes the following commute for any $f\colon A \to B$ in C:

$$\begin{array}{ccc} 1 & \xrightarrow{1_1} & 1 \\ \downarrow{\scriptstyle \gamma_A} & & \downarrow{\scriptstyle \gamma_B} \\ FA & \xrightarrow{Ff} & FB \end{array}$$

But γ_A picks out a particular element a of FA, and γ_B picks out a particular element b of FB. And there is no guarantee that Ff (for each and any f) always sends a to the particular target b. □

(2) Recall the functor $List\colon \mathsf{Set} \to \mathsf{Set}$. $List_{ob}$ sends a set C to the set of finite lists of members of C, and $List_{arw}$ sends a set-function $f\colon A \to B$ to the map from $List(A)$ to $List(B)$ that sends a list $a_0{}^\frown a_1{}^\frown a_2{}^\frown \ldots {}^\frown a_n$ to $fa_0{}^\frown fa_1{}^\frown fa_2{}^\frown \ldots {}^\frown fa_n$. Claim: there is a natural transformation $\alpha\colon 1 \Rightarrow List$, where 1 is the trivial identity functor $1\colon \mathsf{Set} \to \mathsf{Set}$.

Proof. We need a suite of functions α_C that make the following commute for any $f\colon A \to B$ in C:

$$\begin{array}{ccc} A & \xrightarrow{\quad f \quad} & B \\ \Big\downarrow{\scriptstyle \alpha_A} & & \Big\downarrow{\scriptstyle \alpha_B} \\ \mathit{List}(A) & \xrightarrow{\mathit{List}(f)} & \mathit{List}(B) \end{array}$$

For any non-empty C, put α_C to be the function which sends an element of C to the length-one list containing just that element (and sends the empty set to the empty list). We are immediately done. □

What about going in the opposite direction? Can there be a natural transformation from the functor *List* to the trivial functor 1? We'd need a uniform way of choosing from a list in $\mathit{List}(C)$ a specific member of C. Well, how about the function γ_C that chooses the first element of a list in $\mathit{List}(C)$? Won't that make the square always commute?

Nice try! But the snag is that even if C is non-empty, $\mathit{List}(C)$ will contain the empty list – and there is no canonical way of recovering a specific member of C from the empty list. And that thought can be parlayed into a proof that there is no natural transformation from *List* to 1.

(b) Now for a case with rather more significance, but which still requires only relatively elementary ideas.

(3) We are going to set up two functors from CRing, the category of rings where multiplication commutes, to our familiar friend Mon.

One is the functor $F\colon \mathsf{CRing} \to \mathsf{Mon}$ which simply forgets about the ring structure other than multiplication.

The other is the functor we can notate M_n which sends a ring R to the monoid of $n \times n$ matrices with elements from R (with matrix multiplication as the monoid operation, and the unit diagonal matrix as the monoid unit). How does M_n operate on a ring homomorphism $f\colon R \to S$? It applies f to each component of a matrix with elements from R to give us a matrix with elements from S. And since f respects multiplication in R, $M_n(f)\colon M_n(R) \to M_n(S)$ will respect multiplication between matrices, i.e. will be a monoid homomorphism.

Now the claim is that there is a nice natural transformation $M_n \Rightarrow F$. How come? We need a suite of arrows α such that, for every ring homomorphism $f\colon R \to S$, the naturality squares commute:

$$\begin{array}{ccc} M_n(R) & \xrightarrow{M_n(f)} & M_n(S) \\ \Big\downarrow{\scriptstyle \alpha_R} & & \Big\downarrow{\scriptstyle \alpha_S} \\ F(R) & \xrightarrow{\quad F(f) \quad} & F(S) \end{array}$$

OK, what's a nice canonical way of mapping an $n \times n$ matrix of elements of R to an element of $F(R)$ which is (of course) simply an element of R? How do we go in a natural way from matrices of reals, say, to a real number? Take a determinant! So let's put α_R to be the function det_R that sends matrices $M_n(R)$ to their determinants.

And then everything will nicely commute. A ring homomorphism $f\colon R \to S$ respects both addition and multiplication. So if we take a matrix M_R of R-elements and hit all the elements with f to get a matrix M_S of S-elements, it is immediate that $f(det_R(M_R)) = det_S(M_S)$.

In short, then, we have a natural transformation $det\colon M_n \Rightarrow F$.

(c) I'll briefly mention two more cases of natural transformations which aren't isomorphisms and which have mathematical significance (but feel quite free to skip if the background ideas are unfamiliar):

(4) For those who know a bit more group theory, consider the *abelianization* of a group G. Officially, this is the quotient of a group by its commutator subgroup $[G, G]$ (but you can think of it as the 'biggest' Abelian group A for which there is a surjective homomorphism α_G from G onto A). There is then a functor Ab which sends a group G to its abelianization $Ab(G)$, and sends an arrow $f\colon G \to H$ to the arrow $Ab(f)\colon Ab(G) \to Ab(H)$ defined in a fairly obvious way.

We therefore have a pair of functors, $\mathsf{Grp} \underset{Ab}{\overset{1}{\rightrightarrows}} \mathsf{Grp}$, and we can then check that the following diagram always commutes:

$$\begin{array}{ccc} G & \xrightarrow{\quad f \quad} & H \\ \Big\downarrow{\scriptstyle \alpha_G} & & \Big\downarrow{\scriptstyle \alpha_H} \\ Ab(G) & \xrightarrow{\ Ab(f)\ } & Ab(H) \end{array}$$

So we have a natural transformation, but not usually a natural isomorphism, between the functors 1 and Ab.

(5) For those who know rather more topology, we can mention two important functors from topological spaces to groups. One of them we've met before in §27.4, namely the functor $\pi\colon \mathsf{Top}_* \to \mathsf{Grp}$ which sends a space with a basepoint to its fundamental group at the base point. The other functor $H\colon \mathsf{Top} \to \mathsf{AbGrp}$ sends a space to the abelian group which is its first homology group (I'm not going to try to explain that here!). Now these functors aren't yet parallel functors between the same categories. But we can define a functor $H'\colon \mathsf{Top}_* \to \mathsf{Grp}$ which first forgets base points of spaces, then applies H, and then forgets that the relevant groups are abelian. I simply record that it is a very important fact of topology that, in our categorial terms, there is a natural transformation from π to H'.

(d) Can we generalize from these sorts of examples, and motivate a wider 'Eilenberg/Mac Lane Thesis' to the effect that, whenever we have a particularly natural sort of construction taking widgets to wombats we can view it as a natural transformation between suitable functors?

This looks a notably less plausible claim than the more restricted Thesis about isomorphisms that we met in §32.7, because the notion of a natural construction seems quite permissive. In fact, there seem to be clear counterexamples to the wider Thesis.

For example, take the construction that forms the centre of a group (the abelian subgroup of the elements which commute with all the other members of the group). That's surely a natural enough construction. But it is quite easy to show that there can't be a *functor* from Grp to Ab that takes groups to their centres and then behaves functorially on arrows.[1] And that stymies the possibility of thinking of group centres in terms of a suitable natural transformation between functors.

So a universal thesis here would overshoot. But yes, many cases of natural constructions *can* be treated categorially as natural transformations, and that's enough to make the notion of pivotal interest.

33.3 Horizontal composition of natural transformations

(a) We have seen how to compose natural transformations *vertically*. We can, however, also put things together *horizontally* in various ways.

First, there is so-called *whiskering*(!) where we combine a single functor with a natural transformation between two functors to get a new natural transformation. We can neatly represent one sort of combination by 'adding a whisker' on the left of a diagram for a natural transformation as follows:

$\mathsf{C} \xrightarrow{F} \mathsf{D} \underset{K}{\overset{J}{\Downarrow \beta}} \mathsf{E}$ gives rise to $\mathsf{C} \underset{K \circ F}{\overset{J \circ F}{\Downarrow \beta F}} \mathsf{E}$, where the component of βF at C is the same as the component of β at FC – in other words, $(\beta F)_C = \beta_{FC}$.

[1] An argument for enthusiasts. Suppose A is the free group on one generator a, and B is the free group on two generators a, b. The centre of A, $Z(A)$, is the abelian group A itself. The centre of B, $Z(B)$, is the trivial one-object group 1. Consider the homomorphism $f\colon A \to B$ generated by sending a to a, and the homomorphism $g\colon B \to A$ generated by sending both a and b to a. Note that $g \circ f = 1_A$. Suppose there were a functor Grp to Ab which, acting on objects, sends groups G to their centres $Z(G)$. How would it act on homomorphisms? It would need to send $f\colon A \to B$ to $Ff\colon Z(A) \to Z(B)$, i.e. $Ff\colon A \to 1$ (which sends everything to the one object of the trivial group) and send $g\colon B \to A$ to $Ff\colon Z(B) \to Z(A)$, i.e. $Fg\colon 1 \to A$. So $Fg \circ Ff$ is the map $!_A$ which sends everything in A to its unit element. But then we have

$$F1_A = F(g \circ f) = Fg \circ Ff = !_A \neq 1_{Z(A)}$$

So F is not functorial after all, not always sending identity maps to identity maps.

Why does this make sense?

Suppose $f\colon A \to B$ is a C-arrow, and consider the resulting arrow $Ff\colon FA \to FB$ in D. Since β is a natural transformation between the functors $J, K\colon \mathsf{D} \to \mathsf{E}$, we get the commuting naturality square on the left (living in E):

$$\begin{array}{ccc} J(FA) & \xrightarrow{J(Ff)} & J(FB) \\ \big\downarrow{\scriptstyle \beta_{FA}} & & \big\downarrow{\scriptstyle \beta_{FB}} \\ K(FA) & \xrightarrow{K(Ff)} & K(FB) \end{array} \qquad \begin{array}{ccc} (J \circ F)A & \xrightarrow{(J\circ F)f} & (J \circ F)B \\ \big\downarrow{\scriptstyle (\beta F)_A} & & \big\downarrow{\scriptstyle (\beta F)_B} \\ (K \circ F)A & \xrightarrow{(K\circ F)f} & (K \circ F)B \end{array}$$

Putting $(\beta F)_C = \beta_{FC}$ then makes this just the same as the square on the right, which we can now read as giving a natural transformation between $J \circ F$ and $K \circ F$.

Likewise for adding a whisker on the right:

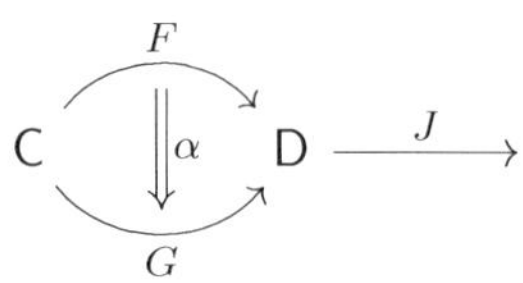

gives rise to

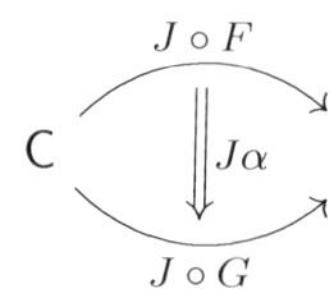

E, where the component of $J\alpha$ at C is the result of applying J to the component of α at C, i.e. $(J\alpha)_C = J(\alpha_C)$.[2]

(b) For future use, by the way, we should note the following mini-result:

Theorem 157. *Whiskering a natural isomorphism yields a natural isomorphism.*

Proof. 'Pre-whiskering' the natural transformation β by F to get βF yields a transformation whose components are (some of the) components of β. So if β's components are all isomorphisms, βF's components must be too.

While 'post-whiskering' α by J to get $J\alpha$ yields a transformation whose components are the result of applying J to the components of α. So if α is an isomorphism, then – since functors preserve isomorphisms – these components of $J\alpha$ are also all isomorphisms. □

(c) Second, we can *horizontally compose* two natural transformations in the following way:

We take

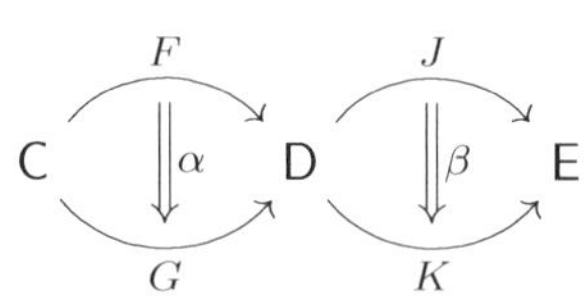

and get

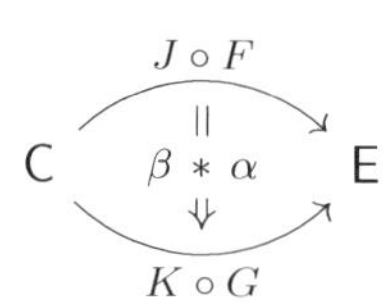

[2]I should note a variant notation here. Since the A-component of the whiskered β-following-F is β_{FA}, this encourages many to prefer the notation 'β_F' for the new natural transformation as opposed to our 'βF'. However, this introduces an unhappy notational asymmetry – as in '$J\alpha$' vs 'β_F' – which I find distracting in some contexts.

How do we define $\beta * \alpha$? Take an arrow $f\colon A \to B$ living in C and form this naturality square for α:

$$\begin{array}{ccc} FA & \xrightarrow{Ff} & FB \\ \downarrow{\scriptstyle \alpha_A} & & \downarrow{\scriptstyle \alpha_B} \\ GA & \xrightarrow{Gf} & GB \end{array}$$

Applying the functor J to the ingredients of that diagram, we get

$$\begin{array}{ccc} J(FA) & \xrightarrow{J(Ff)} & J(FB) \\ \downarrow{\scriptstyle J(\alpha_A)} & & \downarrow{\scriptstyle J(\alpha_B)} \\ J(GA) & \xrightarrow{J(Gf)} & J(GB) \end{array}$$

which also commutes. And since $Gf\colon GA \to GB$ is a map in D, and β is a natural transformation between the parallel functors $J, K\colon \mathsf{D} \to \mathsf{E}$, this too must commute:

$$\begin{array}{ccc} J(GA) & \xrightarrow{J(Gf)} & J(GB) \\ \downarrow{\scriptstyle \beta_{GA}} & & \downarrow{\scriptstyle \beta_{GB}} \\ K(GA) & \xrightarrow{K(Gf)} & K(GB) \end{array}$$

Gluing together those last two commutative diagrams one above the other gives a natural transformation from $J \circ F$ to $K \circ G$, if we set the component of $\beta * \alpha$ at C to be $\beta_{GC} \circ J\alpha_C$.

Three remarks:

(1) That definition for $\beta * \alpha$ looks surprisingly asymmetric. But note that instead of applying J to the initial naturality square for α and then pasting the result above a naturality square for β, we could have similarly applied K to the initial naturality square and pasted the result below another naturality square for β, thus showing that we can alternatively define the natural transformation $J \circ F$ to $K \circ G$ as having the components $K\alpha_C \circ \beta_{FC}$. So symmetry is restored: we get equivalent accounts that mirror each other.

(2) We can think of whiskering as a special case of the horizontal composition of two natural transformations where one of them is the identity natural transformation. For example

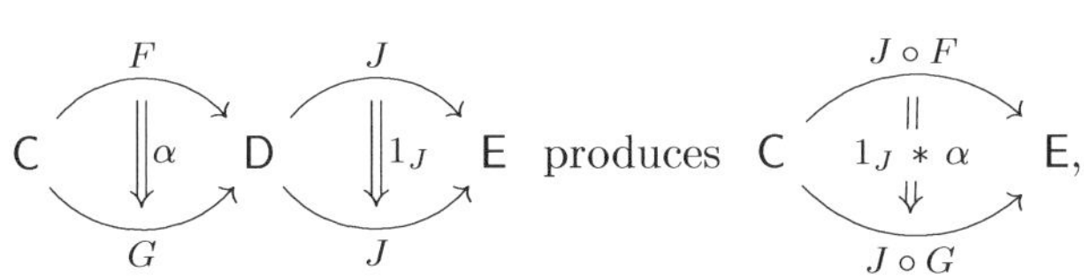

and the component of $1_J * \alpha$ at C is an identity composed with $J\alpha_C$. So this is the same as taking the left-hand natural transformation and simply whiskering with J on the right.

(3) We could now go on to consider the case of horizontally composing a pair of vertical compositions – and show that it comes to the same if we construe the resulting diagram as the result of vertically composing a pair of horizontal compositions. In symbols

$$(\delta \circ \gamma) * (\beta \circ \alpha) = (\delta * \beta) \circ (\gamma * \alpha)$$

But we won't now pause over this, another 'interchange law' – take it as a challenge to draw a diagram and prove it!

33.4 Cones as natural transformations

(a) A *natural transformation* is defined as a suite of arrows from various sources, with each pair of arrows making certain diagrams commute. But now compare: we earlier defined a *cone* as again essentially a suite of arrows – but this time all from the same source, the apex of the cone – with each pair of arrows making certain diagrams commute. Which suggests that we might be able to treat cones as special cases of natural transformations. And we can, giving us a perhaps tidier characterization than Defn. 112. Let's finish the chapter by showing how.

(b) Recall, Defn. 111 (re)defines a diagram of shape J in the category C as a functor $D\colon \mathsf{J} \to \mathsf{C}$.

Recall from §26.2, Ex. (F6) that among such functors from J to C, there are the constant functors $\Delta_X\colon \mathsf{J} \to \mathsf{C}$ which pick out an object X in C, send every J-object to X, and send every J-arrow to 1_X (see §26.2 (F6)).

And now let's ask: what does it take for there to be a natural transformation $\alpha\colon \Delta_X \Rightarrow D$, for some given $D\colon \mathsf{J} \to \mathsf{C}$?

By the definition of a natural transformation, the following must commute for any J-arrow $j\colon K \to L$:

$$\begin{array}{ccc} \Delta_X K & \xrightarrow{\Delta_X j} & \Delta_X L \\ {\scriptstyle\alpha_K}\downarrow & & \downarrow{\scriptstyle\alpha_L} \\ DK & \xrightarrow{Dj} & DL \end{array} \quad = \quad \begin{array}{ccc} X & \xrightarrow{1_X} & X \\ {\scriptstyle\alpha_K}\downarrow & & \downarrow{\scriptstyle\alpha_L} \\ DK & \xrightarrow{Dj} & DL \end{array} \quad = \quad \begin{array}{ccc} & X & \\ {\scriptstyle\alpha_K}\swarrow & & \searrow{\scriptstyle\alpha_L} \\ DK & \xrightarrow{Dj} & DL \end{array}$$

Which makes the α_J (where J runs over objects in J) the legs of a cone over D with a vertex X!

Conversely, the legs of any cone over D with a vertex X assemble into a natural transformation $\alpha\colon \Delta_X \to D$. So that means that cones can be regarded as certain natural transformations.

Or at least, that is so if we think of cones austerely, as a suite of arrows. To recall: we originally defined a cone as some (C, c_j), a vertex C together with some arrows c_j from that vertex satisfying certain conditions. As already noted in §19.1, fn. 1, that mode of presentation redundantly gives the vertex data twice – both explicitly, and then implicitly as the shared source of the arrows. So we lose no information in treating a cone just as the suite of its arrows.

So, with that understanding, we can sum up as a theorem:

Theorem 158. *If D is a diagram of shape J in C, then a cone over D with vertex X is a natural transformation from Δ_X to D (where $\Delta_X \colon \mathsf{J} \to \mathsf{C}$ is the collapse-to-X functor).* □

We will pick up this idea again in §36.7.

34 Isomorphic categories, equivalent categories

Now we have defined natural transformations, and more particularly natural isomorphisms, we are in a position to characterize what it is for categories to be *equivalent* – where equivalent categories 'come to the same', in one good intuitive sense. However, we first define a stronger notion, and then see why it is *too* strong for many purposes.

34.1 Isomorphic categories

(a) When first introducing functors in §26 we quickly met Theorem 129 which (1) tells us that for any category C there is an identity functor $1_{\mathsf{C}} \colon \mathsf{C} \to \mathsf{C}$ which sends C's objects and arrows to themselves, and (2) also tells us that functors $F \colon \mathsf{C} \to \mathsf{D}$ and $G \colon \mathsf{D} \to \mathsf{E}$ always compose to give a functor $GF \colon \mathsf{C} \to \mathsf{E}$.

Hence it makes perfect sense to say that a functor might have a two-sided inverse, and so we get a predictable definition:[1]

Definition 124. A functor $F \colon \mathsf{C} \to \mathsf{D}$ is an *isomorphism*, in symbols $F \colon \mathsf{C} \xrightarrow{\sim} \mathsf{D}$, iff it has a two-sided inverse – meaning that there is a functor $G \colon \mathsf{D} \to \mathsf{C}$ where $GF = 1_{\mathsf{C}}$ and $FG = 1_{\mathsf{D}}$. △

And let's check that we get sensible results with this definition, e.g.

Theorem 159. *If $F \colon \mathsf{C} \xrightarrow{\sim} \mathsf{D}$ is an isomorphism, it is fully faithful and essentially surjective on objects.*

Proof. Take parallel arrows in C, namely $f, g \colon A \to B$. Supposing $Ff = Fg$, then $GFf = GFg$ – where G is F's inverse. So $1_{\mathsf{C}}f = 1_{\mathsf{C}}g$ and hence $f = g$. Therefore F is faithful.

Suppose next we are given an arrow $h \colon FA \to FB$. Put $f = Gh$. Then $Ff = FGh = 1_{\mathsf{D}}h = h$. So every such h in D is the image under F of the associated arrow f in C. So F is full.

[1] NB: To echo fn. 1 of Ch. 32, we are now defining what it is for a functor to be an isomorphism between *categories*, while in the last couple of chapters we have been interested in (natural) isomorphisms between *functors*. Different topics!

Finally take any D-object D, and using F's inverse G again, map that to the C-object GD. Then $F(GD) = D \cong D$. So F is e.s.o. □

The converse doesn't hold, however. We will soon find examples of functors that are fully faithful and e.s.o. but not isomorphisms.

(b) Now that we have the notion of a functor-as-isomorphism in play, this readily prompts a further definition:

Definition 125. The categories C and D are *isomorphic*, in symbols $\mathsf{C} \cong \mathsf{D}$, iff there is an isomorphism $F\colon \mathsf{C} \xrightarrow{\sim} \mathsf{D}$. △

It is immediate that $\cong$ is an equivalence relation, as required. So let's go straight to some examples of isomorphic categories.

(1) Take the toy categories with different pairs of objects which we can diagram as e.g.

$$\circlearrowleft a_1 \longrightarrow b_1 \circlearrowright \qquad\qquad \circlearrowleft a_2 \longrightarrow b_2 \circlearrowright$$

Plainly there is a unique isomorphism that sends the first to the second. And if we don't care about distinguishing copies of structures that are related by a unique isomorphism, then we'll count these as the same in a strong sense. Which to that extent warrants our earlier talk about *the* category $\mathbf{2}$ – e.g. in §5.4, Ex. (C6).

(2) Recall two earlier definitions, of Set_* and $1/\mathsf{Set}$.

The objects of the category of pointed sets Set_* are pairs (X, x), comprising a non-empty set X and x a chosen base point in X. And an arrow $f\colon (X, x) \to (Y, y)$ is a set-function $f\colon X \to Y$ such that $fx = y$.

As for the coslice category $1/\mathsf{Set}$, its objects are all pairs of the form $(X, \vec{x}\colon 1 \to X)$ for any object X in Set (and note, since there is no arrow $\vec{x}\colon 1 \to \emptyset$, X can't be empty here). The arrows from $(X, \vec{x}\colon 1 \to X)$ to $(Y, \vec{y}\colon 1 \to Y)$ are just the set-functions $j\colon X \to Y$ such that $j \circ \vec{x} = \vec{y}$.

Now we said when introducing the construction in §7.3 that $1/\mathsf{Set}$ looks to be in some strong sense 'the same as' the category Set_* of pointed sets. And in fact the categories are isomorphic, as we'll now show.

So take F_{ob} to be the map from objects in $1/\mathsf{Set}$ to objects Set_* which sends $(X, \vec{x}\colon 1 \to X)$ to the pointed set (X, x) where x is the value of the function $\vec{x}$ for its sole argument. And take F_{arw} to send a set function $j\colon X \to Y$ such that $j \circ \vec{x} = \vec{y}$ to the same function treated as an arrow $j\colon (X, x) \to (Y, y)$ which must then preserve base points. It is trivial to check that F is a functor $F\colon 1/\mathsf{Set} \to \mathsf{Set}_*$.

In the other direction, we can define a functor $G\colon \mathsf{Set}_* \to 1/\mathsf{Set}$ which acts on objects by sending (X, x) to $(X, \vec{x})$, where $\vec{x}\colon 1 \to X$ of course maps 1 to the point x, and acts on arrows by sending a basepoint-preserving set-function from X to Y to itself.

It is immediate that these two functors F and G are inverse to each other. Hence, as claimed, $\mathsf{Set}_* \cong 1/\mathsf{Set}$.

(3) We can very similarly show, e.g., that the comma category $(1_{\mathsf{C}} \downarrow X)$ from §29.2 is isomorphic to the slice category C/X. But again, that's too easy to be very interesting, so I leave that as an exercise to check!

So for something giving us a little more to chew on, consider Boolean algebras and the two standard ways of presenting them. In categorial terms, there is a category Bool whose objects are algebras $(B, \neg, \wedge, \vee, 0, 1)$ constrained by the familiar Boolean axioms, and whose arrows are homomorphisms that preserve algebraic structure. But then there is also a category BoolR whose objects are Boolean rings, i.e. rings $(R, +, \times, 0, 1)$ where $x^2 = x$ for all $x \in R$, and whose arrows are ring homomorphisms.

There are familiar ways of marrying up Boolean algebras with corresponding rings and vice versa. Thus if we start from $(B, \neg, \wedge, \vee, 0, 1)$, take the same underlying set and distinguished objects, put

(i) $x \times y =_{\text{def}} x \wedge y$,
(ii) $x + y =_{\text{def}} (x \vee y) \wedge \neg(x \wedge y)$ (exclusive 'or'),

then we get a Boolean ring. And if we apply this same process to two algebras B_1 and B_2, it is elementary to check that it will carry a homomorphism of algebras $f_a \colon B_1 \to B_2$ to a corresponding homomorphism of rings $f_r \colon R_1 \to R_2$.

We can equally easily go from rings to algebras, by putting

(i) $x \wedge y =_{\text{def}} x \times y$,
(ii) $x \vee y =_{\text{def}} x + y + (x \times y)$
(iii) $\neg x =_{\text{def}} 1 + x$.

Note that going from a algebra to the associated ring and back again takes us back to where we started.

In summary, without going into any more details, we can in this way define a functor $F \colon$ Bool $\to$ BoolR, and a functor $G \colon$ BoolR $\to$ Bool which are inverses to each other. So, as we'd surely have expected, the category Bool is isomorphic to the category BoolR.

34.2 Intuitively equivalent but non-isomorphic categories

(a) We've found some nice cases of categories that intuitively 'come to the same' and which are more or less easily seen to be isomorphic in the official sense of Defn. 125. So far, so good.

Now, we could have given the definitions and examples in the last section right back after we first introduced functors. So why have I left the discussion of categories 'coming to the same' until now, after we have the notion of natural isomorphisms between functors in play? Because we can also readily give examples of pairs of categories that again seem morally equivalent but which *aren't* isomorphic. And we'll need our shiny new apparatus to cope with these cases.

To introduce a first example, let's think for a moment about partial functions.

In the general theory of computation, there is no getting away from the central importance of the notion of a partial function. But how should we treat partial functions in logic? Suppose the partial computable function $\varphi\colon \mathbb{N} \to \mathbb{N}$ outputs no value for n (the algorithm defining φ doesn't terminate gracefully for input n). Then the term '$\varphi(n)$' apparently lacks a denotation. *But in standard first-order logic, all terms are assumed to denote* – a sentence with a non-denoting term, on the standard semantics, will lack a truth value. What to do, if we want to formalize our theory?

One option is to bite the bullet, formally allow non-denoting terms, and then go in for some logical revisionism to cope with the truth-value gaps which come along with them.

Another option is to follow Frege and stipulate that apparently empty terms are in fact not empty at all but denote some special rogue object (so there are, strictly speaking, no empty terms and no truth-value gaps, and hence we can preserve standard logic).

So on this second option, presented with what we might naively think of as a partial function $\varphi\colon \mathbb{N} \to \mathbb{N}$, we now treat this as officially being a *total* function – namely $f\colon \mathbb{N} \cup \{\star\} \to \mathbb{N} \cup \{\star\}$, where $\star$ is any convenient non-number, and where $f(n) = \varphi(n)$ when $\varphi(n)$ takes a numerical value and f takes the value $\star$ otherwise. If you like, you can think of $\star$ as coding 'not numerically defined'. On this option, then, since our functions are all officially total, they don't generate non-denoting terms, and we can preserve our standard logic without truth-value gaps.

There is a lot more to be said: but while the debate about the best logical treatment of partial functions is the sort of thing that might grip some philosophically-minded logicians, it does seem of very little general mathematical interest. *And that's exactly the current point.* From a mathematical point of view there surely isn't anything much to choose between the two options. We can think of a world of genuinely *partial* numerical functions $\varphi\colon \mathbb{N} \to \mathbb{N}$, or we can equally think of a corresponding world of *total* functions $f\colon \mathbb{N} \cup \{\star\} \to \mathbb{N} \cup \{\star\}$, with the distinguished point $\star \notin \mathbb{N}$, and $f(\star) = \star$. Take your pick!

More generally, on a larger scale, we can think of a category Pfn whose objects are sets X and whose arrows are (possibly) *partial* functions between them. And there is also the category $\mathsf{Set}_\star$ of pointed sets whose objects are sets with a distinguished base point, and whose arrows are (total) set-functions which preserve base points. And mathematically, shouldn't these come to the same?

However, we can now easily show:

Theorem 160. $\mathsf{Set}_\star$ *is not isomorphic to* Pfn.

We can remark that there *is* an obvious functor $F\colon \mathsf{Set}_\star \to \mathsf{Pfn}$. F sends a pointed set (X, x) to the set $X \setminus \{x\}$, and sends a base-point preserving total function $f\colon (X, x) \to (Y, y)$ to the partial function $\varphi\colon X \setminus \{x\} \longrightarrow Y \setminus \{y\}$, where $\varphi(x) = f(x)$ if $f(x) \in Y \setminus \{y\}$, and is undefined otherwise. But, nice though this is, F isn't an isomorphism (it could send distinct (X, x) and (X', x') to the same target object).

Again, there is a whole family of functors from Pfn to $\mathsf{Set}_\star$ which take a set X and add an element not yet in X to give as an expanded set with the new object as a basepoint. Here's a way of doing this in a uniform way without making independent choices for each X. Define $G\colon \mathsf{Pfn} \to \mathsf{Set}_\star$ as sending a set X to the pointed set $X_* =_{\text{def}} (X \cup \{X\}, X)$, remembering that in standard set theories $X \notin X$. And then let G send a partial function $\varphi\colon X \to Y$ to the total basepoint-preserving function $f\colon X_* \to Y_*$, where $f(x) = \varphi(x)$ if $\varphi(x)$ is defined and $f(x) = Y$ otherwise. G is a natural enough choice, but isn't an isomorphism (it isn't surjective on objects).

Still, those observations don't yet rule out there being *some* pair of functors between $\mathsf{Set}_\star$ and Pfn which are mutually inverse. However, simple cardinality considerations show that there can't be any such pair.

Proof. A functor which is an isomorphism from Pfn to $\mathsf{Set}_\star$ must, among other things, send isomorphisms living in Pfn one-to-one to isomorphisms living in $\mathsf{Set}_\star$, so should preserve the cardinality of isomorphism classes. But the isomorphism class of the empty set in Pfn has just one member, while there is no one-membered isomorphism class in $\mathsf{Set}_\star$. So there can't be an isomorphism between the categories. □

(b) Let's have a rather deeper example. Take again the category FVect whose objects are finite-dimensional vector spaces over the reals and whose arrows are linear maps. And now consider the category Mat whose objects are natural numbers (representing the dimension of a vector space) and whose arrows $M\colon m \to n$ are the $m \times n$ matrices with real-number entries (representing linear maps between spaces). Composition of arrows in Mat is the usual matrix multiplication and the identity arrow on n is the $n \times n$ identity matrix.[2]

Now, it is an entirely familiar thought that linear algebra can be equivalently done either abstractly with linear maps or concretely with matrices – and we move between the two styles as context makes convenient. So in an important sense, the corresponding categories FVect and Mat ought to 'come to the same'.

Well, there are indeed natural enough functors going in each direction. Take a basis for each space: then there is a functor $J\colon \mathsf{FVect} \to \mathsf{Mat}$ which sends a space to its dimension and sends a linear map $f\colon V \to W$ to the corresponding matrix representing f with respect to the chosen bases. And there is a functor $K\colon \mathsf{Mat} \to \mathsf{FVect}$ which sends n to $\mathbb{R}^n$ treated as a vector space, and sends an $n \times m$ matrix to the linear map from $\mathbb{R}^m$ to $\mathbb{R}^n$ which the matrix represents with respect to the standard bases for those spaces. But neither of those functors is an isomorphism, and again cardinality considerations show that there can't be one. Objects m and n in Mat are isomorphic only if they are the same; but there can be isomorphic vector spaces in FVect which are not identical (remember we've only fixed the scalars as being reals: the vectors of an n-dimensional space can vary ad lib).

[2]We can either ban the zero-dimensional case, or allow it and give an ad hoc treatment. The details don't matter.

34.3 Equivalent categories

(a) We've now seen two examples of pairs of categories which are intuitively mathematically equivalent in some strong sense but which aren't isomorphic (according to the natural definition of isomorphism for categories).

We did, however, in the first case note an obvious choice of functors $F\colon \mathsf{Set}_\star \to \mathsf{Pfn}$ and $G\colon \mathsf{Pfn} \to \mathsf{Set}_\star$. Now, since G adds an external base point to a set X, and then F simply removes it again (and the functors do the minimum necessary on arrows), the composite functor $FG\colon \mathsf{Pfn} \to \mathsf{Pfn}$ is actually the identity functor on Pfn. But what about the reverse composition, GF. It can't be the identity functor on $\mathsf{Set}_\star$, or else we'd have isomorphic functors after all. However, GF *does* map $\mathsf{Set}_\star$ to itself in a very simple way. Officially, GF sends the pointed set (X, x) to $(X_x \cup \{X_x\}, X_x)$ where $X_x = X \setminus \{x\}$. But that's to say that GF takes a pointed set (X, x) and methodically replaces the base point with something not already in X. And GF treats arrows to fit. So, applied to some objects and arrows, although GF doesn't take us back to where we started, it should systematically give us back an isomorphic copy. Which suggests that we can expect that GF is naturally isomorphic to the identity functor $\mathsf{Set}_\star$, i.e. $GF \cong 1_{\mathsf{Set}_\star}$.

What about our second example involving FVect and Mat. In this case, the composite KJ will take us from an n-dimensional space V to $\mathbb{R}^n$. But since all n-dimensional spaces are isomorphic, although KJ isn't the identity on FVect, its systematic action again suggests that KJ is naturally isomorphic to the identity, $KJ \cong 1_{\mathsf{FVect}}$. Similarly, depending on the basis for $\mathbb{R}^n$ chosen in defining J, the composite JK needn't be the identity either. It sends an object n to itself; but on arrows, it can systematically send matrices to their image induced by a change of basis. Still, given its methodical operation, we might still expect that $JK \cong 1_{\mathsf{Mat}}$.

(b) Reflection on these cases suggests, then, the following weakening of the definition of isomorphism between categories:

Definition 126. Categories C and D are *equivalent*, in symbols $\mathsf{C} \simeq \mathsf{D}$, iff there are functors $F\colon \mathsf{C} \to \mathsf{D}$ and $G\colon \mathsf{D} \to \mathsf{C}$ that are pseudo-inverses, i.e. $GF \cong 1_{\mathsf{C}}$ and $FG \cong 1_{\mathsf{D}}$.

It can be safely left as an easy exercise to check that equivalence of categories, thus defined, truly is an equivalence relation!

(c) We could now pause to upgrade our hand-waving arguments a moment ago to give a direct proof that Pfn and $\mathsf{Set}_\star$ are equivalent in this sense, and likewise for FVect and Mat.

But we won't do this. Rather, we'll first prove a general result that yields an alternative characterization of equivalence which can often be much easier to apply:[3]

[3] Emily Riehl (2017, p. 31) says that she used to set our next theorem as homework. You are hereby challenged to derive it for yourself before reading the given proof!

Theorem 161. *Assuming a sufficiently strong choice principle, a functor $F\colon \mathsf{C} \to \mathsf{D}$ is part of an equivalence between C and D iff F is faithful, full and essentially surjective on objects.*

Proof. First suppose F is part of an equivalence between C and D, so that there is a functor $G\colon \mathsf{D} \to \mathsf{C}$, where $GF \cong 1_{\mathsf{C}}$ and $FG \cong 1_{\mathsf{D}}$. Then:

(i) For any arrow $f\colon A \to B$ in C, then by hypothesis, the following square commutes (where η is a natural isomorphism between the identity functor 1_{C} and the composite GF),

$$\begin{array}{ccc} A & \xrightarrow{\quad f \quad} & B \\ \big\downarrow{\scriptstyle \eta_A} & & \big\downarrow{\scriptstyle \eta_B} \\ GFA & \xrightarrow{\quad GFf \quad} & GFB \end{array}$$

and hence $\eta_B^{-1} \circ GFf \circ \eta_A = f$. And similarly of course, for any other $g\colon A \to B$, we have $\eta_B^{-1} \circ GFg \circ \eta_A = g$. It immediately follows that if $Ff = Fg$ then $f = g$, i.e. F is faithful. A companion argument, interchanging the roles of C and D, shows that G too is faithful.

(ii) Suppose we are given an arrow $h\colon FA \to FB$, then define f to be $\eta_B^{-1} \circ Gh \circ \eta_A$. But we know from the naturality square that we always have $f = \eta_B^{-1} \circ GFf \circ \eta_A$. So it follows in this case that $GFf = Gh$, and since G is faithful, $h = Ff$. Hence every such h in D is the image under F of some arrow f in C. Therefore F is full.

(iii) Recall, $F\colon \mathsf{C} \to \mathsf{D}$ is e.s.o. iff for any D in D we can find some isomorphic object FC, for C in C. But we know that there is, by assumption, a natural isomorphism $\epsilon\colon FG \Rightarrow 1_{\mathsf{D}}$. So we have a component isomorphism $\epsilon_D\colon FGD \xrightarrow{\sim} D$. Therefore putting $C = GD$ gives the desired result showing that F is e.s.o.

Now for the argument in the other direction. Suppose, then, that $F\colon \mathsf{C} \to \mathsf{D}$ is faithful, full and e.s.o. We need to construct (iv) a corresponding functor $G\colon \mathsf{D} \to \mathsf{C}$, and then a pair of natural isomorphisms (v) $\epsilon\colon FG \Rightarrow 1_{\mathsf{D}}$ and (vi) $\eta\colon 1_{\mathsf{C}} \Rightarrow GF$:

(iv) By hypothesis, F is e.s.o., so by definition every D-object D is isomorphic in D to FC, for some C in C. Hence – and *here* we are invoking an appropriate choice principle – for any given D, we can choose an object C such that there is an isomorphism $\epsilon_D\colon FC \xrightarrow{\sim} D$ in D.

Now define $G_{ob}\colon \mathsf{D} \to \mathsf{C}$ as sending an object D to the chosen C from C (so $GD = C$, and $\epsilon_D\colon FGD \xrightarrow{\sim} D$).

To get a functor, we need the component G_{arw} to act suitably on an arrow $g\colon D \to E$. Now, note that corresponding to g we have a composite arrow $FGD \to FGE$, namely

$$FGD \xrightarrow{\epsilon_D} D \xrightarrow{g} E \xrightarrow{\epsilon_E^{-1}} FGE$$

and since F is full and faithful, there must be some unique $f\colon GD \to GE$ which F sends to that composite. Put $G_{arw}g = f$.

Note, for use in a moment, that this means that (*) given any $g\colon D \to E$, $FGg = \epsilon_E^{-1} \circ g \circ \epsilon_D$, so the left square below commutes, as does the right square (**) for any $f\colon A \to B$ in C (so for any $Ff\colon FA \to FB$):

$$\begin{array}{ccc} FGD & \xrightarrow{FGg} & FGE \\ \downarrow{\scriptstyle \epsilon_D} & & \downarrow{\scriptstyle \epsilon_E} \\ D & \xrightarrow{g} & E \end{array} \qquad\qquad \begin{array}{ccc} FGFA & \xrightarrow{FGFf} & FGFB \\ \downarrow{\scriptstyle \epsilon_{FA}} & & \downarrow{\scriptstyle \epsilon_{FB}} \\ FA & \xrightarrow{Ff} & FB \end{array}$$

We now need to check that G, *with components* G_{ob}, G_{arw}, *is a functor.* So we need to show that G (a) preserves identities and (b) respects composition:

For (a), note that $G_{arw}1_D = e$ where e is the unique arrow from GD to GD such that $Fe = \epsilon_D^{-1} \circ 1_D \circ \epsilon_D = 1_{FGD}$. So $e = 1_{GD}$.

For (b) we need to show that, given D-arrows $g\colon D \to E$ and $h\colon E \to J$, $G(h \circ g) = Gh \circ Gg\colon D \to J$. But note that using (*) we have

$$\begin{aligned} FG(h \circ g) = \epsilon_J^{-1} \circ h \circ g \circ \epsilon_D &= (\epsilon_J^{-1} \circ h \circ \epsilon_E) \circ (\epsilon_E^{-1} \circ g \circ \epsilon_D) \\ &= FG(h) \circ FG(g) = F(G(h) \circ G(g)) \end{aligned}$$

Hence, since $FG(h \circ g) = F(G(h) \circ G(g))$ and F is faithful, $G(h \circ g) = G(h) \circ G(g)$, so G is indeed a functor.

(v) The commuting of the naturality square in (*) for any g shows that components ϵ_D assemble into a natural isomorphism $\epsilon\colon FG \overset{\sim}{\Longrightarrow} 1_{\mathsf{D}}$.

(vi) Note next that we have an isomorphism $\epsilon_{FA}^{-1}\colon FA \xrightarrow{\sim} FGFA$. As F is full and faithful, $\epsilon_{FA}^{-1} = F(\eta_A)$ for some unique $\eta_A\colon A \xrightarrow{\sim} GFA$. Since F is fully faithful, it is conservative, i.e. it reflects isomorphisms (by Theorem 137); hence η_A is also an isomorphism. Also, the naturality diagram

$$\begin{array}{ccc} A & \xrightarrow{f} & B \\ \downarrow{\scriptstyle \eta_A} & & \downarrow{\scriptstyle \eta_B} \\ GFA & \xrightarrow{GFf} & GFB \end{array}$$

always commutes for any arrow $f\colon A \to B$ in C. Why? Because

$$\begin{aligned} F(\eta_B \circ f) = F\eta_B \circ Ff = \epsilon_{FB}^{-1} \circ Ff = \\ FGFf \circ \epsilon_{FA}^{-1} = FGFf \circ F\eta_A = F(GFf \circ \eta_A) \end{aligned}$$

where the middle step applies the naturality square (**). But if $F(\eta_B \circ f) = F(GFf \circ \eta_A)$ then since F is faithful, $\eta_B \circ f = GFf \circ \eta_A$. Hence the η_A are the components of our desired natural isomorphism $\eta\colon 1_{\mathsf{C}} \overset{\sim}{\Longrightarrow} GF$.

So we are done! $\square$

(d) Our useful general Theorem 161 enables us now to very quickly prove the following two equivalence claims without any more hard work:

Theorem 162. Pfn $\simeq$ Set$_\star$

Proof. Define the functor G: Pfn $\to$ Set$_\star$ as before. It sends a set X to a set $X_* =_{\text{def}} X \cup \{X\}$ with basepoint X, and sends a partial function $f\colon X \to Y$ to the total function $f_*\colon X_* \to Y_*$, where for $f_*(x) = f(x)$ if $f(x)$ is defined and $f_*(x) = Y$ otherwise.

G is faithful, as it is easily checked that it sends distinct functions to distinct functions. And it is equally easy to check that G is full, i.e. given any basepoint preserving function between sets X_* and Y_*, there is a partial function f which G sends to it.

But G is essentially surjective on objects. For every pointed set in Set$_\star$ – i.e. every set which can be thought of as the union of a set X with $\{*\}$ where $*$ is an additional basepoint element (not in X) – is isomorphic in Set$_\star$ to the set $X \cup \{X\}$ with X as basepoint.

Hence G is part of an equivalence between Pfn and Set$_\star$. □

Theorem 163. FVect $\simeq$ Mat

Proof. Take the functor J: FVect $\to$ Mat we defined near the beginning of this section. It's faithful as different linear maps get different matrix representations. It's full because every matrix corresponds to a linear map. It's trivially surjective on objects. Hence J is part of an equivalence between FVect and Mat. □

34.4 Why equivalence is the categorially nicer notion

Let's have another very much simpler but instructive example. Recall FinSet is the category of finite sets and functions between them. And FinOrd is the category of finite von Neumann ordinals and functions between them. We then have:

Theorem 164. FinOrd $\simeq$ FinSet

Proof. FinOrd is a full subcategory of FinSet, so the inclusion functor F is fully faithful. F is also essentially surjective on objects: for take any object in FinSet, which is some n-membered set: that is in bijective correspondence (and hence isomorphic in FinSet) with the finite ordinal n. Hence F is part of an equivalence, and FinOrd $\simeq$ FinSet. □

The interesting question here is how should we regard this last result. We saw that defining equivalence of categories in terms of isomorphism would be *too strong*, as it rules out our treating Pfn and Set$_\star$ as in effect equivalent. But now we've seen that defining equivalence of categories as in Defn. 34.3 makes the seemingly very sparse category FinOrd equivalent to the seemingly much more abundant FinSet. Is that a strike against the definition of equivalence, showing it to be *too weak*?

It might help to think of an even simpler toy example. Consider the two categories which we can diagram respectively as follows (with the diagram on the right intended to commute):

$$\bullet \circlearrowright \qquad\qquad \circlearrowleft \bullet \rightleftarrows \star \circlearrowright$$

On the left, we have the category 1. On the right we have a two-object category 2! with arrows in *both* directions between the objects and, since the diagram commutes, those two arrows are inverse to each other and hence are isomorphisms. These two categories are plainly *not* isomorphic, but they *are* equivalent. For one of the inclusion functors $1 \hookrightarrow 2!$ is full and faithful, and it is trivially essentially surjective on objects because each object in the two-object category is isomorphic to the other.

What this second toy example highlights is that our equivalence criterion counts categories as amounting to the same when (putting it very roughly) one is just the same as the other padded out with new objects and enough arrows to make the new objects isomorphic to some old objects.

But on reflection that's fine. Taking a little bit of the mathematical world and bulking it out with copies of the structures it already contains and isomorphisms between the copies won't, for many (most? nearly all?) purposes, give us a real enrichment. Therefore a criterion of equivalence of categories-as-mathematical-universes that doesn't care about surplus isomorphic copies is what we typically need. Hence the results that $1 \simeq 2!$ and Finord $\simeq$ FinSet are arguably welcome features, not bugs, of our account of equivalence.

34.5 Skeletons and evil

(a) Given that two categories can be regarded as being equivalent in an important sense even when one is bulked out with isomorphic extras, shouldn't the usual sort of concern for Bauhaus elegance and lack of redundancy lead us to privilege categories that are as skeletal as possible? Let's say:

Definition 127. The category S is a *skeleton* of the category C if S is a full subcategory of C that contains exactly one object from each class of isomorphic objects of C. A category is *skeletal* if it is a skeleton of some category.

For a toy example, suppose C is a category arising from a preorder – as in §5.4 (C4). Then any skeleton of C will be a poset category. (Check that!)

Theorem 165. *If* S *is a skeleton of the category* C *then* $\mathsf{S} \simeq \mathsf{C}$.

Proof. The inclusion functor $\mathsf{S} \hookrightarrow \mathsf{C}$ is fully faithful, and by the definition of S is essentially surjective on objects. So we can apply Theorem 161. □

Theorem 166. *If* R *and* S *are skeletal categories, then equivalence implies isomorphism, so if* $\mathsf{R} \simeq \mathsf{S}$ *then* $\mathsf{R} \cong \mathsf{S}$.

Proof. By Theorem 161, there must be a functor $F\colon \mathsf{R} \to \mathsf{S}$ which is fully faithful and essentially surjective.

Since S is skeletal, being essentially surjective implies that F is surjective on objects.

F is also injective on objects. For suppose for R-objects C and D, $FC = FD = X$. Since there is an identity C-arrow $1_X\colon FC \to FD$, and F is full, there must be an R-arrow $f\colon C \to D$ such that $Ff = 1_X$. Likewise there must be an R-arrow $g\colon D \to C$ such that $Fg = 1_X$. So $F(g \circ f) = 1_X \circ 1_X = 1_X = F(1_C)$. Hence, since F is faithful, $g \circ f = 1_C$. Similarly $f \circ g = 1_D$. Therefore f and g are isomorphisms between C and D and hence (since R is skeletal) $C = D$.

Since F is bijective on objects, full and faithful, it follows that it is also bijective on arrows. So it is an isomorphism between R and S. □

(b) So how about this for a programme? Take our favoured initial universe of categories, whatever that is. But now slim it down by taking skeletons. Then work with these. And we can now forget bloated non-skeletal categories. And forget too about the notion of equivalence and revert to using the simpler notion of isomorphism, because equivalent skeletal categories are isomorphic. What's not to like?

Well, the trouble is that hardly any categories that occur in the wild (so to speak) are skeletal. And slimming down has to be done by appeal to an axiom of choice as we choose one representative to stand in for a collection of isomorphic objects. In fact the following statements are each equivalent to a version of the Axiom of Choice:

(1) Any category has a skeleton.

(2) A category is equivalent to any of its skeletons

(3) Any two skeletons of a given category are isomorphic.

The required choice of a skeleton will therefore usually be quite artificial – there typically won't be a canonical choice. So any gain in simplicity from concentrating on skeletal categories would be bought at the cost of having to adopt 'unnatural', non-canonical, choices of skeletons. Given that category theory is supposed to be all about natural patterns already occurring in mathematics, this perhaps isn't going to be such a brilliant trade-off after all.

(c) I noted before that caring about whether objects are actually identical as opposed to isomorphic is jokingly said to be 'evil': the same branding is applied more generally to categorial notions that are not invariant under categorial equivalence. So being skeletal is evil. So too is being small:

Theorem 167. *Smallness is not preserved by categorial equivalence.*

In other words, we can have C a small category, $\mathsf{C} \simeq \mathsf{D}$, yet D not small. This is a simple corollary of our observation in §34.3 that if we take a category, inflate it by adding lots of objects and just enough arrows to ensure that these objects are isomorphic to the original objects, then the augmented category is equivalent to the one we started with. For an extreme example, start with the one-object category $\mathbf{1}$, i.e. $\bullet \circlearrowleft$ (that's small)! Now add as new objects e.g. every ordinal, and as new arrows an identity arrow for each ordinal, and also for every ordinal X a pair of arrows $\bullet \rightleftarrows X$ that compose to give identities. Then we get a new pumped-up category $\mathbf{1}^+$ (which is certainly not small). But $\mathbf{1}^+ \simeq \mathbf{1}$.

There is, however, a companion positive result:

Theorem 168. *Local smallness is preserved by categorial equivalence, so isn't evil.*

Proof. An equivalence $\mathsf{C} \underset{G}{\overset{F}{\rightleftarrows}} \mathsf{D}$ requires F and G to be full and faithful functors. So in particular, for any D-objects D, D', there are the same number of arrows between them as between the C-objects GD, GD'. So that ensures that if C has only a set's worth of arrows between any pair of objects, the same goes for D. □

I should add that it is common to take preservation by categorial equivalence as the mark of a truly categorial property.

Definition 128. A property P of categories is called categorial if whenever C satisfies P and $\mathsf{C} \simeq \mathsf{D}$, then D satisfies P.

For example, the properties of being a preorder category or being a groupoid are categorial. Being a partial order or being a group are not categorial.[4] (Why?)

[4]Thus Peter Johnstone in his lectures.

35 Categories of categories

We have seen how structured whatnots equipped with structure-respecting maps between them can be assembled into categories. But we have also now seen that categories too can have structure-respecting maps between *them*, i.e. functors. So can data of *these* two sorts be assembled into further categories?

Yes indeed. Quite unproblematically, there are at least some *categories of categories*.

Going up another level, functors too can have structure-respecting maps between *them*, i.e. natural transformations. And once again, data of these two sorts can also be assembled into further categories, *functor categories*.

Of these two ideas, it's the second one that is going to be by far the more important for us. In this chapter I say a little about the first, if only to calm nerves, and then set the idea aside.

35.1 A definition, and some tame categories of categories

For any category there is an identity functor sending that category to itself. And Theorem 129 tells us that if there are (covariant) functors $F\colon \mathsf{C} \to \mathsf{D}$ and $G\colon \mathsf{D} \to \mathsf{E}$, then they can be composed to give us a (covariant) functor $G \circ F\colon \mathsf{C} \to \mathsf{E}$, and composition is associative.

So the following definition makes good sense:

Definition 129. A category of categories comprises two sorts of data:

(1) *Objects*: some categories, $\mathsf{C}, \mathsf{D}, \mathsf{E}, \ldots,$

(2) *Arrows*: some functors, $F, G, H, \ldots,$ between those categories,

where the arrows (i) include the identity functor on each category, and (ii) also include $G \circ F$ for each included composable pair F and G (where F's target is G's source). △

And we can immediately give some very tame examples:

(1) Trivially, there is a category of categories whose sole object is your favourite category C and whose sole arrow is the identity functor 1_{C}.

(2) Equally trivially, there is a category of categories with just two objects, the categories Mon and Set, and whose arrows are the identity functors

from each object to itself, together with the forgetful functor from Mon to Set.

(3) We noted that every monoid can be thought of as itself being a category, and functors between monoids-as-categories are just monoid homomorphisms – see §5.4 (C3) and §26.2 (F12). So any category of monoids can be regarded as a category of categories. In particular, that goes for Mon.

Which is enough to establish the point of principle, that there are at least *some* (unexciting!) examples of categories of categories. But how far can we go?

35.2 A category of *all* categories?

(a) What about the extreme case? Is there a universal mega-category comprising *all* categories and the functors between them?

Yes, according to some. Thus Tom Leinster cheerfully says "there is a category CAT whose objects are categories and whose maps are functors", and – in case you are in any doubt that this is supposed to be universal – he later says again "the category of *all* categories and functors is written as CAT" (Leinster 2014, p. 18 and p. 77, his emphasis).

But this sort of unqualified talk is going to make logicians pretty nervous. For isn't there a problem of a familiar kind here? Suppose we say:

Definition 130. A category is *normal* iff it is not one of its own objects. △

Typical categories that we've met are normal in this sense: e.g. categories of groups aren't themselves groups. And this much is uncontentious:

Theorem 169. *There is no category whose objects are all and only the normal categories.*

Proof. Suppose that there is a category N whose objects are exactly the normal categories. Now ask, is N normal? If it is, then it is one of the objects of N, so N is not normal. So N can't be normal. But then it is not one of the objects of N, so N is normal after all. Contradiction. □

But arguably, it seems, we can then go on to conclude that

Theorem 170 (?). *There is no category* CAT *of all categories.*

For if there were such an inclusive mega-category, we could separate out from it a subcategory containing just the normal categories.

(b) Of course, that argument echoes Russell's proof that there can be no set of all normal sets (i.e. no set of all the sets which are not members of themselves), together with the familiar further step taking us to the conclusion that there is no universal set (because if there were, we could use a separation principle to carve out from it the set of all normal sets).

Now, to keep ourselves honest, we should note that in the set-theoretic context we can resist the second step, if – and it's a big *if* – we are willing to restrict our separation principle and develop a non-standard set theory.[1] And I suppose we could perhaps similarly try to resist the move from Theorem 169 to Theorem 170 by trying to restrict when we are allowed to carve out subcategories – though it is not very easy to see a principled way of doing this (other than one which leads to a non-standard set theory too).

But let's not get further entangled with the Russellian line of argument here. For I think that there is a rather more basic problem with the idea of a category that is, in an unqualified way, the category of all categories.

(c) Right back in §4.3, I suggested that if we think of a group as some objects equipped with a binary operation obeying the right conditions, where we don't put any restriction on the kind of objects involved, then it is very far from clear that talk of 'all' groups will locate a determinate fixed totality. Which is why I doubted that there is a determinate all-inclusive mega-category of all groups and their homomorphisms.

Well, doesn't the same go for categories? – at least on our Type I definition which places no restrictions at all on what can count as the objects and arrows of a category. If we are not circumscribing in advance the universe where categories live, what good reason is there to suppose that there *is* a definite totality of categories in our generous sense?

I rather suspect, then, that the idea of a determinate category of *all* categories is, for rather boring reasons, a non-starter.

35.3 Cat, Cat and CAT?

When discussing groups we in effect said: 'OK: let's not fret about whether there is a category of *all* groups, whatever that might mean. Instead, let's focus on what happens in some determinate arena which we hope is rich enough to implement copies of all the groups we might care about – and then it can make sense to talk of a category Grp of all the group-implementations living *there* in that arena and the homomorphisms between *them*'. And of course the arena that we suggested to work with was a capacious universe of sets.

Can we make a parallel move here for categories? How about saying this? – 'Again, let's not fret about whether there is a category of *all* categories, whatever that might mean. Instead, let's focus on what happens in some sufficiently expansive but determinate arena. In particular, let's think about those categories that can be implemented in a capacious universe of sets. For this universe ought to implement copies of all the 'naturally occurring' categories that we initially want to think about.'

So, going along with that thought, now consider the following three notions of increasing scope:

[1] See, for example, Forster (1995) for a classic discussion of deviant set theories like Quine's NF, which allows a universal set at the cost of restricting separation.

Definition 131. Cat is the category whose objects are the small categories implemented in our favoured universe of sets and whose arrows are the functors between them.

CAT is the category whose objects are the locally small categories implemented in our favoured universe of sets and whose arrows are the functors between them.

And CAT is, for us, not Leinster's dubious category of all categories, but the category whose objects are the categories implemented in our favoured universe of sets and whose arrows are the functors between them. △

Arguably *these* definitions are unproblematic. Or at least, Russellian problems don't come back to bite us again.

First, the discrete category based on any given set (with objects the members of that set and with just identity arrows) is small. But that implies that there are at least as many small categories as there are sets. Hence the category Cat of small categories has at least as many objects as there are sets, and so is definitely *not* small. Since Cat is unproblematically not small, no Russellian paradox arises for Cat as it did for the putative category of normal categories.

Second, take a one-object category 1, which is certainly locally small. Then a functor from 1 to the locally small Set will map the object of 1 to some particular set: and there will be as many distinct functors $F\colon 1 \to \mathsf{Set}$ as there are sets. In other words, arrows from 1 to Set in CAT are too many to be mapped one-to-one to a set. Hence CAT is definitely *not* locally small. So again no Russellian paradox arises in this case either.

And as for our re-defined CAT, we can take a Russellian argument to show that such a category can't itself be implemented in the original chosen universe of sets with respect to which we defined CAT. Again no problem. If we really want to theorize about such a large category in a set-theoretic style, we'll e.g. need a bigger ambient universe.[2]

In fact, though, I only offer these definitions because you'll find references elsewhere to 'the category of small categories' and the like. But having suggested safe enough versions of such ideas, we'll find that we won't – at least in these introductory-level notes – have real occasion to use them. So let's move on.

[2]Is that too hand-waving for comfort? Then tackle Shulman (2008) for more on options for set theories for coping with very large categories.

36 Functor categories

In the preamble to the last chapter, we noted that the functors between two categories taken together with the natural transformations between those functors give us the data for a new sort of category. This chapter develops the idea.

36.1 Functor categories officially defined

To repeat, Theorem 155 tells us that (1) for any functor there is an identity natural transformation $1_F \colon F \Rightarrow F$ and that (2) if there are natural transformations $\alpha \colon F \Rightarrow G$ and $\beta \colon G \Rightarrow H$, then there is a composite natural transformation $\beta \circ \alpha \colon F \Rightarrow H$, where composition is associative. So take some functors and enough natural transformations between them: then these can constitute a category.

We will be especially interested in the following sort of case:

Definition 132. $[\mathsf{C}, \mathsf{D}]$ – alternatively D^{C} – is the category of covariant functors from C to D, i.e. it is the *functor category* whose objects are *all* the covariant functors $F \colon \mathsf{C} \to \mathsf{D}$, and whose arrows are *all* the natural transformations between those functors.[1] △

We can of course also define the category of contravariant functors from C to D. But we needn't introduce a special notation for this because, as is standard, we will default to talking instead – but equivalently! – about the category $[\mathsf{C}^{op}, \mathsf{D}]$ of covariant functors. So you can take the functor categories we discuss always to be categories of covariant functors.

36.2 Four simple examples

It might help to fix ideas to start by working through four challenges (the first two are very easy):

[1] The alternative notation 'D^{C}' for the category of functors from C to D is of course suggested by the familiar notation 'D^C' for the set of functions from C to D. For example, the functor category we call '$[2, \mathsf{Set}]$' in a moment is often denoted 'Set^2'. (A warning, though, for when you are looking at other texts: that's not to be confused with 'Set^2' when it denotes the product category $\mathsf{Set} \times \mathsf{Set}$.)

(a) Take the discrete category $\overline{2}$ which comprises two objects together with their identity arrows. What is the functor category $[\overline{2}, \mathsf{C}]$?

(b) Next recall the category 2, which has two objects with their identity arrows plus a single arrow between them. What is the functor category $[2, \mathsf{C}]$?

(c) Now take the category 2^+, which again has two objects and *two* parallel arrows between them. Omitting identity arrows, we can diagram this (there is a hint here!) as $E \overset{s}{\underset{t}{\rightrightarrows}} V$. What is the functor category $[2^+, \mathsf{Set}]$?

(d) Suppose M is a particular monoid M treated as a category. What is the functor category $[\mathsf{M}, \mathsf{Set}]$?

(a) For convenience, dub the objects of $\overline{2}$ simply 'A' and 'B'. The objects of $[\overline{2}, \mathsf{C}]$ are functors $F\colon \overline{2} \to \mathsf{C}$ where we can choose *any* pair of objects from C that we like to be FA and FB:

$$\overset{1_A}{\circlearrowleft}\atop A \qquad \overset{1_B}{\circlearrowleft}\atop B \qquad \xrightarrow{\quad F \quad} \qquad \overset{1_{FA}}{\circlearrowleft}\atop FA \qquad \overset{1_{FB}}{\circlearrowleft}\atop FB$$

Which means there is a simple bijective association between the objects F of our functor category and ordered pairs of C-objects.

What about the arrows of $[\overline{2}, \mathsf{C}]$? An arrow between the parallel functors $F, G\colon \overline{2} \to \mathsf{C}$ is a natural transformation α with components

$$\begin{array}{ccc} FA & \qquad & FB \\ \downarrow{\scriptstyle \alpha_A} & & \downarrow{\scriptstyle \alpha_B} \\ GA & & GB \end{array}$$

And since there are no arrows in $\overline{2}$ between A and B, there is nothing more needed to complete a naturality square in C. So there are no constraints on those components of α. Hence a natural transformation from F to G, an arrow of $[\overline{2}, \mathsf{C}]$, is simply *any* pair of C-arrows such that the first goes from FA to GA, and the second from FB to GB.

Putting things together, the objects of our category $[\overline{2}, \mathsf{C}]$ are (in effect) ordered pairs of C-objects; and an arrow between such pairs is a pair of C-arrows (one between the first members of the pairs, one between the second members). But that makes our category come to the same as the category C^2 – where products of categories are defined as in Defn. 23. More officially: $[\overline{2}, \mathsf{C}] \cong \mathsf{C}^2$.

Exercise: describe the inverse pair of functors making an isomorphism between the two categories!

(b) What about the functor category $[2, \mathsf{C}]$?

An object in this category is a functor $F\colon 2 \to \mathsf{C}$. And representing 2 as on the left below, F can send the f to *any* arrow in C we choose to be $F_{arw}f\colon F_{ob}A \to F_{ob}B$. Then, so long as we put $F_{arw}1_A = 1_{FA}$ and $F_{arw}1_B = 1_{FB}$, the components F_{ob} and F_{arw} will give us a functor:

$$\overset{1_A}{\circlearrowleft} A \xrightarrow{\;f\;} \overset{1_B}{\circlearrowleft} B \qquad \xrightarrow{\;F\;} \qquad \overset{1_{FA}}{\circlearrowleft} FA \xrightarrow{\;Ff\;} \overset{1_{FB}}{\circlearrowleft} FB$$

So we have a bijective correlation between functors F, i.e. objects of $[2, \mathsf{C}]$, and the C-arrows.

And what are the arrows in our functor category? A natural transformation from F to the parallel functor G will have as components any two C-arrows, j, k, which make this a commutative square:

$$\begin{array}{ccc} FA & \xrightarrow{\;Ff\;} & FB \\ \big\downarrow{\scriptstyle j} & & \big\downarrow{\scriptstyle k} \\ GA & \xrightarrow{\;Gf\;} & GB \end{array}$$

Thus the arrows of the new category are exactly pairs of C-arrows that make the relevant diagram commute.

So in sum, $[2, \mathsf{C}]$ comes to the same as the arrow category $\mathsf{C}^{\to}$ we met in Defn. 28. In fact the two categories are isomorphic. Exercise: spell out why that's so.

(c) What does a functor F from the mini-category 2^+, i.e. $E \underset{t}{\overset{s}{\rightrightarrows}} V$, to the category Set actually do?

By definition, F sends the object E to a set FE, and sends the object V to a set FV. And it sends the arrow $s \colon E \to V$ to a function $Fs \colon FE \to FV$ and the arrow $t \colon E \to V$ to a function $Ft \colon FE \to FV$. We can then look at the data that F picks out in Set like this. Think of FE as a set of 'edges', and FV as a set of 'vertices'. Then each 'edge' in FE gets assigned a 'vertex' as 'source' by Fs, and also gets assigned a vertex as 'target' by Ft. In this way, 2^+'s image under F can be regarded as a *directed graph.*

OK: now we know what a functor $F \colon 2^+ \to \mathsf{Set}$ does. What about the arrows in the functor category $[2^+, \mathsf{Set}]$? – in other words, what is a natural transformation between functors like F?

By definition, we need a pair of components, φ_E and φ_V, which make the following two diagrams commute:

$$\begin{array}{ccc} FE & \xrightarrow{\;Fs\;} & FV \\ \big\downarrow{\scriptstyle \varphi_E} & & \big\downarrow{\scriptstyle \varphi_V} \\ GE & \xrightarrow{\;Gs\;} & GV \end{array} \qquad \begin{array}{ccc} FE & \xrightarrow{\;Ft\;} & FV \\ \big\downarrow{\scriptstyle \varphi_E} & & \big\downarrow{\scriptstyle \varphi_V} \\ GE & \xrightarrow{\;Gt\;} & GV \end{array}$$

But that makes φ_E and φ_V the components of a graph homomorphism – compare (C25) in §5.7.

So we can think of the functor category $[2^+, \mathsf{Set}]$ as being tantamount to the category Graph of graphs and graph homomorphisms. (Exercise: are these two categories again isomorphic? Explain your answer.)

(d) Start with a monoid $M = (\mathrm{M}, *, e)$. Then, you'll recall, the corresponding category M has a single object $\bullet$ (whatever you like). And any $m \in \mathrm{M}$ counts as an M-arrow $m\colon \bullet \to \bullet$. Composition of arrows $m \circ n$ is defined to be the monoid product $m * n$, and the identity arrow $1_\bullet$ is defined to be the monoid identity e.

What does a functor F from M to Set do? It must send $\bullet$ to some set X, and send an arrow $m\colon \bullet \to \bullet$ to a set-function $f_m\colon X \to X$. For functoriality, $F1_\bullet = 1_X$, and $F(m \circ n) = F(m) \circ F(n)$, i.e. $f_{m*n} = f_m \circ f_n$. So F specifies an M-set (X, f_M) in the sense of §5.7, Ex. (C21).

And what's a natural transformation between the functor F giving the M-set (X, f_M) and the functor G giving the M-set (Y, g_M)? It will be an arrow $j\colon X \to Y$ making the following commute for any m:

$$\begin{array}{ccc} F\bullet & \xrightarrow{Fm} & F\bullet \\ \downarrow{\scriptstyle j} & & \downarrow{\scriptstyle j} \\ G\bullet & \xrightarrow{Gm} & G\bullet \end{array} \qquad \text{which is} \qquad \begin{array}{ccc} X & \xrightarrow{f_m} & X \\ \downarrow{\scriptstyle j} & & \downarrow{\scriptstyle j} \\ Y & \xrightarrow{g_m} & Y \end{array}$$

In other words, if the operation f_m on X sends x to x', then the corresponding operation g_m on Y will send $j(x)$ to $j(x')$. But this means that the natural transformation j counts, by definition, as an arrow in the category of M-sets.

So, in sum, the functor category $[\mathsf{M}, \mathsf{Set}]$ is just the category M-Set in thin disguise.

36.3 On issues of size

I should say something about the size of functor categories. Here's a simple result:

Theorem 171. *If* C *is small and* D *is locally small, then the functor category* $[\mathsf{C}, \mathsf{D}]$ *is locally small.*

Proof. We need to show that there is only a set's worth of $[\mathsf{C}, \mathsf{D}]$-arrows, i.e. natural transformations, between any two functors $F, G\colon \mathsf{C} \to \mathsf{D}$.

By hypothesis there is only a set's worth of C-objects C, and hence a set's worth of pairs of D-objects FC and GC. And then by hypothesis there is only a set's worth of D-arrows from any FC to GC. Hence there is altogether only a set's worth of arrows to choose from in selecting the components to build a natural transformation $\alpha\colon F \Rightarrow G$. So there can only be a set's worth of such natural transformations. □

But evidently we can't use this line of argument if C isn't small and has more than a set's worth of objects. And indeed, if C and D are only limited to being *locally* small (by the standard of our current ambient universe of sets), a putative functor category $[\mathsf{C}, \mathsf{D}]$ need no longer be locally small (by the same standard). So it won't count as a kosher category, at least according to a definition that builds in the requirement of local smallness. What to do?

Officially, we will avoid trouble by following the conventional line and focusing on functor categories [C, D] where C is small, and where D is at least locally small, so our functor category is too. However, I suggest we don't regard this as a restriction on the categories C and D that can feature so much as an injunction to work with a generous enough ambient universe of sets when dealing with functor categories (see the discussion at the end of §30.1).

Unofficially, however, the line continues to be: at our introductory level, we won't actually fuss too much about such issues of size!

36.4 Functor categories and limits

(a) Our examples so far might already suggest, to put it very roughly, that if D is a rich category, so too is [C, D]. And this is fairly straightforward to confirm in the case of limits. Assume, if you want to be pernickety, that C is small and D is locally small, and then we have:

Theorem 172. *If the category* D *has all limits of shape* J*, so does any functor category* [C, D].

As a special case, since Set *has all finite limits, so does any functor category* [C, Set].

Proof sketch for the special case: [C, Set] *has all finite limits.* With an eye to deploying Theorem 97 about the existence of limits, we prove three mini-results.

(1) A terminal object for [C, Set] is provided by the functor $\overline{1}\colon \mathsf{C} \to \mathsf{Set}$, which acts like this:

$$\begin{array}{lrcl} \overline{1}\colon & X & \longmapsto & \{\bullet\} \\ & j\colon X \to Y & \longmapsto & 1_1\colon \{\bullet\} \to \{\bullet\} \end{array}$$

where as usual $\{\bullet\}$ is your favourite singleton. It is trivial to show that every functor $F\colon \mathsf{C} \to \mathsf{Set}$ has a unique natural transformation to $\overline{1}$, and hence that $\overline{1}$ is terminal in [C, Set].

(2) A product-object for the functors $F, G\colon \mathsf{C} \to \mathsf{Set}$ is provided by the functor $F \times G\colon \mathsf{C} \to \mathsf{Set}$, where

$$\begin{array}{lrcl} F \times G\colon & X & \longmapsto & FX \times GX \\ & j\colon X \to Y & \longmapsto & Fj \times Gj\colon FX \times GX \to FY \times GY \end{array}$$

and $Fj \times Gj$ acts component-wise.

In addition, this product object can be equipped with two projection arrows, where the arrows of the category are natural transformations. We just do the obvious thing, and put $\pi_1\colon F \times G \Rightarrow F$ to be the natural transformation whose component $(\pi_1)_X\colon FX \times GX \to FX$, for any C-object X, sends a pair $\langle x, x' \rangle \in FX \times GX$ to $x \in FX$; and similarly for $\pi_2\colon F \times G \Rightarrow G$.

It is then easily checked that, so defined, $(F \times G, \pi_1, \pi_2)$ has the defining universal property of a product. So [C, Set] has all products.

(3) What about equalizers? Suppose $\alpha, \beta\colon F \Rightarrow G$ are parallel natural transformations between functors $F, G\colon \mathsf{C} \to \mathsf{Set}$. We want to show that there is always a functor E and natural transformation $e\colon E \Rightarrow F$, such that (E, e) has the universal property of an equalizer.

For every C-object X, the components $\alpha_X, \beta_X : FX \to GX$ are set functions; and we essentially want to simultaneously equalize all these components. And the obvious choice for E is the functor

$$\begin{array}{llll} E\colon & X & \longmapsto & \{x \in FX \mid \alpha_X(x) = \beta_X(x)\} \\ & j\colon X \to Y & \longmapsto & Fj|_{EX}\colon EX \to EY \end{array}$$

We just need to check that $Fj|_{EX}$, the restriction of Fj to EX, really does send any member of EX to a member of EY. But, given the assumption that α_X and β_X agree on any member $x \in EX$, the naturality squares tell us that $\alpha_Y \circ Fj(x) = Gj \circ \alpha_X(x) = Gj \circ \beta_X(x) = \beta_Y \circ Fj(x)$, so $Fj(x) \in EY$.

And with E now defined, I'll leave it to you to complete the construction by (i) defining a matching e as a natural transformation whose components are inclusion functions, and (ii) showing (E, e) has the universal property of an equalizer.

So $[\mathsf{C}, \mathsf{Set}]$ has an initial object, binary products and equalizers: therefore, by Theorem 97, it has all finite limits. □

(b) We won't actually need the more general claim in Theorem 172. I will outline a proof strategy, but you can cheerfully skip!

Proof strategy for the general claim. We need to show that for any diagram of shape J in our functor category, i.e. for any $D\colon \mathsf{J} \to [\mathsf{C}, \mathsf{D}]$, we can construct a limit cone over D.

We again proceed 'pointwise' (as e.g. with products above). So first fix on a 'point' of C, I mean a C-object X. Then let's define D_X as sending a J-object J to $DJ(X)$ (i.e. we apply the functor $DJ\colon \mathsf{C} \to \mathsf{D}$ to X and get a D-object). And let D_X send a J-arrow $j\colon J \to K$ to $Dj_X\colon D_X J \to D_X K$ (Dj is a natural transformation from the functor $DJ\colon \mathsf{C} \to \mathsf{D}$ to the functor $DK\colon \mathsf{C} \to \mathsf{D}$, and we are taking its X-component).

This defines a functor $D_X : \mathsf{J} \to \mathsf{D}$, i.e. a diagram of shape J in D. So, by the assumption that D has all limits of J, there is a limit $(L_X, \lambda_{X,J})$ over D_X, with (for each J) a leg $\lambda_{X,J}$ targeting $D_X J$.

Now vary the C-object X, and we use the ingredients of the various limit cones $(L_X, \lambda_{X,J})$ to build a limit cone over D.

First, then, we need a functor $L\colon \mathsf{C} \to \mathsf{D}$ as the vertex of the cone, and we can define it like this:

$$\begin{array}{llll} L\colon & X & \longmapsto & L_X \\ & f\colon X \to Y & \longmapsto & L_f\colon L_X \to L_Y \end{array}$$

where L_f is the unique arrow that makes this commute for all J:

$$\begin{array}{ccc} L_X & \xrightarrow{L_f} & L_Y \\ \Big\downarrow{\scriptstyle \lambda_{X,J}} & & \Big\downarrow{\scriptstyle \lambda_{Y,J}} \\ DJ(X) & \xrightarrow{DJ(f)} & DJ(Y) \end{array}$$

(This makes sense because L_X together with the various composite arrows $DJ(f) \circ \lambda_J^X : L^X \to DJ(Y)$ will assemble, as we vary J, into a cone over D_Y. Therefore, since L_Y is the vertex of a limit cone over D_Y, there will indeed be a unique arrow $L_X \to L_Y$ making everything commute.)

We now need to equip our vertex functor L with 'legs', arrows which will be natural transformations between L and the functors which are D-images of J-objects J. But we have the ingredients of natural transformations $L \Rightarrow DJ$ to hand: they will be assembled from components $\lambda_{X,J} : L_X \to DJ(X)$. We then just need to check this does give us a limit cone.[2] □

36.5 Presheaf categories

(a) Special cases apart – think about our simple examples in §36.2 – the functor categories $[\mathsf{C}, \mathsf{D}]$ and $[\mathsf{C}^{op}, \mathsf{D}]$ will differ in important ways. And there are contexts where the second will be the one we want to think about.

Let's have an example (if you will forgive some hand-waving). Suppose we are dealing with a topological space X. Then we can be interested in assigning data to each open set $U \subseteq X$. And we will want to do this in such a way that the data assigned to more inclusive sets coheres with the data assigned to subsets. What do I mean?

Suppose U, V are open sets of X, and suppose F is our function assigning data to open sets. Then, in nice cases, if $V \subseteq U$, then there will be a map $\rho_V^U : FU \to FV$ that tells us how to 'restrict' the data on U to the smaller open set V (if the data are functions defined on the open sets, then ρ_V^U could literally be the map that takes a function on U to its restriction on V). And for coherence, we evidently want this for a start: if $W \subseteq V \subseteq U$ then the restriction map ρ_W^U from the data FU to the data FW should be the composite $\rho_W^V \circ \rho_V^U$.

Now put that in categorial terms. Suppose $\mathcal{O}(X)$ is the category whose objects are open sets of X and where there is an arrow $V \to U$, an inclusion function, if and only if $V \subseteq U$. And suppose, for generality's sake, that our data is coded by sets living in Set. Then our data-assigning rule F in the first place sends objects of $\mathcal{O}(X)$ to objects of Set. But F also sends an $\mathcal{O}(X)$-arrow, an inclusion $i : V \to U$, to a Set-arrow Fi (i.e. ρ_V^U): $FU \to FV$, subject to the condition that $F(i \circ j) = Fj \circ Fi$. So, assuming F sends identity arrows to identity arrows, we get a contravariant functor $F : \mathcal{O}(X) \to \mathsf{Set}$, or equivalently – a covariant functor $F : \mathcal{O}(X)^{op} \to \mathsf{Set}$.

[2] Enthusiasts can find a full proof in e.g. Leinster (2014, Theorem 6.2.5 and its corollary).

Of course, sticking for a moment with the same context where we are interested in assigning data to each open set U in a space, we'll not only want to respect restriction-to-subsets, but also want the data assigned to U to be coherent with the data assigned to smaller open sets which together can be patched together to cover U. However, we won't pause to spell out that additional patching condition. We'll merely note that a data assignment which satisfies both the restriction and the patching conditions is said to be a *sheaf*. And that's why, now generalizing from the topological case,

Definition 133. A functor $F\colon \mathsf{C}^{op} \to \mathsf{Set}$ – a data assignment satisfying at least the restriction condition – is said to be a *presheaf* on C.

So the functor category $[\mathsf{C}^{op}, \mathsf{Set}]$ is a *presheaf category*, the category of presheaves on C: it is often notated simply $\widehat{\mathsf{C}}$.[3] △

(b) We will see in the next chapter why, starting from a category C we are often more interested in the category of contravariant functors from C to Set, i.e. $[\mathsf{C}^{op}, \mathsf{Set}]$, then in the companion category of covariant functors $[\mathsf{C}, \mathsf{Set}]$. But, for now, let's note a general result which applies to both types of functor category:

Theorem 173. *(i) For any small category* C*, the presheaf category* $[\mathsf{C}^{op}, \mathsf{Set}]$ *is Cartesian closed and has a subobject classifier. Equivalently, since every category is the opposite of its opposite, (ii) for any small category* C*, the functor category* $[\mathsf{C}, \mathsf{Set}]$ *is Cartesian closed and has a subobject classifier.*

Theorem 172 already tells us that functor categories $[\mathsf{C}, \mathsf{Set}]$, and so presheaf categories too, have all finite limits. So to prove our new theorem, it remains to show these categories also have all exponentials and a subobject classifier. I am not minded to delay over the proofs.[4]

36.6 Hom-functors from functor categories

We move on. Having introduced the idea of a functor category, the discussion now takes another twist, as we spiral up to a new level of abstraction. In the rest of this chapter, we go on to look at some examples of new functors *which act on functor categories*.

(a) To prepare for our next pair of examples, let's introduce some new notation.

We've used $[\mathsf{C}, \mathsf{D}]$ to denote the category of functors from C to D. So, given a pair of objects in that category, i.e. a pair of functors $F, G\colon \mathsf{C} \to \mathsf{D}$, then our earlier notational convention would make $[\mathsf{C}, \mathsf{D}](F, G)$ denote the hom-set of

[3] As remarked at the outset of the chapter, it is usual to default to talking about the category of covariant functors $[\mathsf{C}^{op}, \mathsf{Set}]$ rather than talking equivalently about the category of contravariant functors $[\mathsf{C}, \mathsf{Set}]$.

[4] You can find the requisite proofs in e.g. Goldblatt (1984, pp. 205–210). But the arguments aren't elegant, and the ideas used don't seem to hook up illuminatingly with later discussions in this book; which is why I've decided – on balance – to leave the proofs for completists to follow up.

arrows from F to G in the functor category, i.e. the set of natural transformations from F to G.

However, I find that notation a bit cumbersome, and it will do no harm to emphasize that the hom-set here is indeed a set of natural transformations: so I prefer to use this more self-explanatory notation:

Definition 134. $Nat(F, G)$ will denote the set of natural transformations from F to G.[5] △

(b) Now, where there are hom-sets like $Nat(F, G)$, there will be corresponding hom-functors. And here they are – we just apply the definitions built into Theorems 148 and 149, and adapt to our current notational style:

Definition 135. $Nat(F, -)\colon [\mathsf{C}, \mathsf{D}] \to \mathsf{Set}$ is the covariant functor which

(1) sends a $[\mathsf{C}, \mathsf{D}]$-object such as the functor $G\colon \mathsf{C} \to \mathsf{D}$ to the corresponding set $Nat(F, G)$; and
(2) sends a $[\mathsf{C}, \mathsf{D}]$-arrow such as a natural transformation $\gamma\colon G \Rightarrow H$ to the function $\gamma \circ -$, which takes an arrow α in $Nat(F, G)$ and returns the arrow $\gamma \circ \alpha$ in $Nat(F, H)$.

And likewise, $Nat(-, G)\colon [\mathsf{C}, \mathsf{D}] \to \mathsf{Set}$ is the contravariant functor which

(1) sends a $[\mathsf{C}, \mathsf{D}]$-object such as the functor $F\colon \mathsf{C} \to \mathsf{D}$ to the corresponding set $Nat(F, G)$; and
(2) sends a $[\mathsf{C}, \mathsf{D}]$-arrow such as a natural transformation $\gamma\colon F \Rightarrow E$ to the function $- \circ \gamma$, which takes an arrow α in $Nat(E, G)$ and returns the arrow $\alpha \circ \gamma$ in $Nat(F, G)$. △

It will again help to fix ideas to check carefully that this all makes good sense: as I said, we'll need hom-functors of these kinds later.

(c) Another quick example. Start again with the functor category $[\mathsf{C}, \mathsf{D}]$ and this time also pick an object A in C. Then there is an evaluation functor that looks at what is in $[\mathsf{C}, \mathsf{D}]$ and evaluates it at A:

Definition 136. The functor $eval_A\colon [\mathsf{C}, \mathsf{D}] \to \mathsf{D}$ sends any functor $F\colon \mathsf{C} \to \mathsf{D}$ to FA and sends any natural transformation $\alpha\colon F \Rightarrow G$ to $\alpha_A\colon FA \to GA$. △

Once more, check that $eval_A$ (which we'll need later) really is functorial.

36.7 Categories of diagrams and limit functors

(a) To repeat: in the terminology of Defn. 111, we can think of the objects of the functor category $[\mathsf{J}, \mathsf{C}]$, i.e. functors $D\colon \mathsf{J} \to \mathsf{C}$, as diagrams of shape J in C.

[5] Riehl (2017, e.g. p. 57) prefers 'Hom(F, G)'; I'm following Mac Lane (1997, e.g. p. 61). Might the collection of natural transformations be 'too large' to be a set? Perhaps. But again let's not fret about such issues of size right now: it will turn out that when we later want to talk about such hom-sets, it will often be in contexts where they are safely set-sized.

And then the arrows of the category are natural transformations between the functors.

Those diagrams-as-functors include constant functors such as $\Delta_X \colon \mathsf{J} \to \mathsf{C}$ (recall, that's the functor that picks a single object X from the category C and then sends every object of J to X, and sends every arrow in J to 1_X). And what is a natural transformation between two such constant functors, $\alpha \colon \Delta_X \Rightarrow \Delta_Y$? By definition, we require that for every J-arrow $j \colon A \to B$, the naturality square on the left commutes in C:

$$\begin{array}{ccc} \Delta_X A & \xrightarrow{\Delta_X j} & \Delta_X B \\ \downarrow{\scriptstyle \alpha_A} & & \downarrow{\scriptstyle \alpha_B} \\ \Delta_Y A & \xrightarrow{\Delta_Y j} & \Delta_Y B \end{array} \qquad\qquad \begin{array}{ccc} X & \xrightarrow{1_X} & X \\ \downarrow{\scriptstyle \alpha_A} & & \downarrow{\scriptstyle \alpha_B} \\ Y & \xrightarrow{1_Y} & Y \end{array}$$

But the square on the left is just the square on the right. That will commute so long as every component of α is the same, i.e. is some fixed arrow $f \colon X \to Y$. Call the resulting natural transformation α_f.

And now note that this sets us up for the following definition:

Definition 137. Given a category C, the *diagonal functor* $\Delta_{\mathsf{J}} \colon \mathsf{C} \to [\mathsf{J}, \mathsf{C}]$ acts as follows:

$$\begin{array}{rrcl} \Delta_{\mathsf{J}}\colon & X & \longmapsto & \Delta_X \colon \mathsf{J} \to \mathsf{C} \\ & f \colon X \to Y & \longmapsto & \alpha_f \colon \Delta_X \to \Delta_Y \end{array} \qquad \triangle$$

The 'diagonal' label is standard: the (rather weak?) motivation is that in particular case where J is the discrete two-object category $\overline{2}$, we get the functor $\Delta_{\overline{2}}$ which is tantamount to the original binary diagonal functor $\Delta \colon \mathsf{C} \to \mathsf{C} \times \mathsf{C}$ which we met in §26.3.

(b) Let's now assume that we are dealing with a category C which has all limits of shape J, i.e. every diagram $D \colon \mathsf{J} \to \mathsf{C}$ has a limit.

Then we can aim to define a functor $\underset{\leftarrow \mathsf{J}}{\mathit{Lim}} \colon [\mathsf{J}, \mathsf{C}] \to \mathsf{C}$ whose object component sends a diagram D living in the functor category $[\mathsf{J}, \mathsf{C}]$ to the vertex L for some chosen limit cone over D in C.

Three initial comments:

i. It is standard to use the notation '$\underset{\leftarrow \mathsf{J}}{\mathit{Lim}}$' for this functor. But in fact plain 'Lim' will serve us well enough when J is fixed by context and when we are explicitly dealing with limits rather than colimits (which are conventionally notated using $\underset{\to \mathsf{J}}{\mathit{Lim}}$ with the arrow in the other direction).

ii. More importantly, note that we do need to do some choosing here! This functor is not entirely 'naturally' or 'canonically' defined: in the general case, limits over D are only unique up to isomorphism, so we will have to select a particular limit whose vertex is to be $\mathit{Lim}\, D$, the value of our functor for input D.

iii. But to get a kosher functor, we also need to suitably define *Lim*'s component which acts on arrows. This must send an arrow in $[\mathsf{J}, \mathsf{C}]$, i.e. a natural transformation $\alpha\colon D \Rightarrow D'$ between the $[\mathsf{J}, \mathsf{C}]$ objects D and D', to an arrow in C from $Lim\, D$ to $Lim\, D'$. How can it do this in, well, a natural way?

To answer that question, suppose that we do have a natural transformation $\alpha\colon D \Rightarrow D'$. By hypothesis there are limit cones over D and D', respectively $(Lim\, D, \lambda_J)$ and $(Lim\, D', \lambda'_J)$. So now take any arrow $d\colon K \to L$ living in J and consider the following diagram:

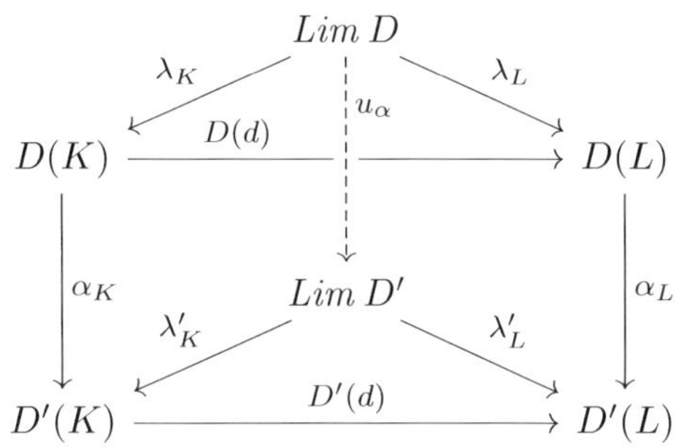

The top triangle commutes because $(Lim\, D, \lambda_J)$ is a limit. The lower square commutes by the naturality of α. Therefore the outer pentagon commutes and so, generalizing over objects J in J, $(Lim\, D, \alpha_J \circ \lambda_J)$ is a cone over D'. But then *this* cone must factor uniquely through D''s limit cone $(Lim\, D', \lambda'_J)$ via some unique $u_\alpha\colon Lim\, D \to Lim\, D'$.

The map $\alpha \mapsto u_\alpha$ is therefore a plausible candidate for *Lim*'s action on arrows; and this assignment is easily verified to yield a functor. In summary:

Definition 138. Assuming every diagram D of shape J has a limit in C, then a functor $Lim\colon [\mathsf{J}, \mathsf{C}] \to \mathsf{C}$

(1) sends an object D in $[\mathsf{J}, \mathsf{C}]$ to the vertex $Lim\, D$ of some chosen limit cone $(Lim\, D, \lambda_J)$ over D,

(2) sends an arrow $\alpha\colon D \Rightarrow D'$ in $[\mathsf{J}, \mathsf{C}]$ to the arrow $u_\alpha\colon Lim\, D \to Lim\, D'$ where for all J in J, $\lambda'_J \circ u_\alpha = \alpha_J \circ \lambda_J$. △

(c) Let's note a couple of simple theorems about limit functors. First, the diagram above can be recycled to show

Theorem 174. *Assuming limits of the relevant shape exist, if we have a natural isomorphism $D \cong D'$, then $Lim\, D \cong Lim\, D'$.*

Proof. Because we have a natural isomorphism $D \cong D'$, we can show as above both that there is a unique $u\colon Lim\, D \to Lim\, D'$ and symmetrically that there is a unique $u'\colon Lim\, D' \to Lim\, D$. These compose to give us a map $u' \circ u\colon Lim\, D \to Lim\, D$ which must be $1_{Lim\, D}$ by the now familiar argument (the limit cone with vertex $Lim\, D$ can factor through itself by both $u' \circ u$ and $1_{Lim\, D}$, but by definition there is only one way for the limit cone to factor through itself). Likewise, $u \circ u' = 1_{Lim\, D'}$. So u is an isomorphism. □

Theorem 175. *Suppose that* C *has all limits of shape* J*, so* Lim *is well-defined. Then for any limit over* $D\colon \mathsf{J} \to \mathsf{C}$ *which the functor* $F\colon \mathsf{C} \to \mathsf{D}$ *preserves,* $F(Lim\, D) \cong Lim(FD)$*. In brief:* F *commutes with* Lim*.*

Proof. If F preserves a limit cone over $D\colon \mathsf{J} \to \mathsf{C}$ with vertex $Lim\, D$, then F sends that limit cone to a limit cone over FD (i.e. $F \circ D$) with vertex $F(Lim\, D)$. But that vertex must be isomorphic to the vertex of any other limit cone over FD. So it must be isomorphic to whatever has been chosen to be $Lim(FD)$. □

37 The Yoneda Embedding

"The Yoneda Lemma is perhaps the single most used result in category theory" (that's from Steve Awodey 2010, p. 185). Again, the Lemma "is arguably the most important result in category theory" (that's from Emily Riehl 2017, p. 57). It is difficult, though, to illustrate such claims at the level of these notes.[1] Still, given the Lemma's claimed importance, I should at least try to say *something* introductory.

Now, it has also been said that "the level of abstraction in the Yoneda Lemma means that many people find it quite bewildering" (that's from Tom Leinster 2000, p. 1). So here is the challenge, then: to make the Lemma as unbewildering as possible. I'll take things in gentle stages, over two chapters.

What happens in this chapter? We met hom-functors in Chapter 31. We then introduced natural transformations in Chapter 33. Now we bring things together and look at natural transformations between hom-functors. And very quickly we will arrive at what we can think of as a restricted version of the Yoneda Lemma. We will also meet the related Yoneda Embedding theorem, and the Yoneda Principle. These initial technicalities, I hope you'll be able to agree, are actually quite straightforward. At the end of the chapter, I say something about their significance.

37.1 Natural transformations between hom-functors

(a) So, down to business! Here is a covariant hom-functor from the (locally small) category C to Set:

$$\begin{array}{llll} \mathsf{C}(A,-)\colon & X & \longmapsto & \mathsf{C}(A,X) \\ & j\colon X \to Y & \longmapsto & (j \circ -)_A\colon \mathsf{C}(A,X) \to \mathsf{C}(A,Y), \end{array}$$

where $(j \circ -)_A$ sends an arrow $g\colon A \to X$ to the arrow $j \circ g\colon A \to Y$. And here's another covariant hom-functor from C to Set:

$$\begin{array}{llll} \mathsf{C}(B,-)\colon & X & \longmapsto & \mathsf{C}(B,X) \\ & j\colon X \to Y & \longmapsto & (j \circ -)_B\colon \mathsf{C}(B,X) \to \mathsf{C}(B,Y), \end{array}$$

[1] And we might wonder if Awodey and Riehl exaggerate. Compare, say, Birgit Richter's excellent *From Categories to Homotopy Theory* (2020). The first part is a brisk introduction to category theory; the second part turns to homotopy theory. In that second part where categorial ideas are put to work, Yoneda is mentioned just once, in an incidental remark.

where $(j \circ -)_B$ sends an arrow $h\colon B \to X$ to the arrow $j \circ h\colon B \to Y$.

Now, recall from §31.2 that we can think of a functor like $\mathsf{C}(A, -)$ as encapsulating how A sees its world; and then a question arises about how A's view meshes with the view from another object B encapsulated by $\mathsf{C}(B, -)$. Dropping the metaphor, we want a story about how a hom-functor $\mathsf{C}(A, -)$ can be related to the hom-functor $\mathsf{C}(B, -)$ by some natural transformation(s).

We'll also want a story about natural transformations between hom-functors of the second, contravariant, flavour. But here I will just spell out the covariant story, leaving its companion as an exercise in dualizing (you just need to keep a beady eye on the direction of arrows).

(b) By definition, if α is to be a natural transformation from $\mathsf{C}(A, -)$ to $\mathsf{C}(B, -)$, its components must be such that the following diagram commutes, for any given C-arrow $j\colon X \to Y$:

$$\begin{array}{ccc} \mathsf{C}(A, X) & \xrightarrow{\mathsf{C}(A,j)} & \mathsf{C}(A, Y) \\ \Big\downarrow{\scriptstyle \alpha_X} & & \Big\downarrow{\scriptstyle \alpha_Y} \\ \mathsf{C}(B, X) & \xrightarrow{\mathsf{C}(B,j)} & \mathsf{C}(B, Y) \end{array}$$

Here $\mathsf{C}(A, j)$ is standard shorthand for the result of applying the functor $\mathsf{C}(A, -)$ to the C-arrow j. And that result, as we've reminded ourselves, is $(j \circ -)_A$. Similarly $\mathsf{C}(B, j)$ is $(j \circ -)_B$.

By definition, then, the component α_X must send any arrow $g\colon A \to X$ to some arrow $B \to X$. And the very easiest way of doing this is to fix on some arrow $f\colon B \to A$, and put $\alpha_X = (- \circ f)_X$, which sends any arrow $g\colon A \to X$ to the corresponding composite arrow $g \circ f\colon B \to X$.

Define α_Y, for any Y, similarly. In other words, put $\alpha_Y = (- \circ f)_Y$, which sends an arrow $h\colon A \to Y$ to the composite $h \circ f\colon B \to Y$.

And lo and behold, this easy first guess at suitable α_X, α_Y, etc., makes our diagram commute! Take any arrow $g\colon A \to X$ in $\mathsf{C}(A, X)$, and chase it round the diagram in both directions, and we end up with the same result. Thus:

$$\begin{array}{ccc} g\colon A \to X & \xmapsto{(j \circ -)_A} & j \circ g\colon A \to Y \\ \Big\downarrow{\scriptstyle (- \circ f)_X} & & \Big\downarrow{\scriptstyle (- \circ f)_Y} \\ g \circ f\colon B \to X & \xmapsto{(j \circ -)_B} & j \circ g \circ f\colon B \to Y \end{array}$$

Generalizing: for any X, Y, and $j\colon X \to Y$, our first diagram always commutes when α's components are defined like α_X, α_Y. So α is a natural transformation.

Let's sum this up, and also introduce some new notation:

Theorem 176. *Suppose C is a locally small category, and $\mathsf{C}(A, -)$, $\mathsf{C}(B, -)$ are hom-functors (for objects A, B in C).*

Then, given an arrow $f\colon B \to A$, there exists a corresponding natural transformation which we will now notate $\mathsf{C}(f, -)\colon \mathsf{C}(A, -) \Rightarrow \mathsf{C}(B, -)$, where for each

X, the component $\mathsf{C}(f,-)_X\colon \mathsf{C}(A,X) \to \mathsf{C}(B,X)$ *sends an arrow* $g\colon A \to X$ *to* $g \circ f\colon B \to X$. □

Note: if f in our theorem is an isomorphism, then each component of our natural transformation $(- \circ f)$ has an inverse (i.e. $- \circ f^{-1}$) so is an isomorphism. Therefore the induced transformation $\mathsf{C}(f,-)$ will be a natural isomorphism.

(c) To check understanding and for future use, show the following:

Theorem 177. *Given a locally small category* C *including objects* A, B, C*, and given* C*-arrows* $f\colon B \to A$ *and* $g\colon C \to B$*, then*

(1) $\mathsf{C}(f \circ g, -) = \mathsf{C}(g,-) \circ \mathsf{C}(f,-)$.

(2) $\mathsf{C}(f,-)_A 1_A = f$.

(3) $\mathsf{C}(1_A,-) = 1_{\mathsf{C}(A,-)}$.

Not that this is much of a challenge! – we just have to apply definitions. So:

Proof of (1). By the definition of $\mathsf{C}(f \circ g,-)$, a component $\mathsf{C}(f \circ g,-)_X$ sends any arrow $k\colon A \to X$ to $k \circ (f \circ g)$. However, $\mathsf{C}(f,-)_X$ sends k to $k \circ f$, and $\mathsf{C}(g,-)_X$ sends that on to $(k \circ f) \circ g$. Hence, component by component, $\mathsf{C}(f \circ g,-)$ acts in the same way as $\mathsf{C}(g,-) \circ \mathsf{C}(f,-)$, which makes them the same natural transformation. □

Proof of (2). By definition $\mathsf{C}(f,-)_A$ sends any $k\colon A \to A$ to $k \circ f\colon B \to A$. So in particular it sends 1_A to f. □

Proof of (3). By definition, $\mathsf{C}(1_A,-)_X$ sends any $k\colon A \to X$ to $k \circ 1_A\colon A \to X$ – i.e. it sends k to itself.

And what is $1_{\mathsf{C}(A,-)}$? I haven't said! But the notation indicates the identity arrow on the object $\mathsf{C}(A,-)$, where the relevant category must be a functor category including such hom-functors as objects. But then the identity object on such an object will be the identity natural transformation from $\mathsf{C}(A,-)$ to itself. So what does the X-component of that identity natural transformation do to an arrow such as $k\colon A \to X$? It will send the arrow to itself. Which shows that the X-components of $\mathsf{C}(1_A,-)$ and $1_{\mathsf{C}(A,-)}$ agree on their actions; and that holds for all X and so the transformations are identical. □

(d) The result (2) has an immediate corollary that we can add to our earlier main theorem:

Theorem 176 (cont'd). *If* $f, f'\colon B \to A$ *are distinct arrows, then the corresponding natural transformations* $\mathsf{C}(f,-)$ *and* $\mathsf{C}(f',-)$ *are also distinct.*

Proof. We know from the result just proved that

$$\mathsf{C}(f,-)_A 1_A = f \neq f' = \mathsf{C}(f',-)_A 1_A$$

Hence the A-components of $\mathsf{C}(f,-)$ and $\mathsf{C}(f',-)$ can't be the same, hence the natural transformations can't overall be the same either. □

(e) So far, we haven't had to think hard at all! We asked 'what does it take to get a natural isomorphism between hom-functors', immediately spotted *one* trick that will work, and very easily proved some results about it. The next question to ask is: can *all* possible natural transformations between the hom-functors $\mathsf{C}(A,-)$ and $\mathsf{C}(B,-)$ be produced by the same trick, i.e. are they all generated from arrows $f\colon B \to A$ in the way described in Theorem 176?

If a natural transformation $\alpha\colon \mathsf{C}(A,-) \Rightarrow \mathsf{C}(B,-)$ *is* already given as being of the form $\mathsf{C}(f,-)$ for some $f\colon B \to A$, then we know in this case that $f = \mathsf{C}(f,-)_A 1_A = \alpha_A 1_A$. It would be very nice if the same idea always works, so that we have:

Theorem 178. *Suppose* C *is a locally small category, and consider the hom-functors* $\mathsf{C}(A,-)$ *and* $\mathsf{C}(B,-)$*, for objects* A, B *in* C*. Then, if there is a natural transformation* $\alpha\colon \mathsf{C}(A,-) \Rightarrow \mathsf{C}(B,-)$*, there is a unique arrow* $f\colon B \to A$ *such that* $\alpha = \mathsf{C}(f,-)$*, namely* $f = \alpha_A(1_A)$.

And that is indeed right. We just have to think about what happens when we chase 1_A round a naturality square involving the component α_A (what else?). Where does 1_A live? – in the hom-set $\mathsf{C}(A,A)$. So the obvious thing to do is look again at the sort of square we met before, but now putting $X = A$. Off we go:

Proof. Since α is a natural transformation, the following diagram in particular must commute, for any X and any $j\colon A \to X$,

$$\begin{array}{ccc} \mathsf{C}(A,A) & \xrightarrow{\mathsf{C}(A,j)} & \mathsf{C}(A,X) \\ \downarrow{\scriptstyle \alpha_A} & & \downarrow{\scriptstyle \alpha_X} \\ \mathsf{C}(B,A) & \xrightarrow{\mathsf{C}(B,j)} & \mathsf{C}(B,X) \end{array}$$

As before, $\mathsf{C}(A,j)$ is the map that (among other things) sends an arrow $h\colon A \to A$ to the arrow $j \circ h\colon A \to X$, and $\mathsf{C}(B,j)$ sends an arrow $k\colon B \to A$ to the arrow $j \circ k\colon B \to X$.

So now chase that identity arrow 1_A round the diagram from the top left to bottom right nodes. The top route sends it to $\alpha_X(j)$. The bottom route sends it to $j \circ (\alpha_A(1_A))$, which equals $\mathsf{C}(\alpha_A(1_A),-)_X(j)$ (check how we set up the notation in Theorem 176).

Since our square always commutes we have $\alpha_X(j) = \mathsf{C}(\alpha_A(1_A),-)_X(j)$, for all objects X and for all arrows $j\colon A \to X$. Thus the X-components of α and $\mathsf{C}(\alpha_A(1_A),-)$ agree on their application to all arrows $j\colon A \to X$, hence must be the same. Since X was arbitrary, that means all the components of α and $\mathsf{C}(\alpha_A(1_A),-)$ are the same.

Hence those natural transformations are identical, which is what we need for the existence part of our theorem – we've found an f, i.e. $\alpha_A(1_A)$, such that $\alpha = \mathsf{C}(f,-)$.

Look at it this way: *fixing just one bit of data – about what (the relevant component of) α does to 1_A – fixes the whole natural transformation* by the requirement that the naturality squares all commute.

Finally, suppose both f and f' are such that $\alpha = \mathsf{C}(f, -) = \mathsf{C}(f', -)$. Then by Theorem 177 (2)

$$f = \mathsf{C}(f, -)_A(1_A) = \mathsf{C}(f', -)_A(1_A) = f'$$

which shows f's uniqueness. □

(f) The theorems so far in this section have been about covariant hom-functors. Predictably, there are dual results for contravariant hom-functors $\mathsf{C} \to \mathsf{Set}$ (or equivalently, covariant hom-functors $\mathsf{C}^{op} \to \mathsf{Set}$).

Here's a summary theorem, whose proof can be left as a routine exercise in dualization – just pay attention to the direction of arrows:

Theorem 179. *Suppose* C *is a locally small category, and* $\mathsf{C}(-, A)$, $\mathsf{C}(-, B)$ *are contravariant hom-functors (for objects* A, B *in* C*). Then:*

(1) If there exists an arrow $f \colon A \to B$, *there is a natural transformation* $\mathsf{C}(-, f) \colon \mathsf{C}(-, A) \Rightarrow \mathsf{C}(-, B)$, *where for each* X, *the component* $\mathsf{C}(-, f)_X \colon \mathsf{C}(X, A) \to \mathsf{C}(X, B)$ *sends an arrow* $j \colon X \to A$ *to* $f \circ j \colon X \to B$.

(2) $\mathsf{C}(-, g \circ f) = \mathsf{C}(-, g) \circ \mathsf{C}(-, f)$.

(3) Different arrows $f, f' \colon A \to B$ *give rise to different corresponding natural transformations* $\mathsf{C}(-, f)$, $\mathsf{C}(-, f')$.

(4) If there is a natural transformation $\alpha \colon \mathsf{C}(-, A) \Rightarrow \mathsf{C}(-, B)$, *there is a unique arrow* $f \colon A \to B$ *such that* $\alpha = \mathsf{C}(-, f)$, *namely* $f = \alpha_A(1_A)$. □

37.2 The Restricted Yoneda Lemma

(a) We now have all we need to prove the Restricted Yoneda Lemma (that's my non-standard label: its rationale becomes clear in the next chapter). The key idea is simply this: in proving Theorems 176 and 178, we have shown that the arrows $B \to A$ of a locally small category C line up one-to-one with natural transformations $\mathsf{C}(A, -) \Rightarrow \mathsf{C}(B, -)$.

In other words, in the notation of Defn. 134, we have shown that *there is a bijection between the hom-set* $\mathsf{C}(B, A)$ *and the collection* $Nat(\mathsf{C}(A, -), \mathsf{C}(B, -))$. And ah-ha! – since we are dealing with a locally small category, $\mathsf{C}(B, A)$ is by assumption set-sized; therefore, since the collection $Nat(\mathsf{C}(A, -), \mathsf{C}(B, -))$ is the same size, we can happily treat that as a set too.

Likewise Theorem 179 tells us that its arrows $A \to B$ line up one-to-one with natural transformations $\mathsf{C}(-, A) \Rightarrow \mathsf{C}(-, B)$. In other words, there is a bijection between the hom-set $\mathsf{C}(A, B)$ and the set $Nat(\mathsf{C}(-, A), \mathsf{C}(-, B))$.

But bijections between sets, of course, count as isomorphisms in Set. So we have established the following key theorem:

Theorem 180 (The Restricted Yoneda Lemma). *Suppose* C *is a locally small category, and* A, B *are objects of* C. *Then* $Nat(\mathsf{C}(A,-), \mathsf{C}(B,-)) \cong \mathsf{C}(B, A)$ *and* $Nat(\mathsf{C}(-,A), \mathsf{C}(-,B)) \cong \mathsf{C}(A, B)$. □

(b) It is worth spelling out the justification for our theorem in a slightly different style.

Fix the objects A and B. Then we've shown that there is a function $\mathcal{X}_{AB}$ with source $\mathsf{C}(B, A)$ and target $Nat(\mathsf{C}(A,-), \mathsf{C}(B,-))$, which sends an arrow $f\colon B \to A$ to $\mathsf{C}(f,-)$. And there is a function $\mathcal{E}_{AB}$ in the reverse direction, from $Nat(\mathsf{C}(A,-), \mathsf{C}(B,-))$ to $\mathsf{C}(B, A)$, which sends a natural transformation $\alpha\colon \mathsf{C}(A,-) \Rightarrow \mathsf{C}(B,-)$ to $\alpha_A(1_A)$.

Then we immediately have:

(1) Given any $f\colon B \to A$,

$$(\mathcal{E}_{AB} \circ \mathcal{X}_{AB})f = \mathcal{E}_{AB}(\mathsf{C}(f,-)) = \mathsf{C}(f,-)_A(1_A) = f.$$

where the last identity is from Theorem 177. But f was arbitrary. Whence $\mathcal{E}_{AB} \circ \mathcal{X}_{AB} = 1$ (that's the identity on $\mathsf{C}(B, A)$).

(2) Given any $\alpha\colon \mathsf{C}(A,-) \Rightarrow \mathsf{C}(B,-)$,

$$(\mathcal{X}_{AB} \circ \mathcal{E}_{AB})\alpha = \mathcal{X}_{AB}(\alpha_A(1_A)) = \mathsf{C}(\alpha_A(1_A),-) = \alpha$$

where the last identity is from Theorem 178. But α was arbitrary. Whence $\mathcal{X}_{AB} \circ \mathcal{E}_{AB} = 1$ (that's the identity on $Nat(\mathsf{C}(A,-), \mathsf{C}(B,-))$).

Having a two-sided inverse, $\mathcal{X}_{AB}$ is therefore an isomorphism, and we have half our last theorem again.

The proof of the other half is dual. There is a function $\mathcal{Y}_{AB}$, with source $\mathsf{C}(A, B)$ and target $Nat(\mathsf{C}(-,A), \mathsf{C}(-,B))$, which sends an arrow $f\colon A \to B$ to $\mathsf{C}(-,f)$. And there is a function we'll again notate $\mathcal{E}_{AB}$ from $Nat(\mathsf{C}(-,A), \mathsf{C}(-,B))$ to $\mathsf{C}(A, B)$, which sends a natural transformation α to $\alpha_A(1_A)$. As before, we can show these two functions are inverses. And so $\mathcal{Y}_{AB}$ is also an isomorphism, and we have the other half of our theorem.

37.3 The Yoneda Embedding, the Yoneda Principle

(a) A moment ago we fixed objects A and B and defined $\mathcal{X}_{AB}$ as a map from arrows $f\colon B \to A$ to natural transformations $\mathsf{C}(f,-)\colon \mathsf{C}(A,-) \Rightarrow \mathsf{C}(B,-)$.

But there is of course nothing special about the particular objects A and B here. So we can in fact think of a more general operation on arrows $\mathcal{X}$ which, now for *any* arrow f at all living in C, sends it to a corresponding natural transformation $\mathsf{C}(f,-)$.

Similarly, we can define an operation on objects which we will also label $\mathcal{X}$ that takes any C-object A and sends it to the corresponding hom-functor $\mathsf{C}(A,-)$.

That double use of '$\mathcal{X}$' for an operation on objects and operation on arrows promises that the two components assemble into a functor! Which they do.

For consider: hom-functors like $\mathsf{C}(X,-)$ are objects of the functor category $[\mathsf{C},\mathsf{Set}]$. And natural transformations like $\mathsf{C}(f,-)$ are arrows in that same category. So, the $\mathcal{X}$ operation on objects and the $\mathcal{X}$ operation on arrows are of the right types to be components of a contravariant functor $\mathcal{X}\colon \mathsf{C} \to [\mathsf{C},\mathsf{Set}]$ (contravariant, of course, because an arrow $f\colon B \to A$ is sent to a natural transformation $\mathsf{C}(f,-)\colon \mathsf{C}(A,-) \Rightarrow \mathsf{C}(B,-)$).

And we can easily confirm that the two key conditions for functoriality are satisfied. First, identities are preserved:

$$\mathcal{X}(1_A) = \mathsf{C}(1_A,-) = 1_{\mathsf{C}(A,-)} = 1_{\mathcal{X}(A)}.$$

Second, composition is respected. In other words, for any composable f, g in C,

$$\mathcal{X}(g \circ f) = \mathsf{C}(g \circ f,-) = \mathsf{C}(f,-) \circ \mathsf{C}(g,-) = \mathcal{X}(f) \circ \mathcal{X}(g),$$

reversing the order of composition as is required for a contravariant functor.

So to summarize this important result, and state its dual:

Theorem 181. *For any locally small category* C, *there is a contravariant functor* $\mathcal{X}\colon \mathsf{C} \to [\mathsf{C},\mathsf{Set}]$ *which operates as follows:*

$$\begin{array}{rrcl} \mathcal{X}: & A & \longmapsto & \mathsf{C}(A,-) \\ & f\colon B \to A & \longmapsto & \mathsf{C}(f,-)\colon \mathsf{C}(A,-) \Rightarrow \mathsf{C}(B,-). \end{array}$$

Dually, there is a covariant functor $\mathcal{Y}\colon \mathsf{C} \to [\mathsf{C}^{op},\mathsf{Set}]$ *which works like this:*

$$\begin{array}{rrcl} \mathcal{Y}: & A & \longmapsto & \mathsf{C}(-,A) \\ & f\colon A \to B & \longmapsto & \mathsf{C}(-,f)\colon \mathsf{C}(-,A) \Rightarrow \mathsf{C}(-,B). \end{array}$$ □

By the way, as I mentioned in passing in the previous chapter, the presheaf category $[\mathsf{C}^{op},\mathsf{Set}]$ is often briskly notated as $\widehat{\mathsf{C}}$. So in that idiom, we'd notate the second functor simply as $\mathcal{Y}\colon \mathsf{C} \to \widehat{\mathsf{C}}$.

(b) It is immediate that the functors $\mathcal{X}$ and $\mathcal{Y}$ behave nicely:

Theorem 182. $\mathcal{X}$ *and* $\mathcal{Y}$ *are fully faithful and injective on objects.*

Proof. Let's work through the second case. By definition, $\mathcal{Y}$ is full just in case, for any C-objects A, B, and any natural transformation $\alpha\colon \mathsf{C}(-,A) \Rightarrow \mathsf{C}(-,B)$, there is an arrow $f\colon A \to B$ in C such that $\alpha = \mathcal{Y}f = \mathsf{C}(-,f)$. Which is given in Theorem 179.

By definition, $\mathcal{Y}$ is faithful just in case, for any C-objects A, B, and any pair of arrows $f, g\colon A \to B$ in C, then if $\mathsf{C}(-,f) = \mathsf{C}(-,g)$ then $f = g$. But that also follows immediately from Theorem 179.

So the only new claim is that $\mathcal{Y}$ is injective on objects, meaning that if $\mathcal{Y}(A) = \mathcal{Y}(B)$ then $A = B$. Suppose then that we are given $\mathcal{Y}(A) = \mathcal{Y}(B)$, i.e. $\mathsf{C}(-,A) = \mathsf{C}(-,B)$. Then for any object C we'll have $\mathsf{C}(C,A) = \mathsf{C}(C,B)$. But that can't be so if $A \neq B$, since hom-sets on different pairs of objects must be disjoint – as we've set things up, no arrow $g\colon C \to A$ can equal some $h\colon C \to B$, having distinct targets. □

The situation, then, is this. The functor $\mathcal{Y}$ injects a one-to-one copy of the C-objects into the objects of the functor category $[\mathsf{C}^{op}, \mathsf{Set}]$; and then it fully and faithfully matches up the arrows between C-objects with arrows between the corresponding objects in $[\mathsf{C}^{op}, \mathsf{Set}]$. In other words:

Theorem 183 (The Yoneda Embedding). *The image of* C *under the functor* $\mathcal{Y}$ *is an isomorphic copy of* C*, embedded inside the presheaf category* $[\mathsf{C}^{op}, \mathsf{Set}]$ *as a full O.* □

Note: it is customary to emphasize this result rather than its dual involving $\mathcal{X}$. To be sure, $\mathcal{X}$ embeds a copy of C^{op} in the functor category $[\mathsf{C}, \mathsf{Set}]$ – but we are much more likely to be interested in finding copies of the category C we start off with rather than a copy of its opposite.

(c) For the last of our initial trio of Yoneda results, here's an important corollary of what's gone before:[2]

Theorem 184 (The Yoneda Principle). *For any objects* A, B *in the locally small category* C*,* $A \cong B$ *iff* $\mathcal{X}A \cong \mathcal{X}B$ *and also iff* $\mathcal{Y}A \cong \mathcal{Y}B$*.*

Proof. I'll prove the second claim, with the dual result left as an exercise.

Suppose $A \cong B$. Then there is an isomorphism $f\colon A \xrightarrow{\sim} B$. So there is a natural transformation $\mathsf{C}(-, f)\colon \mathsf{C}(-, A) \Rightarrow \mathsf{C}(-, B)$, which by the remark after Theorem 176 is a natural isomorphism. So in our alternative notation, $\mathcal{Y}f\colon \mathcal{Y}A \overset{\sim}{\Rightarrow} \mathcal{Y}B$. Hence $\mathcal{Y}A \cong \mathcal{Y}B$.

Now suppose $\mathcal{Y}A \cong \mathcal{Y}B$. So there exists a natural isomorphism $\alpha\colon \mathsf{C}(-, A) \overset{\sim}{\Rightarrow} \mathsf{C}(-, B)$. By Theorem 179, α is $\mathsf{C}(-, f)$ for some $f\colon A \to B$, i.e. is $\mathcal{Y}f$. But $\mathcal{Y}$ is fully faithful. So Theorem 137 tells us that since $\mathcal{Y}f$ is an isomorphism, so is f. Hence $A \cong B$.

That shows $A \cong B$ iff $\mathcal{Y}A \cong \mathcal{Y}B$. □

37.4 Yoneda meets Cayley

Now, as I said in the preamble, the more interesting *applications* of the results so far come rather too far downstream to explore at this point. But I can perhaps usefully make some elementary remarks about their significance.

(a) We have seen before that it can be instructive to test-drive categorial ideas on toy cases such as ordered collections of objects considered as categories or monoids considered as categories. Let's take the second sort of case.

[2]The association of all three with the name of the Japanese mathematician Nobuo Yoneda (1930–1996) is perhaps *rather* tenuous.

For what it is worth, the story goes that a key idea was in a 1954 paper by Yoneda on homology. And then Saunders Mac Lane and Yoneda met in Paris the same year, were talking (at the Gare du Nord!) about Yoneda's paper and his ideas, and Mac Lane baptized the general Yoneda Lemma. Though it seems that the first explicit appearance of the Lemma in print, in the restricted form, is in a 1960 paper by Grothendieck (which may well have been an independent rediscovery).

More specifically, let's look at those monoids which are groups, considered as one-object categories all of whose arrows are isomorphisms (see §8.7). To spell that out, take a group comprising some objects G (no more than a set's worth, for present purposes) equipped with a suitable binary operation $\star$ and with the object e distinguished as the group identity. Then the corresponding category G has the following data:

(1) The sole object of G: choose whatever object you like, dub it '$\bullet$'.

(2) The arrows of G: any object g of the group counts as an arrow $g\colon \bullet \to \bullet$. The composite $h \circ g$ of the two arrows $h, g\colon \bullet \to \bullet$ is $h \star g\colon \bullet \to \bullet$. And the identity arrow $1_\bullet$ is the group identity e.

G is locally small since its sole potential hom-set $\mathsf{G}(\bullet, \bullet)$ is the set whose members are the group objects G (or are set-indices for those objects, if we're fussy).

We can therefore apply our Yoneda Embedding results, Theorems 181 and 182. Consider then the version which tells us that there is a fully faithful functor $\mathcal{Y}$ which embeds G into the category $[\mathsf{G}^{op}, \mathsf{Set}]$, where

(1) for the G-object $\bullet$, $\mathcal{Y}\bullet = \mathsf{G}(-, \bullet)$.
(2) For any G-arrow $g\colon \bullet \to \bullet$, $\mathcal{Y}g = \mathsf{G}(-, g)\colon \mathsf{G}(-, \bullet) \Rightarrow \mathsf{G}(-, \bullet)$.

And what does that mean?

Note first that the arrows $\mathcal{Y}g$ (one for each $g \in$ G) form a group. Functoriality gives the required associativity and ensures $\mathcal{Y}e$ behaves as the identity element; further $\mathcal{Y}g \circ \mathcal{Y}g^{-1} = \mathcal{Y}(g \circ g^{-1}) = \mathcal{Y}e$ and likewise $\mathcal{Y}g^{-1} \circ \mathcal{Y}g = \mathcal{Y}e$, so we have the required group inverses.

And what are the elements of this group? By definition, $\mathsf{G}(-, g)$ sends an arrow $x\colon \bullet \to \bullet$ to $g \circ x\colon \bullet \to \bullet$. In other words, $\mathcal{Y}g$ acts by sending any group-object x to the group-object $g \star x$. That gives us a permutation of G – because if $g \star x = g \star x'$ then $g^{-1} \star g \star x = g^{-1} \star g \star x'$ and hence $x = x'$.

In sum, G's isomomorphic image under the Yoneda functor $\mathcal{Y}$ is a category which has a single object, and whose arrows form a group of permutations of the objects G. And *that* is just a group of permutations of the objects G treated as a category.

In other words, Yoneda tells us that any group $(\mathrm{G}, \star, e)$, when thought of as a category, is isomorphic to a group of permutations of its objects G (a subgroup of *all* G's permutations), when thought of as a category. But – deleting the reference to categories – *that's just Cayley's theorem in group theory.*

(b) Let's not get over-excited! It would be quite misleading to say that we have arrived at a new, distinctively categorial, proof of Cayley's theorem. For the key part of the argument above is that the permutation functions $x \mapsto g \star x$, for the various $g \in$ G, form a group – and that's the essence of the usual elementary proof of Cayley's theorem. So what new insight might we get out of the categorial detour?

Well, compare our earlier treatment of products. Pre-categorially, we are familiar with a bunch of constructions which look, intuitively, to involve more or less the same idea – consider, for example, forming a product of two groups, forming the product of two lattices, taking the meet of two elements *in* a lattice, forming a logical product of two propositions in logic, and so on. Then the nice thing about the categorial account of products is that it enables us to see all these and more as instances of the very same construction. Similarly here. There are a variety of pre-categorial results that are intuitively in the same ballpark as Cayley's theorem, telling us about how structures of one kind can be isomorphically embedded into other structures. There are some algebraic cousins, such as e.g. the result that a ring can be embedded into the endomorphism ring of its underlying abelian group. Then there are results such as that a partial ordering of objects can be mirrored in a collection of subsets of those objects ordered by inclusion. It turns out that the Yoneda embedding theorem applied to relevant categories reveals such results as again instances of the very same construction. And that will be an insight worth having.

37.5 Putting the Yoneda Principle to work

(a) Theorem 184 – the Yoneda Principle, as we are calling it – can be technically helpful. Because in some cases where we want to show that A is isomorphic to B it is actually rather easier to first spot a strategy for showing the relevant $\mathcal{Y}A$ and $\mathcal{Y}B$ are isomorphic, and then we can get our desired result by appealing to the Principle.

Here's an example. Back in Chapter 18, I outlined a brute-force proof of Theorem 77 (2): for all A, B, C in a Cartesian closed category C, $(A^B)^C \cong A^{B\times C}$. We can now box a bit more cleverly.

Assume C is a Cartesian closed category that is locally small. Theorem 75 establishes (*) there is a bijection between arrows $X \to Z^Y$ and $X \times Y \to Z$. Hence we straightforwardly have

$$\begin{aligned}\mathsf{C}(X, (A^B)^C) &\cong \mathsf{C}(X \times C, A^B)\\ &\cong \mathsf{C}((X \times C) \times B, A)\\ &\cong \mathsf{C}(X \times (B \times C), A)\\ &\cong \mathsf{C}(X, A^{(B\times C)})\end{aligned}$$

where we are applying (*) three times, and the remaining line depends on the isomorphism of the multi-products $(X \times C) \times B$ and $X \times (B \times C)$.

Now, not only are these isomorphisms, but intuitively they should each be provable without making special assumptions about X (or assumptions about A, B, or C), and without having to make arbitrary choices along the way. So our isomorphisms ought morally be natural in X.

That means we ought to be able to show $\mathsf{C}(-, (A^B)^C) \cong \mathsf{C}(-, A^{(B\times C)})$, or in other words, $\mathcal{Y}(A^B)^C \cong \mathcal{Y}A^{(B\times C)}$.

And then an application of the Yoneda Principle gives us $(A^B)^C \cong A^{B\times C}$, as desired. Neat!

(b) However, there was some hand-waving at the penultimate step of our argument there. For how do we actually *prove* that those four isomorphisms in the displayed chain are natural in X? Well, for the applications of (*), note the proof in §32.4 (4) that the functors $\mathsf{C}(-, Z^Y)$ and $\mathsf{C}(- \times Y, Z)$ are naturally isomorphic. And the other step where we re-order products can be shown to involve a natural isomorphism too (compare our first example in §32.1).

So OK: we *can* fill in the needed details, then, without too much difficulty. And structuring the proof of our result as a smart argument via Yoneda makes it relatively easy to spot what's needed. But it is perhaps worth also pointing out that the work done in filling out those details and proving the naturality-in-X of each of the isomorphisms in the chain will give us along the way the ingredients we need to fill out a direct brute-force proof that $(A^B)^C \cong A^{B \times C}$ along the lines sketched in Chapter 18, a proof which never mentions Yoneda.

37.6 The philosophical content of the Yoneda Principle?

Putting technicalities aside, there is something perhaps rather more interesting about the Yoneda Principle: it reflects a certain categorial perspective.

Recall, we can think of arrows from various objects X to a given object A as probing A from various perspectives. In special cases, probing from a limited selection of objects will reveal everything we need to know about A. For example, in the category Set, probing from a singleton $\{\star\}$ is enough, as the members of a set A are bijectively associated with arrows $\{\star\} \to A$. In general, however, we will need to probe from more sources: for example, in Grp, probes from a singleton source can only hit a group's identity element. But if we probe from *all* other objects X, that must suffice to fix the object A up to isomorphism – in other words, if we take all the arrows $X \to A$, for varying X, you get all the categorially relevant information about A. Why so?

Well, as we noted before, the functor $\mathsf{C}(-, A)$ wraps up all that perspectival information, telling you for each X what the corresponding hom-set of arrows $X \to A$ comprises. Then the Yoneda Principle tells us that if we have pinned down $\mathsf{C}(-, A)$ (i.e. $\mathcal{Y}A$) up to isomorphism – pinned down how A is seen by its world – then we have pinned down A up to isomorphism. For $\mathsf{C}(-, A) \cong \mathsf{C}(-, B)$ indeed implies $A \cong B$.

As we might say, then: you know an object, categorially speaking, by knowing how its friends see it.[3]

[3]Or here's another folklore image. If you think of the objects of a category as particles and the arrows as ways to smash one particle into another, then the Yoneda Principle tells us that if you know all about the interactions when you smash various particles X into the mystery particle A, you know everything there is to know about A. Note though: this sort of point, while often described as giving the 'philosophical' content of the Yoneda Lemma, does not actually require the full Lemma we meet in the next chapter.

38 The Yoneda Lemma

In the previous chapter, we proved what I called the Restricted Yoneda Lemma and showed that the Yoneda Embedding functor really is an embedding. And the proofs of those initial results are, I hope, not at all bewildering. It is only a bit of an exaggeration to say that we asked what the natural transformations between two hom-functors might look like, and then followed our noses, doing the obvious things. And in fact the theorems obtained so far are often the ones that are actually needed when a result is proved 'by the Yoneda Lemma'.

Still, having got this far, it is worth pressing on and proving the full-power, unrestricted, Yoneda Lemma. What does this involve?

38.1 Onwards to the full Yoneda Lemma!

Assume once more that C is locally small. Then here again is one half of our restricted Theorem 180:

> Let F be the hom-functor $\mathsf{C}(B, -) \colon \mathsf{C} \to \mathsf{Set}$. Then there is an isomorphism between $Nat(\mathsf{C}(A, -), F)$ and FA.

Now, to get from that to the full Yoneda Lemma takes two more stages:

(1) *Generalizing on F.* We look again at the ingredients of the proof of the restricted version and ask 'Did we essentially depend on the fact that the second functor in the story, now notated simply 'F', was actually a hom-functor $\mathsf{C}(B, -)$ for some B?'

Inspection reveals that we didn't. So we have the more general result that for *any* functor $F \colon \mathsf{C} \to \mathsf{Set}$, and any C-object A, there is an isomorphism between $Nat(\mathsf{C}(A, -), F)$ and FA.

(2) *Confirming it's all natural.* Our proof of this general result – like the proof of the original Restricted Lemma – provides a recipe for constructing the required isomorphism that doesn't involve any arbitrary choices, and doesn't depend on any special features of A or F.

In an *intuitive* sense, then, we've constructed a natural isomorphism between the objects $Nat(\mathsf{C}(A, -), F)$ and FA. And so hopefully we should be able to show that these objects are naturally isomorphic in the *official categorial* sense of Defn. 122.

In short, we will get from the Restricted Yoneda Lemma to the full-dress Yoneda Lemma by generalizing a construction, and then recasting in category-theoretic terms our intuitive judgement of the naturality of our construction. Neither stage involves anything conceptually very difficult. It is forgivable to skip the proof details, though you might want to grasp the general strategy for each stage.

And needless to say, it's a two-for-one deal: once we have done all the work of strengthening one half of the Restricted Lemma, we can leave strengthening the dual half as an exercise.

38.2 The generalizing move: the Core Lemma

(a) We continue working in a locally small category C. And let's restate some of what we already know, again temporarily using 'F' to abbreviate '$\mathsf{C}(B, -)$':

(i) F sends a C-object A to the set $FA = \mathsf{C}(B, A)$, and there is a bijection between elements of FA and natural transformations $\mathsf{C}(A, -) \Rightarrow F$ – this bijection sends $f\colon B \to A$ in FA to the transformation whose X-component maps an arrow $g\colon A \to X$ to $g \circ f\colon B \to X$.

(ii) F sends a C-arrow $g\colon A \to X$ to a function Fg, where this takes a C-arrow $f\colon B \to A$ to the C-arrow $g \circ f\colon B \to X$. In short, $Fg(f) = g \circ f$. (That's from Theorem 148, re-lettered.)

(iii) Hence, putting (i) and (ii) together, we have: there's a bijection which sends an element f in FA to the natural transformation whose X-component maps $g\colon A \to X$ to $Fg(f)$.

But now note: (iii) at least makes sense for *any* functor $F\colon \mathsf{C} \to \mathsf{Set}$. Why? Because FA, where A is a C-object, is a set. And Fg, where g is a C-arrow will be a Set-arrow, i.e. a (total!) set-function. Hence function Fg then can be applied to any element of the set FA.

(b) That last thought gives us the clue as to how to prove the following generalization of (half of) the Restricted Lemma:

Theorem 185 (The Core Yoneda Lemma). *For any object A of the locally small category C, and any functor $F\colon \mathsf{C} \to \mathsf{Set}$, $Nat(\mathsf{C}(A, -), F) \cong FA$.*[1]

Proof, step 1: Defining $\mathcal{X}_{AF}$, a candidate bijection. Pick an element e from the set FA (this now need not be a function in the general case, so it would perhaps be misleading to still use the notation 'f').

And taking up the idea in (iii), define $\alpha^e_X : \mathsf{C}(A, X) \to FX$ as the function that maps any C-arrow $g\colon A \to X$ to $Fg(e)$. Similarly, define $\alpha^e_Y : \mathsf{C}(A, Y) \to FY$ as the function that maps any C-arrow $h\colon A \to Y$ to $Fh(e)$; and so on.

[1] Again my new label is non-standard. Some call *this* result the Yoneda Lemma, plain and simple: see e.g. Adámek et al. (2009, p. 88), Barr and Wells (1985, p.26), Grandis (2018, p. 44). But it is more usual to take unqualified talk of the Lemma to refer to the full result we eventually arrive at as our Theorem 188.

Then it is immediate that a diagram like the following commutes for any $j\colon X \to Y$:

$$\begin{array}{ccc} \mathsf{C}(A,X) & \xrightarrow{\mathsf{C}(A,j)} & \mathsf{C}(A,Y) \\ \downarrow{\scriptstyle \alpha^e_X} & & \downarrow{\scriptstyle \alpha^e_Y} \\ FX & \xrightarrow{\quad Fj \quad} & FY \end{array}$$

The upper route takes a C-arrow $g\colon A \to X$ to $j \circ g\colon A \to Y$. And α^e_Y sends that on to $F(j \circ g)(e)$ which equals $Fj(Fg(e))$ by functoriality. While the lower route takes g to $Fg(e)$ to $Fj(Fg(e))$. So we get the same result either way.

Hence, as defined, the components $\alpha^e_X, \alpha^e_Y, \ldots$ assemble into a natural transformation $\alpha^e\colon \mathsf{C}(A,-) \Rightarrow F$. Great!

So we have brought into play a nice function $\mathcal{X}_{AF}\colon FA \to Nat(\mathsf{C}(A,-), F)$ which sends an element e of FA to the natural transformation α^e. It just remains to show that this function is bijective (as was its analogue in §37.2). □

Proof step 2: Showing $\mathcal{X}_{AF}$ is surjective. We want to prove that every natural transformation $\alpha\colon \mathsf{C}(A,-) \Rightarrow F$ is some α^e generated by an element e in FA. Given the proof of Theorem 178, we know exactly how to do that.

We start by noting that, given any natural transformation $\alpha\colon \mathsf{C}(A,-) \Rightarrow F$, the following diagram in particular must commute, for any Y and any $j\colon A \to Y$:

$$\begin{array}{ccc} \mathsf{C}(A,A) & \xrightarrow{\mathsf{C}(A,j)} & \mathsf{C}(A,Y) \\ \downarrow{\scriptstyle \alpha_A} & & \downarrow{\scriptstyle \alpha_Y} \\ FA & \xrightarrow{\quad Fj \quad} & FY \end{array}$$

Now chase the identity arrow 1_A round the diagram from the top left to bottom right nodes. The top route sends it first to j and then on to $\alpha_Y(j)$. The bottom route sends it to $Fj(\alpha_A(1_A))$ – which, by definition, equals $\alpha^e_Y j$ for $e = \alpha_A(1_A)$. Since the diagram commutes, $\alpha_Y(j) = \alpha^e_Y j$, and since this holds for any j, we have $\alpha_Y = \alpha^e_Y$.

But Y was arbitrary, so the equality holds for all components, therefore $\alpha = \alpha^e$ when $f = \alpha_A(1_A)$. □

Proof step 3: Showing $\mathcal{X}_{AF}$ is injective. We use the same pattern of argument as for Theorem 176 (cont'd), except that where we previously used the fact that $\mathsf{C}(f,-)_A 1_A = f$, we now use the fact that $\alpha^e_A(1_A) = e$. And why is that a fact? By definition, α^e_A sends an arrow $g\colon A \to A$ to $Fg(e)$. So $\alpha^e_A(1_A)$ yields $F1_A(e)$. But the functoriality of F ensures that $F1_A$ is an identity function.

So if $e \neq e'$ we have $\alpha^e_A(1_A) = e \neq e' = \alpha^{e'}_A(1_A)$, and hence $\alpha^e \neq \alpha^{e'}$ □

So we are done: we've got a bijection $\mathcal{X}_{AF}$ between the sets $Nat(\mathsf{C}(A,-), F)$ and FA, showing they are isomorphic, and the Core Lemma is in the bag.

Of course, there's a companion dual result which can safely be left as an exercise:[2]

Theorem 185 (cont'd). *For any object A of the locally small category* C, *and any functor* $F\colon \mathsf{C}^{op} \to \mathsf{Set}$, $Nat(\mathsf{C}(-,A),F) \cong FA$. □

(c) Let's take up an idea from the discussion after Theorem 180. We'll show that $\mathcal{X}_{AF}\colon FA \to Nat(\mathsf{C}(A,-),F)$ has a two-sided inverse $\mathcal{E}_{AF}\colon Nat(\mathsf{C}(A,-),F) \to FA$ where that is the function which sends a natural transformation $\alpha\colon \mathsf{C}(A,-) \Rightarrow F$ to the element $\alpha_A(1_A)$. Proceeding more or less as before,

(1) Given any element e of FA,

$$(\mathcal{E}_{AF} \circ \mathcal{X}_{AF})e = \mathcal{E}_{AF}\alpha^e = \alpha^e_A(1_A) = e.$$

But e was arbitrary. Whence $\mathcal{E}_{AF} \circ \mathcal{X}_{AF} = 1$ (that's the identity on FA).

(2) Given any $\alpha\colon \mathsf{C}(A,-) \Rightarrow F$,

$$(\mathcal{X}_{AF} \circ \mathcal{E}_{AF})\alpha = \mathcal{X}_{AF}(\alpha_A(1_A)) = \alpha^{\alpha_A(1_A)} = \alpha$$

(for the last identity, see the end of the proof step 2 above). But α was arbitrary. Whence $\mathcal{X}_{AF} \circ \mathcal{E}_{AF} = 1$ (that's the identity on $Nat(\mathsf{C}(A,-),F)$).

Having a two-sided inverse, $\mathcal{X}_{AF}$ is therefore (as we know!) an isomorphism with inverse $\mathcal{E}_{AF}$, a point we'll need in a moment.

There is of course a dual story to be told about an isomorphism $\mathcal{Y}_{AF}\colon FA \to Nat(\mathsf{C}(-,A),F)$ which sends an element e of FA to a suitable $\alpha^e\colon \mathsf{C}(-,A) \Rightarrow F$. But I'll leave it as another challenge to fill in the details.

38.3 Making it all natural

(a) So where have we got to?

To concentrate again on half the story, Theorem 185 tells us that – when C is a locally small category, A is any object in that category, and $F\colon \mathsf{C} \to \mathsf{Set}$ is some functor – then FA is isomorphic to $Nat(\mathsf{C}(A,-),F)$. I called that the Core Yoneda Lemma. And of course, the earlier Restricted Lemma Theorem 180 is what we get when we restrict to the cases where F has the form $\mathsf{C}(B,-)$.

Now, our proof of Theorem 185 didn't depend on any special facts about A or F, and didn't depend on any arbitrary choices. So, at least in an *intuitive* sense, we have found a natural isomorphism. But when we find an intuitively natural isomorphism, what I called the Eilenberg/Mac Lane Thesis in §32.7 enjoins us to see this as a natural isomorphism in the official *categorial* sense, arising from

[2] I'll state this dual in the conventional way, in terms of a covariant functor $F\colon \mathsf{C}^{op} \to \mathsf{Set}$. But you may well find it easier to keep track of which arrows go in which direction, and avoid dancing between C and C^{op}, if you set out the proof thinking in terms of the equivalent contravariant functor $F\colon \mathsf{C} \to \mathsf{Set}$.

a natural isomorphism between functors. And so this is going to be the final stage of the argument taking us to the full Yoneda Lemma. We want to show how the intuitively natural isomorphism we've found between the objects FA and $Nat(\mathsf{C}(A,-)), F)$ can be seen as arising, in fact in two ways, from natural isomorphisms between functors – making it officially both 'natural in A' and 'natural in F'. Though, as often, checking official naturality is rather tedious.

(b) A quick but important observation before continuing.

At the very beginning of the previous chapter, I noted Awodey's remark about the Yoneda Lemma being perhaps the most used result in category theory. And taking a look at e.g. a characteristic pair of important monographs that go rather beyond entry level, Borceux (1994) and Mac Lane and Moerdijk (1992), we do find multiple invocations of one or other of the interrelated Yoneda results.

However, I think it is correct to say that the appeal is most frequently either to our easy Yoneda Embedding result, Theorem 183, or to the simple existence of some isomorphism of the kind reported in the Core Theorem 185. The further fact that such an isomorphism between objects can officially be seen as arising from a natural isomorphism between functors is often not needed.

(c) Still, let's get down to work again. Here's a straight application of our earlier Defn. 122: two objects in Set are said to be *naturally isomorphic in A* if they are the images of the same object A under a couple of naturally isomorphic functors $F, G\colon \mathsf{C} \to \mathsf{Set}$.

So to show that, in particular, FA and $Nat(\mathsf{C}(A,-), F)$ are naturally isomorphic in A, we want to find a functor $G\colon \mathsf{C} \to \mathsf{Set}$ such that $GA = Nat(\mathsf{C}(A,-), F)$, and G is naturally isomorphic to F. How can we construct a suitable G?

Well, here are two functors we already know about:

(1) Theorem 181 defined the contravariant functor $\mathcal{X}\colon \mathsf{C} \to [\mathsf{C}, \mathsf{Set}]$ which
  (i) sends an object A to the hom-functor $\mathsf{C}(A,-)$; and
  (ii) sends a C-arrow $f\colon A \to B$ to the corresponding natural transformation $\mathsf{C}(f,-)\colon \mathsf{C}(B,-) \Rightarrow \mathsf{C}(A,-))$.

(2) Defn. 135 defined another contravariant functor, $Nat(-, F)\colon [\mathsf{C}, \mathsf{Set}] \to \mathsf{Set}$, which
  (i) sends a functor $\mathsf{C}(A,-)$ to the corresponding set $Nat(\mathsf{C}(A,-), F)$; and
  (ii) sends a natural transformation $\mathsf{C}(f,-)\colon \mathsf{C}(B,-) \Rightarrow \mathsf{C}(A,-)$ to the function which takes a natural transformation $\alpha\colon \mathsf{C}(A,-) \Rightarrow F$ and outputs $\alpha \circ \mathsf{C}(f,-)\colon \mathsf{C}(B,-) \Rightarrow F$).

Hence if we put $G = Nat(-, F) \circ \mathcal{X}$ we'll get a covariant functor $G\colon \mathsf{C} \to \mathsf{Set}$ (because contravariant functors compose to form a covariant functor by the mini-Theorem 130). And, as required, $GA = Nat(\mathsf{C}(A,-), F)$.

Then, to prove GA and FA are naturally isomorphic in A, we'll establish the following:

Theorem 186. *As defined, the functors $G = Nat(-, F) \circ \mathcal{X}$ and F are naturally isomorphic.*

And it wouldn't be quite absurd to set this as a challenge to prove for yourself. You'll need to check a suitable naturality square. But what will be the components of the candidate natural isomorphism taking us from the likes of GA (i.e. $Nat(\mathsf{C}(A,-),F)$) to FA? Well, what else but $\mathcal{E}_{AF}$ as defined in the previous section? Then you just need to prove that the square commutes. Try it before reading on!

Proof. Given some arrow $j\colon A \to B$, consider the following square:

$$\begin{array}{ccc} GA = Nat(\mathsf{C}(A,-),F) & \xrightarrow{\quad Gj \quad} & GB = Nat(\mathsf{C}(B,-),F) \\ \Big\downarrow{\scriptstyle \mathcal{E}_{AF}} & & \Big\downarrow{\scriptstyle \mathcal{E}_{BF}} \\ FA & \xrightarrow{\quad Fj \quad} & FB \end{array}$$

Take any $\alpha\colon \mathsf{C}(A,-) \Rightarrow F$ in GA. Then, applying our definitions, we have:

(1) $\mathcal{E}_{BF} \circ Gj\alpha = \mathcal{E}_{BF}(\alpha \circ \mathsf{C}(j,-)) = (\alpha \circ \mathsf{C}(j,-))_B(1_B) = \alpha_B \circ \mathsf{C}(j,-)_B(1_B) = \alpha_B(j)$.

(2) But also $Fj \circ \mathcal{E}_{AF}(\alpha) = Fj \circ \alpha_A(1_A) = \alpha_B \circ \mathsf{C}(A,j)(1_A) = \alpha_B(j)$ (for the middle equation we note that $Fj \circ \alpha_A = \alpha_B \circ \mathsf{C}(A,j)$ by a naturality square for α).

Our diagram will therefore always commute, and hence there is a natural isomorphism $\mathcal{E}_F\colon G \Rightarrow F$ with components $(\mathcal{E}_F)_A = \mathcal{E}_{AF}$ for each A in C, and we are done. □

(d) That captures in categorial terms the intuition that the isomorphism between FA and $Nat(\mathsf{C}(A,-),F)$ depends in a natural way on A. Now for the companion intuition that it depends in a natural way on F too. Keeping A fixed, we want to prove $Nat(\mathsf{C}(A,-),F) \cong FA$ naturally in F.

So we want to show that our isomorphism arises again from a natural isomorphism between two functors which we might initially notate as $Nat(\mathsf{C}(A,-),-)$ and $-A$, when these functors are applied to F.

And in fact, we have fleetingly met the *second* of these functors in a different notation in Defn. 136: it is the functor $eval_A\colon [\mathsf{C},\mathsf{Set}] \to \mathsf{Set}$ which sends any functor $F\colon \mathsf{C} \to \mathsf{Set}$ to FA and sends any natural transformation $\alpha\colon F \Rightarrow G$ to its component $\alpha_A\colon FA \to GA$.

While what about the *first* functor, $Nat(\mathsf{C}(A,-),-)$? It is another covariant functor from $[\mathsf{C},\mathsf{Set}] \to \mathsf{Set}$.[3] Hence we now want to prove the following:

Theorem 187. *As defined, the functors $Nat(\mathsf{C}(A,-),-)$ and $eval_A$ are naturally isomorphic.*

[3]Why so? Any $Nat(J,-)$ defined in terms of a functor $J\colon \mathsf{E} \to \mathsf{F}$ is a hom-functor from $[\mathsf{E},\mathsf{F}]$ to Set. So yes, in the present case $Nat(\mathsf{C}(A,-),-)$ defined in terms of the functor $\mathsf{C}(A,-)\colon \mathsf{C} \to \mathsf{Set}$ will give us a hom-functor from $[\mathsf{C},\mathsf{Set}]$ to Set!

Again, here's a challenge: prove that before reading on ...

Proof. Given any $\gamma\colon F \Rightarrow G$, consider the following diagram,

$$\begin{array}{ccc} Nat(\mathsf{C}(A,-),F) & \xrightarrow{Nat(\mathsf{C}(A,-),\gamma)} & Nat(\mathsf{C}(A,-),G) \\ \Big\downarrow{\scriptstyle \mathcal{E}_{AF}} & & \Big\downarrow{\scriptstyle \mathcal{E}_{AG}} \\ eval_A(F) = FA & \xrightarrow{eval_A(\gamma)} & eval_A(G) = GA \end{array}$$

Take any $\alpha\colon \mathsf{C}(A,-) \Rightarrow F$, and recall that $Nat(\mathsf{C}(A,-),\gamma)$ sends α to $\gamma \circ \alpha$. Then we have:

(1) $\mathcal{E}_{AG} \circ Nat(\mathsf{C}(A,-),\gamma)(\alpha) = \mathcal{E}_{AG}(\gamma \circ \alpha) = (\gamma \circ \alpha)_A(1_A) = \gamma_A(\alpha_A(1_A))$.

(2) But also $eval_A(\gamma) \circ \mathcal{E}_{AF}(\alpha) = \gamma_A(\alpha_A(1_A))$.

Hence the diagram always commutes. Therefore there is a natural isomorphism $\mathcal{E}_A\colon K \Rightarrow eval_A$ with components $(\mathcal{E}_A)_F = \mathcal{E}_{AF}$ for each F from $[\mathsf{C}, \mathsf{Set}]$. □

38.4 Putting everything together

Our last two theorems have duals – which I'll leave you to state and prove. But taking those as read, we can now combine all the ingredients from the last three theorems ...

Cue drum-roll!

... and at last we get the fully caffeinated Lemma:

Theorem 188 (Full Yoneda Lemma). *For any locally small category* C, *object* A *in* C, *and covariant functor* $F\colon \mathsf{C} \to \mathsf{Set}$, $Nat(\mathsf{C}(A,-),F) \cong FA$, *both naturally in* A *and naturally in* F.

Likewise for any contravariant functor $F\colon \mathsf{C} \to \mathsf{Set}$ *(equivalently, covariant functor* $F\colon \mathsf{C}^{op} \to \mathsf{Set}$*),* $Nat(\mathsf{C}(-,A),F) \cong FA$, *both naturally in* A *and naturally in* F. □

So at last we get there![4]

[4] I have gone for what might be called mid-level notational brevity. Thus some would write the likes of $Hom_{[\mathsf{C}^{op},\mathsf{Set}]}(Hom_{\mathsf{C}}(-,A),F)$ where I would more briskly (and, in context, I hope more readably) have $Nat(\mathsf{C}(-,A),F)$.

Others go in the opposite direction, opting for additional compression; where I have written $\mathsf{C}(A,-)$ and $\mathsf{C}(-,A)$ for, respectively, the covariant and contravariant hom-functors, you will often find simply h^A and h_A.

39 Representables and universal elements

Hom-functors have very nice properties like preserving limits or colimits (see e.g. Theorem 150). Other functors that are naturally isomorphic to hom-functors will share these nice properties (see e.g. Theorem 153). It is an obvious next move to say something about this wider class of well-behaved functors.

That, then, is the task of this chapter. However, at our introductory level, hom-functors will remain our prime concern. And you might well want to skim or even skip over most of the increasingly abstract material here.

At the end of the chapter, however, I do briefly pick up an interesting issue left hanging in §19.6.

39.1 Representable functors

(a) You can assume throughout this chapter that we are dealing with locally small categories. And we first need to fix some standard terminology.

Definition 139. Let $F : \mathsf{C} \to \mathsf{Set}$ be a covariant set-valued functor. If, for some C-object A there is a natural isomorphism $\psi\colon F \overset{\sim}{\Longrightarrow} \mathsf{C}(A, -)$, then (A, ψ) is said to be a *representation* of F, with A its *representing object.*[1]

If F has a representation, i.e. is naturally isomorphic to a hom-functor, it is said to be *representable.*

Dually, let F now be a contravariant set-valued functor. If there is a natural isomorphism $\psi\colon F \overset{\sim}{\Longrightarrow} \mathsf{C}(-, A)$, then (A, ψ) is again said to be a representation of F, with A its representing object. Again, F is representable if it has a representation. △

(b) We can immediately state two important results.

Theorem 189. *If $F\colon \mathsf{C} \to \mathsf{Set}$ is represented by both A and B, then $A \cong B$.*

Proof. If we have $\mathsf{C}(A, -) \cong F \cong \mathsf{C}(B, -)$ then, in the notation of Theorem 184, $\mathcal{X}A \cong \mathcal{X}B$ and hence $A \cong B$ by the Yoneda Principle. □

Theorem 190. *A representable functor $F\colon \mathsf{C} \to \mathsf{Set}$ preserves all limits that exist in the small category* C.

[1] I guess that we might instead have expected the hom-functor $\mathsf{C}(A, -)$ to be called a representation of F.

Proof. By hypothesis there is a hom-functor $\mathsf{C}(A,-)$ such that $\mathsf{C}(A,-) \cong F$. By Theorem 153, F preserves whatever limits $\mathsf{C}(A,-)$ preserves. But by Theorem 150 $\mathsf{C}(A,-)$ preserves all limits that exist in C. □

And as a corollary of that theorem, in the special case where C has all limits of shape J (so the functor $Lim = \underset{\leftarrow \mathsf{J}}{Lim}$ is well-defined and we can apply Theorem 175), we have

Theorem 191. *If $F\colon \mathsf{C} \to \mathsf{Set}$ is representable, then for any diagram $D\colon \mathsf{J} \to \mathsf{C}$, $F(Lim\, D) \cong Lim(FD)$.* □

39.2 Two elementary examples

Hom-functors themselves are by definition representable functors. But what other kinds of examples are there?

(a) First, a toy case. Take the trivial identity functor from Set to Set:

$$\begin{array}{rrcl} 1_{\mathsf{Set}}\colon & X & \longmapsto & X \\ & f\colon X \to Y & \longmapsto & f\colon X \to Y. \end{array}$$

Is this representable? Well, if it is to be representable by a covariant hom-functor $\mathsf{Set}(S,-)$ for some S, we would need a suite of isomorphisms $\psi_X, \psi_Y, \ldots$, such that this square always commutes for any f:

$$\begin{array}{ccc} X & \xrightarrow{\quad f \quad} & Y \\ \Big\downarrow{\scriptstyle \psi_X} & & \Big\downarrow{\scriptstyle \psi_Y} \\ \mathsf{Set}(S,X) & \xrightarrow{\quad f \,\circ\, - \quad} & \mathsf{Set}(S,Y) \end{array}$$

But how can we ensure that the members of X are in bijection with the arrows from S to X? By taking S to be some singleton $1 = \{\bullet\}$ (any will do). Then, if we put ψ_X to be the bijective function which sends $x \in X$ to the corresponding arrow $\vec{x}\colon 1 \to X$, and similarly for ψ_Y, etc., our square will trivially commute. Then these ψ_Z are components of a natural isomorphism $\psi\colon 1_{\mathsf{Set}} \overset{\sim}{\Longrightarrow} \mathsf{Set}(1,-)$.

Which gives us the following mini-theorem:

Theorem 192. *The identity functor $1_{\mathsf{Set}}\colon \mathsf{Set} \to \mathsf{Set}$ is representable, and is represented by a singleton* 1.

That's not surprising. As we put it before, we can think of a hom-functor $\mathsf{C}(A,-)$ as encapsulating A's view of the category C (the view as seen via the arrows from A). In particular, then, $\mathsf{Set}(1,-)$ encapsulates how a singleton sees the category of sets via arrows from that singleton. But we know that the various arrows from a singleton to any given set can in turn target all the members of that set. So $\mathsf{Set}(1,-)$ encapsulates a complete view of Set. No wonder it comes to the same as the identity functor on Set.

(b) Let's next return to the very first functor we met back in §26.2, the forgetful functor $F\colon \mathsf{Mon} \to \mathsf{Set}$ which sends any monoid living in Mon to its underlying set, and sends a monoid homomorphism to the same function thought of as an arrow in Set.

Suppose we have monoids $M = (\underline{M}, *, e)$ and $M' = (\underline{M'}, \star, d)$, where underlining indicates the underlying set of the monoid. Then in our shorthand:[2]

$$F\colon \quad \begin{array}{rcl} M & \longmapsto & \underline{M} \\ f\colon M \to M' & \longmapsto & f\colon \underline{M} \to \underline{M'} \end{array}$$

And now let's ask: is there a representing monoid R such that the hom-functor $\mathsf{Mon}(R, -)$ is naturally isomorphic to the forgetful F?

Applying the usual definition, the hom-functor $\mathsf{Mon}(R, -)$ sends a monoid M to $\mathsf{Mon}(R, M)$. And it sends a monoid homomorphism $f\colon M \to M'$ to the set-function $f \circ -$ which maps an arrow $g\colon R \to M$ in $\mathsf{Mon}(R, M)$ to the arrow $f \circ g\colon R \to M'$ in $\mathsf{Mon}(R, M')$. And if this functor $\mathsf{Mon}(R, -)$ is to be naturally isomorphic with the forgetful functor F, there will have to be an isomorphism ψ with a component at each monoid M such that, for any $f\colon M \to M'$ in Mon, the following diagram commutes in Set:

$$\begin{array}{ccc} \underline{M} & \xrightarrow{\quad f \quad} & \underline{M'} \\ \Big\downarrow{\scriptstyle \psi_M} & & \Big\downarrow{\scriptstyle \psi_{M'}} \\ \mathsf{Mon}(R, M) & \xrightarrow{\quad f \,\circ\, - \quad} & \mathsf{Mon}(R, M') \end{array}$$

Now, for this to work, we need to choose our representing monoid R such that (for any monoid M) there is a bijection between $\underline{M}$ and $\mathsf{Mon}(R, M)$. And presumably, for the needed generality, R will have to be a 'boring' monoid without too much distinctive structure. That severely limits the possible options.

First shot: take the simplest such boring monoid, the one-element monoid 1. But a moment's reflection shows that this can't work as a candidate for R (typically $\underline{M}$ has many members, $\mathsf{Mon}(1, M)$ can have only one, so there won't be an isomorphism between them).

Second shot: take the next simplest unstructured monoid, the free monoid with a single generator ('free' in the sense that there are no further conditions on the monoid other than it *is* a monoid and every element bar the identity is the result of applying the monoid operations as many times as we like to the generating element). We can usefully think of this monoid as $N = (\mathbb{N}, +, 0)$ whose generator is 1, and whose every element other than the identity element 0 is a sum of 1s. Now consider a homomorphism from N to M. $0 \in \mathbb{N}$ has to be sent to the identity element e in M. And once we fix where $1 \in \mathbb{N}$ gets sent to, namely some $m \in \underline{M}$, that determines where every element of $\mathbb{N}$ goes (since

[2]Remember, a monoid homomorphism in Mon just *is* a function between the relevant underlying sets satisfying certain conditions.

every non-zero $\mathbb{N}$ element $1 + 1 + 1 + \ldots + 1$ will be sent to a corresponding $\underline{M}$-element $m * m * m * \ldots * m$).

So consider $\psi_M \colon M \to \mathsf{Mon}(N, M)$ which maps m to the unique homomorphism $\overline{m} \colon N \to M$ that sends $1 \in \mathbb{N}$ to $m \in \underline{M}$. ψ_M is evidently bijective – each homomorphism from N to M is some $\overline{m}$ for one and only one m in M. Hence ψ_M is an isomorphism in Set. Define $\psi_{M'}$ similarly.

And now it is easily seen that our diagram always commutes, for any homomorphism $f \colon (\underline{M}, *, e) \to (\underline{M}', \star, d)$. Chase an element $m \in \underline{M}$ round the diagram. The route via the north-east node gives us $m \longmapsto fm \longmapsto \overline{fm}$, the other route gives us $m \longmapsto \overline{m} \longmapsto f \circ \overline{m}$. But $f \circ \overline{m} = \overline{fm}$. Why? Because $f \circ \overline{m}$ is the composite map which takes e.g. $3 \in \mathbb{N}$ to the result of f applied to $m * m * m$, i.e. takes 3 to $fm \star fm \star fm$ – but that's what $\overline{fm}$ does.

Since the diagram always commutes, this means in turn that the maps ψ_M assemble into a natural isomorphism $\psi \colon F \overset{\sim}{\Longrightarrow} \mathsf{Mon}(N, -)$. Hence, in summary:

Theorem 193. *The forgetful functor $F \colon \mathsf{Mon} \to \mathsf{Set}$ is representable, and is represented by N, the free monoid on one generator.*

Again this isn't too surprising. Our hom-functor $\mathsf{Mon}(N, -)$ encapsulates N's view of the category of monoids. Now, N can survey all the elements of $\underline{M}$ by running though the homomorphisms $f \colon N \to M$ and looking at all the various possible values $f(1)$ in turn. But homomorphic probes from N won't give us any distinctive information about how the monoid operation works on those elements of $\underline{M}$. Which is why $\mathsf{Mon}(N, -)$ is tantamount to the forgetful functor.

By the way, being representable, and therefore naturally isomorphic to a hom-functor, it follows that the forgetful $F \colon \mathsf{Mon} \to \mathsf{Set}$ preserves limits (by Theorems 150 and 153). But we knew that already (from Theorem 142).

39.3 More examples of representables

(a) Unsurprisingly, there are analogous representation theorems for other forgetful functors. For instance, although we won't pause over the proofs, we have:

Theorem 194. *(1) The forgetful functor $F \colon \mathsf{Grp} \to \mathsf{Set}$ is representable, and is represented by Z, the group of integers under addition.*

(2) The forgetful functor $F \colon \mathsf{Ab} \to \mathsf{Set}$ is representable, and is also represented by Z.

(3) The forgetful functor $F \colon \mathsf{Vect} \to \mathsf{Set}$ (where Vect is the category of vector spaces over the reals) is representable, and is represented by R, the reals treated as a vector-space.

(4) The forgetful functor $F \colon \mathsf{Top} \to \mathsf{Set}$ is representable, and is represented by the one-point topological space, call it S_0.

And given such examples, you might be tempted to conjecture that *all* such forgetful functors into Set are representable. But not so. Consider FinGrp, the category of finite groups. Then

Theorem 195. *The forgetful functor $F\colon \mathsf{FinGrp} \to \mathsf{Set}$ is not representable.*

Proof. Suppose a putative representing group R has r members, and take any group G with $g > 1$ members, where g is coprime with r. Then it is well known that the only group homomorphism from R to G is the trivial one that sends everything to the identity in G. But then the underlying set of G can't be in bijective correspondence with $\mathsf{FinGrp}(R, G)$ as would be required for a naturality square showing that R represented F. □

(b) Let's take another pair of examples. We first need to recall definitions from §26.6:

(i) The *covariant powerset functor* $P\colon \mathsf{Set} \to \mathsf{Set}$ maps a set X to its powerset $\mathcal{P}X$ and maps a set-function $f\colon X \to Y$ to the function which sends $U \in \mathcal{P}X$ to its image $f[U] \in \mathcal{P}Y$.

(ii) The *contravariant powerset functor* $\overline{P}\colon \mathsf{Set} \to \mathsf{Set}$ again maps a set to its powerset, and maps a set-function $f\colon Y \to X$ to the function which sends $U \in \mathcal{P}X$ to its inverse image $f^{-1}[U] \in \mathcal{P}Y$.

Theorem 196. *The contravariant powerset functor $\overline{P}$ is represented by the set $2 = \{0, 1\}$; but the covariant powerset functor P is not representable.*

Proof. As yet, we don't have any general principles about representables and non-representables which we can invoke to prove theorems such as this. So again we will just need to plough through by applying definitions and seeing what we get.

If the contravariant functor $\overline{P}$ is to be representable, then there must be a representing set R and a natural isomorphism ψ with components such that, for all set functions $f\colon Y \to X$, the following diagram always commutes:

$$\begin{array}{ccc} \overline{P}X & \xrightarrow{\ \overline{P}f\ } & \overline{P}Y \\ \Big\downarrow{\scriptstyle \psi_X} & & \Big\downarrow{\scriptstyle \psi_Y} \\ \mathsf{Set}(X, R) & \xrightarrow{\ \mathsf{Set}(f, R)\ } & \mathsf{Set}(Y, R) \end{array}$$

Now $\mathsf{Set}(X, R)$ is the set of set-functions from X to R, whose cardinality is $|R|^{|X|}$; and the cardinality of $\overline{P}X$, i.e. $\mathcal{P}$, is $2^{|X|}$. So that forces R to be a two-membered set: let's pick the set $2 = \{0, 1\}$.

$\mathsf{Set}(X, 2)$ is then the set of characteristic functions for subsets of X, i.e. the set of functions $c_U\colon X \to \{0, 1\}$ where $c_U(x) = 1$ iff $x \in U \subseteq X$. So the obvious next move is to take $\psi_X\colon \overline{P}X \to \mathsf{Set}(X, 2)$ to be the isomorphism that sends a set $U \in \mathcal{P}X$ to its characteristic function c_U.

With this choice, the diagram always commutes. Chase the element $U \in \overline{P}X$ around. The route via the north-east node takes us from $U \subseteq X$ to $f^{-1}[U] \subseteq Y$ to its characteristic function. i.e. the function that maps $y \in Y$ to 1 iff $f(y) \in U$. Meanwhile, the route via the south-west node takes us first from $U \subseteq X$ to c_U, and then we apply $\mathsf{Set}(f, 2)$, which maps $c_U \colon X \to 2$ to $c_U \circ f \colon Y \to 2$, which again is the function that maps $y \in Y$ to 1 iff $f(y) \in U$. Which establishes the first half of the theorem.

For the second half of the theorem, note that if we try to run a similar argument for the covariant functor P, we'd need to find a representing set R' such that PX and $\mathsf{Set}(R', X)$ are always in bijective correspondence. But $\mathsf{Set}(R', X)$ is the set of set-functions from R' to X, whose cardinality is $|X|^{|R'|}$, while the cardinality of PX is $2^{|X|}$. And there is no choice of R' which will make these equal for varying X. □

(c) Let me add an interesting third example. Take a category C and pick two C-objects X and Y. Then we can define a contravariant functor from C to Set which we will notate $\mathsf{C}(-, X) \times \mathsf{C}(-, Y)$ and which works as follows:

$$\mathsf{C}(-, X) \times \mathsf{C}(-, Y)\colon \quad \begin{array}{rcl} S & \longmapsto & \mathsf{C}(S, X) \times \mathsf{C}(S, Y) \\ f \colon S \to S' & \longmapsto & \mathsf{C}(f, X) \times \mathsf{C}(f, Y) \end{array}$$

So (1) Our functor sends the object S to the Cartesian product of the hom-sets $\mathsf{C}(S, X)$ and $\mathsf{C}(S, Y)$.

Next, recall that $\mathsf{C}(f, X)$ is the function $(- \circ f)_X$ which sends an arrow $j \colon S' \to X$ to $j \circ f \colon S \to X$. Likewise $\mathsf{C}(f, Y)$ is the function $(- \circ f)_Y$ which sends an arrow $k \colon S' \to Y$ to $k \circ f \colon S \to Y$. Then (2) $\mathsf{C}(f, X) \times \mathsf{C}(f, Y)$ is the function which acts component-wise on a pair in $\mathsf{C}(S, X) \times \mathsf{C}(S, Y)$, by applying $(- \circ f)_X$ to the first component and $(- \circ f)_Y$ to the second component.

You might like to check that (1) and (2) do define a functor!

Now let's ask: is our functor representable? Is there a C-object O such that there is a natural isomorphism $\psi \colon \mathsf{C}(-, O) \xRightarrow{\sim} \mathsf{C}(-, X) \times \mathsf{C}(-, Y)$? In other words, is there a suite of bijections $\psi_S \colon \mathsf{C}(S, O) \xrightarrow{\sim} \mathsf{C}(S, X) \times \mathsf{C}(S, Y)$, one for each S, which makes the relevant naturality squares commute?

Suppose that there is. So for any $f \colon S \to O$ there is in particular a commuting naturality square like this (noting the direction of arrows in a naturality square for contravariant functors, and remembering what $\mathsf{C}(f, O)$ and $\mathsf{C}(f, X) \times \mathsf{C}(f, Y)$ are):

$$\begin{array}{ccc} \mathsf{C}(O, O) & \xrightarrow{\quad (- \circ f)_O \quad} & \mathsf{C}(S, O) \\ \Big\downarrow{\scriptstyle \psi_O} & & \Big\downarrow{\scriptstyle \psi_S} \\ \mathsf{C}(O, X) \times \mathsf{C}(O, Y) & \xrightarrow{\ (- \circ f)_X \times (- \circ f)_Y\ } & \mathsf{C}(S, X) \times \mathsf{C}(S, Y) \end{array}$$

Now take any pair of arrows $j \colon S \to X$ and $k \colon S \to Y$. By hypothesis, since ψ_S is an isomorphism, there is some unique $f \colon S \to O$ such that ψ_S sends f to that pair. And let's chase the identity arrow 1_O round the square when f is so defined. On the north-east route, by definition,

$$1_O \longmapsto f \longmapsto \langle j, k \rangle$$

And suppose ψ_O sends 1_O to the pair of arrows $\pi_1 \colon O \to X$ and $\pi_2 \colon O \to Y$. Then taking the south-west route, we have

$$1_O \longmapsto \langle \pi_1, \pi_2 \rangle \longmapsto \langle \pi_1 \circ f, \pi_2 \circ f \rangle.$$

Since our diagram commutes, it follows that $j = \pi_1 \circ f$ and $k = \pi_2 \circ f$. In other words, our arbitrarily chosen arrows $j \colon S \to X$ and $k \colon S \to Y$ factor through the arrows $\pi_1 \colon O \to X$ and $\pi_2 \colon O \to Y$ respectively via a uniquely fixed arrow f. Hence (O, π_1, π_2) form a categorial product of X with Y. Which is rather neat!

To summarize, we have:

Theorem 197. *Suppose that the functor* $\mathsf{C}(-, X) \times \mathsf{C}(-, Y)$ *is represented by* (O, ψ)*. Then the representing object* O *is the object of a product of* X *and* Y *in* C*, with* $\psi_O(1_O)$ *giving the pair of projection arrows for the product.* □

39.4 Universal elements

(a) To repeat, we say that a pair (A, ψ) is a representation of the covariant functor $F \colon \mathsf{C} \to \mathsf{Set}$ if and only if there is a natural isomorphism $\psi \colon F \overset{\sim}{\Longrightarrow} \mathsf{C}(A, -)$.

But in talking about such a natural isomorphism we are of course back in Yoneda territory. And in proving the Core Yoneda Lemma (Theorem 185) we showed that every such natural isomorphism – now taken in the direction from $\mathsf{C}(A, -)$ to F – is of the form α^e where e is a unique element of FA and for any C-object X, the component $\alpha^e_X \colon \mathsf{C}(A, X) \to FX$ is the bijection that maps any C-arrow $f \colon A \to X$ to $Ff(e)$.[3]

Obviously, it doesn't make any difference whether we talk of a functor F as being represented by (1) A together with an isomorphism $\psi \colon F \overset{\sim}{\Longrightarrow} \mathsf{C}(A, -)$, or as being represented by (2) A together with an isomorphism $\alpha \colon \mathsf{C}(A, -) \overset{\sim}{\Longrightarrow} F$. And the same data could then equally well be presented by giving (3) A and the element e from FA such that $\alpha = \alpha^e$.

So we might have expected to find the likes of (A, e) also being described as a representation of F. But that isn't the conventional jargon. Instead, we find this:

Definition 140. A *universal element* of the functor $F \colon \mathsf{C} \to \mathsf{Set}$ is a pair (A, e), where A is a C-object and $e \in FA$, and where for each C-object X and $x \in FX$, there is a unique map $f \colon A \to X$ such that $Ff(e) = x$. △

Our discussion so far therefore gives us

Theorem 198. *A functor* $F \colon \mathsf{C} \to \mathsf{Set}$ *is representable by the object* A *iff it has a universal element* (A, e). □

[3] A reminder: our usual convention is for lower-case letters to denote arrows in general or functions in particular: but here we are using lower-case 'e' for a member of the set FA, which won't in general be a (set-)function.

The story for contravariant functors will be the same, except that the map f will go the other way about, $f\colon X \to A$. Here's an application, picking up from the discussion at the end of the last section, and neatly re-packaging the same result:

Theorem 199. *(O, π_1, π_2) is a product of X with Y in C iff $(O, \langle \pi_1, \pi_2 \rangle)$ is a universal element of the contravariant functor $F = \mathsf{C}(-, X) \times \mathsf{C}(-, Y)$.*

Proof. A universal element of F is an object O and an element e of FO, i.e. an element of $\mathsf{C}(O, X) \times \mathsf{C}(O, Y)$, i.e. a pair of arrows $e = \langle \pi_1 \colon O \to X, \pi_2 \colon O \to Y \rangle$.

And (matching the lettering used in the previous section) it is required that for any C-object S, and any pair of arrows $\langle j\colon S \to X, k\colon S \to Y \rangle$, there is a unique map $f\colon S \to O$ such that $Ff(e) = \langle j, k \rangle$.

But $Ff = \mathsf{C}(f, X) \times \mathsf{C}(f, Y)$ is the function that acts component-wise on a pair $e = \langle \pi_1, \pi_2 \rangle$, by applying $(- \circ f)_X$ to π_1 and $(- \circ f)_Y$ to π_2.

So it is required that for any $\langle j, k \rangle$, there is a unique f such that $\pi_1 \circ f = j$, $\pi_2 \circ f = k$. Which is exactly the condition for (O, π_1, π_2) being a product of X with Y. □

(b) Why '*universal*' element? Because the definition invokes a universal mapping property: (A, e) is a universal element iff for every ... there is a unique map such that ....

Now, recall the case of products which we defined by a universal mapping property; in §11.3 we then showed products to be terminal objects in a category of wedges. Dually, of course, coproducts are also defined by a universal mapping property; in this case they are initial objects in a category of corners. Similarly in other cases, we showed that constructions defined by universal mapping properties could be identified (up to isomorphism) as the initial or terminal objects in some appropriate categories. We can play the same game again. So first let's define a suitable category:

Definition 141. $\mathsf{Elts}_{\mathsf{C}}(F)$, the *category of elements of the functor* $F\colon \mathsf{C} \to \mathsf{Set}$, has the following data:

(1) Objects are elements of the functor, i.e. pairs (A, e), where A is a C-object and $e \in FA$.
(2) An arrow from (A, e) to (A', e') is a C-arrow $f\colon A \to A'$ such that $Ff(e) = e'$.
(3) The identity arrow on (A, e) is 1_A.
(4) Composition of arrows is induced by composition of C-arrows. △

It is easily checked that this *is* a category.[4] And note that a category of pairs of objects like (A, e) invites a more straightforward definition than a category of pairs of an object and a natural transformation like (A, ψ) – what would the arrows of the second sort of category be? Which gives a reason for working with

[4]Alternative symbolism for the category includes variations on '$\int_{\mathsf{C}} F$'.

'elements' rather than 'representations' even if they do ultimately encode the same data.[5]

(c) Here is another way of thinking of this category of elements. Let 1 be some singleton in Set; and recycling notation in the usual kind of way let 1 be the trivial functor from the one-object category **1** to Set that sends the sole object of **1** to 1. Then what is the comma category $(1 \downarrow F)$ where as before $F\colon \mathsf{C} \to \mathsf{Set}$? Applying the account of such comma categories given in §29.2, the objects of this category are pairs $(A, \vec{e})$ where A is an object in C and $\vec{e}\colon 1 \to FA$ is an arrow in Set. And the arrows of the category from $(A, \vec{e})$ to $(A', \vec{e'})$ is a C-arrow $f\colon A \to A'$ such that $\vec{e'} = Ff \circ \vec{e}$.

But *that* is just the definition of $\mathsf{Elts}_\mathsf{C}(F)$ except that we have traded in the requirement that e is *member* of FA for the requirement that $\vec{e}$ is an *arrow* $1 \to FA$. But as we well know by now, members of a set are in bijective correspondence with such arrows from a fixed singleton, and from a categorial perspective we can treat members as such arrows. Therefore:

Theorem 200. *For a given functor* $F\colon \mathsf{C} \to \mathsf{Set}$, *the category* $\mathsf{Elts}_\mathsf{C}(F)$ *is (isomorphic to) the comma category* $(1 \downarrow F)$ *where* 1 *is terminal in* Set.

(d) Having defined a category $\mathsf{Elts}_\mathsf{C}(F)$ for elements of $F\colon \mathsf{C} \to \mathsf{Set}$ to live in, we can now ask: how do we distinguish universal elements from other elements categorially?

The answer is immediate from Defn. 140, which in our new terminology immediately implies:

[5] But why '*elements*'? After all, functors don't in any straightforward sense have elements. For what it is worth, this is the best I can do to make the terminology look sensible.

Suppose, as a first step, that we are given a category C whose objects *are* sets (perhaps with some additional structure on them) and whose arrows are functions between sets. Then we might be interested in a derived category which digs inside the sets which are C's objects, and looks at their elements.

Now perhaps we don't want the derived category to forget about which sets in C have which elements as members. Then a natural way to go would be to say that the objects of the derived category are all the pairs (A, e) for some C-object (i.e. set) A and $e \in A$. And then given elements $e \in A$, $e' \in A$, whenever there is a C-arrow $f\colon A \to A'$ such that $f(e) = e'$, we'll say that f is also an arrow from (A, e) to (A', e') in our new category. In a sense, *this* kind of derived 'category of elements' unpacks what's going on inside the original category C.

However, in the general case, C's objects need not be sets, so need not have elements in the ordinary sense. But a functor $F\colon \mathsf{C} \to \mathsf{Set}$ gives us an image or diagram of C inside Set, and of course the objects in the resulting diagram of C *do* have elements. So we can consider the category of elements (ordinary sense) of F's-diagram-of-C, which – following the suggested template in – has as objects all the pairs (FA, e) for A a C-object and $e \in FA$. And then given elements $e \in FA$, $e' \in FA'$, whenever there is a Set-arrow $Ff\colon FA \to FA'$ such that $Ff(e) = e'$, we'll say that Ff is also an arrow from (FA, e) to (FA, e') in our new category.

Now, we can streamline that. Instead of taking the objects to be pairs (FA, e) take them simply to be pairs (A, e) (but where, still, $e \in FA$). And instead of talking of the arrow $Ff\colon FA \to FA'$ we can instead talk more simply of $f\colon A \to A'$ (but where, still, $Ff(e) = e'$). And with that streamlining – lo and behold! – we are back with the category $\mathsf{Elts}_\mathsf{C}(F)$, which is isomorphic to the category of elements of F's-diagram-of-C, and which – as convention has it – we'll call the category of elements of F, for short.

Theorem 201. *An object $I = (A, e)$ in $\mathsf{Elts_C}(F)$ is a universal element for F iff, for every object E in $\mathsf{Elts_C}(F)$ there is exactly one morphism $f\colon I \to E$, so I is initial in $\mathsf{Elts_C}(F)$.*

But initial objects are unique up to unique isomorphism. Which, recalling what isomorphisms in $\mathsf{Elts_C}(F)$ are, implies

Theorem 202. *If (A, e) and (A', e') are universal elements for $F\colon \mathsf{C} \to \mathsf{Set}$, then there is a unique C-isomorphism $f\colon A \to A'$ such that $Ff(e) = e'$.*

39.5 Limits and exponentials as universal elements

Let's finish the chapter by making some connections between our categorial gadgets old and new.

(a) We will use the notation $Cone_{\mathsf{C}}(C, D)$ for the set of cones over some diagram D with vertex C – and we will assume that we are in a category C which is small enough for any $Cone_{\mathsf{C}}(C, D)$ to be treated as a set living in Set.

Dropping the subscript, we can now define a contravariant functor $Cone(-, D)\colon \mathsf{C} \to \mathsf{Set}$ as follows.

(i) $Cone(-, D)$ sends an object C to $Cone(C, D)$ (which could be the empty set),

(ii) $Cone(-, D)$ sends an arrow $f\colon C' \to C$ to $Cone(f, D)\colon Cone(C, D) \to Cone(C', D)$, which takes a cone (C, c_j) and sends it to the cone $(C', c_j \circ f)$.

It is easily checked that this really is a functor.

We now apply the definition of universal elements, tweaked for the contravariant case (so universal elements are terminal in the relevant category of elements). Then a universal element of the functor $Cone(-, D)$ is a pair $(L, (L, \lambda_j))$, where L is in C and (L, λ_j) is in $Cone(L, D)$, the set of cones over D with vertex L. And moreover, we require that for each C-object C and each cone (C, c_j), there is a unique map $f\colon C \to L$ such that $Cone(f)(L, \lambda_j) = (C, c_j)$, which requires $\lambda_j \circ f = c_j$ for each relevant j. *But that's just to say that (L, λ_j) is a limit cone*! Hence:

Theorem 203. *In small enough categories, there is a limit over the diagram D if and only if the functor $Cone(-, D)$ has a universal element (and so is represented by the object in the universal element).*[6]

We can also note that

Theorem 204. *If C has limits of shape J, then for each $D\colon \mathsf{J} \to \mathsf{C}$,*

$$Cone(C, Lim D) \cong Lim(\mathsf{C}(C, -) \circ D).$$

[6] An aside for the record. In the light of Theorem 158, given a diagram $D\colon \mathsf{J} \to \mathsf{C}$, we have another way of thinking of $Cone_{\mathsf{C}}(C, D)$. It is, in our notation, $Nat(\Delta_C, D)$ – or, as some would prefer to refer to it, $[\mathsf{J}, \mathsf{C}](\Delta_C, D)$. So the functor $Cone(-, D)$ can be – and sometimes is – written as some variant of $Nat(\Delta-, D)$.

Proof. Theorem 84 tells us that cones over D with vertex C correspond one-to-one with C-arrows from C to the vertex of the limit over D. In other words, $Cone(C, D) \cong \mathsf{C}(C, Lim D)$.

But $\mathsf{C}(C, -)$ is a representable functor so we can apply Theorem 191, hence $\mathsf{C}(C, Lim D) \cong Lim(\mathsf{C}(C, -) \circ D)$.

Putting those two bijections together gives us our result. (And although we won't prove it, this is as you would predict, all natural in C and D.) □

(b) Consider next the contravariant functor $\mathsf{C}(- \times B, C)$ which we met in §32.4 Ex. (4). This sends an object A in C to the hom-set of arrows from $A \times B$ to C. And it sends an arrow $f \colon A' \to A$ to the map $- \circ f \times 1_B$ (i.e. to the map that takes an arrow $j \colon A \times B \to C$ and yields the arrow $j \circ f \times 1_B \colon A' \times B \to C$).

Now apply the definition of universal element for the contravariant case. Then a universal element of $\mathsf{C}(- \times B, C)$ is a pair (E, ev), with E in C and ev in $\mathsf{C}(E \times B, C)$, such that for every A and every $g \in \mathsf{C}(A \times B, C)$, there is a unique $\widetilde{g} \colon A \to E$ such that $\mathsf{C}(- \times B, C)(\widetilde{g})(ev) = g$, i.e. $ev \circ \widetilde{g} \times 1_B = g$.

But a pair (E, ev) with those properties is exactly the exponential (C^B, ev). Hence

Theorem 205. *The exponential (C^B, ev), when it exists in C, is a universal element of $\mathsf{C}(- \times B, C)$.*

Since exponentials are therefore also terminal objects in an associated category of elements, they too must be unique up to a unique appropriate isomorphism, giving us another proof of Theorem 72.

(c) By the end of §19.6 we had noted that limits and colimits are defined by a universal mapping property and that exponentials are also defined by a universal mapping property. But exponentials aren't limits or colimits. Which raised the question: is there an abstract categorial notion which subsumes both limits/colimits *and* exponentials, and which perhaps might be offered as formally regimenting the intuitive idea of being defined by a universal mapping property?

Well, we do now have a candidate answer in terms of the idea of being initial or terminal in a suitable category of elements (or equivalently being initial or terminal in a suitable comma category).

But does this abstract categorial notion actually illuminate the intuitive idea of a universal mapping property in the way that (say) the categorial notion of product illuminates the intuitive notion of a product? That seems debatable to me.

40 Galois connections

We will have more to say about functors and limits (and about representables too and about how everything interrelates) after we have introduced our next important Big Idea from core category theory – namely, the idea of pairs of *adjoint functors* and the *adjunctions* they form.

Now, the multi-faceted story about adjoints in general can initially seem puzzlingly complex, and it is easy to get lost in the details. So the plan here is to start by looking first at a very restricted class of cases, namely adjunctions between two posets-as-categories. Or rather, to keep things as down-to-earth as possible, we discuss them in this chapter in their elementary, pre-categorial, guise as so-called Galois connections.

40.1 Posets: some probably unnecessary reminders

For the record, and to fix local notation:

Definition 142. A poset consists of a set C equipped with a partial order $\preccurlyeq$ – i.e., for all $x, y, z \in C$, (i) $x \preccurlyeq x$, (ii) if $x \preccurlyeq y$ and $y \preccurlyeq z$ then $x \preccurlyeq z$, (iii) if $x \preccurlyeq y$ and $y \preccurlyeq x$ then $x = y$. (We will, as appropriate, recruit '$\sqsubseteq$', '$\leqslant$', '$\subseteq$' as other symbols for partial orders.)

If $(C, \preccurlyeq)$ is a poset and $X \subseteq C$, then m is a *maximum* of X (with respect to the inherited order $\preccurlyeq$) iff $m \in X$ and $(\forall x \in X)\, x \preccurlyeq m$.

Suppose $C = (C, \preccurlyeq)$ and $D = (D, \sqsubseteq)$ are two posets, and the map $F \colon C \to D$ is a function between the sets C and D. Then

(1) F is *monotone* iff, for all $x, y \in C$, if $x \preccurlyeq y$ then $Fx \sqsubseteq Fy$;
(2) F is an *order-embedding* iff, for all $x, y \in C$, $x \preccurlyeq y$ just in case $Fx \sqsubseteq Fy$;
(3) F is an *order-isomorphism* iff F is a surjective order-embedding.

Posets are deemed isomorphic when there is an order-isomorphism between them. △

Three quick comments about this:

(i) Talk of sets here is conventional. We could use plurals or non-committal, eliminable, talk of collections instead (see again §3.1) – as we would need to do if we wanted to cover cases where the ordered objects in question might be too many to form a set on standard stories. But I don't want to labour this sort of point yet again; so to avoid distractions, I'll here stick to the conventional set idiom (and cheerfully not worry too much about issues of 'size').

(ii) There is of course a related notion of a strict poset defined in terms of a strict partial order $\prec$, where $x \prec y$ iff $x \preccurlyeq y \wedge x \neq y$ for some partial order $\preccurlyeq$. It is a matter of convenience whether we concentrate on the one flavour of poset or the other, and you will already be familiar with a variety of examples of 'naturally occurring' posets of both flavours.

(iii) 'F' here denotes a function. But I'm capitalizing rather than using 'f' so that things look the same when we later upgrade the function to a functor!

And now a very elementary composite theorem:

Theorem 206. *(1) Maxima are unique when they exist.*

(2) Order-embeddings are injective.

(3) Order-isomorphisms are bijective, and have unique inverses which are also order-isomorphisms.

(4) Monotone maps compose to give monotone maps, and composition is associative. Likewise for order-embeddings and order-isomorphisms.

(5) If $F[C]$ is the image of C under F, an order-embedding $F\colon (C, \preccurlyeq) \to (D, \sqsubseteq)$ is an order-isomorphism from $(C, \preccurlyeq)$ to $(F[C], \sqsubseteq)$.

(6) Any partially ordered collection is order-isomorphic to an inclusion poset, i.e. a collection of sets ordered by inclusion.

Proof. For (1) we note that if $m, m' \in X$ are both maxima, $m' \preccurlyeq m$ and similarly $m \preccurlyeq m'$ and hence $m = m'$.

For (2) we suppose $Fx = Fy$ and hence both $Fx \sqsubseteq Fy$ and $Fy \sqsubseteq Fx$, and then note that if F is an embedding, $x \preccurlyeq y$ and $y \preccurlyeq x$, and hence $x = y$.

(3) to (5) are immediate. For (6) take $(C, \preccurlyeq)$, and for each $y \in C$, form the set π_y containing it and its $\preccurlyeq$-predecessors, so $\pi_y = \{x \in C \mid x \preccurlyeq y\}$. Let Π be the set of π_y for $y \in C$. Define $F\colon (C, \preccurlyeq) \to (\Pi, \subseteq)$ by putting $Fx = \pi_x$. Then F is easily seen to be a bijection, and also $x \preccurlyeq y$ iff $\pi_x \subseteq \pi_y$. So F is an order-isomorphism.[1] □

40.2 A first example of a Galois connection

We'll now rather informally describe a nice logical example of a Galois connection: and let's choose notation with an eye to smoothing the transitions to later generalizations.

Let C be the set whose members are the various *sets of sentences* of some suitable formal language L (the details of L won't matter too much); and let $\preccurlyeq$ simply be set-inclusion. So we can think of C as collecting together *theories*

[1] Category theorist's joke: prove (6) using Yoneda.

couched in the language L, with these theories then partially ordered from less specific (saying less) to more specific (saying more).

Now let D be the set whose members are all the various sets of L-structures, so each $d \in D$ is a set of potential models for theories couched in L;[2] and this time we will take $\sqsubseteq$ to be the *converse* of inclusion. So the sets of potential models are also partially ordered from less specific (more alternatives) to more specific (a narrower range).

There are then two very natural maps between the resulting 'syntactic' and 'semantic' posets:

i. $F\colon (C, \preccurlyeq) \to (D, \sqsubseteq)$ sends a theory c from C to d among D, where d is the set of models of c (i.e. d is the set containing each model on which all the sentences in c are true).

ii. $G\colon (D, \sqsubseteq) \to (C, \preccurlyeq)$ sends a set of models d to the set c containing the sentences that are true on every model in d.

Put it this way: F is the 'find the models' function. It takes a bunch of L-sentences and returns all its models, the set of L-structures where the sentences in the bunch are all true. In the other direction, G is the equally natural 'find the agreed truths' function. It takes a bunch of L-structures and returns the set of L-sentences that are true across all of those structures.

In general F and G will not be inverse to each other. But the mapping functions do interrelate in the following nice ways:

(1) F and G are monotone.

And for all $c \in C$, $d \in D$,

(2) $c \preccurlyeq GFc$ and $FGd \sqsubseteq d$,

(3) $Fc \sqsubseteq d$ iff $c \preccurlyeq Gd$.

Why so? For (1) we note that if $c \preccurlyeq c'$, i.e. if the theory c' is more informative than c, then c' will be true of a narrower range of possible models than c, so Fc' is included in Fc, and hence $Fc \sqsubseteq Fc'$, so F is monotone. Likewise for G.

For the first half of (2) we note that if we start with a bunch of sentences c, look at the models where they are all true together, and then look at the sentences true in all those models together, we'll get back original sentences in c plus all their consequences (where consequence is defined in terms of preservation of truth with respect to the relevant structures).

For the other half of (2) we note that if we start from a collection of models d, find the sentences true in all of them, and then look at the models for those sentences, we must get back at least the models we started with, maybe more. (Remember, $\sqsubseteq$ is the converse of inclusion!)

For (3) we note that if the models where all the sentences of c are true include all those in d then the theory c must be included in the set of sentences true in all the models in d, and vice versa.

[2] To sidestep issues of size, take it that L-structures all live in some big-enough set.

We might also note that it follows from (1) to (3) that

(4) $FGFc = Fc$ and $GFGd = Gd$.

That's because (2) tells us that (i) $FGFc \sqsubseteq Fc$. While (3) tells us $c \preccurlyeq GFc$ iff $Fc \sqsubseteq Fc$, hence we have $c \preccurlyeq GFc$, and hence (ii) $Fc \sqsubseteq FGFc$ since F is monotone. Together, (i) and (ii) give us half of (4), and we get the other half similarly. And again, this is as it should be. For the first half of (4) we note that the models of a set of sentences c together with their consequences are just the models of the original set c. Similarly for the other half.

So in summary: we have here a pair of posets $(C, \preccurlyeq)$, $(D, \sqsubseteq)$ and a pair of functions $F\colon C \to D$ and $G\colon D \to C$ for which conditions (1) to (4) hold. And a pair of functions between posets for which these conditions hold is a *Galois connection*. In a famous Dialectica paper 'Adjointness in foundations' (1969a), F. William Lawvere writes of "the familiar Galois connection between sets of axioms and classes of models, for a fixed [signature]". A set of axioms in the wide sense that Lawvere is using is just any old set of sentences from the right signature. So we've explained what Lawvere was referring to.

40.3 Galois connections defined

As we'll see in the next section, the conditions (1) to (4) on our functions between posets are not independent. The first two together imply the third and fourth, and the third by itself implies the rest. Simply because it is prettier, then, we plump in this section for an official definition just in terms of the third condition (which we re-label):

Definition 143. Suppose that $(C, \preccurlyeq)$ and $(D, \sqsubseteq)$ are posets, and let $F\colon C \to D$ and $G\colon D \to C$ be a pair of functions such that

$$\text{(Gal)} \qquad Fc \sqsubseteq d \text{ iff } c \preccurlyeq Gd \text{ (for all } c \in C,\ d \in D).$$

Then F and G form a *Galois connection* between C and D. When this holds, we write $F \dashv G$, and F is said to be the *left adjoint* of G, and G the *right adjoint* of F.[3] △

Note: F counts as the *left* adjoint not because it here happens to be written on the left of the (symmetric!) biconditional in (Gal), but because it is on the left of one of the order signs. Likewise for the *right* adjoint G.

The first discussion of a version of such a connection – and hence the name – is to be found in Évariste Galois's work in what has come to be known as Galois theory, a topic well beyond our purview here. And there are plenty of other serious mathematical examples (e.g. from number theory, abstract algebra and

[3] Talk of adjoints here seems to have been originally borrowed from the old theory of Hermitian operators, where in e.g. a Hilbert space with inner product $\langle\cdot,\cdot\rangle$ the operators A and A^* are said to be adjoint when we have, generally, $\langle Ax, y\rangle = \langle x, A^*y\rangle$. The formal analogy is evident.

topology) of two posets with a Galois connection between them. But we don't want to get bogged down in unnecessary mathematics at this early stage; so for the moment let's just give some simple cases, to add to our informally described motivating example in the last section:

(G1) Suppose F is an order-isomorphism between $(C, \preccurlyeq)$ and $(D, \sqsubseteq)$: then F^{-1} is an order-isomorphism in the reverse direction. Take $c \in C, d \in D$: then $Fc \sqsubseteq d$ iff $F^{-1}Fc \preccurlyeq F^{-1}d$ iff $c \preccurlyeq F^{-1}d$. Hence $F \dashv F^{-1}$.

(G2) Take the posets $(\mathbb{N}, \leqslant)$ and $(\mathbb{Q}^+, \leqslant)$ comprising the naturals and the non-negative rationals in their standard orders. Let $I\colon \mathbb{N} \to \mathbb{Q}^+$ be the injection function which maps a natural number to the corresponding rational integer, and let $F\colon \mathbb{Q}^+ \to \mathbb{N}$ be the 'floor' function which maps a non-negative rational to the greatest natural number less than or equal to it. Then $I \dashv F$ is a Galois connection, with I the left adjoint. Likewise if $C\colon \mathbb{Q}^+ \to \mathbb{N}$ is the 'ceiling' function which maps a rational to the smallest natural which is at least as big, then $C \dashv I$ is a Galois connection with I the right adjoint.[4]

(G3) Let $f\colon X \to Y$ be some function between two sets X and Y. It induces a function $F\colon \mathcal{P}X \to \mathcal{P}Y$ between their powersets which sends $A \subseteq X$ to $f[A]$, and another function $F^{-1}\colon \mathcal{P}Y \to \mathcal{P}X$ which sends $B \subseteq Y$ to its pre-image under f, $F^{-1}[B] = \{x \in X \mid f(x) \in B\}$. Then $F \dashv F^{-1}$ is a Galois connection between the inclusion posets $(\mathcal{P}X, \subseteq)$ and $(\mathcal{P}Y, \subseteq)$.

(G4) Take any poset $(C, \preccurlyeq)$, and let 1 be a one object poset such as $(\{0\}, =)$. Let $F\colon (C, \preccurlyeq) \to 1$ be the only possible function, the one which sends everything to 0. Then F has a right adjoint $G\colon 1 \to (C, \preccurlyeq)$ if and only if, for any $x \in C$, $Fx = 0$ iff $x \preccurlyeq G0$. So F has a right adjoint just in case C has a maximum, and then G sends 1's only element to it. Dually, F has a left adjoint $E\colon 1 \to (C, \preccurlyeq)$ just in case $(C, \preccurlyeq)$ has a minimum, and then the left adjoint E sends 1's only element to *that*.

(G5) Let's have a (very!) simple result from topology. Recall, a topological space is standardly treated as a set X equipped with a topology, a set of 'open' sets $\mathcal{O}$ – where that is a family of subsets of X including $\emptyset$ and X and closed under unions and finite intersections. Given a space $(X, \mathcal{O})$, we can now consider two related posets, $(\mathcal{P}X, \subseteq)$ and $(\mathcal{O}, \subseteq)$.

By definition $\mathcal{O} \subseteq \mathcal{P}X$, and the inclusion function $F\colon \mathcal{O} \to \mathcal{P}X$ which sends $A \in \mathcal{O}$ to the same set as a member of $\mathcal{P}X$ is trivially monotone.

And now consider $G\colon \mathcal{P}X \to \mathcal{O}$ which sends a set $B \subseteq X$ to its (topological) interior, i.e. to the largest member of $\mathcal{O}$ which is a subset of B – this is the union of all the $O \in \mathcal{O}$ such that $O \subseteq B$, a union which will itself be in $\mathcal{O}$.

Then it is immediate that $F \dashv G$ – because $FA \subseteq B$ if and only if $A \subseteq GB$.

[4] So the adjoint functions in these cases map an object in one poset to its 'best fit' in the other – on some understanding of 'best'. Adjoint functions are sometimes said, more generally, to provide a solution to an optimization problem.

We can, as you'd expect, do the same trick for closed sets. Let $(\mathcal{P}X, \subseteq)$ be as before, and let $(\mathcal{C}, \subseteq)$ be the poset of closed sets of X. This time, let $F' \colon \mathcal{C} \to \mathcal{P}X$ be the inclusion function and let $G' \colon \mathcal{P}X \to \mathcal{C}$ send a subset of X to its closure. Then $G' \dashv F'$.

(G6) For our next example we return to elementary logic. Choose a favourite logical proof-system S – it could be classical or intuitionistic, or indeed any other logic, so long as it has a normally-behaved conjunction and conditional connectives and a sensible deducibility relation. Let $\alpha \vdash \beta$ notate, as usual, that there is a formal S-proof from premiss α to conclusion β. Then let $|\alpha|$ be the equivalence class of wffs of the system interderivable with α. Take E to be set of all such equivalence classes, and put $|\alpha| \preccurlyeq |\beta|$ in E iff $\alpha \vdash \beta$. Then it is easily checked that $(E, \preccurlyeq)$ is a poset.

Now consider the following two functions between $(E, \preccurlyeq)$ and itself. Fix γ to be some S-wff. Then let F send the equivalence class $|\alpha|$ to the class $|(\gamma \wedge \alpha)|$, and let G send $|\alpha|$ to the class $|(\gamma \to \alpha)|$.

Given our normality assumption, $\gamma \wedge \alpha \vdash \beta$ if and only if $\alpha \vdash \gamma \to \beta$. Hence $|\gamma \wedge \alpha| \preccurlyeq |\beta|$ iff $|\alpha| \preccurlyeq |\gamma \to \beta|$. That is to say $F|\alpha| \preccurlyeq |\beta|$ iff $|\alpha| \preccurlyeq G|\beta|$. Hence we have a Galois connection $F \dashv G$ between $(E, \preccurlyeq)$ and itself and, in a slogan, 'Conjunction is left adjoint to conditionalization'.

(G7) Our last example for the moment is another example from elementary logic. Let S now be a first-order logic, and consider the set of S-wffs with at most the variables $\vec{x}$ free.

We will write $\varphi(\vec{x})$ for a formula in this class, $|\varphi(\vec{x})|$ for the class of formulae interderivable with $\varphi(\vec{x})$, and $E_{\vec{x}}$ for the set of such equivalence classes of formulae with at most $\vec{x}$ free. Using $\preccurlyeq$ as in the last example, $(E_{\vec{x}}, \preccurlyeq)$ is a poset for any choice of variables $\vec{x}$.

We now consider two maps between the posets $(E_{\vec{x}}, \preccurlyeq)$ and $(E_{\vec{x},y}, \preccurlyeq)$. In other words, we are going to be moving between (equivalence classes of) formulae with at most $\vec{x}$ free, and (equivalence classes of) formulae with at most $\vec{x}, y$ free – where y is a new variable not among the $\vec{x}$.

First, since every wff with at most the variables $\vec{x}$ free also has at most the variables $\vec{x}, y$ free, there is a trivial map $F \colon E_{\vec{x}} \to E_{\vec{x},y}$ that sends the class of formulas $|\varphi(\vec{x})|$ in $E_{\vec{x}}$ to the same class of formulas which is also in $E_{\vec{x},y}$.

Second, we define the companion map $G \colon E_{\vec{x},y} \to E_{\vec{x}}$ that sends $|\varphi(\vec{x}, y)|$ in $E_{\vec{x},y}$ to $|\forall y \varphi(\vec{x}, y)|$ in $E_{\vec{x}}$.

Then $F \dashv G$, i.e. we have another Galois connection, because

$$F(|\varphi(\vec{x})|) \preccurlyeq |\psi(\vec{x}, y)| \quad \text{iff} \quad |\varphi(\vec{x})| \preccurlyeq G(|\psi(\vec{x}, y)|).$$

For this simply reflects the familiar logical rule that

$$\varphi(\vec{x}) \vdash \psi(\vec{x}, y) \text{ iff } \varphi(\vec{x}) \vdash \forall y \psi(\vec{x}, y),$$

so long as y is not free in $\varphi(\vec{x})$. Hence universal quantification is right-adjoint to a certain trivial inclusion operation.

And we can exactly similarly show that existential quantification is left-adjoint to the same operation.

So, with these various examples on the table, let's list some morals!

(i) Our first example shows that Galois connections are at least as plentiful as order-isomorphisms: and such an isomorphism will have a right adjoint and left adjoint which are the same (i.e. both are the isomorphism's inverse).

(ii) The second and fourth cases show that posets that aren't order-isomorphic can still be Galois connected.

(iii) The third case shows that posets can have many Galois connections between them (as any $f\colon X \to Y$ generates a connection between the inclusion posets on the powersets of X and Y).

(iv) The fourth example gives a case where a function has both a left and a right adjoint that are different.

(v) The fourth and seventh cases give a couple of illustrations of how a significant construction (taking maxima, forming a universal quantification respectively) can be regarded as adjoint to some quite trivial operation.

(vi) The sixth example, like the third, shows that even when the relevant posets are isomorphic (in the sixth case because they are identical!), there can be a pair of functions which aren't isomorphisms but which also go to make up a Galois connection.

(vii) And the last two examples, like the motivating example in the previous section, illustrate why Galois connections might be of some interest to logicians.

40.4 Galois connections re-defined

Here again, in brisker form, is our initial definition of a Galois connection. Assume $(C, \preccurlyeq)$ and $(D, \sqsubseteq)$ are posets, and let $F\colon C \to D$ and $G\colon D \to C$ be a pair of functions between them. Then:

Definition 143. F and G form a Galois connection between C and D, in symbols $F \dashv G$, iff (Gal) $Fc \sqsubseteq d$ iff $c \preccurlyeq Gd$ (for all $c \in C$, $d \in D$). △

And now here is an alternative definition, given the same background assumption:

Definition 144. F and G form a Galois connection between C and D, in symbols $F \dashv G$, iff

(1) F and G are both monotone,

(2) $c \preccurlyeq GFc$ and $FGd \sqsubseteq d$ (for all $c \in C$, $d \in D$). △

We need, of course, to confirm that these claimed alternative definitions do come to the same. But that's easy. Given the same background assumptions, we have:

Theorem 207. *(Gal) is true iff (1) F and G are both monotone, and (2) $c \preccurlyeq GFc$ and $FGd \sqsubseteq d$ (for all $c \in C$, $d \in D$).*

Proof. (*If*) Assume conditions (1) and (2) both hold. And suppose $Fc \sqsubseteq d$. Since by (1) G is monotone, $GFc \preccurlyeq Gd$. But by (2) $c \preccurlyeq GFc$. Hence by transitivity $c \preccurlyeq Gd$. That establishes one half of (Gal), and the proof of the other half is dual.

(*Only if*) Suppose (Gal) is true. Then in particular, $Fc \sqsubseteq Fc$ iff $c \preccurlyeq GFc$. Since $\sqsubseteq$ is reflexive, $c \preccurlyeq GFc$. Similarly for the other half of (2).

Now, suppose also that $c \preccurlyeq c'$. Then since we've just shown $c' \preccurlyeq GFc'$, we have $c \preccurlyeq GFc'$. But by (Gal) we have $Fc \sqsubseteq Fc'$ iff $c \preccurlyeq GFc'$. Whence, $Fc \sqsubseteq Fc'$ and F is monotone. Similarly for the other half of (1). □

One comment about this. Note that we could replace clause (2) in our alternative definition with the equivalent clause

(2′) (i) If $c \preccurlyeq c'$, then both $c \preccurlyeq c' \preccurlyeq GFc'$ and $c \preccurlyeq GFc \preccurlyeq GFc'$; and
(ii) if $d \sqsubseteq d'$, then both $FGd \sqsubseteq d \sqsubseteq d'$ and $FGd \sqsubseteq FGd' \sqsubseteq d'$.

For (2′) implies (2); conversely (1) and (2) imply (2′). We mention this variant on our alternative definition of Galois connections for later use.

40.5 Some basic results about Galois connections

(a) We now have a pair of equivalent definitions of Galois connections, and a small range of elementary examples. In this section we start by proving a couple of theorems that show that such connections behave as you would hope, in two different respects.

First, we show that inside a Galois connection, a left adjoint uniquely fixes its right adjoint, and vice versa:

Theorem 208. *If we have Galois connections $F \dashv G$, $F \dashv G'$ between the posets $(C, \preccurlyeq)$ and $(D, \sqsubseteq)$, then $G = G'$. Likewise, if $F \dashv G$, $F' \dashv G$ are both Galois connections between the same posets, then $F = F'$.*

Proof. We prove the first part. $F \dashv G'$ implies, in particular, that for any $d \in D$, $FGd \sqsubseteq d$ iff $Gd \preccurlyeq G'd$.

But by the alternative definition, applied to the connection $F \dashv G$, we have $FGd \sqsubseteq d$. So we can infer that $Gd \preccurlyeq G'd$.

By symmetry, $G'd \preccurlyeq Gd$. But d was arbitrary, so $G = G'$. □

Careful! This theorem does not say that, for any $F\colon C \to D$ there must exist a unique corresponding $G\colon D \to C$ such that $F \dashv G$ (compare §40.3 Ex. (G4)). Nor does it say that when there is a Galois connection between two posets, it is unique (our toy examples have shown that that is false too). The claim is only that, if you are given a particular function between posets, then there is at most one candidate for being its right adjoint, and similarly at most one candidate for being its left adjoint.

Second, we confirm that Galois connections combine nicely in the no doubt predictable way:

Theorem 209. *Suppose there is a Galois connection $F \dashv G$ between the posets $(C, \preccurlyeq)$ and $(D, \sqsubseteq)$, and a connection $H \dashv K$ between the posets $(D, \sqsubseteq)$ and $(E, \subseteq)$. Then there is a Galois connection $HF \dashv GK$ between $(C, \preccurlyeq)$ and $(E, \subseteq)$.*

Proof. Take any $c \in C, e \in E$. Then, using the first connection, we have $Fc \sqsubseteq Ke$ iff $c \preccurlyeq GKe$. And by the second connection, we have $HFc \subseteq e$ iff $Fc \sqsubseteq Ke$.

Hence $HFc \subseteq e$ iff $c \preccurlyeq GKe$. Therefore $HF \dashv GK$. □

(b) Given that adjoint functions determine each other, we naturally seek an explicit definition of one in terms of the other. Here it is:

Theorem 210. *If $F \dashv G$ is a Galois connection between the posets $(C, \preccurlyeq)$ and $(D, \sqsubseteq)$, then*

(1) Gd = the $\preccurlyeq$-maximum of $\{c \in C \mid Fc \sqsubseteq d\}$,

(2) Fc = the $\sqsubseteq$-minimum of $\{d \in D \mid c \preccurlyeq Gd\}$.

Proof. We argue for (1), leaving the dual (2) to take care of itself. Fix on an arbitrary $d \in D$ and for brevity, put $\Sigma = \{c \in C \mid Fc \sqsubseteq d\}$.

The alternative definition tells us that for any d, $FGd \sqsubseteq d$. So $Gd \in \Sigma$.

Now suppose $c \in \Sigma$, i.e. $Fc \sqsubseteq d$. Then since G is monotone, $GFc \preccurlyeq Gd$. But again the alternative definition tells us that $c \preccurlyeq GFc$. Therefore $c \preccurlyeq Gd$.

That shows Gd is both a member of and an upper bound for Σ, i.e. is a maximum for Σ. □

(c) Recall the posets $(\mathbb{N}, \leqslant)$ and $(\mathbb{Q}^+, \leqslant)$, the injection map $I \colon \mathbb{N} \to \mathbb{Q}^+$, and the floor function $F \colon \mathbb{Q}^+ \to \mathbb{N}$ which maps a non-negative rational to the largest natural less than or equal to it. Then we remarked before that $I \dashv F$. Now we note that $F \dashv I$ is false. Indeed, there can be no connection of the form $F \dashv G$ between $(\mathbb{Q}^+, \leqslant)$ and $(\mathbb{N}, \leqslant)$. For $Fq = 1$ iff $1 \leqslant q < 2$, and hence $\{q \in \mathbb{Q}^+ \mid Fq \leqslant 1\}$ has no maximum, Therefore, by the previous theorem, there can be no right adjoint to F.

Generalizing, we should perhaps highlight the following:

Theorem 211. *Galois connections are not necessarily symmetric. That is to say, given $F \dashv G$ is a Galois connection between the posets C and D, it does not follow that $G \dashv F$ is a connection between D and C.*

40.6 Isomorphisms and closures

It is quite surprising how much structure we can get out of the simple condition (Gal). This section mentions a couple more results.

(a) We know that a pair of posets which have a Galois connection between them needn't be isomorphic overall. But this next theorem says that they will typically contain an interesting pair of isomorphic sub-posets.

Theorem 212. *If $F \dashv G$ is a Galois connection between the posets $(C, \preccurlyeq)$ and $(D, \sqsubseteq)$, then $(G[D], \preccurlyeq)$ and $(F[C], \sqsubseteq)$ are order-isomorphic.*[5]

Proof. We show that F restricted to $G[D]$ provides the desired isomorphism.

Note first that if $c \in G[D]$, then $Fc \in F[C]$. So F as required sends elements of $G[D]$ to elements of $F[C]$. Moreover every element of $F[C]$ is Fu for some $u \in G[D]$. For if $d \in F[C]$, then for some c, $d = Fc = FGFc = Fu$ where $u = GFc \in G[D]$.

So F restricted to $G[D]$ is onto $F[C]$. It remains to show that it is an order-embedding. We know that F will be monotone, so what we need to prove is that, if $c, c' \in G[D]$ and $Fc \sqsubseteq Fc'$, then $c \preccurlyeq c'$.

But if $Fc \sqsubseteq Fc'$, then by the monotonicity of G, $GFc \preccurlyeq GFc'$. Recall, though, that $c, c' \in G[D]$ are fixed points of GF. Hence $c \preccurlyeq c'$ as we want. □

(b) For our second result, we need the idea of a closure function on a poset – i.e. a function T which, roughly speaking, maps a poset 'upwards' to a subposet which then stays fixed under further applications of T. Officially:

Definition 145. Suppose $(C, \preccurlyeq)$ is a poset; then a *closure function* on C is a function $T \colon C \to C$ such that, for all $c, c' \in C$,

(1) $c \preccurlyeq Tc$;

(2) if $c \preccurlyeq c'$, then $Tc \preccurlyeq Tc'$, i.e. T is monotone;

(3) $TTc = Tc$, i.e. T is idempotent. △

Theorem 213. *If $F \vdash G$ is a Galois connection between $(C, \preccurlyeq)$ and another poset, then $T = GF$ is a closure function for C.*

Proof. We quickly check that the three conditions for closure apply. (1) is given by definition. (2) is immediate as GF is a composition of monotone functions. And for (3), we know from the argument in §40.2 that $FGF = F$, and hence $GFGF = GF$. □

What this shows is that the existence of a Galois connection $F \dashv G$ between the 'home' poset $(C, \preccurlyeq)$ and some 'foreign' poset $(D, \sqsubseteq)$ imposes a condition here at home on $(C, \preccurlyeq)$ – namely, there has to be a closure function on that poset. And the reverse is also true:

Theorem 214. *If there is a closure function T for the poset $(C, \preccurlyeq)$, then we can find another poset such that there is a Galois connection between the two.*

[5] Life is too short to fuss about notationally distinguishing the order relation $\preccurlyeq$ defined on C from that relation's restriction to $G[D]$.

Proof. Consider the poset $(T[C], \preccurlyeq)$. Mildly abusing notation, let $T\colon (C, \preccurlyeq) \to (T[C], \preccurlyeq)$ be the closure function but now thought of as having codomain $T[C]$. And let $I\colon (T[C], \preccurlyeq) \to (C, \preccurlyeq)$ be the inclusion function which maps an element of $T[C]$ to itself.

Suppose $c \in C$, and $d \in T[C]$ so for some $c' \in C$, $d = Tc'$. Then

(i) if $Tc \preccurlyeq d$, then since $c \preccurlyeq Tc$, we have $c \preccurlyeq d$ hence $c \preccurlyeq Id$;
(ii) if $c \preccurlyeq Id$, then $c \preccurlyeq d$, so $c \preccurlyeq Tc'$, so $Tc \preccurlyeq TTc'$, so $Tc \preccurlyeq Tc'$, i.e. $Tc \preccurlyeq d$.

Hence $T \dashv I$. □

40.7 Syntax and semantics briefly revisited

(a) Finally, we'll note one rather characteristic way in which Galois connections can naturally arise.

Theorem 215. *Given sets X and Y, with R a binary relation between their members, form the posets $(\mathcal{P}X, \subseteq)$ and $(\mathcal{P}Y, \supseteq)$,*

i. define $F\colon (\mathcal{P}X, \subseteq) \to (\mathcal{P}Y, \supseteq)$ by putting $FA = \{b \mid (\forall a \in A)aRb\}$ for $A \subseteq X$; and
ii. similarly define $G\colon (\mathcal{P}Y, \supseteq) \to (\mathcal{P}X, \subseteq)$ by putting $GB = \{a \mid (\forall b \in B)aRb\}$ for $B \subseteq Y$.

(So the idea is simply that F sends A to the set of objects R-related to A's members. Similarly for G.) Then $F \dashv G$.

Proof. We just have to prove that principle (Gal) holds, i.e. for given $A \subseteq X$, $B \subseteq Y$, then $FA \supseteq B$ iff $A \subseteq GB$.

But simply by applying definitions we see $FA \supseteq B$ iff $(\forall b \in B)(\forall a \in A)aRb$ iff $(\forall a \in A)(\forall b \in B)aRb$ iff $A \subseteq GB$. □

Let's say that a Galois connection produced in this way is *relation-generated.* Galois's original example was of this kind.

(b) And our original motivating example is relation-generated too. Let's briefly return to it.

Fixing again on a suitable language L, Let X be the set of L-sentences, let Y be the set of L-structures, and let R be the relation between $x \in X$ and $y \in Y$ such that xRy iff $y \vDash x$.

Then $(\mathcal{P}X, \subseteq)$, and $(\mathcal{P}Y, \supseteq)$ are the posets from §40.2 again. Then our last theorem tells us that there is a Galois connection between these posets.

Here, from the theorem, F is defined by putting $Fc = \{b \mid (\forall a \in c)b \vDash a\}$ for any $c \subseteq X$. So, as before, F is the 'find the models' function.

Similarly, from the theorem, G is defined by putting $Gd = \{a \mid (\forall b \in d)b \vDash a\}$ for $d \subseteq Y$. So, as before, G is the 'find-the-agreed-truths' function.

So Theorem 215 immediately gives us back the same Galois connection $F \dashv G$ between the 'syntax' poset and 'semantics' poset for a language L that we introduced in §40.2. And we can now turn the handle and apply all the implications of a Galois connection that we've been deriving.

For example, we know that for any set c of L-sentences, $FGFc = Fc$. What does that tell us? We noted in §40.2 that GFc is the set of semantic consequences of c. So $FGFc = Fc$ says that the set of models of the semantic consequences of c is the same as the set of models of c.

OK, that's most certainly not exciting logical *news*! The same goes for the interpretation of the other implications of our Galois connection here. But deriving novel results is not the name of the game. The point rather is this. We take the fundamental *true-of* relation which can obtain between an L-sentence and an L-structure: this immediately generates a certain Galois connection $F \dashv G$ between two naturally ordered 'syntactic' and 'semantic' posets. Looking at the implications of such a connection, we come to see some familiar old logical ideas as actually exemplifying quite general order-theoretic patterns which recur elsewhere. Which is rather neat.

41 Adjunctions introduced

Recall that lovely quotation from Tom Leinster which I gave at the very outset of Part I:

> Category theory takes a bird's eye view of mathematics. From high in the sky, details become invisible, but we can spot patterns that were impossible to detect from ground level. (Leinster 2014, p. 1)

Perhaps the most dramatic example of patterns that category theory newly reveals are those that involve *adjunctions*. Indeed, Goldblatt claims

> The isolation and explication of the notion of adjointness is perhaps the most profound contribution that category theory has made to the history of general mathematical ideas. (Goldblatt 1984, p. 438)

As Mac Lane famously puts it, the slogan is

> Adjoint functors arise everywhere. (Mac Lane 1997, p. vii)

In the previous chapter, we were looking at what, in hindsight, turns out to be a restricted version of the phenomenon. But category theory enables us to generalize radically.[1]

41.1 Adjoint functors: a first definition

(a) Take a poset $(C, \preccurlyeq)$ and now consider the corresponding category C. So the objects of C are the members of C, and there is a unique C-arrow $A \to A'$ if and only if $A \preccurlyeq A'$ (for any $A, A' \in C$). Similarly let D be the category corresponding to the poset $(D, \sqsubseteq)$, whose objects are the members of D, and which has a unique arrow $B \to B'$ if and only if $B \sqsubseteq B'$ (for any $B, B' \in D$).

Now, we've defined a Galois connection between the posets $(C, \preccurlyeq)$ and $(D, \sqsubseteq)$ as a pair of functions $F\colon C \to D$ and $G\colon D \to C$ such that

$$\text{(Gal)} \qquad FA \sqsubseteq B \text{ iff } A \preccurlyeq GB \text{ (for all } A \in C, B \in D).$$

[1] In this chapter we define adjunctions and give elementary examples. In the next two chapters we can prove some basic results about them. But at the level of these introductory notes, you'll have to take it somewhat on trust that adjunctions do keep cropping up in significant applications of category theory e.g. in algebraic topology.

But we know that (Gal) implies that F and G are monotone – see Theorem 207. And monotone functions between posets give rise to functors between the corresponding categories – see §26.2, Ex. (F13).

Let's spell that out again. The monotone function $F\colon C \to D$ gives rise to a corresponding functor $F\colon \mathsf{C} \to \mathsf{D}$ whose object component sends a C-object A to the D-object FA and whose arrow component sends a C-arrow $A \to A'$ to the unique arrow D-arrow $FA \to FA'$. Similarly, of course, the monotone $G\colon D \to C$ gives rise to a corresponding functor $G\colon \mathsf{D} \to \mathsf{C}$.

So our Galois-connected *functions* F, G in opposite directions between the posets $(C, \preccurlyeq)$ and $(D, \sqsubseteq)$ give rise to a corresponding pair of *functors* F, G in opposite directions between the corresponding poset categories C and D. And given (Gal), these two functors will have the following connecting property:

> There is a unique arrow $FA \to B$ in D iff there is a corresponding unique arrow $A \to GB$ in C (for all C-objects A and D-objects B).

In other words, (Gal) for the posets $(C, \preccurlyeq)$ and $(D, \sqsubseteq)$ becomes the following for the corresponding categories C, D:

> There is a bijection, i.e. an isomorphism in Set, between the hom-sets $\mathsf{D}(FA, B)$ and $\mathsf{C}(A, GB)$ (for all C-objects A and D-objects B).

Moreover, this isomorphism arises systematically from the underlying Galois connection, in an entirely uniform way without making arbitrary choices, for any A, B. And we now know how to capture that informal claim in category-theoretic terms – that isomorphism will be *natural* in A and in B.

In summary then: Suppose that the functions $F\colon C \to D$ and $G\colon D \to C$ are left and right adjoints in a Galois connection $F \dashv G$ between posets $(C, \preccurlyeq)$ and $(D, \sqsubseteq)$. And suppose that $F\colon \mathsf{C} \to \mathsf{D}$ and $G\colon \mathsf{D} \to \mathsf{C}$ are the corresponding functors between the corresponding categories C and D. Then

$$\mathsf{D}(FA, B) \cong \mathsf{C}(A, GB)$$

naturally in A and naturally in B (for all A in C and B in D).

(b) Now – sounding the trumpets – here's the key new move. We radically generalize from that special case of poset categories:

Definition 146. Suppose C and D are categories and $F\colon \mathsf{C} \to \mathsf{D}$ and $G\colon \mathsf{D} \to \mathsf{C}$ are functors. Then F *is left adjoint to* G *and* G *is right adjoint to* F, notated $F \dashv G$, iff

$$\mathsf{D}(FA, B) \cong \mathsf{C}(A, GB)$$

naturally in A and naturally in B (for all A in C and B in D). △

If we want notation that makes explicit where adjoint functors are going, we can write $F \dashv G\colon \mathsf{C} \to \mathsf{D}$, or we can diagram the situation like this:

$$\mathsf{C} \underset{G}{\overset{F}{\underset{\longleftarrow}{\overset{\longrightarrow}{\bot}}}} \mathsf{D}.$$

One immediate comment. Here, and onwards through our discussions of adjunctions, we'll take it that there is no problem in talking about the relevant hom-*sets* – either because the categories are small enough, or because we are taking a relaxedly inclusive line on what counts as set-like collections.

(c) Let's quickly state a first couple of key theorems. First, recall Theorem 208 that told us that if we have Galois connections $F \dashv G$, $F \dashv G'$ between the posets $(C, \preccurlyeq)$ and $(D, \sqsubseteq)$, then $G = G'$. And likewise, if $F \dashv G$, $F' \dashv G$ are both Galois connections between the same posets, then $F = F'$. We now have the following fairly predictable analogue result for adjoints more generally:

Theorem 216. *If a functor has an adjoint, it is unique up to natural isomorphism. If $F \dashv G$ and $F \dashv G'$ then $G \cong G'$. If $F \dashv G$ and $F' \dashv G$ then $F \cong F'$.*

Second, recall Theorem 209 that tells us that Galois connections compose: if there is a Galois connection $F \dashv G$ between the posets $(C, \preccurlyeq)$ and $(D, \sqsubseteq)$, and a connection $H \dashv K$ between the posets $(D, \sqsubseteq)$ and $(E, \subseteq)$, then there is a Galois connection $HF \dashv GK$ between C and E. The obvious generalization again applies:

Theorem 217. *Given* $\mathsf{C} \underset{G}{\overset{F}{\underset{\longleftarrow}{\overset{\longrightarrow}{\bot}}}} \mathsf{D}$ *and* $\mathsf{D} \underset{K}{\overset{H}{\underset{\longleftarrow}{\overset{\longrightarrow}{\bot}}}} \mathsf{E}$ *, then* $\mathsf{C} \underset{GK}{\overset{HF}{\underset{\longleftarrow}{\overset{\longrightarrow}{\bot}}}} \mathsf{E}$.

But let's leave the proofs of these technicalities until the next chapter. Before then, we want to get enough examples of adjoint pairs onto the table; we need to explain what the naturality requirement in our first official definition comes to; we also want to introduce an alternative definition.

41.2 Examples

All the instances of Galois connections between posets which we looked at in the last chapter can of course now be re-purposed as examples of adjunctions between corresponding poset categories. You should pause here to think through a few of those cases in their new guise.

(a) Done? Then we'll proceed to look at some different sorts of examples of adjunctions. And to speed things along, we will proceed informally. We mostly won't actually prove that the bijections between the relevant hom-sets in our various examples are natural in the technical sense; rather, we will take it as enough to find a bijection that can clearly be set up in a systematic and intuitively natural way, without arbitrary choices.

For a warm-up exercise, we start with a particularly easy case:

(A1) Consider any (non-empty!) category C and the one object category $\mathbb{1}$ (comprising the sole object $\bullet$ and its identity arrow). There is a unique constant functor $\Delta_\bullet \colon \mathsf{C} \to \mathbb{1}$. Questions: when does $\Delta_\bullet$ have a right adjoint $G \colon \mathbb{1} \to \mathsf{C}$? what about a left adjoint?

If G is to be a right adjoint, we require

$$1(\Delta_\bullet A, \bullet) \cong \mathsf{C}(A, G\bullet)$$

to hold for any C-object A. So we require

$$1(\bullet, \bullet) \cong \mathsf{C}(A, G\bullet).$$

But the hom-set on the left contains just the identity arrow. And that can only be in bijection with the hom-set on the right, for each A, if there is always a *unique* arrow $A \to G\bullet$, i.e. if $G\bullet$ is terminal in C (and then the bijection is intuitively natural, as no arbitrary choices can be involved in setting it up).

In sum, $\Delta_\bullet$ has a right adjoint $G\colon 1 \to \mathsf{C}$ just in case G sends 1's unique object to C's terminal object: no terminal object, no right adjoint.

Dually, $\Delta_\bullet$ has a left adjoint if and only if C has an initial object.

This toy example reminds us of what we have already seen in the special case of Galois connections, namely that a functor may or may not have a right adjoint, and independently may or may not have a left adjoint, and if both adjoints exist they may be different.

But let's also note that we have here a first indication that adjunctions and (co)limits can interact in interesting ways: in this case, we could *define* terminal and initial objects for a category C in terms of the existence of right and left adjoints to the functor $\Delta_\bullet\colon \mathsf{C} \to 1$.

(b) Another example involving a limit:

(A2) For any category C there is a binary diagonal functor $\Delta\colon \mathsf{C} \to \mathsf{C}^2$ which sends a C-object A to the pair $\langle A, A\rangle$, and sends a C-arrow f to the pair of arrows $\langle f, f\rangle$. What would it take for this functor to have a right adjoint $G\colon \mathsf{C}^2 \to \mathsf{C}$?

We'd need

$$\mathsf{C}^2(\Delta A, \langle B, C\rangle) \cong \mathsf{C}(A, G\langle B, C\rangle)$$

naturally in A and in $\langle B, C\rangle$. But the left-hand hom-set is the set of C^2-arrows from $\langle A, A\rangle$ to $\langle B, C\rangle$, which is the set of pairs of C-arrows $f_1\colon A \to B$, $f_2\colon A \to C$. So how can we get such pairs of arrows from A lined up in a natural way one-to-one with single C-arrows from A?

Assume that C has all products: then we can match pairs of arrows $f_1\colon A \to B$, $f_2\colon A \to C$ with the mediating arrows $\langle\!\langle f_1, f_2\rangle\!\rangle\colon A \to B \times C$ in the corresponding product diagrams. So, take G to be the product functor $\otimes\colon \mathsf{C}^2 \to \mathsf{C}$ (which exists if C has all products, see §26.4). Then we'll get our desired natural isomorphism:

$$\mathsf{C}^2(\Delta A, \langle B, C\rangle) \cong \mathsf{C}(A, \otimes\langle B, C\rangle),$$

showing that, assuming C has products, $\otimes$ is right adjoint to Δ.

There is also, as you ought to expect by now, a dual result: assuming C has coproducts, the functor that sends pairs of objects to their coproduct is left adjoint to Δ. (Pause to think this through.)

(A3) Here next is a radical generalization.

Suppose C has all limits of shape J. Now, in §36.7 we defined the diagonal functor $\Delta_{\mathsf{J}} \colon \mathsf{C} \to [\mathsf{J}, \mathsf{C}]$ and the limit functor $Lim \colon [\mathsf{J}, \mathsf{C}] \to \mathsf{C}$. And earlier results show that, for any C in C and for any D in $[\mathsf{J}, \mathsf{C}]$ (i.e., any diagram-as-functor $D \colon \mathsf{J} \to \mathsf{C}$) we have a bijection

$$[\mathsf{J}, \mathsf{C}](\Delta_{\mathsf{J}} C, D) \cong \mathsf{C}(C, Lim D).$$

Why so?

(i) $\Delta_{\mathsf{J}} C$ is the constant functor Δ_C. And Theorem 158 tells us the $[\mathsf{J}, \mathsf{C}]$-arrows from Δ_C to D, i.e. the natural transformations from Δ_C to D, are the cones over D with vertex C.

(ii) While the definition of Lim tells us that $\mathsf{C}(C, Lim D)$ is the set of arrows from C to the vertex of our chosen limit over D.

(iii) But Theorem 84 tells us the cones over D with vertex C correspond one-to-one with C-arrows from C to the vertex of the limit over D.

Moreover, once we've fixed on the particular limit involved in defining Lim, there are no further arbitrary choices involved: our bijection is natural in C and D.

Hence, assuming all along that C has all limits of shape J, we have an adjunction $\Delta_{\mathsf{J}} \dashv Lim$.

(Exercise: What is the dual case?)

(A4) Suppose C is a category with exponentiation (and hence with products). Then, in a slogan, exponentiation by B is right adjoint to taking the product with B.

To see this, we define a pair of functors from C to itself. First, there is the functor $- \times B \colon \mathsf{C} \to \mathsf{C}$ which sends an object A to the product $A \times B$, and sends an arrow $f \colon A \to A'$ to $f \times 1_B \colon A \times B \to A' \times B$.

Second, there is the functor $(-)^B \colon \mathsf{C} \to \mathsf{C}$ which sends an object C to C^B, and sends an arrow $f \colon C \to C'$ to $\overline{f \circ ev} \colon C^B \to C'^B$, as defined in §26.4, (F17).

But $\mathsf{C}(A \times B, C) \cong \mathsf{C}(A, C^B)$ naturally in A and C (for a partial proof see §32.4(4)). Hence $(- \times B) \dashv (-)^B$.

(A5) Consider the familiar category Set of sets and total functions, and then the category Pfn of sets and partial functions. There's an inclusion functor $F \colon \mathsf{Set} \to \mathsf{Pfn}$ (remember that, as always, a total function counts as a limiting case of a partial function, so we do indeed have an inclusion functor here)! And in the other direction we have the 'totalizing' functor G which sends a set X to the augmented set $X + \star$ (with $\star$ some novel addition), and sends a partial function $\varphi \colon X \to Y$ to the total function

$f\colon X + \star \to Y + \star$, where $f(\star) = \star$, $f(x) = y$ when $\varphi(x)$ is defined and takes the value y, and $f(x) = \star$ otherwise.[2]

It is then immediate that there is a naturally arising isomorphism $\mathsf{Pfn}(FX, Y) \cong \mathsf{Set}(X, GY)$, and hence $F \dashv G$.

(c) Now for a whole family of important examples involving adjoints of forgetful functors.

(A6) Let's next consider the forgetful functor $U\colon \mathsf{Top} \to \mathsf{Set}$ which sends each topological space to its underlying set of points, and sends any continuous function between topological spaces to the same function thought of as a set-function. Questions: does this have a left adjoint? a right adjoint?

If U is to have a left adjoint $F\colon \mathsf{Set} \to \mathsf{Top}$, then for any set S and for any topological space $(T, \mathcal{O})$ – with T a set of points and $\mathcal{O}$ a topology (a suitable collection of open sets) – we require

$$\mathsf{Top}(FS, (T, \mathcal{O})) \cong \mathsf{Set}(S, U(T, \mathcal{O})) = \mathsf{Set}(S, T),$$

where the bijection here needs to be a natural one.

Now, on the right we have the set of *all* functions $f\colon S \to T$. So that needs to be in bijection with the set of all *continuous* functions from FS to $(T, \mathcal{O})$. How can we ensure this holds in a systematic way, for any S and $(T, \mathcal{O})$? Well, suppose that for any S, F sends S to the topological space $(S, \mathcal{D})$ which has the discrete topology (i.e. all subsets of S count as open). It is a simple exercise to show that *every* function $f\colon S \to T$ then counts as a continuous function $f\colon (S, \mathcal{D}) \to (T, \mathcal{O})$. So the functor F that assigns a set the discrete topology will be left adjoint to the forgetful functor – and so by Theorem 216 will be, up to isomorphism, *the* left adjoint.

Similarly, the functor $G\colon \mathsf{Set} \to \mathsf{Top}$ that assigns a set the indiscrete topology (the only open sets are the empty set and S itself) is the right adjoint to the forgetful functor U.

(A7) Here's another case of a forgetful functor, this time the functor $U\colon \mathsf{Mon} \to \mathsf{Set}$ which forgets about monoidal structure. Does U have a left adjoint $F\colon \mathsf{Set} \to \mathsf{Mon}$?

If $M = (\underline{M}, *, e)$ is a monoid and X some set, we will need

$$\mathsf{Mon}(FX, M) \cong \mathsf{Set}(X, UM) = \mathsf{Set}(X, \underline{M}).$$

But the hom-set on the right contains all possible functions set-functions from X to the monoid's underlying set $\underline{M}$. How can these be in one-to-one correspondence with the monoid homomorphisms from FX to M?

[2]It's a matter of taste exactly how you want to treat $X + \star$ – one trick would be to take that as alternative notation for $X \cup \{X\}$, and read other occurrences of $\star$ to match. (And compare the discussion in §34.2.)

Suppose FX is some monoid with a lot of structure (over and above the minimum required to be a monoid). Then there may be few monoid homomorphisms from FX to M. Therefore, if there are to be *lots* of such monoid homomorphisms, one for each $f\colon X \to \underline{M}$, then FX will surely need to have minimal structure. Which suggests going for broke and considering the limiting case of the least structured monoid built on X, i.e. the *free* monoid on X which we met back in §26.5, Ex. (F18).

So we want the functor F to send a set X to $FX = (List(X), ^\frown, \emptyset)$ – where, recall, the objects of $(List(X), ^\frown, \emptyset)$ are finite lists of X-elements (including the null sequence) and its monoid operation is concatenation.

Now consider the correlation between a function $f\colon X \to \underline{M}$ and the corresponding function $\hat{f}\colon FX \to M$ which sends the empty sequence of X-elements to the unit of M, and sends the finite sequence $x_1^\frown x_2^\frown x_3^\frown \ldots ^\frown x_n$ to the M-element $fx_1 * fx_2 * fx_3 * \ldots * fx_n$. So defined, $\hat{f}$ respects the unit and the monoid operation and so is a monoid homomorphism.

Evidently that correlation is injective: different functions f will have different correlates $\hat{f}$ (if for some x, $fx \neq gx$, then $\hat{f}\langle x\rangle \neq \hat{g}\langle x\rangle$, where $\langle x\rangle$ is the one-object sequence whose member is x).

The correlation is surjective too. Take any homomorphism $h\colon FX \to M$, and for each $x \in X$, let fx be the member of $\underline{M}$ such that $h\langle x\rangle = fx$. Varying x, that defines a function $f\colon X \to \underline{M}$ such that $h = \hat{f}$.

So we have a naturally arising bijection between functions $f\colon X \to \underline{M}$ and the corresponding functions $\hat{f}\colon FX \to M$. Which establishes that, as required, $\mathsf{Mon}(FX, M) \cong \mathsf{Set}(X, \underline{M}) = \mathsf{Set}(X, UM)$, naturally in X and M.

In sum, the free functor F which takes a set to the free monoid on that set is left adjoint to the forgetful functor U which sends a monoid to its underlying set: $F \dashv U$. And by the yet-to-be-proved Theorem 216 again, this free functor is, up to isomorphism, the unique left adjoint to the forgetful functor.

This example involving monoids is actually typical of a whole cluster of cases. A left adjoint of the forgetful functor from some class of algebraic structures to their underlying sets is characteristically provided by the non-trivial functor that takes us from a set to a free structure on that set of the relevant algebraic kind. Thus we have, for example,

(A8) The forgetful functor $U\colon \mathsf{Grp} \to \mathsf{Set}$ has as a left adjoint the functor $F\colon \mathsf{Set} \to \mathsf{Grp}$ which sends a set to the free group on that set (i.e. the group obtained from a set S by adding just enough elements for it to become a group while imposing no constraints beyond those required to ensure we do have a group).

What about *right* adjoints to our last two forgetful functors?

(A9) We will later show that the forgetful functor $U\colon \mathsf{Mon} \to \mathsf{Set}$ has no right adjoint by a neat proof in §43.5. But here's a more hand-waving argument.

U would have a right adjoint $G\colon \mathsf{Set} \to \mathsf{Mon}$ just in case $\mathsf{Set}(\underline{M}, S) = \mathsf{Set}(UM, S) \cong \mathsf{Mon}(M, GS)$, for all monoids $M = (\underline{M}, *, e)$ and sets S. But this requires the monoid homomorphisms from M to GS always to be in bijection with the set-functions from $\underline{M}$ to S. But that's not possible (consider keeping the sets $\underline{M}$ and S fixed, but changing the possible monoid operations with which $\underline{M}$ is equipped).

A similar argument shows that the forgetful functor $U\colon \mathsf{Grp} \to \mathsf{Set}$ has no right adjoint.

There are, however, examples of 'less forgetful' algebraic functors which have both left and right adjoints:

(A10) Take the functor $U\colon \mathsf{Grp} \to \mathsf{Mon}$ which forgets about group inverses but keeps the monoidal structure. This has a left adjoint $F\colon \mathsf{Mon} \to \mathsf{Grp}$ which converts a monoid to a group by adding inverses for elements (and otherwise making no more assumptions than are needed to get a group). U also has a right adjoint $G\colon \mathsf{Mon} \to \mathsf{Grp}$ which rather than adding elements subtracts them by mapping a monoid to the submonoid of its invertible elements (which can be interpreted as a group).

Let's check the second of those claims. We have $U \dashv G$ so long as

$$\mathsf{Mon}(UK, M) \cong \mathsf{Grp}(K, GM),$$

for any monoid M and group K, and in a natural way. Now we just remark that every element of K-as-a-monoid is invertible and a monoid homomorphism sends invertible elements to invertible elements. Hence a monoid homomorphism from K-as-a-monoid to M will in fact also be a group homomorphism from K to the submonoid-as-a-group GM.

And here is another forgetful functor with a right adjoint:

(A11) Recall the functor $F\colon \mathsf{Set} \to \mathsf{Rel}$ which 'forgets' that arrows are functional (see §26.2, Ex. (F4)). And now we introduce a powerset functor $P\colon \mathsf{Rel} \to \mathsf{Set}$ defined as follows:

a) P sends a set A to its powerset $\mathcal{P}A$, and

b) P sends a relation R in $A \times B$ to the function $f_R\colon \mathcal{P}A \to \mathcal{P}B$ which sends $X \subseteq A$ to $Y = \{b \mid (\exists x \in X)Rxb\} \subseteq B$.

Claim: $F \dashv P$. Why? We observe that there is a (natural!) one-to-one correlation between a relation R in $A \times B$ and a function $f\colon A \to \mathcal{P}B$ where $f(x) = \{y \mid Rxy\}$ and so Rxy iff $y \in f(x)$. This gives us a natural enough bijection $\mathsf{Rel}(FA, B) \cong \mathsf{Set}(A, PB)$, for any A, B.

(d) Our examples so far *are* all pretty elementary. But once we delve a bit further into areas of algebra, geometry, topology, etc., we find that more exciting adjunctions keep cropping up. So, for an algebraic example of another case where the adjoint of a trivial functor is something much more substantial, we have:

(A12) The inclusion functor from the category of abelian groups into the category of groups has a left adjoint which assigns to every group G its abelianization $G/[G, G]$ (see §33.2 Ex. (4)).

For a nice topological example (again stated without proof):

(A13) The inclusion functor from KHaus, the category of compact Hausdorff spaces, into Top has a left adjoint, namely the Stone-Čech compactification functor.

But perhaps we already have enough to be going on with.[3]

41.3 Naturality

(a) We said: $F \dashv G \colon \mathsf{C} \to \mathsf{D}$ when (for all C-objects A and D-objects B)

$$\mathsf{D}(FA, B) \cong \mathsf{C}(A, GB)$$

holds *naturally* in A and B. Let's now be more explicit about what the official naturality requirement comes to in this case.

By assumption there is a whole suite of bijections $\varphi_{AB} \colon \mathsf{D}(FA, B) \to \mathsf{C}(A, GB)$ (one for each choice of A, B). For naturality in A, we need to consider what happens when we vary the occupant of the A-place: keeping B fixed, we want $\varphi_{AB}, \varphi_{A'B}, \varphi_{A''B}, \ldots$, to assemble into a natural isomorphism $\varphi_B \colon \mathsf{D}(F-, B) \xRightarrow{\sim} \mathsf{C}(-, GB)$.

Similarly for naturality in B: $\varphi_{AB}, \varphi_{AB'}, \varphi_{AB''}, \ldots$, keeping A fixed, should assemble into a natural isomorphism between hom-functors $\varphi_A \colon \mathsf{D}(FA, -) \xRightarrow{\sim} \mathsf{C}(A, G-)$.

Let's look at the second case. If $\varphi_{AB}, \varphi_{AB'}, \varphi_{AB''}, \ldots$ are to assemble into a natural isomorphism then we require that, for any choice of B, B' and arrow $h \colon B \to B'$, a naturality square of the usual kind always commutes:

$$\begin{array}{ccc} \mathsf{D}(FA, B) & \xrightarrow{\mathsf{D}(FA, h)} & \mathsf{D}(FA, B') \\ \downarrow{\scriptstyle \varphi_{AB}} & & \downarrow{\scriptstyle \varphi_{AB'}} \\ \mathsf{C}(A, GB) & \xrightarrow{\mathsf{C}(A, Gh)} & \mathsf{C}(A, GB') \end{array}$$

But how does the covariant hom-functor $\mathsf{D}(FA, -)$ operate on $h \colon B \to B'$? As we saw in §31.1, it sends h to $h \circ -$, i.e. to that function which composes h with

[3] If you want more examples, see e.g. Mac Lane (1997, pp. 87–88) for a list, and Riehl (2017, §4.1) for another.

an arrow from $\mathsf{D}(FA, B)$ to give an arrow in $\mathsf{D}(FA, B')$. Similarly, $\mathsf{C}(A, G-)$ will send h to $Gh \circ -$.

So consider an arrow $d\colon FA \to B$ living in $\mathsf{D}(FA, B)$. The naturality square now tells us that for any $h\colon B \to B'$, $\varphi_{AB'}(h \circ d) = Gh \circ \varphi_{AB} d$.[4]

(b) Let's introduce a standard bit of notation which is used to indicate the action of the bijections φ_{AB} and their inverses on the relevant hom-sets:

Definition 147. Given $F \dashv G\colon \mathsf{C} \to \mathsf{D}$, whose adjunction is φ, then φ_{AB} sends an arrow $d\colon F(A) \to B$ to its *transpose* $\overline{d}\colon A \to GB$; likewise the inverse bijection φ_{AB}^{-1} sends an arrow $c\colon A \to GB$ to its transpose $\overline{c}\colon F(A) \to B$. △

An alternative notation uses $d^\flat$ for our $\overline{d}$ and $c^\sharp$ for our $\overline{c}$, and this notation might well be thought to be preferable in principle since transposing by 'sharpening' and 'flattening' are different operations. But the double use of the overlining notation is pretty standard, and is quite slick.

Evidently, transposing twice takes us back to where we started: $\overline{\overline{c}} = c$ and $\overline{\overline{d}} = d$.

(c) Deploying this new notation, we have shown the first part of the following theorem follows from the naturality requirement. And the second part (exercise!) follows by a dual argument, in which some arrows get reversed because the relevant hom-functors in this case are contravariant.

Theorem 218. *Given $F \dashv G\colon \mathsf{C} \to \mathsf{D}$, then*

(1) for any $d\colon FA \to B$ and $h\colon B \to B'$, $\overline{h \circ d} = Gh \circ \overline{d}$,

(2) for any $c\colon A \to GB$ and $k\colon A' \to A$, $\overline{c \circ k} = \overline{c} \circ Fk$, so $\overline{\overline{c} \circ Fk} = c \circ k$. □

There is also a nice converse to this theorem. Given functors $F\colon \mathsf{C} \to \mathsf{D}$ and $G\colon \mathsf{D} \to \mathsf{C}$ such that there is always a bijection φ_{AB} between $\mathsf{D}(FA, B)$ and $\mathsf{C}(A, GB)$ then, if conditions (1) and (2) hold with transposes defined as before, the various φ_{AB} will assemble into natural transformations, so that $\mathsf{D}(FA, B) \cong \mathsf{C}(A, GB)$ holds naturally in A and in B, and hence $F \dashv G$.

41.4 An alternative definition

(a) To recap, in §41.1 we parlayed the idea of a Galois connection between posets into the idea of a connection between functors which act on posets-as-categories: then we simply generalized to give us the definition of an adjunction $F \dashv G$ between functors acting on other sorts of category too. In sum, $F \dashv G\colon \mathsf{C} \to \mathsf{D}$ holds when $\mathsf{D}(FA, B) \cong \mathsf{C}(A, GB)$ naturally in A and B.

In §41.2 we gave some easy examples of adjunctions to add to examples based on Galois connections.

[4] Let's have a helpful convention here: I'll use d, d' for arrows living in D, and hence belonging to homsets such as $\mathsf{D}(FA, B)$, while I use c, c' for arrows living in C, and hence belonging to homsets such as $\mathsf{C}(A, GB)$.

Then in §41.3 we paused to spell out what the naturality clause means in our initial definition of an adjunction.

So far, so straightforward. Now, the definition in §41.1 was inspired by our original definition of a Galois connection in §40.3. But we gave an alternative definition of such connections in §40.4. This too can be generalized to give a second definition of adjunctions. In this section we show how.

Of course, we are eventually going to prove the two definitions are equivalent. But you need to know both. While our second characterization initially looks more complicated, it brings out something more of the structural richness of adjunctions, and it is very useful in applications.

(b) Recall: a Galois connection between the posets $(C, \preccurlyeq)$ and $(D, \sqsubseteq)$, according to the tweaked version of our alternative definition in §40.4, comprises a pair of functions $F\colon C \to D$ and $G\colon D \to C$ such that (for any $A, A' \in C$ and $B, B' \in D$):

(1) F and G are both monotone,

(2′) (i) if $A \preccurlyeq A'$, then $A \preccurlyeq A' \preccurlyeq GFA'$ and $A \preccurlyeq GFA \preccurlyeq GFA'$,
(ii) if $B \sqsubseteq B'$, then $FGB \sqsubseteq B \sqsubseteq B'$ and $FGB \sqsubseteq FGB' \sqsubseteq B'$.

And as before, let C be the category corresponding to the poset $(C, \preccurlyeq)$, while D corresponding to $(D, \sqsubseteq)$. Also as before, the monotone functions F, G between the posets give rise to functors F, G between the corresponding categories. And now note that, the composite monotone function $GF\colon C \to C$ gives rise to a composite functor $GF\colon \mathsf{C} \to \mathsf{C}$, and likewise the composite monotone function $FG\colon D \to D$ gives rise to a functor $FG\colon \mathsf{D} \to \mathsf{D}$.

Now, the first part of (2′) corresponds to the claim that the following diagram always commutes in the poset category C, for any A, A':

$$\begin{array}{ccc} A & \longrightarrow & A' \\ \big\downarrow & & \big\downarrow \\ GFA & \longrightarrow & GFA' \end{array}$$

In a poset category, arrows from a given source to a given target are unique, so the two composites from A to GFA' must be equal.

So let's define η_X to be the arrows $X \to GFX$, one for each C-object X. And let $1_{\mathsf{C}}\colon \mathsf{C} \to \mathsf{C}$ be the identity functor. Then our commutative diagram can be revealingly redrawn as follows:

$$\begin{array}{ccc} 1_{\mathsf{C}}A & \longrightarrow & 1_{\mathsf{C}}A' \\ \big\downarrow{\scriptstyle \eta_A} & & \big\downarrow{\scriptstyle \eta_{A'}} \\ GFA & \longrightarrow & GFA' \end{array}$$

This commutes for all A, A'. *Which is just to say that the η_X assemble into a natural transformation $\eta\colon 1_{\mathsf{C}} \Rightarrow GF$.*

Likewise, the second part of (2′) corresponds to the claim that there is a natural transformation $\varepsilon\colon FG \Rightarrow 1_{\mathsf{D}}$.

(c) This gives us the two key ingredients for an alternative definition for an adjunction between functors $F\colon \mathsf{C} \to \mathsf{D}$ and $G\colon \mathsf{D} \to \mathsf{C}$: we will require there to be a pair of natural transformations $\eta\colon 1_{\mathsf{C}} \Rightarrow GF$ and $\varepsilon\colon FG \Rightarrow 1_{\mathsf{D}}$.

But we can extract a bit more from the motivating case of Galois connections. For note that (1) and (2′) imply

(3) (i) $FA \sqsubseteq FGFA \sqsubseteq FA$,
(ii) $GB \preccurlyeq GFGB \preccurlyeq GB$,

(Check that!)

Hence, now moving again from the poset to the corresponding poset category C, we will have an arrow $FA \to FGFA$ (which must be equal to $F\eta_A$, since arrows between a source and target are unique in a poset category), and also have an arrow $FGFA \preccurlyeq FA$ (which must be equal to ε_{FA}). Then the categorial analogue of (3.i) is the claim that the following diagram commutes for each A:

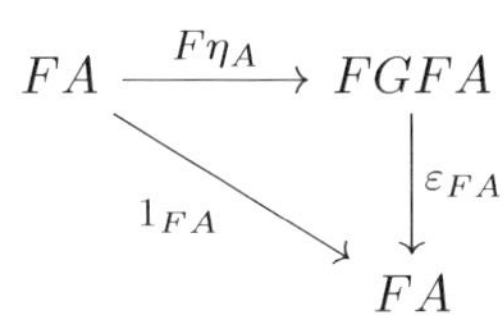

Or what comes to the same, in the functor category $[\mathsf{C}, \mathsf{D}]$ this next diagram commutes[5]

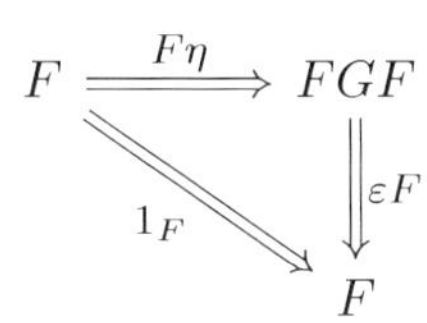

For remember 'whiskering', discussed in §33.3: the various components $F\eta_A$ assemble into a natural transformation $F\eta$, and the components ε_{FA} assemble into a natural transformation εF. And then recall from §36.1 that 'vertical' composition of natural transformations between functors is defined component-wise. So, for each A,

$$(\varepsilon F \circ F\eta)_A = \varepsilon_{FA} \circ F\eta_A = 1_{FA} = (1_F)_A,$$

[5]Notational fine print: our convention has been to use single arrows to represent arrows inside particular categories, and double arrows to represent natural transformations between functors across categories.

We are now dealing with natural transformations thought of as arrows within a particular functor category. Some therefore use single arrows in the usual way when drawing diagrams in functor categories; some use double arrows to remind us that the local arrows are natural transformations (between functors relating some other categories). The first way is perhaps more principled and it produces cleaner diagrams; but I'm jumping the second way, which can provide a helpful reminder of what is going on (as in e.g. the contrasting diagrams on p. 406).

where 1_F is the natural transformation whose component at A is 1_{FA}. Since all components are equal, the left-most and right-most natural transformations in that equation are equal and our diagram in the functor category commutes.

Exactly similarly, from (3.ii) we infer that $G\varepsilon_B \circ \eta_{GB} = 1_{GB}$. In other words, the next diagram commutes in $[\mathsf{D}, \mathsf{C}]$:

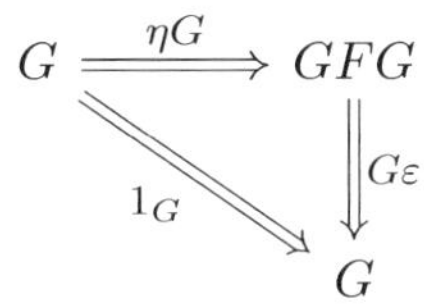

(d) And *now*, let's generalize from the special case of poset categories. We can put everything together to give us our second equally standard definition for adjoint functors:

Definition 148. Suppose C and D are categories and $F\colon \mathsf{C} \to \mathsf{D}$ and $G\colon \mathsf{D} \to \mathsf{C}$ are functors. Then *F is left adjoint to G and G is right adjoint to F*, notated $F \dashv G$, iff there are natural transformations $\eta\colon 1_{\mathsf{C}} \Rightarrow GF$ and $\varepsilon\colon FG \Rightarrow 1_{\mathsf{D}}$ such that

(i) $\varepsilon_{FA} \circ F\eta_A = 1_{FA}$ for all A in C, and $G\varepsilon_B \circ \eta_{GB} = 1_{GB}$ for all B in D;

(ii) or equivalently, the following *triangle identities* hold in the functor categories $[\mathsf{C}, \mathsf{D}]$ and $[\mathsf{D}, \mathsf{C}]$ respectively:

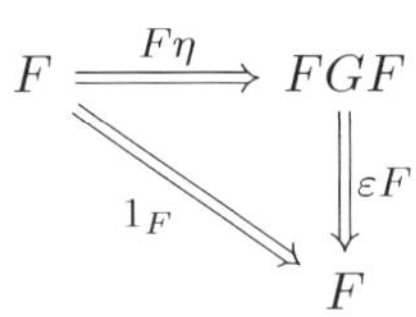

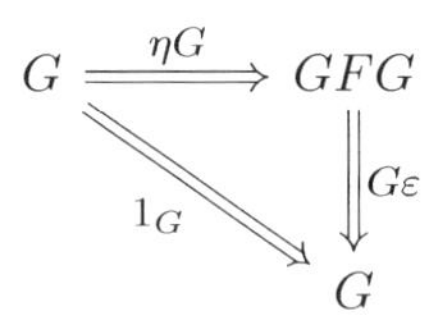

Note, the transformations η and ε are standardly called the *unit* and *co-unit* of the adjunction. △

Of course, we have to show that our two definitions of adjunctions *are* definitions of the same relationship. In other words, we need to prove:

Theorem 219. *For given functors $F\colon \mathsf{C} \to \mathsf{D}$ and $G\colon \mathsf{D} \to \mathsf{C}$, $F \dashv G$ holds by Defn. 146 iff it holds by Defn. 148.*

But let's leave the proof of this technicality (along with the proofs of Theorems 217 and 216) for the following chapter, so we can concentrate in this chapter on headline news.

41.5 Isomorphism, equivalence, adjointness

We got to our second general definition of an adjunction by starting from our second definition of those special adjunctions which are Galois connections. But there is another line of thought taking us in the same general direction.

For consider, first, this rephrasing of our earlier definition of an *isomorphism* between categories:

Definition 125* The categories C and D are isomorphic iff there are functors $F\colon \mathsf{C} \to \mathsf{D}$ and $G\colon \mathsf{D} \to \mathsf{C}$ such that we have the *identities* $1_{\mathsf{C}} = GF$ and $FG = 1_{\mathsf{D}}$. △

And then next, recall that in order to define an *equivalence* of categories, we weakened the requirement that F and G are inverses to give a definition which can be rephrased like this:

Definition 126* The categories C and D are equivalent iff there are functors $F\colon \mathsf{C} \to \mathsf{D}$ and $G\colon \mathsf{D} \to \mathsf{C}$ such that there is a pair of *natural isomorphisms* $\eta\colon 1_{\mathsf{C}} \overset{\sim}{\Longrightarrow} GF$ and $\varepsilon\colon FG \overset{\sim}{\Longrightarrow} 1_{\mathsf{D}}$. △

And now a further weakening: we arrive at our second definition of adjunctions by no longer requiring natural isomorphisms but making do with natural transformations satisfying certain side conditions:

Definition 148 There is an adjunction between the categories C and D iff there is an adjoint pair of functors $F\colon \mathsf{C} \to \mathsf{D}$ and $G\colon \mathsf{D} \to \mathsf{C}$ – i.e. there is a pair of *natural transformations* $\eta\colon 1_{\mathsf{C}} \Rightarrow GF$ and $\varepsilon\colon FG \Rightarrow 1_{\mathsf{D}}$ (where $\varepsilon_{FA} \circ F\eta_A = 1_{FA}$ for all A in C, and $G\varepsilon_B \circ \eta_{GB} = 1_{GB}$ for all B in D). △

An isomorphism is a fortiori an equivalence. But since the pair of isomorphisms making an equivalence need not satisfy the triangle identities, they needn't immediately give us an adjunction as they stand. However, taking such an equivalence defined as in Defn 126* and fixing one of the natural isomorphisms, we can always tinker (if necessary) with the other to get a corresponding adjunction. More carefully, we have

Theorem 220. *If there is an equivalence between C and D constituted by a pair of functors $F\colon \mathsf{C} \to \mathsf{D}$ and $G\colon \mathsf{D} \to \mathsf{C}$ and a pair of natural isomorphisms $\eta\colon 1_{\mathsf{C}} \overset{\sim}{\Longrightarrow} GF$ and $\varepsilon\colon FG \overset{\sim}{\Longrightarrow} 1_{\mathsf{D}}$, then there is an adjunction $F \dashv G$ with unit η and new co-unit ε' (defined in terms of η and ε). There is also an adjunction $F \dashv G$ with co-unit ε and new unit η' (again defined in terms of η and ε).*

Look at it this way. A particular equivalence between categories might locate the 'wrong' transformations to make an adjunction; but we can always massage the ingredients of the equivalence to get something that *does* work as an adjunction. The reverse doesn't hold (for a start, we can't expect asymmetric cases – where $F \dashv G$ holds but not $G \dashv F$ – to give rise to equivalences).

Once more, we'll leave the proof of our new theorem until the next chapter. But taking the theorem as given, our three notions of isomorphism, equivalence and adjunction can now be seen as giving rise to progressively weaker connections between categories. (Of course, if we go on to drop the added condition from Defn. 148, the parenthetical clause about triangle identities, we get an even weaker relation between F and G. But this turns out to have less mathematical interest.)

42 Five basic theorems

In the preceding chapter, I explained what it takes to get an adjunction between two functors, and gave some examples of adjoint pairs. In fact I gave two definitions, mirroring the two alternative definitions of Galois connections – and I merely announced that the two definitions are equivalent. So the first order of business for this chapter must be to prove this equivalence.

We will also need to prove three other theorems that I stated without proof in the last chapter, and I add a new fifth result, Theorem 221.

Nearly everything in this chapter counts as 'filling in technical details' rather than 'introducing some new big idea'. The only real novelty is the brief observation after Theorem 221 that we could in principle define adjunctions in a pair of other ways. So you might well want to skim and skip through *very* quickly on a first reading: the details can get pretty convoluted![1]

42.1 Two definitions again

(a) Suppose $F\colon \mathsf{C} \to \mathsf{D}$ and $G\colon \mathsf{D} \to \mathsf{C}$ are functors. When is F left-adjoint to G and G right-adjoint to F? When does $F \dashv G$ hold? We gave two accounts:

Definition 146. $F \dashv G$ if and only if $\mathsf{D}(FA, B) \cong \mathsf{C}(A, GB)$ naturally in A in C and in B in D.

Definition 148. $F \dashv G$ if and only if there are natural transformations $\eta\colon 1_{\mathsf{C}} \Rightarrow GF$ and $\varepsilon\colon FG \Rightarrow 1_{\mathsf{D}}$ such that $\varepsilon_{FA} \circ F\eta_A = 1_{FA}$ for all C-objects A, and $G\varepsilon_B \circ \eta_{GB} = 1_{GB}$ for all D-objects B.

And these definitions, I claimed, come to the same. In other words, we have:

Theorem 219. *$F \dashv G$ holds by Defn. 146 iff it holds by Defn. 148.*

Proof (If). Suppose there are natural transformations $\eta\colon 1_{\mathsf{C}} \Rightarrow GF$ and $\varepsilon\colon FG \Rightarrow 1_{\mathsf{D}}$ for which the so-called triangle identities hold.

Take any d in $\mathsf{D}(FA, B)$. Then $\eta_A\colon A \to GFA$ and $Gd\colon GFA \to GB$ compose. And so we can coherently define an arrow $\varphi_{AB}\colon \mathsf{D}(FA, B) \to \mathsf{C}(A, GB)$ by putting $\varphi_{AB}\, d = Gd \circ \eta_A$.

[1] And indeed my proof-versions are longer than in some (many?) other presentations. Because, for better or worse, I *do* spell out the not-always-obvious details, and in two cases give more than one proof.

Likewise, we can define $\psi_{AB}\colon \mathsf{C}(A, GB) \to \mathsf{D}(FA, B)$ by putting $\psi_{AB}\, c = \varepsilon_B \circ Fc$ for any $c\colon A \to GB$.

We can now show that the pair φ_{AB} and ψ_{AB} will be mutually inverse. Take any $d\colon FA \to B$. Then

$$\begin{aligned}
\psi_{AB}(\varphi_{AB}\, d) &= \psi_{AB}(Gd \circ \eta_A) && \text{by definition of } \varphi \\
&= \varepsilon_B \circ F(Gd \circ \eta_A)) && \text{by definition of } \psi \\
&= \varepsilon_B \circ FGd \circ F\eta_A && \text{by functoriality of } F \\
&= d
\end{aligned}$$

where the last step is by the following commuting diagram, combining a naturality square and a triangle identity:

$$\begin{array}{ccccc}
FA & \xrightarrow{F\eta_A} & FGFA & \xrightarrow{FGd} & FGB \\
 & \searrow^{1_{FA}} & \downarrow{\scriptstyle \varepsilon_{FA}} & & \downarrow{\scriptstyle \varepsilon_B} \\
 & & FA & \xrightarrow{\quad d \quad} & B
\end{array}$$

Hence, since d was arbitrary, $\psi_{AB} \circ \varphi_{AB} = 1$. And note, by the way, how we *do* need to appeal here to the added triangle equality, and can't rely only on functoriality and the naturality of ε.

Likewise we can show $\varphi_{AB} \circ \psi_{AB} = 1$ (exercise!). Therefore the components φ_{AB} are isomorphisms, proving that $\mathsf{D}(FA, B) \cong \mathsf{C}(A, GB)$.

So to complete the proof we 'just' need to show that this isomorphism is natural in A and B. And morally it should be, as we made no arbitrary choices along the way. But I suppose we ought to spell out a proof of naturality.

OK: now keep A fixed: then I claim that, as we vary B, the various components φ_{AB} assemble into a natural isomorphism φ_A from the hom-functor $\mathsf{D}(FA, –)$ to the hom-functor $\mathsf{C}(A, G(–))$. To show this, note that the usual sort of naturality square for hom-functors – the sort we met in §37.1 – commutes for every $h\colon B \to B'$:

$$\begin{array}{ccc}
\mathsf{D}(FA, B) & \xrightarrow{h\ \circ\ -} & \mathsf{D}(FA, B') \\
\downarrow{\scriptstyle \varphi_{AB}} & & \downarrow{\scriptstyle \varphi_{AB'}} \\
\mathsf{C}(A, GB) & \xrightarrow{Gh\ \circ\ -} & \mathsf{C}(A, GB')
\end{array}$$

Why? Because for every f in $\mathsf{D}(FA, B)$ we have

$$\varphi_{AB'}(h \circ f) = G(h \circ f) \circ \eta_A = Gh \circ (Gf \circ \eta_A) = Gh \circ \varphi_{AB}(f)$$

which holds in virtue of our definitions and the functoriality of G.

Now keep B fixed: then by a parallel argument (exercise!), as we vary A, the various components φ_{AB} assemble into a natural isomorphism $\varphi_B\colon \mathsf{D}(F(–), B) \Rightarrow \mathsf{C}(–, GB)$ between the two contravariant functors.

Hence, as we wanted, $\mathsf{D}(FA, B) \cong \mathsf{C}(A, GB)$ naturally in A and in B. $\square$

Proof (Only if). Suppose $\mathsf{D}(FA, B) \cong \mathsf{C}(A, GB)$ naturally in A and in B. We need to define a unit η and co-unit ε for the adjunction, and show that they satisfy the triangle identities.

Now, as we noted in §41.3, the naturality in B means that we have a natural isomorphism $\varphi_A\colon \mathsf{D}(FA, -) \cong \mathsf{C}(A, G-)$. But we've met natural transformations between covariant hom-functors before in §37.1. And the proof of Theorem 178 indicates that the whole transformation φ_A is fixed by what it does to relevant identity arrow, in this case 1_{FA}. So let's look at that.

As a particular case, then, we have $\mathsf{D}(FA, FA) \cong \mathsf{C}(A, GFA)$, naturally in A. And the FA-component of φ_A here sends 1_{FA} to an arrow with the right source and target that we will (in hope!) call $\eta_A\colon A \to GFA$.

We now show that, as we vary A, the components η_A assemble into a natural transformation $\eta\colon 1_{\mathsf{C}} \Rightarrow GF$. So consider the following two diagrams:

$$\begin{array}{ccc} FA & \xrightarrow{Fc} & FA' \\ \downarrow{\scriptstyle 1_{FA}} & & \downarrow{\scriptstyle 1_{FA'}} \\ FA & \xrightarrow{Fc} & FA' \end{array} \qquad\qquad \begin{array}{ccc} A & \xrightarrow{c} & A' \\ \downarrow{\scriptstyle \eta_A} & & \downarrow{\scriptstyle \eta_{A'}} \\ GFA & \xrightarrow{GFc} & GFA' \end{array}$$

The diagram on the left (living in D) commutes for all $c\colon A \to A'$: i.e. $Fc \circ 1_{FA} = 1_{FA'} \circ Fc$. The bijection from D-arrows to C-arrows transposing across the adjunction must preserve identities. So $\overline{Fc \circ 1_{FA}} = \overline{1_{FA'} \circ Fc}$. But by the first of the naturality requirements in §41.3, $\overline{Fc \circ 1_{FA}} = GFc \circ \overline{1_{FA}} = GFc \circ \eta_A$. And by the other naturality requirement, $\overline{1_{FA'} \circ Fc} = \overline{\overline{\eta_{A'}} \circ Fc} = \eta_{A'} \circ c$. So we have $GFc \circ \eta_A = \eta_{A'} \circ c$ and the diagram above on the right also commutes for all c. Hence the components η_X do assemble into a natural transformation.

Similarly the same bijection forming the adjunction but taken in the opposite direction sends 1_{GB} to its transpose $\varepsilon_B\colon FGB \to B$, and the components ε_B assemble into a natural transformation from FG to 1_{D}.

But do η and ε satisfy the triangle identities? Consider these two diagrams:

$$\begin{array}{ccc} A & \xrightarrow{\eta_A} & GFA \\ \downarrow{\scriptstyle \eta_A} & & \downarrow{\scriptstyle 1_{GFA}} \\ GFA & \xrightarrow{1_{GFA}} & GFA \end{array} \qquad\qquad \begin{array}{ccc} FA & \xrightarrow{F\eta_A} & FGFA \\ \downarrow{\scriptstyle 1_{FA}} & & \downarrow{\scriptstyle \varepsilon_{FA}} \\ FA & \xrightarrow{1_{FA}} & FA \end{array}$$

The diagram on the left in C trivially commutes. Transpose it into D via the adjunction, and we find that the diagram on the right must also commute. Therefore $\varepsilon_{FA} \circ F\eta_A = 1_{FA}$ for all A in C – which gives us one of the triangle identities. The other identity we get dually. And we are done. □

42.2 Another definition?

(a) It immediately follows from what we've just shown that, if $F \dashv G\colon \mathsf{C} \to \mathsf{D}$, then the adjunction's unit $\eta\colon 1_{\mathsf{C}} \Rightarrow GF$ has the following 'universal mapping property': for any $c\colon A \to GB$ in C there is a unique associated $d\colon FA \to B$ in D such that $c = Gd \circ \eta_A$.

Why? For existence, put $d = \psi_{AB}c$, with ψ_{AB} as before. Then

$$Gd \circ \eta_A = G\psi_{AB}c \circ \eta_A = G(\varepsilon_B \circ Fc) \circ \eta_A = G\varepsilon_B \circ GFc \circ \eta_A = \\ G\varepsilon_B \circ \eta_{GB} \circ c = 1_{GB} \circ c = c$$

where the equation at the split depends on our earlier result $GFc \circ \eta_A = \eta_{A'} \circ c$, but replacing $c\colon A \to A'$ with $c\colon A \to GB$.

For uniqueness, note that the adjunction implies any $d'\colon FA \to B$ will be equal to $\psi_{AB}c'$ for some $c'\colon A \to GB$. And if $Gd \circ \eta_A = Gd' \circ \eta_A$, then by the previous argument $c = c'$ and hence $d = d'$.

It is worth noting that we can also prove the converse here. Suppose we have functors $F\colon \mathsf{C} \to \mathsf{D}$ and $G\colon \mathsf{D} \to \mathsf{C}$, and a natural transformation $\eta\colon 1_{\mathsf{C}} \Rightarrow GF$ such that for any $d\colon A \to GB$ in C there is a unique $c\colon FA \to B$ for which $d = Gc \circ \eta_A$. Then $F \dashv G$.

Why? Define $\varphi_{AB}\colon \mathsf{D}(FA, B) \to \mathsf{C}(A, GB)$ by putting $\varphi_{AB}c = Gc \circ \eta_A$. By the same proof as for Theorem 219, when we keep A fixed, the various components φ_{AB} assemble into a natural transformation $\varphi_A\colon \mathsf{D}(FA, -) \Rightarrow \mathsf{C}(A, G-)$. And when we keep B fixed, the various components φ_{AB} assemble into a natural transformation $\varphi_B\colon \mathsf{D}(F-, B) \Rightarrow \mathsf{C}(-, GB)$. Further, by the uniqueness clause, the components φ_{AB} are bijections, so the natural transformations are actually natural isomorphisms. Therefore $\mathsf{D}(FA, B) \cong \mathsf{C}(A, GB)$ naturally in A and in B.

Putting all this together, then, we get the first half of the following theorem (with the dual half left as another exercise for enthusiasts):

Theorem 221. *Given functors $F\colon \mathsf{C} \to \mathsf{D}$ and $G\colon \mathsf{D} \to \mathsf{C}$, then $F \dashv G$ iff (i) there is a natural transformation $\eta\colon 1_{\mathsf{C}} \Rightarrow GF$, for which (ii) for any $c\colon A \to GB$ there is a unique $d\colon FA \to B$ such that $c = Gd \circ \eta_A$.*

And dually, we also have $F \dashv G$ iff (i') there is a natural transformation $\varepsilon\colon FG \Rightarrow 1_{\mathsf{D}}$, for which (ii') for any $d\colon FA \to B$ there is a unique $c\colon A \to GB$ such that $d = \varepsilon_B \circ Fc$. □

(b) Evidently, we could have recruited either half of this latest theorem as the basis of a further alternative definition for $F \dashv G$. In fact, this is the *first* definition of an adjunction given by Awodey (2010, §9.1). I'm not sure that's the best way to start; but the theorem is certainly useful in applications.

Go back to one of our first examples of an adjunction in §41.2. We noted that – assuming that C has all products – the product functor $\otimes\colon \mathsf{C}^2 \to \mathsf{C}$ is right adjoint to the binary diagonal functor $\Delta\colon \mathsf{C} \to \mathsf{C}^2$. We can now prove a converse result:

Theorem 222. *If the binary diagonal functor $\Delta\colon \mathsf{C} \to \mathsf{C}^2$ has a right adjoint, then C has all products.*

Proof. Suppose $\Delta \dashv G$. By definition, we want to show that, given any X, Y, there is an object O and two projection arrows $\pi_1 : O \to X$ and $\pi_2 : O \to Y$ such that, for any object S and arrows $f_1 : S \to X$ and $f_2 : S \to Y$, there is a unique mediating arrow $u : S \to O$ such that $f_1 = \pi_1 \circ u$ and $f_2 = \pi_1 \circ u$ (those objects and arrows all in C of course).

Now, the pair of arrows $\langle f_1, f_2 \rangle$ is, by definition, a C^2-arrow from ΔS (i.e. $\langle S, S \rangle$) to $\langle X, Y \rangle$. And our assumed adjunction associates that arrow with a unique C-arrow $u : S \to O$ (for $O = G\langle X, Y \rangle$), such that $\langle f_1, f_2 \rangle = \varepsilon_O \circ \Delta u$, where $\varepsilon : \Delta G \Rightarrow 1_{\mathsf{C}^2}$ is the co-unit of the adjunction.

So what is the component of ε at $O = \langle X, Y \rangle$? It is an arrow in C^2 from $\Delta G\langle X, Y \rangle$ to $1_{\mathsf{C}^2} \circ \langle X, Y \rangle$, in other words an arrow from $\langle O, O \rangle$ to $\langle X, Y \rangle$. But such an arrow in C^2 is a pair of arrows $\pi_1 : O \to X, \pi_2 : O \to Y$ (which is fixed independently of f_1, f_2).

Hence we have $\langle f_1, f_2 \rangle = \langle \pi_1, \pi_2 \rangle \circ \langle u, u \rangle$. However, composition of arrows in C^2 is defined component-wise. So that means $f_1 = \pi_1 \circ u$ and $f_2 = \pi_1 \circ u$. Which was to be proved. □

We can use a similar argument to show e.g. that if $\Delta_{\mathsf{J}} : \mathsf{C} \to [\mathsf{J}, \mathsf{C}]$ has a right adjoint, then C has all limits of shape J.

42.3 Uniqueness and composition

(a) Let's now return to two other theorems that we stated without proof in the last chapter. First:

Theorem 216. *If a functor has an adjoint, it is unique up to natural isomorphism. If $F \dashv G$ and $F \dashv G'$ then $G \cong G'$. If $F \dashv G$ and $F' \dashv G$ then $F \cong F'$.*

I'll outline a proof using Yoneda here. Then, because you may have so far skipped over the chapters on Yoneda, I'll fully spell out a different line of proof in the next chapter.

Proof sketch. We assume $F \dashv G : \mathsf{C} \to \mathsf{D}$ and $F \dashv G' : \mathsf{C} \to \mathsf{D}$, and aim to show $G \cong G'$.

By assumption, $\mathsf{C}(A, GB) \cong \mathsf{D}(FA, B)$ naturally in A and in B. So, in particular, there is (i) a natural isomorphism $\mathsf{C}(-, GB) \xRightarrow{\sim} \mathsf{D}(F-, B)$. Likewise, $\mathsf{D}(FA, B) \cong \mathsf{C}(A, G'B)$ again naturally in A and in B. So, in particular, there is (ii) a natural isomorphism $\mathsf{D}(F-, B) \xRightarrow{\sim} \mathsf{C}(-, G'B)$.

But Theorem 152 tells us that natural isomorphisms compose. So from (i) and (ii) we can conclude that $\mathsf{C}(-, GB) \cong \mathsf{C}(-, G'B)$. Or equivalently $\mathcal{Y}(GB) \cong \mathcal{Y}(G'B)$. And hence by the Yoneda Principle, i.e. by our Theorem 184, $GB \cong G'B$. But (*) since we started off with everything natural in B, the isomorphism here will also be natural in B. Meaning that $G \cong G'$, as we want. □

To complete this proof sketch – which at least tells us why our theorem *ought* to be true! – we strictly speaking need to confirm (*). But rather than do that, I will as promised give another proof in the next chapter.

(b) Moving on: Theorem 209 told us that if there is a Galois connection $F \dashv G$ between the posets $(C, \preccurlyeq)$ and $(D, \sqsubseteq)$, and a connection $H \dashv K$ between the posets $(D, \sqsubseteq)$ and $(E, \subseteq)$, then there is a Galois connection $HF \dashv GK$ between $(C, \preccurlyeq)$ and $(E, \subseteq)$. And now we can show that, as announced in the last chapter, the same holds for adjunctions more generally:

Theorem 217. *Given* $\mathsf{C} \underset{G}{\overset{F}{\underset{\longleftarrow}{\overset{\longrightarrow}{\bot}}}} \mathsf{D}$ *and* $\mathsf{D} \underset{K}{\overset{H}{\underset{\longleftarrow}{\overset{\longrightarrow}{\bot}}}} \mathsf{E}$, *then* $\mathsf{C} \underset{GK}{\overset{HF}{\underset{\longleftarrow}{\overset{\longrightarrow}{\bot}}}} \mathsf{E}$.

Proof via homsets. Since $H \dashv K$, we have $\mathsf{E}(HFA, C) \cong \mathsf{D}(FA, KC)$, naturally in A – by the argument for Theorem 154(3)[2] – and also naturally in C.

Also, since $F \dashv G$, we have $\mathsf{D}(FA, KC) \cong \mathsf{C}(A, GKC)$, naturally in A and in C.

So by Theorem 154(2), $\mathsf{E}(HFA, C) \cong \mathsf{C}(A, GKC)$ naturally in A and in C. Hence $HF \dashv GK$ □

That was quick and easy. But there is perhaps some additional fun to be had – at least for those who like a bit of diagram-wrangling – by working through another argument. Others can skip.

Proof by units and co-units. Since $F \dashv G$, there is a pair of natural transformations $\eta\colon 1_{\mathsf{C}} \Rightarrow GF$ and $\varepsilon\colon FG \Rightarrow 1_{\mathsf{D}}$, satisfying the usual triangle identities.

Since $H \dashv K$, there are natural transformations $\eta'\colon 1_{\mathsf{D}} \Rightarrow KH$ and $\varepsilon'\colon HK \Rightarrow 1_{\mathsf{E}}$, again satisfying the triangle identities.

We now define two more natural transformations by composition,

$$\eta'' : \quad 1_{\mathsf{C}} \xRightarrow{\ \eta\ } GF \xRightarrow{G\eta' F} GKHF$$

$$\varepsilon'' : \quad HFGK \xRightarrow{H\varepsilon K} HK \xRightarrow{\ \varepsilon'\ } 1_{\mathsf{E}}$$

To show $HF \dashv GK$ it suffices to check that η'' and ε'' also satisfy the triangle identities.

Consider, then, the following diagram:

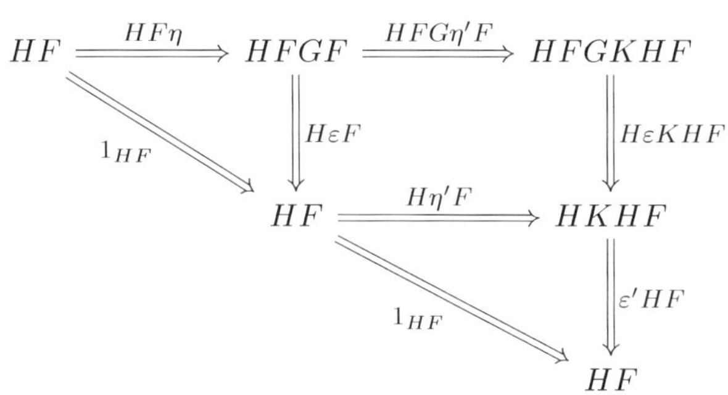

[2] Because by definition $\mathsf{E}(HFA, C) \cong \mathsf{D}(FA, KC)$ naturally in FA; so then put FA for KB in the proof of Theorem 154.

'Whiskering' the triangle identity $\varepsilon F \circ F\eta = 1_F$ by H shows that the top left triangle commutes. And whiskering the identity $\varepsilon' H \circ H\eta' = 1_H$ on the other side by F shows that the bottom triangle commutes.

Further, the square commutes. For by either the naturality of ε or the naturality of η', the following square commutes in the functor category:

$$\begin{array}{ccc} FG & \xRightarrow{FG\eta'} & FGKH \\ \Big\Downarrow{\scriptstyle\varepsilon} & & \Big\Downarrow{\scriptstyle\varepsilon KH} \\ 1 & \xRightarrow{\eta'} & KH \end{array}$$

And whiskering again gives the commuting square in the big diagram. (Exercise: check the claims about whiskering and the naturality square.)

So the whole big diagram commutes, and in particular the outer triangle commutes. But that tells us that $\varepsilon'' HF \circ HF\eta'' = 1_{HF}$ – which is one of the desired triangle identities for η'' and ε''.

The other identity follows similarly. □

42.4 Equivalences and adjunctions again

There's another unproved theorem remaining from the previous chapter. I said:

Theorem 220. *If there is an equivalence between* C *and* D *constituted by a pair of functors* $F\colon \mathsf{C} \to \mathsf{D}$ *and* $G\colon \mathsf{D} \to \mathsf{C}$ *and a pair of natural isomorphisms* $\eta\colon 1_{\mathsf{C}} \xRightarrow{\sim} GF$ *and* $\varepsilon\colon FG \xRightarrow{\sim} 1_{\mathsf{D}}$*, then there is an adjunction* $F \dashv G$ *with unit* η *and new co-unit* ε' *(defined in terms of* η *and* ε*). There is also an adjunction* $F \dashv G$ *with co-unit* ε *and new unit* η' *(again defined in terms of* η *and* ε*).*

In other words, we can take an equivalence defined as in Defn 126*, fix one of the natural transformations, but tinker (if necessary) with the other one so as to get an adjunction. The devil is in the details. I'll prove one half of the theorem.

Proof. Define the natural transformation ε' by composition as follows:

$$\varepsilon'\colon FG \xRightarrow{FG\varepsilon^{-1}} FGFG \xRightarrow{(F\eta G)^{-1}} FG \xRightarrow{\ \varepsilon\ } 1_{\mathsf{B}}$$

Since η and ε are isomorphisms, and by Theorem 157 whiskering natural isomorphisms yields another natural isomorphism, the inverses mentioned here must exist.

I claim that $F \dashv G$ with unit η and counit ε'. So we 'just' need to establish that, with ε' so defined, we get the usual triangle identities $\varepsilon'_{FA} \circ F\eta_A = 1_{FA}$ for all C-objects A, and also get $G\varepsilon'_B \circ \eta_{GB} = 1_{GB}$ for all D-objects B.

Therefore, firstly, for any A, we need the composite arrow (*)

$$FA \xrightarrow{F\eta_A} FGFA \xrightarrow{(FG\varepsilon^{-1})_{FA}} FGFGFA \xrightarrow{(F\eta G)^{-1}_{FA}} FGFA \xrightarrow{\varepsilon_{FA}} FA$$

to equal the identity arrow on FA (recall, the component of a 'vertical' composite of natural transformations for FA is the composite of the components of the individual transformations).

We begin by noting that, for any C-object A, the following square commutes by the naturality of η:

$$\begin{array}{ccc}
A & \xrightarrow{\eta_A} & GFA \\
\big\downarrow{\scriptstyle \eta_A} & & \big\downarrow{\scriptstyle \eta_{GFA}} \\
GFA & \xrightarrow{GF\eta_A} & GFGFA
\end{array}$$

So we have $\eta_{GFA} \circ \eta_A = GF\eta_A \circ \eta_A$. But since η_A is an isomorphism, it is epic (right-cancellable), so we have $\eta_{GFA} = GF\eta_A$ for all A. Similarly, we have $\varepsilon^{-1}_{FGB} = (FG\varepsilon^{-1})_B$ for all D-objects B.

Now consider, then, the following diagram:

$$\begin{array}{ccc}
FA & \xrightarrow{F\eta_A} & FGFA \\
\big\downarrow{\scriptstyle (\varepsilon^{-1})_{FA}} & & \big\downarrow{\scriptstyle (\varepsilon^{-1})_{FGFA} = (FG\varepsilon^{-1})_{FA}} \\
FGFA & \xrightarrow{FGF\eta_A} & FGFGFA \\
\big\downarrow{\scriptstyle 1_{FGFA}} & \swarrow{\scriptstyle (F\eta G)^{-1}_{FA}} & \\
FGFA & & \\
\big\downarrow{\scriptstyle \varepsilon_{FA}} & & \\
FA & &
\end{array}$$

The top square commutes, being a standard naturality square. (Fill in the schema of Defn. 123 by putting the natural transformation $\alpha = \varepsilon^{-1}\colon 1_{\mathsf{D}} \to FG$, and put f to be the function $F\eta_A\colon FA \to FB$.) And the triangle below commutes because $FGF\eta_A = F\eta_{GFA}$ from the equation above and $F\eta_{GFA} = (F\eta G)_{FA}$ (since $\eta_{GFA} = (\eta G)_{FA}$), so the arrows along two sides are simply inverses, and therefore compose to the identity.

The whole diagram therefore commutes. The arrows on the longer circuit from top-left to bottom form the composite (*). The arrows on the direct route from top to bottom compose to the identity 1_{FA}. The composites are equal and hence we have established that the first triangle identity holds.

The second triangle identity holds by a similar argument. Hence $F \dashv G$. Hooray! □

43 Adjunctions explored

Let's quickly gather our thoughts again. In Chapter 41 we introduced adjunctions, defining them in two ways, and stating some of their very basic properties. In Chapter 42 we did the formal work of actually proving that adjunctions have the claimed properties. The only new idea we met is that adjunctions can also be defined in a further way, in terms of a universal mapping property we can read off Theorem 221.

This chapter now pushes on the discussion by making some intriguing connections between adjunctions, comma categories, representables, limits, and more.

43.1 Adjunctions and comma categories

(a) Back in Chapter 29 we introduced the idea of comma categories. And at the end of §29.2, we looked at a particular case, where – changing labels – A is an object of some category C, and $G\colon \mathsf{D} \to \mathsf{C}$ is a functor, and the resulting comma category $(A \downarrow G)$ has the following data:

(1) The objects are pairs (B, c), where B is a D-object, and c is any C-arrow $c\colon A \to GB$.

(2) An arrow $f\colon (B, c) \to (B', c')$ is a D-arrow $f\colon B \to B'$ such that this triangle commutes in C:

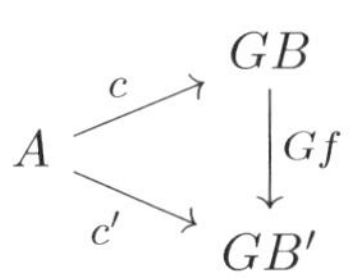

The objects of $(A \downarrow G)$ therefore involve the members of hom-sets like $\mathsf{C}(A, GB)$. But this edges us into now familiar territory: can we link up an adjunction arising from $\mathsf{D}(FA, B) \cong \mathsf{C}(A, GB)$ with some sort of linkage between F and $(A \downarrow G)$?

Yes! – a moment's reflection shows we can make a very neat connection:

Theorem 223. *If $F \dashv G\colon \mathsf{C} \to \mathsf{D}$ is an adjunction with unit η, then for any C-object A, (FA, η_A) is an initial object of $(A \downarrow G)$.*

Proof. We need to show that for every (B, c) in $(A \downarrow G)$ there is a unique arrow $d\colon (FA, \eta_A) \to (B, c)$. In other words, we need to show that there is a unique $d\colon FA \to B$ such that this commutes:

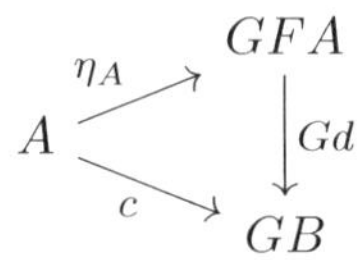

But that's the universal mapping property established in Theorem 221! □

(b) We get a nice converse result too.

Theorem 224. *Suppose $G\colon \mathsf{D} \to \mathsf{C}$ is a functor. If the derived comma category $(A \downarrow G)$ has an initial object for every* C*-object A, then G has a left adjoint.*

Proof. Suppose for each C-object A, the comma category $(A \downarrow G)$ has an initial object: and – in a spirit of hope! – let's write that initial object as (FA, η_A).

We will now define a functor $F\colon \mathsf{C} \to \mathsf{D}$ which, on objects, sends A to FA as just defined. And on arrows, we define F as sending an arrow $c\colon A \to A'$ to $Fc\colon FA \to FA'$ where Fc is the arrow making this commute:

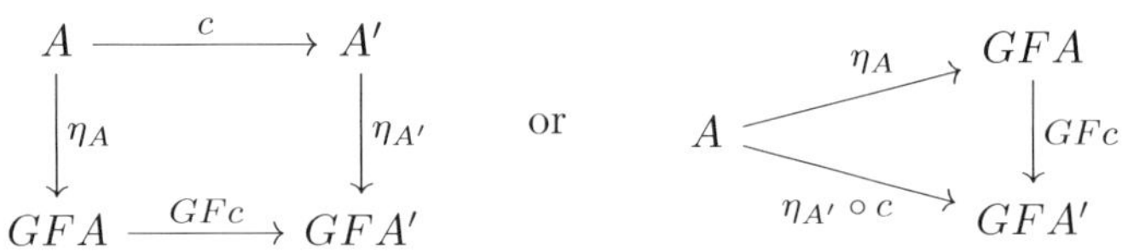

Fc exists and is unique because (FA, η_A) is initial. Then, to show functoriality, consider:

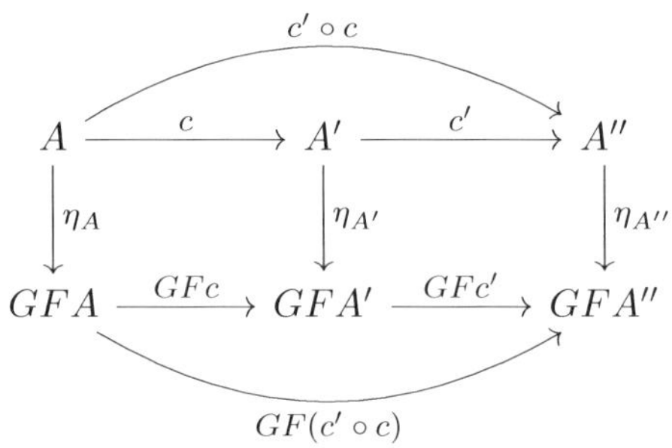

$F(c' \circ c)$ is by definition the unique arrow making the outer rectangle commute. But since $GFc' \circ GFc$, i.e. $G(Fc' \circ Fc)$, also makes that rectangle commute, we have $F(c' \circ c) = Fc' \circ Fc$.

Where have we got to? We have defined a functor F, and looking again at the previous left-hand commuting square we have shown that $\eta_A, \eta_{A'}, \ldots$ assemble into a natural transformation $\eta\colon 1_{\mathsf{C}} \Rightarrow GF$. And by the assumption that (FA, η_A) is initial in $(A \downarrow G)$, we know that for every $c\colon A \to GB$ there is a unique $d\colon FA \to B$ such that $c = Gd \circ \eta_A$.

So we can now appeal to Theorem 221 to conclude that $F \dashv G$. □

(c) Theorem 210 told us how to define e.g. the left adjoint of a function in a Galois connection (if it exists). We now have a nice analogue.

If a functor $G\colon \mathsf{D} \to \mathsf{C}$ has a left adjoint, then there is an initial object for $(A \downarrow G)$ for all C-objects A; and then the proof of the last theorem tells us how to define G's left adjoint F in terms of it. Which is neat.

Of course, we will get dual results for everything in the section, swapping around left and right adjoints. How will the story go? Another exercise!

(d) We can now use Theorem 223 to give a snappy proof of our earlier

Theorem 216. *If a functor has an adjoint, it is unique up to natural isomorphism. If $F \dashv G$ and $F \dashv G'$ then $G \cong G'$. If $F \dashv G$ and $F' \dashv G$ then $F \cong F'$.*

Proof. Suppose we have both $F \dashv G\colon \mathsf{C} \to \mathsf{D}$ with unit η and $F' \dashv G\colon \mathsf{C} \to \mathsf{D}$ with unit η'.

By Theorem 223, for any C-object A, both (FA, η_A) and $(F'A, \eta'_A)$ are initial objects of $(A \downarrow G)$. So, as initial objects, there must be an isomorphism α_A between them, which by the definition of arrows in $(A \downarrow G)$ means we have an isomorphism $\alpha_A\colon FA \to F'A$ in D such that $\eta'_A = G\alpha_A \circ \eta_A$. Likewise, of course, we have an isomorphism $\alpha_{A'}\colon FA' \to F'A'$ such that $\eta'_{A'} = G\alpha_{A'} \circ \eta_{A'}$.

It is then easy to show that $\alpha_A, \alpha_{A'}, \ldots$ assemble into a natural isomorphism $\alpha\colon F \overset{\sim}{\Longrightarrow} F'$, and therefore – as claimed – $F \cong F'$.[1] □

(e) Or was that *too* snappy? Maybe, to complete the argument, I should give the book-keeping details. So take any arrow $c\colon A \to A'$, and form the square

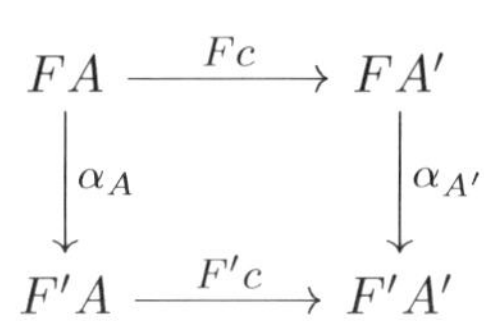

If α is to be a natural isomorphism, we need this square always to commute. And we show it commutes by checking that both (i) $\alpha_{A'} \circ Fc$ and (ii) $F'c \circ \alpha_A$ are arrows from (FA, η_A) to $(F'A', \eta'_{A'} \circ c)$ in $(A \downarrow G)$ – and hence must be equal since (FA, η_A) is initial in the comma category.

But for (i) and (ii) to be arrows with the announced source and target, we need the following two triangles to commute, again by definition of what it takes to be an arrow in $(A \downarrow G)$:

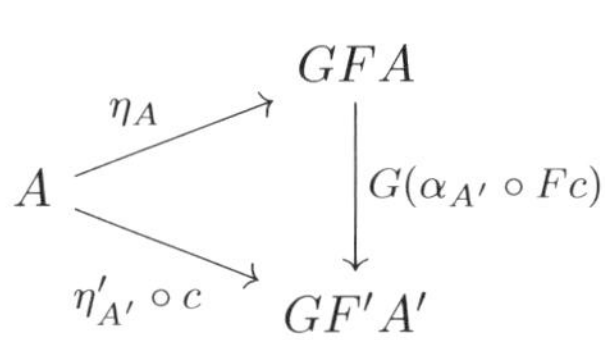

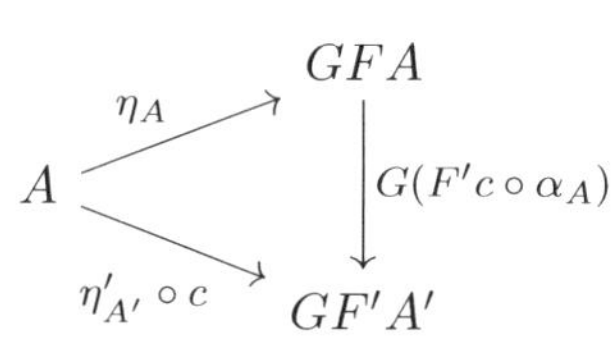

[1] I learnt this basic proof idea, as so much else, from Peter Johnstone's famed Cambridge 'Part III' course on category theory. I particularly recall this as seeming very pleasingly neat at the time! The book-keeping was left as an exercise. Thanks to Izaak van Dongen for importantly correcting my garbled notes.

The left-hand triangle commutes because

$$G(\alpha_{A'} \circ Fc) \circ \eta_A = G\alpha_{A'} \circ GFc \circ \eta_A = G\alpha_{A'} \circ \eta_{A'} \circ c = \eta'_{A'} \circ c$$

and the right-hand triangle commutes because

$$G(F'c \circ \alpha_A) \circ \eta_A = GF'c \circ G\alpha_A \circ \eta_A = GF'c \circ \eta'_A = \eta'_{A'} \circ c$$

with the inner equations in each case appealing to earlier results (exercise: confirm that!). So we are done.

43.2 Adjunctions and fully faithful functors

Let's continue with a simple question. If there is an adjunction $F \dashv G$, then what properties such as being full or faithful must F or G have?

Here's one basic result (a nicely testing exercise to prove – try it, before reading on):

Theorem 225. *Suppose $F \dashv G\colon \mathsf{C} \to \mathsf{D}$ is an adjunction with co-unit ε. Then*

(1) G is faithful iff ε is 'pointwise epic' (i.e. its component ε_B is epic for each D-object B).
(2) G is fully faithful iff ε is an isomorphism.

Proof (1). Given the adjunction, the proof of Theorem 219 tells us that for any D-objects B, B', there is an isomorphism $\psi_{GB,GB'} = \varepsilon_{B'} \circ F\colon \mathsf{C}(GB, GB') \xrightarrow{\sim} \mathsf{D}(FGB, B')$. So, as we vary g, we get a bijection between arrows $Gg\colon GB \to GB'$ and arrows $\varepsilon_{B'} \circ FGg$. But this naturality square for ε

$$\begin{array}{ccc} FGB & \xrightarrow{\;FGg\;} & FGB' \\ \downarrow{\scriptstyle \varepsilon_B} & & \downarrow{\scriptstyle \varepsilon_{B'}} \\ B & \xrightarrow{\;\;g\;\;} & B' \end{array}$$

tells us that $\varepsilon_{B'} \circ FGg = g \circ \varepsilon_B$. So we have a bijection between arrows Gg and arrows $g \circ \varepsilon_B$.

Assume G is faithful. So if $g \neq g'$, then $Gg \neq Gg'$, hence $g \circ \varepsilon_B \neq g' \circ \varepsilon_B$. From which it immediately follows that ε_B is epic.

For the converse, assume ε_B is epic. Then if $Gg = Gg'$, $g \circ \varepsilon_B = g' \circ \varepsilon_B$, so $g = g'$, so G is faithful. □

Proof (2) $\Rightarrow$. Now assume G is fully faithful. Since our adjunction must have a unit $\eta\colon 1_{\mathsf{C}} \Rightarrow GF$, for any D object B there must in particular be a C-arrow $\eta_{GB}\colon GB \to GFGB$. Now, $\eta_{GB} = Gd_B$ for a unique $d_B\colon B \to FGB$ (such a d_B must exist because G is full and must be unique because G is faithful). We'll now show that this d_B is a two-sided inverse to ε_B.

Consider then the following naturality square:

$$\begin{array}{ccc} FGB & \xrightarrow{FGd_B} & FGFGB \\ \downarrow{\scriptstyle \varepsilon_B} & & \downarrow{\scriptstyle \varepsilon_{FGB}} \\ B & \xrightarrow{d_B} & FGB \end{array}$$

But now note that

$$\varepsilon_{FGB} \circ FGd_B = \varepsilon_{FGB} \circ F\eta_{GB} = 1_{FGB}$$

with the first equation holding by the definition of d_B and the second equation being one of the triangle identities. So it follows, given our commuting square, that $d_B \circ \varepsilon_B = 1_{FGB}$. But then, trivially, $\varepsilon_B \circ d_B \circ \varepsilon_B = \varepsilon_B$ – and hence, since ε_B is epic, $\varepsilon_B \circ d_B = 1_B$.

So ε_B has d_B as a two-sided inverse, which proves that ε_B is an isomorphism and hence (since B was arbitrary) that ε is an isomorphism. □

Proof (2) $\Leftarrow$. For the converse result suppose that ε is an isomorphism so that each component ε_B is too. Since ε_B is then epic, we already know that G is faithful, so we just need to show that it is also full. In other words, given any $f\colon GB \to GB'$, there is an arrow $g\colon B \to B'$ such that $f = Gg$.

The obvious construction to try for g, given what's already on the table, is the composite

$$B \xrightarrow{\varepsilon_B^{-1}} FGB \xrightarrow{Ff} FGB' \xrightarrow{\varepsilon_{B'}} B'.$$

And indeed we can show that $Gg = f$.

We need two mini-lemmas. (i) First note that $G\varepsilon_B^{-1}\colon GB \to GFGB$ in C gets sent by the adjunction's isomorphism $\psi_{GB,GFGB} = \varepsilon_{FGB} \circ F$ to its transpose $\varepsilon_{FGB} \circ F(G\varepsilon_B^{-1}) = \varepsilon_{FGB} \circ \varepsilon_{FGB}^{-1} = 1_{FGB}$. But by definition 1_{FGB} is the transpose of η_{GB}. Whence, since transposition involves an isomorphism, $G\varepsilon_B^{-1} = \eta_{GB}$.

(ii) Second, note that whiskering ε with G gives us a natural isomorphism $G\varepsilon\colon GFG \xrightarrow{\sim} G$. And then we have, in particular, this naturality square:

$$\begin{array}{ccc} GFGB & \xrightarrow{GFf} & GFGB' \\ \downarrow{\scriptstyle G\varepsilon_B} & & \downarrow{\scriptstyle G\varepsilon_{B'}} \\ GB & \xrightarrow{f} & GB' \end{array}$$

Which gives us

$$Gg = G(\varepsilon_{B'} \circ Ff \circ \varepsilon_B^{-1}) = G\varepsilon_{B'} \circ GFf \circ \eta_{GB} = f \circ G\varepsilon_B \circ \eta_{GB} = f \circ 1_{GB} = f$$

with the first equation by definition, the second by functoriality and (i), the third equation by (ii), and the fourth by a triangle identity. □

The proof of (2) wasn't straightforward – was I mean to suggest tackling it as an exercise? Well, Leinster (2014, p. 57) does the same. And he adds a supplementary exercise: he notes that an adjunction satisfying the equivalent conditions of (2) is called a *reflection* and he asks you to consider which examples of adjunctions that you know about are reflections.

But we won't pursue that last question here. Instead I'll just add that, almost needless to say, there is a matching pair of dual results we can add:

Theorem 225 (cont'd). *Suppose* $F \dashv G\colon \mathsf{C} \to \mathsf{D}$ *is an adjunction with unit* η. *Then*

(i) F *is faithful iff* η *is 'pointwise monic'.*
(ii) F *is fully faithful iff* η *is an isomorphism.* □

43.3 Adjunctions and representables

(a) Changing tack, let's now think about adjunctions and representable functors. Here's a connection. If $F \dashv G\colon \mathsf{C} \to \mathsf{D}$, then $\mathsf{D}(FA, B) \cong \mathsf{Set}(A, GB)$ naturally in B, i.e. there is a natural isomorphism $\varphi_A\colon \mathsf{D}(FA, -) \cong \mathsf{C}(A, G-)$. So it is immediate that each functor $\mathsf{C}(A, G-)\colon \mathsf{C} \to \mathsf{D}$ is represented by the corresponding object FA. But that's not very exciting.

(b) But suppose we have a (covariant) functor $G\colon \mathsf{D} \to \mathsf{Set}$. This is the sort of functor that *could* itself be representable in the sense of Defn. 139. And it *could* have a left or right adjoint. So now what connections can we make?

Well, first we have an easy result:

Theorem 226. *If* $G\colon \mathsf{D} \to \mathsf{Set}$ *has a left adjoint, it is representable.*

Proof. By assumption, there is a functor $F\colon \mathsf{Set} \to \mathsf{D}$ such that for any set A and D-object B, $\mathsf{D}(FA, B) \cong \mathsf{Set}(A, GB)$, naturally in A and in B.

And by definition, a D-object X together with a natural isomorphism ψ represents G if and only if $\psi\colon G \overset{\sim}{\Longrightarrow} \mathsf{D}(X, -)$.

Choose $A = 1$, your favourite singleton; and put $X = F1$. Then, $\mathsf{Set}(1, GB) \cong \mathsf{D}(F1, B)$ naturally in B. But of course, $GB \cong \mathsf{Set}(1, GB)$ naturally in B (as we noted in §32.6). So putting these facts together, $GB \cong \mathsf{D}(F1, B)$ naturally in B. Hence, G is representable by $F1$ together with the natural isomorphism $\psi\colon G \overset{\sim}{\Longrightarrow} \mathsf{D}(F1, -)$. □

(c) Is there a converse to this last theorem? If a functor G from some category to Set is representable, must it have a left adjoint?

Let's consider what we can learn from looking at (almost) the simplest case; so take the hom-functor $G = \mathsf{Set}(2, -)$, where 2 is your favourite two-membered set.[2] What would it take for this representable functor to have a left adjoint F?

We need, for any A, B, $\mathsf{Set}(FA, B) \cong \mathsf{Set}(A, \mathsf{Set}(2, B))$. On the right we have, in effect, a set of functions mapping A once to B via 0 and once to B via 1. The obvious way of getting a one-to-one matching set of arrows from FA to B

[2] I owe this presentational suggestion to Patrick Stevens.

is then to suppose that F gives us two copies of A, again one tagged 0, the other tagged 1. In other words, we want F to send A to the disjoint union $A+A$, or in other notation $\coprod_{i\in 2} A_i$ (with each $A_i = A$) Make F behave sensibly on arrows so that we have a genuine functor here, and our bijection can then quite easily be shown to be natural in A and B, as needed for an adjunction.

Then generalize, in two steps. First, take the trivially representable hom-functor $G = \mathsf{Set}(X, -)$ where X is now any set you choose. Then the same basic idea works. This will have a left adjoint F where F sends an object X to the disjoint union of X-indexed copies of A, i.e. $\coprod_{i\in X} A_i$ (with each $A_i = A$). Of course, in Set that coproduct is always available. Then, second, suppose we are dealing with any other category D where generalized coproducts like that are always available. Then again the same construction will work. So, hand-waving and for once not pausing to nail down details, we have a result along the lines: if D has enough coproducts, then any representable functor $G\colon \mathsf{D} \to \mathsf{Set}$ will have a left adjoint.

43.4 Right adjoints preserve limits ('RAPL')

(a) Here again is the key definition we need to recall:

Definition 114. A functor $G\colon \mathsf{D} \to \mathsf{C}$ *preserves limits of shape* J iff, for any diagram $D\colon \mathsf{J} \to \mathsf{D}$, if (L, λ_J) is a limit cone over D, then $(GL, G\lambda_J)$ is a limit cone over $G \circ D\colon \mathsf{J} \to \mathsf{C}$.

And we immediately have the following result:

Theorem 227. *Any set-valued functor $G\colon \mathsf{D} \to \mathsf{Set}$ that is a right adjoint preserves all limits that exist in* D *(at least if that's a small category).*

Proof. By Theorem 226, G is a representable functor, so by Theorem 190 preserves all limits that exist in D. □

Which was easy, given what we had already shown. But we now want to prove that there is nothing special here about set-valued functors. We have, quite generally,

Theorem 228. *If the functor $G\colon \mathsf{D} \to \mathsf{C}$ is a right adjoint, it preserves all limits that exist in* D*. Dually, if the functor $F\colon \mathsf{C} \to \mathsf{D}$ is a left adjoint, it preserves all colimits that exist in* C*.*

And we can prove this in various ways. But I think there is a lot to be said for tackling the problem head-on, appealing just to definitions and some very basic principles about (co)limits and adjoints. So, for the first part:

Proof. Suppose that G has the left adjoint $F\colon \mathsf{C} \to \mathsf{D}$; and suppose also that the diagram $D\colon \mathsf{J} \to \mathsf{D}$ has a limit cone (L, λ_J) in D. Then $(GL, G\lambda_J)$ is certainly a cone over GD in C. But what we need to show is that this is a *limit* cone.

So consider the following diagrams, built up in the stages I'm about to describe:

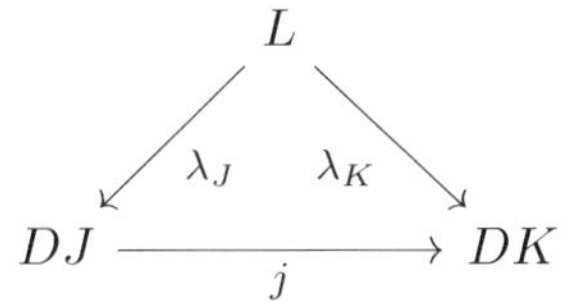

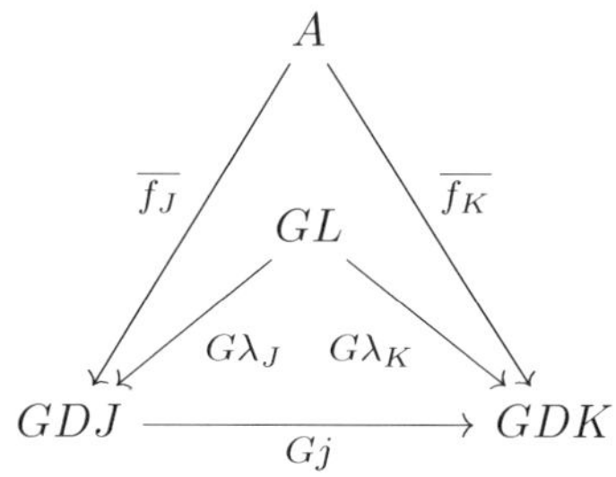

(1) We start on the left, in the category D with the limit cone over D with vertex L. I've diagrammed just part of the cone, the two legs λ_J and λ_K with targets DJ and DK respectively (for J-objects J and K). By assumption, for any J-arrow $j\colon K \to L$, $\lambda_K = j \circ \lambda_J$.

(2) We now use the functor G to send that cone to the category C, and we get the cone with vertex GL we want to prove is a limit cone.

(3) Choose any other cone over GD, and let its vertex be A. Now, the legs of the cone are C-arrows $A \to GDJ$. But under the adjunction $F \dashv G$, such arrows correspond one-to-one to their transposes $FA \to DJ$. So we can specify the legs of the cone as transposes of D-arrows f_J and f_K etc. Which we'll now do, getting as far as the diagram above on the right.

(4) So now go back to D and consider the result of adding the object FA with all the arrows f_J, f_K, etc., as on the left below:

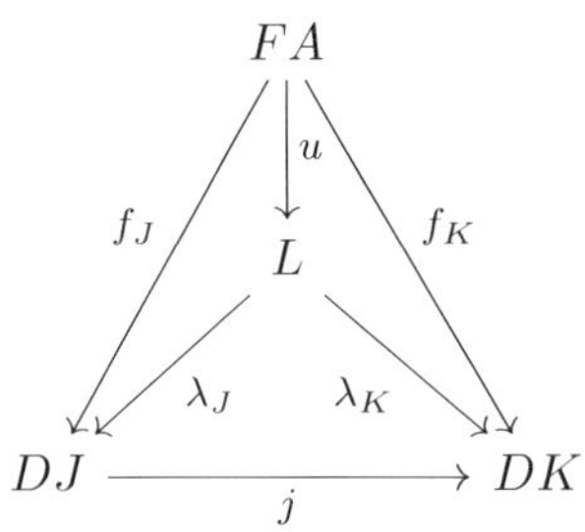

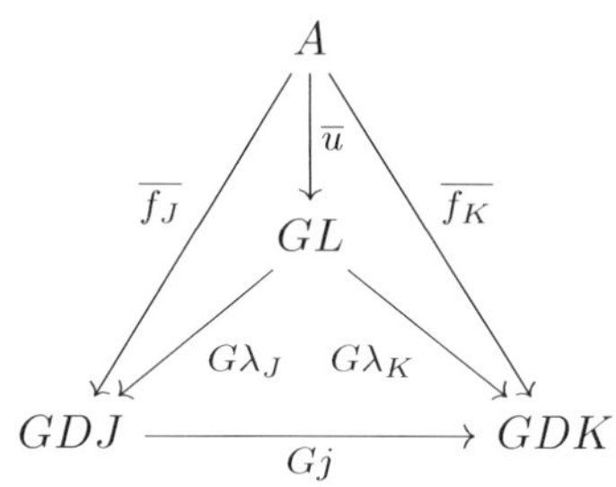

I claim that we've here drawn another cone. Why? Note that $\overline{f_K} = Gj \circ \overline{f_J}$ (since, by assumption, we have a cone on the right with vertex A). But Theorem 218 (1) tells us that $Gj \circ \overline{f_J} = \overline{j \circ f_J}$. Hence $\overline{f_K} = \overline{j \circ f_J}$ and therefore $f_K = j \circ f_J$, showing that FA with the legs f_J, f_K etc. really is a cone.

(5) Since that is a cone in D, it factors through the limit (L, λ_J) via a unique arrow $u\colon FA \to L$. Now, this has a transpose $\overline{u}\colon A \to GL$ under the adjunction. I claim that the cone in C with vertex A factors uniquely through the cone with vertex GL via this arrow $\overline{u}$. And since the cone with vertex A was arbitrarily chosen, that will prove that the cone with vertex GL is a limit cone, as claimed.

(6) So first we need to show that $(A, \overline{f_J})$ factors through $(GL, G\lambda_J)$ via $\overline{u}$, i.e. for any J, $\overline{f_J} = G\lambda_J \circ \overline{u}$.

The adjunction $F \dashv G$ means that $\mathsf{D}(FA, L) \cong \mathsf{C}(A, GL)$ naturally in L. Which means in turn – see §41.3 – that the following square commutes, for any $\lambda_J \colon L \to D_J$,

$$\begin{array}{ccc}
\mathsf{D}(FB, L) & \xrightarrow{\lambda_J \circ -} & \mathsf{D}(FB, D_J) \\
\downarrow & & \downarrow \\
\mathsf{C}(B, GL) & \xrightarrow{G\lambda_J \circ -} & \mathsf{C}(B, GD_J)
\end{array}$$

where the vertical arrows are components of the natural transformation that sends an arrow to its transform. Chase the arrow $u \colon FA \to L$ round the diagram in both directions and we get $G\lambda_J \circ \overline{u} = \overline{\lambda_J \circ u} = \overline{f_J}$.

(7) It remains to confirm $\overline{u}$'s uniqueness. Suppose that $(A, \overline{f_J})$ factors through $(GL, G\lambda_J)$ by some $\overline{v}$. Then for all J, $\overline{f_J} = G\lambda_J \circ \overline{v}$. We show as before that $f_J = \lambda_J \circ v$, whence (FB, f_J) factors through (L, λ_J) via n. By the uniqueness of factorization, $v = u$ again. And we are done. □

I've chunked up that proof into seven stages; but we simply did the more or less obvious thing at each stage, moving between to and fro between the categories C and D either by applying the appropriate one of F and G or by looking at transposes. And this nicely reveals the nuts and bolts of the underlying mechanism by which a right adjoint preserves limits.[3]

And there is of course a dual argument showing why left adjoints preserve colimits.

43.5 Limit (non)preservation: a few examples

Right adjoints preserve limits and dually left adjoints preserve colimits. So now let's see a few elementary examples of (co)limit preservation – and also some

[3]The only pause for thought in writing this up was in choosing which way round – 'f_J' vs '$\overline{f_J}$' – to label a pair of transposes to make things fit nicely with Theorem 218.

Leinster offers the following more abstract line of proof for RAPL (see his 2014, p. 158, and for a close variant, see Awodey 2010, pp. 225–6). Using the notation '$\mathsf{D}(X, D)$' as shorthand for the composite functor $\mathsf{D}(X, -) \circ D$, we have

$$\begin{aligned}
\mathsf{C}(A, G\mathit{Lim}D) &\cong \mathsf{D}(FA, \mathit{Lim}D) \\
&\cong \mathit{Lim}\,\mathsf{D}(FA, D) \\
&\cong \mathit{Lim}\,\mathsf{C}(A, GD) \\
&\cong \mathit{Cone}(A, GD).
\end{aligned}$$

The first line is by the assumed adjunction $F \dashv G$. The next line is by Theorem 191. The next line relies on adjointness again (with a bit of care). And then the last line comes from Theorem 204. Therefore, Leinster concludes, "$G\mathit{Lim}D$ represents $\mathit{Cone}(-, G \circ D)$; that is, it is a limit of $G \circ D$" (now appealing to Theorem 203).

But have we quite got there? We've shown that the right adjoint G takes the vertex of a limit for D to the vertex of a limit of GD; but we officially need to show that G preserves not only vertices of limits but preserves whole limit cones. This further step is supposed to fall out of the naturality of our sequence of congruences in A and D: however, confirming that naturality takes more digging. So in this case I do find my given, much more pedestrian, line of proof at least as illuminating!

examples where we can argue from non-preservation to the non-existence of adjoints.

(1) Back in §28.2 we noted that the forgetful functor $U\colon \mathsf{Mon} \to \mathsf{Set}$ preserves limits. But we now have another proof: U has a left adjoint (by §41.2, Ex. (A7)) i.e. it *is* a right adjoint, so must preserve limits.

There are other examples of this kind, involving a forgetful functor $U\colon \mathsf{Alg} \to \mathsf{Set}$, where Alg is a category of sets equipped with some algebraic structure for U to ignore. Such a forgetful U standardly has a left adjoint, so must preserve whatever limits exist in the relevant Alg.

Further, a left-adjoint to U must preserve existing colimits in Set. But Set has *all* colimits; in order to preserve so much, the left adjoint construction needs to be as free from constraints as possible. As we saw.

(2) Consider Example (A4) of §41.2 again. If C is a category with exponentiation, and hence with products, there is this pair of adjoint functors from C to itself: $(- \times B) \dashv (-)^B$.

Since the functor $(-)^B$ is a right adjoint, it preserves such limits as exist in C. So it will, in particular, preserve terminal objects. Hence 1^B must be terminal, so $1^B \cong 1$. (Compare Theorem 74.)

Similarly for products. To spell this out, consider the diagram $A\colon 2 \to \mathsf{C}$ (where as before 2 is a discrete two object category with objects $0, 1$). A limit over the diagram A in C is a product $(A_1 \times A_2)$ (with its projection arrows, but let's not fuss about them). The claim that $(-)^B$ preserves limits entails that $(A_1 \times A_2)^B$ is the limit over the diagram $(-)^B \circ A\colon 2 \to \mathsf{C}$, which is the product $(A_1^B \times A_2^B)$. So $(A_1 \times A_2)^B \cong (A_1^B \times A_2^B)$. (Compare Theorem 77.)

Since the functor $- \times B$ is a left adjoint, it preserves such colimits as exist in C. Assume C also has coproducts. Then, by a similar argument, $(A_1 + A_2) \times B \cong (A_1 \times B) + (A_2 \times B)$.

(3) Take the discussion in §40.3, Ex. (G7) where we looked at the Galois connection between two functions between posets of equivalence classes of wffs, with the left adjoint an 'add a dummy variable' map, and the right adjoint provided by applying a universal quantifier. This carries over to an adjunction of functors between certain poset categories. Since quantification is a right adjoint, it preserves limits, and in particular preserves products, which are conjunctions in this category. Which reflects the familiar logical truth that $\forall x(Px \wedge Qx) \equiv (\forall x Px \wedge \forall x Qx)$.

(4) Claim: the forgetful functor $F\colon \mathsf{Grp} \to \mathsf{Set}$ has no right adjoint. Proof: then one-object group is initial in Grp; but a singleton is not initial in Set; so there is a colimit which F doesn't preserve and it therefore cannot be a left adjoint.

(5) The proof of Theorem 133 tells us that the forgetful functor $F\colon \mathsf{Mon} \to \mathsf{Set}$ fails to preserve all epimorphisms. By Theorem 134 this implies that F doesn't preserve all pushouts, and hence doesn't preserve all colimits.

Hence this forgetful functor too can't be a left adjoint. Compare the arm-waving argument to the same conclusion in §41.2. Ex. (A9).

43.6 An afterword on monads

We began this group of chapters by looking at a special case of adjunctions, namely Galois connections. And near the end of that initial discussion, we noted that a Galois connection $F \dashv G$ between the 'home' poset $(C, \preccurlyeq)$ and some 'foreign' poset $(D, \sqsubseteq)$ imposes a condition at home on $(C, \preccurlyeq)$ – namely, there has to be a closure function on that poset (see 40.6 and Theorem 213).

Is something along the same lines true of adjunctions in general? Suppose we have an adjunction $F \dashv G\colon \mathsf{C} \to \mathsf{D}$. Then what does this tell us about structures that we must already be able to find at home in C by itself?

(a) We know that, given $F \dashv G\colon \mathsf{C} \to \mathsf{D}$,

(i) there must be an endofunctor $T\colon \mathsf{C} \to \mathsf{C}$, such that

(ii) there is a natural transformation $\eta\colon 1_{\mathsf{C}} \Rightarrow T$.

That follows, of course, by putting $T = GF$, and taking η to be the unit of the adjunction. The functor T will also feature in another natural transformation. For don't forget the co-unit $\varepsilon\colon FG \Rightarrow 1_{\mathsf{D}}$. This can be whiskered left and right to give a natural transformation $G\varepsilon F\colon GFGF \Rightarrow GF$, i.e.

(iii) there is a natural transformation which we can abbreviate $\mu\colon TT \Rightarrow T$.

So what more do we know about this triple (T, η, μ)?

Well, first recall the two triangle identities for our adjunction, diagrammed on the left below. And then post-whisker the ingredients of the first triangle with G, and pre-whisker the second one with F, giving us the diagram on the right which must commute too:

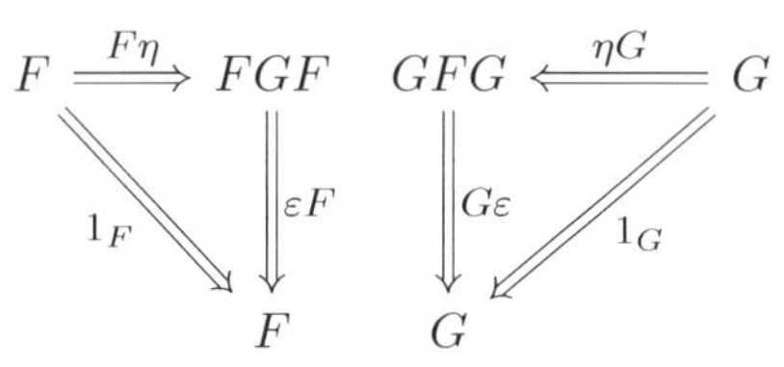

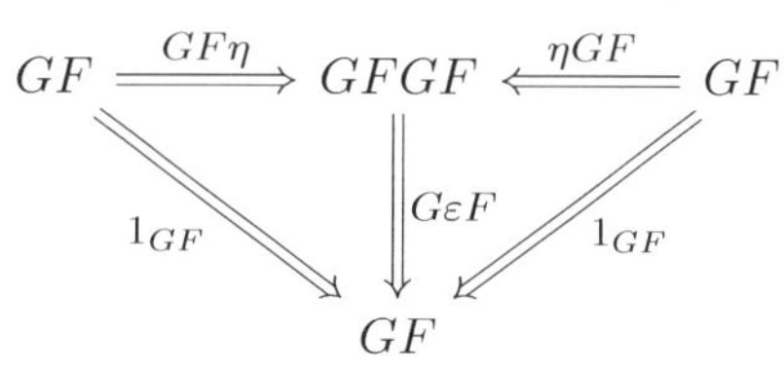

In other words, the following diagram commutes – call it (T1):

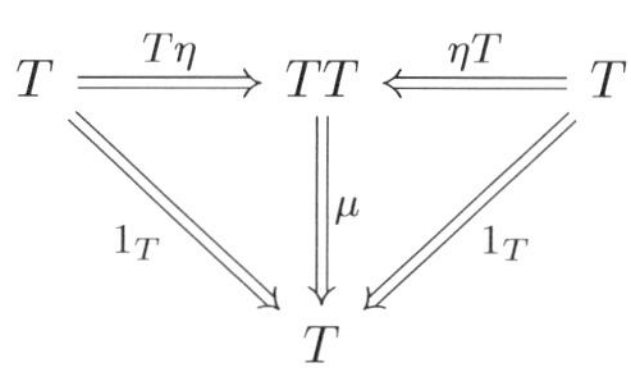

Next, let's take again the natural transformation $\varepsilon\colon FG \Rightarrow 1_{\mathsf{D}}$, and think about how this operates on its own component $\varepsilon_{FA}\colon FGFA \to FA$ (tricksy, eh?). We get the naturality square on the left below. And hitting everything with the functor G, this means the square on the right commutes too:

$$\begin{array}{ccc} FGFGFA & \xrightarrow{FG\varepsilon_{FA}} & FGFA \\ \downarrow{\scriptstyle \varepsilon_{FGFA}} & & \downarrow{\scriptstyle \varepsilon_{FA}} \\ FGFA & \xrightarrow{\varepsilon_{FA}} & FA \end{array} \qquad \begin{array}{ccc} GFGFGFA & \xrightarrow{GFG\varepsilon_{FA}} & GFGFA \\ \downarrow{\scriptstyle G\varepsilon_{FGFA}} & & \downarrow{\scriptstyle G\varepsilon_{FA}} \\ GFGFA & \xrightarrow{G\varepsilon_{FA}} & GFA \end{array}$$

But A was arbitrary, so this tells us that the following left-hand square of natural transformations commutes. In other words, we get the commuting diagram on the right – call it (T2):

$$\begin{array}{ccc} GFGFGF & \xRightarrow{GFG\varepsilon F} & GFGF \\ \Downarrow{\scriptstyle G\varepsilon FGF} & & \Downarrow{\scriptstyle G\varepsilon F} \\ GFGF & \xRightarrow{G\varepsilon F} & GF \end{array} \qquad \begin{array}{ccc} TTT & \xRightarrow{T\mu} & TT \\ \Downarrow{\scriptstyle \mu T} & & \Downarrow{\scriptstyle \mu} \\ TT & \xRightarrow{\mu} & T \end{array}$$

OK: so we now have some reason for highlighting triples (T, η, μ) for which (T1) and (T2) hold. Let's give them a name:

Definition 149. If $T\colon \mathsf{C} \to \mathsf{C}$ is a functor, and $\eta\colon 1_{\mathsf{C}} \Rightarrow T$ and $\mu\colon TT \Rightarrow T$ are natural transformations such that the diagrams (T1) and (T2) commute, i.e. such that

(1) $\mu \circ T\eta = 1_T = \mu \circ \eta T$, and
(2) $\mu \circ T\mu = \mu \circ \mu T$,

then the triple (T, η, μ) is called a *monad* in the category C. △

Why *monad*? Look again at §14.2, where we saw how to define groups categorially inside Set. (G1) and (G2), the first two diagrams there, suffice to define *monoids* in Set (since for a monoid we don't need the additional diagram (G3) giving us inverses). Run together the two versions of triple products in (G1) and we immediately arrive at two diagrams for monoids with very close parallels to the two diagrams (T1) and (T2) characterizing monads. Indeed, monads *are* monoids in a suitable category of endofunctors. And this explains where the transformation μ is referred to as the *multiplication* of the monad; η is, naturally enough, the *unit*.

We saw in §40.6 that Galois connections give rise to closure functions, which we notated by 'T'. And now we have seen that

Theorem 229. *If $F \dashv G$ is an adjunction between* C *and another category, then there is a monad (T, η, μ) in* C.

(b) If adjunctions are everywhere, as the slogan has it, and if adjunctions always give rise to monads, then monads are everywhere too. And monads – originally called 'standard constructions' or simply 'triples' – were in fact first investigated independently of adjunctions, as early as the late 1950s. There are significant examples which arise 'in the wild', in algebra and topology. However, I'm just going to give a couple of very elementary examples of monads. So consider:

(M1) Recall Example (A5) from §41.2. We located an adjunction $F \dashv G\colon \mathsf{Set} \to \mathsf{Pfn}$ where (i) F is simply an inclusion functor, while G totalizes functions by sending a set X to the augmented set $X + \star$, and sending a partial function $\varphi\colon X \to Y$ to the total function $f\colon X + \star \to Y + \star$, where $f\star = \star$, $f(x) = y$ when $\varphi(x)$ is defined and takes the value y, and $f(x) = \star$ otherwise. Now,

(i) let's put $T = GF$, so T is the endofunctor that sends X to $X + \star$, and sends a function $f\colon X \to Y$ to the function $f^+\colon X + \star \to Y + \star$ such that $f^+\star = \star$ and otherwise f^+ agrees with f.

And what accompanying natural transformations give us a monad?

(ii) Evidently, we'll want the unit's components $\eta_X\colon X \to TX$ simply to be inclusion functions.

(iii) As for components of the multiplication, let $\mu_X\colon X + \star + \star \to X + \star$ be the identity on X while sends each of the two augmenting elements in $X + \star + \star$ to the single augmenting element of $X + \star$.

It is the simplest of direct checks to confirm that we have well-defined two natural transformations and that (T, η, μ) is a monad.

(M2) Let's now take the familiar adjunction $F \dashv U\colon \mathsf{Set} \to \mathsf{Mon}$, where U is the forgetful functor sending a monoid to its underlying set, and F is the functor sending a set to X the free monoid on that set – where we take that monoid to be the standard construct $(\mathit{List}(X), ^\frown, \emptyset)$. This adjunction induces a monad on Set:

(i) Its endofunctor $L = UF\colon \mathsf{Set} \to \mathsf{Set}$ sends a set X to the set of finite lists of members of X, and sends a function $f\colon X \to Y$ to the function $Lf\colon LX \to LY$ which maps the list $\langle x_0, x_1, x_2, \ldots, x_n\rangle$ to $\langle fx_0, fx_1, fx_2, \ldots, fx_n\rangle$.

And what are the corresponding natural transformations η and μ? Recall, $\eta_X : X \to UFX$ is the transpose under the adjunction of 1_{FX} (and likewise $\eta_Y : FUY \to Y$ is the transpose of 1_{UY}). Looking at the discussion in §41.2, we can read off that

(ii) η has components like $\eta_X : X \to LX$, which sends each element $x \in X$ to the one-element list $\langle x\rangle \in LX$.

(iii) μ flattens lists of lists: so it has components like $\mu_X : LLX \to LX$, which sends an LL-element such as

$$\langle\langle x_{00}, x_{01}, x_{02}, \ldots, x_{0m}\rangle, \langle x_{10}, x_{11}, x_{12}, \ldots, x_{1n}\rangle, \ldots, \langle x_{k0}, x_{k1}, x_{k2}, \ldots, x_{ks}\rangle\rangle$$

to the corresponding L-element

$$\langle x_{00},\ x_{01},\ x_{02}, \ldots,\ x_{0m},\ x_{10}, x_{11},\ x_{12}, \ldots,\ x_{1m}, \ldots,\ x_{k0},\ x_{k1},\ x_{k2}, \ldots,\ x_{ks}\rangle.$$

Thus defined, (L, η, μ) is a monad.

But we don't need to rely on the generation of the triple from the adjunction to see that. It again is a very easy exercise to check directly that η_X

and μ_X do assemble into natural transformations, and that the resulting natural transformations satisfy the conditions diagrammed by (T1) and (T2).[4]

43.7 Further questions

I'll mention a couple of particular issues we've left hanging in this chapter.

(a) Right adjoints preserve limits. But is there a converse result? If a functor preserves limits must it be a right adjoint?

Well, there are some significant theorems of the following shape:

Suppose $G\colon \mathsf{D} \to \mathsf{C}$, where G, C and D satisfy certain special conditions: then G is a right adjoint if and only if it preserves limits.

However, it is not straightforward either to motivate the special conditions that give us significant results of this form or to usefully apply them in practice.[5] And so – a judgement call – I've decided that these various so-called Adjoint Functor Theorems sit over the boundary of what belongs in these introductory, entry-level, notes on category theory, and I'll give no more details here. I have to draw the line somewhere!

(b) As noted a moment back, we earlier showed that Galois connections give rise to closure functions. But we also proved a reverse result: Theorem 213 tells us that given a closure function T for the poset $(C, \preccurlyeq)$, then we can find another poset such that there is a Galois connection between the two. So another question: is there similarly a reverse result for monads?

Again yes: not only do adjunctions give rise to monads, but also any monad can be seen as arising from an adjunction (in fact, in more than one way). But proving this, and exploring the new ideas about algebras for monads that we meet in doing so, also takes us beyond entry level, say I!

So this is as good a point as any to end Part II of these notes.[6]

---

[4]Our two elementary examples of monads aren't exactly thrilling. However, the first does record some of the functional apparatus we might use for gracefully handling partial functions (by adding a default point to signal 'undefined'); and the second records some functional apparatus we might want for gracefully working with lists. But handling partial functions and handling lists are *very* basic programming tasks – so we get, perhaps, the merest hint of how it could be that the categorial notion of a monad might turn out be of interest to theoretical computer science.

[5]Peter Johnstone once set an exam question 'Write an essay on (a) the usefulness, or (b) the uselessness, of the Adjoint Functor Theorems'.

[6]If you do want to know a little more about (a) the adjoint functor theorems and (b) monads and their algebras, a good introduction is given in Julia Goedecke's notes (2013, pp. 30–47). For more on (a) see Awodey (2010, pp.239–244), Leinster (2014, pp. 159–164), Riehl (2017, pp. 144–151), or Roman (2017, pp. 135–141). For more on (b) see Awodey (2010, Ch. 10) and Riehl (2017, Ch. 5).

Interlude

Part I (Chapters 2–25) of these notes looked inside categories, exploring various familiar constructions such as forming subobjects, products, quotients, exponentials, as they now appear in their categorial guise. In Part II (Chapters 26–43) we said something about the functors that can map a construction in one category to the same sort of construction in another category, and we went on to talk about maps between functors, and more. We eventually encountered some distinctive novelties such as the Yoneda Lemma and the concept of adjunctions.

All this has unfolded in a pretty conventional way, probably distinguished only by its gentle pace.[1] The modest goal has simply been to introduce enough ideas, and do it accessibly enough, to give you a solid foundation from which to tackle more sophisticated treatments of category theory and its applications provided elsewhere.

There is a history to be told about the way in which the central ideas of category theory first emerge in the nineteen forties and fifties, starting from explorations of relations between topology and algebra, and initially percolating very slowly.[2] Within a few decades, however, the ideas have become thoroughly mainstream and the topologist J. Peter May can write

> A great deal of modern mathematics, by no means just algebraic topology, would quite literally be unthinkable without the language of categories, functors, and natural transformations introduced by Eilenberg and Mac Lane in their 1945 paper. It was perhaps inevitable that some such language would have appeared eventually. It was certainly not inevitable that such an early systematization would have proven so remarkably durable and appropriate; it is hard to imagine that this language will ever be supplanted. (May 1999, p. 666)

---

[1]Or painfully slow dawdle, according to taste! I have touched on less than half the topics in Peter Johnstone's famed Cambridge course: you can find links to student lecture notes at logicmatters.net/categories.

[2]For an interesting, reasonably brief and readable account, see e.g. Jean-Pierre Marquis (2006, in particular the preamble and §§1 and 2). It is doubtful, however, whether digging much further into the background of early homology theory and abstract algebra is really worth the effort. Though I suppose I should mention that Marquis has also written a long but rather dense book on the history and philosophy of category theory (2008). So too has Ralf Krömer (2007).

That's more than enough reason to engage with category theory and to want to know more; and where you go next will now depend on your particular mathematical interests.

We could stop these entry-level notes at this point. However, I have added a short Part III.

Back at the outset, to get our categorial story under way, I proposed that we initially think of the groups and homomorphisms between them which assemble into the category Grp as living in a universe of sets, understood in a standard sort of way. Similarly for other categories like Top or Pos. So we had in mind the familiar, pre-categorial, idea of mathematical structures as sets equipped with properties, relations and functions which can themselves be thought of as implemented as sets. But having set sail from this safe anchorage, now that our category theory is launched underway, can it give us a better account of mathematical structures, one that is better attuned to mathematical practice? After all, it is an equally familiar truism that much mathematics proceeds in cheerful ignorance of the alleged conventional set-theoretic underpinnings of the structures it studies.

In the final chapters of the book, making up Part III, I'll explore one theme. We will descend from the giddier heights of categorial abstraction to say a little more about an especially interesting family of categories, namely the toposes which have enough gadgetry to provide generous arenas in which we can reconstruct significant amounts of mathematics. Topos theory quickly becomes an extraordinarily rich field of modern mathematics, uniting ideas from algebraic geometry, topology, higher-order logic, and more; nearly all that is way beyond our scope here. But having come this far, it would be a pity not to take a first quick peek over the fence at one small corner of the field, though our discussion will still remain at a decidedly elementary level. So one goal of the final chapters is the again modest one of saying just enough to entice you into further explorations of topos theory and categorial logic. But we will also eventually meet the topos described by ETCS (the Elementary Theory of the Category of Sets), which was first mentioned in §4.3. This arguably provides an alternative to conventional set-theoretic foundations for an account of mathematical structures.

44 On elementary toposes

Let's briefly recall where we had reached by the end of Part I. We had seen that if a category has (1) all finite limits then we can, for example, construct products of any two objects in the category and construct inverse images and other pullbacks too. If our category has (2) all finite colimits then we can, for example, form all quotients. If our category also has (3) all exponentials, we get objects B^A which behave like a function space that collects the arrows from A to B. If a category has (4) a subobject classifier, it will have arrows that behave like characteristic functions. If a category has all of (1) to (4), this enables more constructions – for example, we get an analogue of powersets.

In sum, especially if it also has a natural numbers object, a category with (1) to (4) will provide an arena in which we can implement a lot of 'ordinary' mathematical constructions. Let's pursue this idea.

44.1 Defining an elementary topos

Here, then, is a standard definition:

Definition 150. A category is an *(elementary) topos* if and only if

(1) it is finitely complete,
(2) it is finitely cocomplete,
(3) it has all exponentials,
(4) it has a subobject classifier. △

Why that qualifier *elementary*? It marks a contrast with the earlier notion of a *Grothendieck* topos which has its origins in hard-core algebraic geometry. It would take us much too far afield even to begin to explicate that richer notion. So for now, I'll simply note that the idea of a topos does come in both a stronger (and significantly scarier) and a weaker (and notably more friendly) version. Every Grothendieck topos is an elementary topos, but not vice versa.

There really should be nothing at all puzzling about the weaker idea with which we are going to be concerned: to repeat, it is simply the concept of a category which combines a number of easily-explained and now-familiar features. Henceforth in these notes, take unqualified talk of a topos to be referring to an elementary one.

44.2 A note on our definition

We will take Defn. 150 as our official definition of a topos. But it does turn out to involve a redundancy, since we have:

Theorem 230. *A category that is finitely complete, has all exponentials, and has a subobject classifier, is also finitely cocomplete.*

In other words, condition (2) for being a topos is automatically satisfied if the other three conditions are satisfied. However, this is *not* elementary to prove.[1] Even proving the special case of the existence of an initial object given (1), (3) and (4) isn't simple. So, to get things underway, let's keep condition (2) explicitly built in from the start.

But the fact that (2) *can* be deduced from the other conditions explains why you'll often find an elementary topos defined, more briskly, as being a (properly) Cartesian closed category with a subobject classifier.[2] And in the light of our finite completeness theorems, Theorems 97 and 99, we will have the following result, useful in establishing that a candidate category really is a topos:

Theorem 231. *If a category has a terminal object, all binary products and equalizers (or equivalently, a pullback for every corner), has all exponentials and a subobject classifier, then it is a topos.*

44.3 A few initial examples

(a) Let's start by noting that – as with so many definitions of types of structure – Defn. 150 admits a degenerate case. It is easily checked that

(T1) The category $\mathbf{1}$ comprising a single object and its identity arrow forms a topos.

And echoing Defn. 76, we can say a bit more generally:

Definition 151. A *degenerate* topos is a category where any object has just one isomorphism to itself and one isomorphism to any other object, and there are no other arrows. △

At the other extreme, here's the paradigm example of a non-degenerate topos:

(T2) Set is a topos.

[1] The result was originally announced by C. J. Mikkelsen in a 1972 conference talk, and a published proof was then given by Paré (1974). That proof requires a perhaps surprising amount of apparatus beyond what we have so far introduced, as will become evident if you look at e.g. the version in Mac Lane and Moerdijk (1992, §IV.5).

[2] For the idea of being *properly* Cartesian closed see Defn. 75 and its footnote. And given what we said at the end of Chapter 24, we can also define an elementary topos as a category with finite limits and a power object for every object, as is done by Barr and Wells (1985, p. 75).

Or at least, this is so given that we are making sufficiently conventional assumptions about our default category of sets. Which is not a trivial point. For recall, I did remark in passing in §17.3 that the deviant set theory NF doesn't have all exponentials; so it follows that the category of NF sets doesn't straightforwardly form a topos.[3] But OK: let's continue to assume we take a sufficiently standard line about sets.

(b) Next, though, note that even if we just look at the *finite* sets, all the needed constructions of finite limits and so on are still available. Hence

(T3) FinSet is a topos.

Further, any finite set is isomorphic to a finite ordinal; so all the 'unique-up-to-isomorphism' categorial constructions that we can do with finite sets can be done using only finite ordinals. Therefore we also have

(T4) FinOrd is a topos.

And there is a sense in which FinOrd is the minimal non-trivial topos, for it can be shown that every non-degenerate topos has an isomorphic copy of FinOrd sitting inside it.[4]

Putting aside the degenerate cases, then, every topos has an infinite number of non-isomorphic objects. However, as we've remarked before, having an infinite number of objects doesn't imply having an infinite object. And an axiom of infinity in the form of the assumption that there is a natural numbers object is *not* conventionally built into the definition of a topos.

(c) We know that e.g. Set_* is not even Cartesian closed. But some other categories that are derived in simple ways from the category of sets *are* toposes.[5] For a start, recall what we know about the category M_2. It has an initial object (§9.1(8)), binary products (Theorem 57) and equalizers (Theorem 69), so it has all finite limits. It also has exponentials (§17.3(7)) and a subobject classifier (Theorem 113). Hence

(T5) M_2 is a topos.

And we can generalize. Referring back to §5.7 Example (C21), we have

(T6) For any monoid M, the corresponding category M-Set is a topos.

Further, we can fairly straightforwardly show

(T7) The arrow category $\mathsf{Set}^{\rightarrow}$ not only has a subobject classifier but satisfies the other conditions for being a topos.

[3]See again a very brief but clear explanation by Randall Holmes: tinyurl.com/holmesev. But there is a twist to the story: if we take the so-called 'strongly cantorian sets', which can be regarded as the 'small' sets of the NF universe, then these *do* form a topos: see Forster et al. (2019).

[4]Hint: any topos will contain $1, 1+1, 1+1+1, \ldots$ and, degenerate cases aside, these objects are pairwise non-isomorphic.

[5]Ah, a vexed question: is it one topos and many toposes? or many topoi? I'm with team Johnstone (see his 1997, p. xx) in preferring the first.

(T8) The slice category Set/X is also a topos for any set X.[6]

(d) What about categories that aren't so directly derived from Set? We know Mon and Grp are not Cartesian closed (see §18.2), so cannot be toposes. Pos is Cartesian closed, but it isn't a topos because it doesn't have a subobject classifier (see §23.4). Top is neither Cartesian closed (not so easy to prove) nor does it have a subobject classifier (an easy result, also noted in §23.4). However,

(T9) Graph is a topos.

Theorem 113 tells us that this category has a subobject classifier. Obviously it has a terminal object too – it's the graph with a single vertex and single loop. So it remains to confirm that Graph has binary products, equalizers, and exponentials – something for enthusiasts to try their hands at, perhaps.

We could continue. There is, for example, this centrally important case:

(T10) The category of sheaves over a topological space is a topos.

But it would again take us *much* too far afield to try to explain the real significance of this to those who are unfamiliar with the relevant topological ideas. Likewise for some other important cases. So let's rest content with our initial list of more elementary examples.

44.4 Toposes beget toposes

I should certainly note, though, a couple of key theorems about two very general ways of getting new toposes from old.

The first we have met before – we can restate Theorem 173 like this:

Theorem 173*. *If* C *is a small category, then the functor category* [C, Set] *is a topos, as is the presheaf category* [C^{op}, Set].[7]

Putting this theorem together with our observation in §36.2 that the categories M-Set and Graph are tantamount to functor categories of the kind [C, Set], we then get proofs of the claims (T6) and (T9) from the previous section (at least for the set-sized cases).

The second general theorem I want to mention generalizes the claim that any slice category Set/X is a topos:

Theorem 232. *If* E *is a topos and X is one of its objects, the slice category* E/X *is also a topos.*

This result is quite often called 'the fundamental theorem' of topos theory. Again, this is not outrageously difficult to prove. In each case, we 'just' have to go

[6]For more on Set$^{\rightarrow}$ and Set/X, as well as about versions of M-Set, see Goldblatt (1984, pp. 86–93).

[7]The role of the assumption of smallness-relative-to-Set, recall, is not so much to put a restriction on C as to ensure that our universe Set is capacious enough.

through, checking that the derived categories have the required products, exponentials, subobject classifiers, etc. However, the devil is in the details, and we don't need the details here.[8] It is perhaps enough that we've made the point that categories that are rich enough to be toposes, hence rich enough to make available a whole swathe of mathematical constructions, are indeed many and various.

[8]Enthusiasts can find a proof in e.g. McLarty (1992, §17.3), though you will need to backtrack in that book if you are to follow all the moves in the proof.

45 Four useful theorems

This chapter states four theorems about toposes which aren't intrinsically very exciting, but which turn out to be useful in what follows. You are entirely welcome to note the results, take them on trust, and skip forward. But I have added proofs/proof-sketches: these are modestly revealing about how various features of a topos can interact to deliver a theorem.

45.1 A couple of remarks about future theorems

(a) In Parts I and II, I gave worked-out proofs of more or less every announced result. But here in Part III I am rather more often going to simply state, giving pointers to proofs available elsewhere. On the plus side, this helps us avoid getting bogged down in some distractingly complicated arguments. On the minus side, it will frustrate those who share my 'completist' turn of mind to be sent hither and thither to fill in the gaps. It is of course a judgement call how to strike a balance.

(b) It is worth noting that where I *do* prove results of the form 'If E is an elementary topos, then …', the given argument may not in fact require the category in question to have all the properties of a topos. Now, I could have tried to state the minimal required assumptions in each case. But in practice this again turns out to be unhelpfully distracting. So instead, I'll leave it as a perhaps instructive exercise accompanying each result – ask yourself, theorem by theorem, 'which aspects of the assumption that E is a topos are being relied on *here*?'.

45.2 Three theorems stated

(a) Recall this definition and its motivation from back in §22.3:

Definition 91. An *image* of an arrow $f\colon X \to Y$ is a subobject of Y, namely $(I, m\colon I \rightarrowtail Y)$, such that (i) f factors through the monic arrow m, i.e. for some $e\colon X \to I$, $f = m \circ e$, and (ii) for any monic m', if f also factors through m' so does m (through some mediating arrow u). △

To diagram the situation:

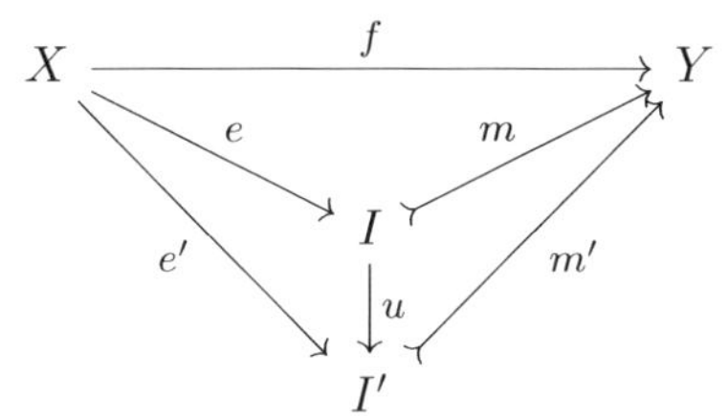

△

And now we can state our first theorem, which generalizes to any topos the simple property of Set which we noted in §8.6:

Theorem 233. *In a topos, any arrow $f\colon X \to Y$ can be factored into an epimorphism $e\colon X \twoheadrightarrow I$ followed by a monomorphism $m\colon I \rightarrowtail Y$, where (I, m) is the image of f.*

Here is the construction we need, developing a hint from §20.5(d). Since we are in a topos, there is a pushout for any pair of arrows. In particular, we have a pushout as on the left, for the arrow $f\colon X \to Y$ taken twice:

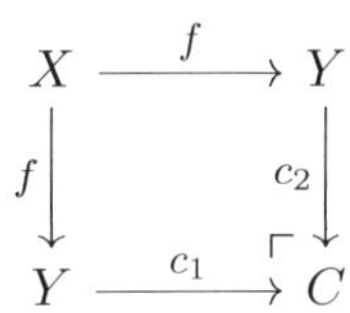

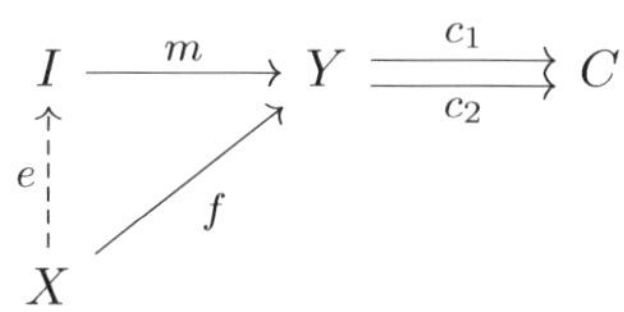

But every parallel pair of arrows gets an equalizer in a topos, so we have in particular an equalizer for the pushout's arrows c_1, c_2, an equalizer which – in anticipation – we will label (I, m). And by the universal property of an equalizer, there must be a unique $e\colon X \to I$ such that the fork with handle f and prongs c_1, c_2 factors through the equalizer, as diagrammed on the right. Then $f = m \circ e$: and I claim this is our desired epi-mono factorization.

So we need to show (1) (I, m) is an image of f, and (2) $e\colon X \to I$ is epic. Both parts involve only some modest diagram chases. So here's a first challenge: try them for yourself before reading §45.3.

(b) Here's a predictable accompanying theorem:

Theorem 234. *In a topos, epi-mono factorizations are unique up to isomorphism. In other words, if $f\colon X \to Y$ factors as $e\colon X \twoheadrightarrow I$ followed by $m\colon I \rightarrowtail Y$ and as $e'\colon X \twoheadrightarrow I'$ followed by $m'\colon I' \rightarrowtail Y$, then there is an isomorphism making this diagram commute, and $m \equiv m'$:*

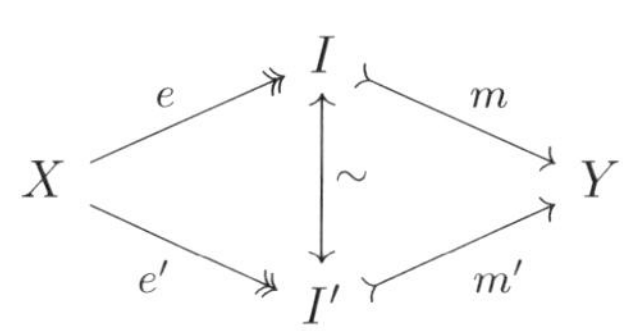

The proof-strategy again makes play with a pushout. So consider the following diagram:

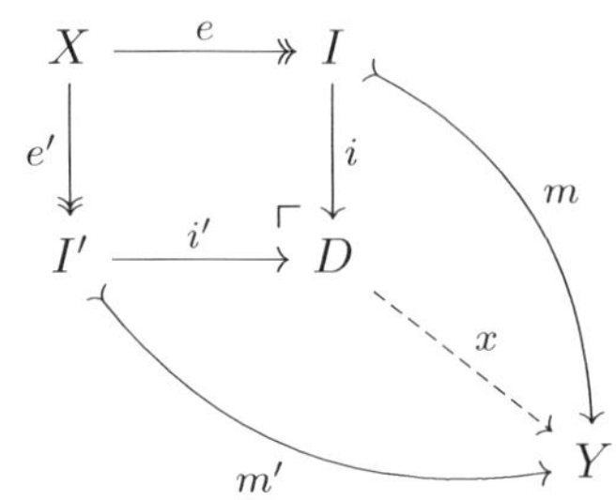

where the corner $I' \to D \leftarrow I$ is the pushout from the wedge $I' \leftarrow X \to I$, and so there is a unique $x\colon D \to Y$ making the diagram commute. We want to show that i and i' are isomorphisms, from which it follows that I and I' are isomorphic. But we can show that i and i' are isomorphisms by showing that (1) they are both monic, and (2) they are both epic – for recall that, in a topos, epic monics are isomorphisms. And if there is an isomorphism $I \xrightarrow{\sim} I'$, then m and m' factor through each other to give $m \equiv m'$.

So here's another challenge: complete the proof.

(c) These last two theorems highlight a respect in which every topos is Set-like: all arrows have epi-mono factorizations, unique up to isomorphism. By contrast, as we already know, there are respects in which toposes can be very different from Set, in particular in the structure of their subobject classifiers (which need not be two-element structures).

But there is an important result about classifiers that *does* obtain across the board:

Theorem 235. *In a topos, the arrow* $[\![\top, \bot]\!]$ *as defined by the coproduct diagram*

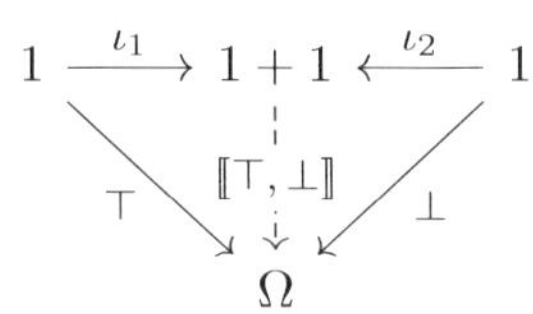

is always monic.

In the case of the degenerate topos, there is nothing to prove (why?). So suppose we are in a non-degenerate topos, where $\top \neq \bot$ (by Theorem 111). Then if the objects in our topos are anything like sets-with-some-structure, the two distinct elements of the local $1+1$ must get sent by a function-like $[\![\top, \bot]\!]$ to distinct elements of the local Ω, making the arrow injective. So, hand-waving, we can reasonably expect our theorem to hold.

I will outline a proof in §45.4, where we will need to invoke a fourth theorem as a lemma en route.

45.3 Two proofs

You can certainly skip these proofs for Theorems 233 and 234, which is why I have separated them off as optional extras. But they are mildly interesting, like e.g. the proof of Theorem 24 about power objects, in showing how features of a topos (e.g. having pushouts, having equalizers, having subobjects) can interact in possibly surprising ways.

Theorem 233: Proof that (I, m) is an image of f. Start again from the diagram

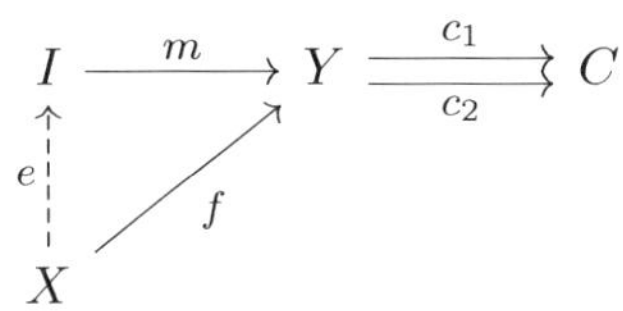

We know m is monic, since any equalizer is monic by the easy Theorem 67. So it remains to show that if f also factors through some monic $m' \colon I' \to Y$, then m likewise factors through m'.

Suppose then that $f = m' \circ e'$ with m' monic. Now, Theorem 116 tells us that m', being monic, is an equalizer of two arrows to Ω, in fact $c_1' = \top_Y$ and $c_2' = \chi_{m'}$. So we have a commuting diagram as on the left:

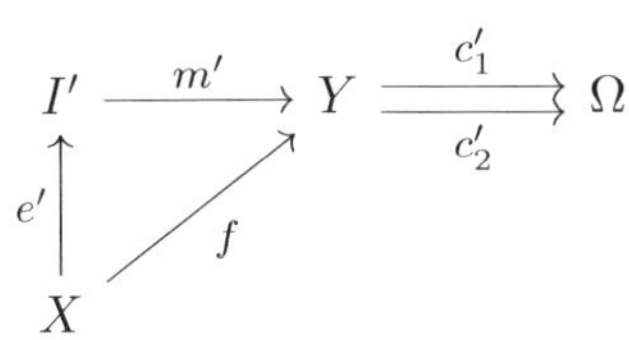

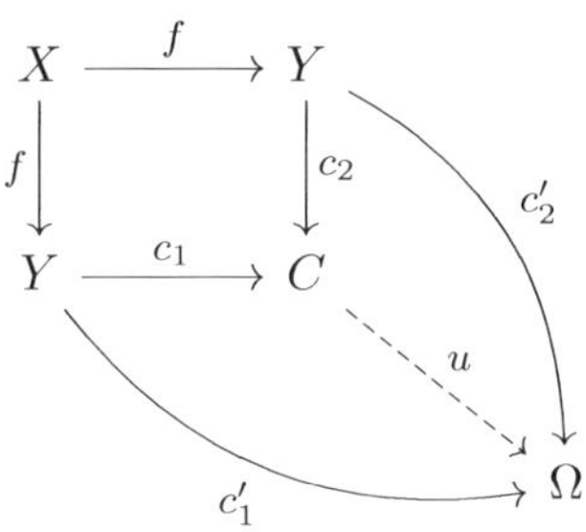

Hence there is a unique u making the diagram on the right above commute, since the upper square is a pushout. Therefore we have

$$c_1' \circ m = u \circ c_1 \circ m = u \circ c_2 \circ m = c_2' \circ m$$

with the first and third equations from the definition of u, and with the second equation reflecting m's definition as equalizing c_1, c_2.

But then, since m' equalizes c_1', c_2' there must in particular be a unique v making this commute:

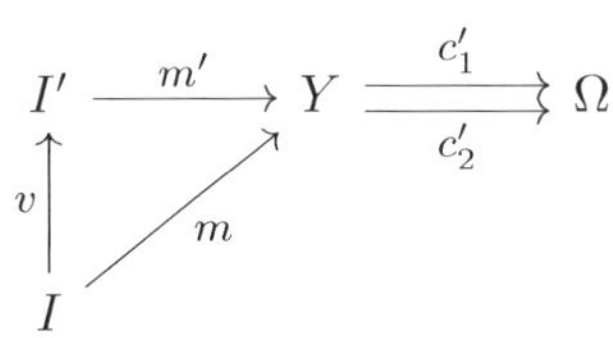

And we are done! We've shown that if f factors through some monic m' then there is some v such that $m = m' \circ v$ □

Theorem 233: Proof that $e \colon X \to I$ is epic. Applying the previous result, we can factor the arrow e itself through its image!

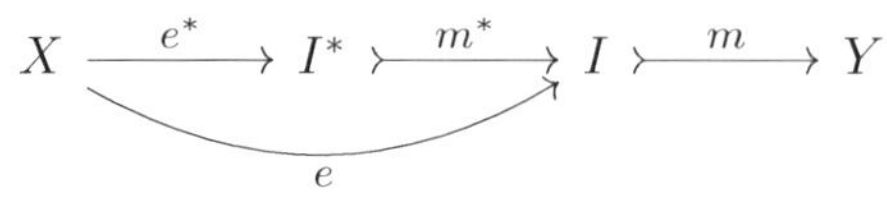

Since monics compose, $m \circ m^*$ is a monic, and since f now factors through this monic, it follows – again from the previous result – that m also must factor through $m \circ m^*$, so for some w, $m = m \circ m^* \circ w$.

A simple little dance then immediately tells us that m^* must be an isomorphism. (If you insist: since $m = m \circ m^* \circ w$ and m is monic, we can cancel on the left to get $1_I = m^* \circ w$. But then $m^* \circ w \circ m^* = m^*$ so, since m^* is also monic, we can cancel on the left again to get $w \circ m^* = 1_{I^*}$. Therefore m^* has a two-sided inverse, and is an isomorphism.)

Now, since the factorization of e through its image follows the same pattern as our factorization of f through *its* image, we have again a pushout square on the left, and then m^* is an equalizer of d_1 and d_2 as on the right:

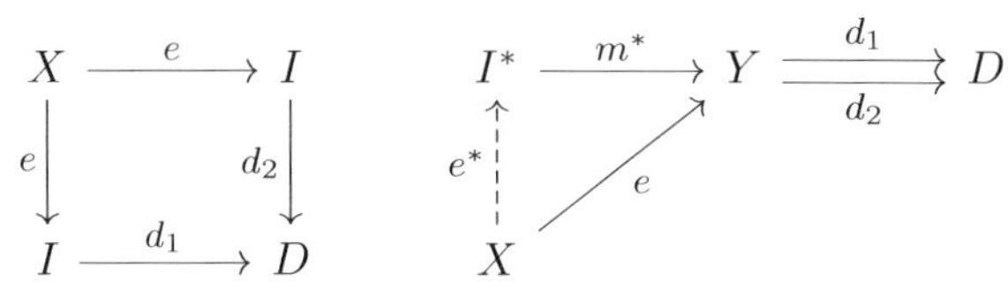

By assumption, then, $d_1 \circ m^* = d_2 \circ m^*$. Since m^* is an isomorphism, it is immediate that $d_1 = d_2$.

So now assume that there are parallel arrows j, k such that $j \circ e = k \circ e$.

Then, given our upper square in the next diagram is a pushout, there is a (unique) arrow x making it all commute:

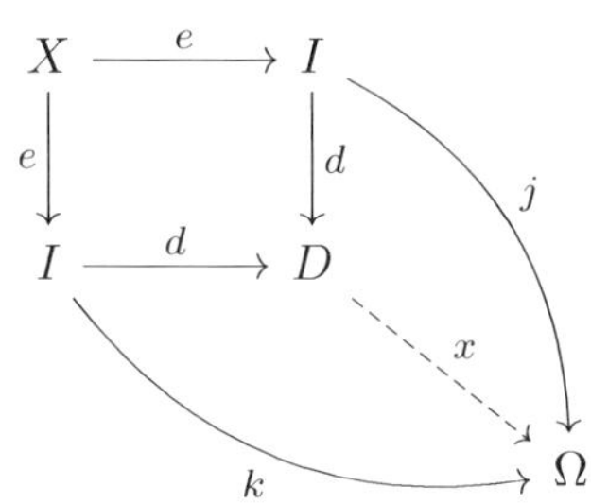

Hence $j = x \circ d = k$. Therefore, as we wanted, e is an epimorphism. □

Thm. 234: Proof. Start from the pushout diagram

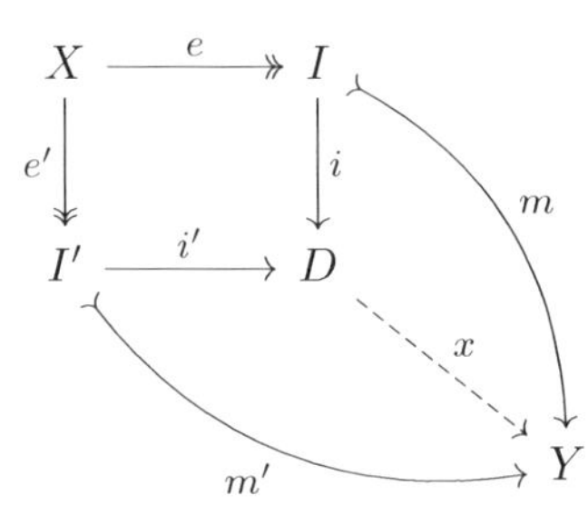

It is immediate that (1) i and i' are monic. Because the arrow m is monic by assumption, and $m = x \circ i$. We just apply Theorem 16 to conclude i is monic. Similarly for i'.

To show that (2) i and i' are epic it is enough to show that $i \circ e$, which equals $i' \circ e'$, is epic. We can then apply Theorem 16 again to conclude i and i' are epic.

So suppose, for arrows $f_1, f_2 \colon D \to E$,

$$f_1 \circ i \circ e = f_1 \circ i' \circ e' = f_2 \circ i \circ e = f_2 \circ i' \circ e'$$

We need to prove $f_1 = f_2$.

Our supposition tells us that the outer paths commute in the left-hand diagram below, and hence by the universal property of the pushout, there is a unique arrow $D \to E$ making the whole diagram commute, which must be f_1. Likewise, the unique arrow $D \to E$ making the right-hand diagram commute is f_2.

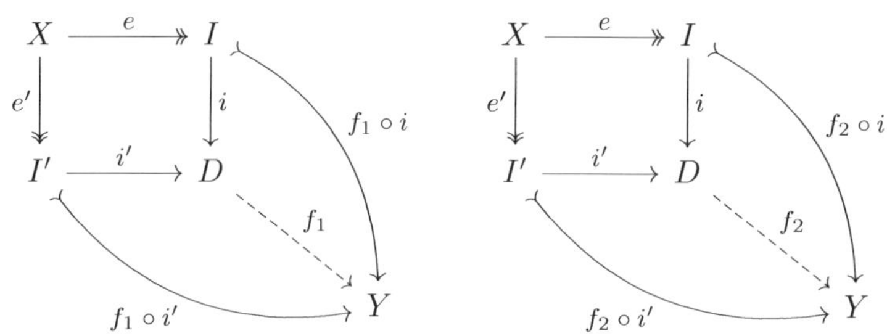

But our supposition, together with the fact that e is epic, tells us that $f_1 \circ i = f_2 \circ i$; and similarly, because e' is epic, $f_1 \circ i' = f_2 \circ i'$. But that means the outer parts of the two diagrams are exactly the same. Hence $f_1 = f_2$. □

45.4 A fourth theorem, and a proof sketch

(a) We need a lemma, both to give a proof sketch for our stated Theorem 235 but also for the crucial later Theorem 253. It tells us about how pullbacks and coproducts interact in a topos:

Theorem 236. *In a topos, if the first two diagrams below are pullbacks, so is the third:*

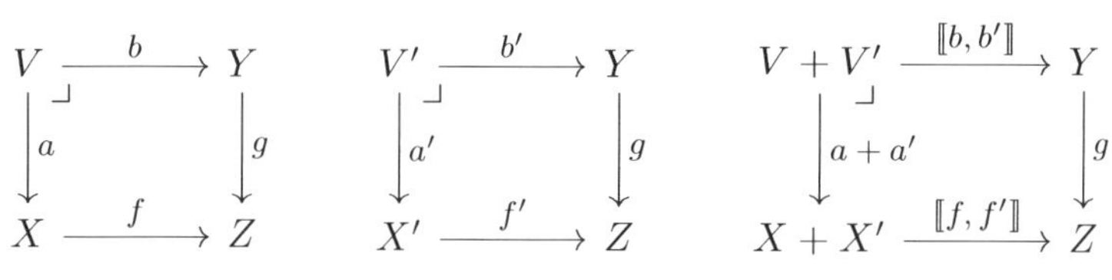

(where the '+' in '$a + a'$' symbolizes the dual of $\times$ from §13.3)

A proof strategy. For a clue about why this should be true, suppose we write the V-vertices of our three diagrams using the product-like notation of §20.1.

Then $V = X \times_Z Y$, $V' = X' \times_Z Y$, and we would expect the vertex of the third pullback, given its opposite corner, to be $(X + X') \times_Z Y$. But now suppose that the product-like $\times_Z$ distributes over coproducts so that $(X + X') \times_Z Y \cong (X \times_Z Y) + (X' \times_Z Y) = V + V'$, and we have recovered the vertex our theorem wants.

But we know that in a topos, products distribute over coproducts (see §43.5). And we know that pullbacks are product-like to the extent of actually being products in an associated slice-category (see Theorem 94). So we should be able to move from our home topos to the appropriate slice topos (using Theorem 232), thereby taking our product-like pullbacks to true products; we distribute a true product over a coproduct in the slice topos; and then we can carry the result back to the home topos as a result about pullbacks. That way, with the details filled in, we'll arrive at our claimed result.[1] □

(b) We'll henceforth allow ourselves, then, to appeal to the lemma Theorem 236. And then I can give

A proof sketch for Theorem 235. We want to show that the arrow $[\![\top, \bot]\!]$ as defined by the coproduct diagram

$$\begin{array}{ccccc} 1 & \xrightarrow{\iota_1} & 1+1 & \xleftarrow{\iota_2} & 1 \\ & {\scriptstyle \top}\searrow & \Big\downarrow{\scriptstyle [\![\top,\bot]\!]} & \swarrow{\scriptstyle \bot} & \\ & & \Omega & & \end{array}$$

is always monic.

Step (1) Use our lemma to show that these two squares are pullbacks (for the trick, see Goldblatt 1984, pp. 119-120, whose proof idea I have borrowed):

$$\begin{array}{ccc} 1 & \xrightarrow{\iota_1} & 1+1 \\ {\scriptstyle 1_1}\Big\downarrow & & \Big\downarrow{\scriptstyle [\![\top,\bot]\!]} \\ 1 & \xrightarrow{\top} & \Omega \end{array} \qquad \begin{array}{ccc} 1 & \xrightarrow{\iota_2} & 1+1 \\ {\scriptstyle 1_1}\Big\downarrow & & \Big\downarrow{\scriptstyle [\![\top,\bot]\!]} \\ 1 & \xrightarrow{\bot} & \Omega \end{array}$$

Step (2) We use our lemma again to conclude that this is a pullback:

$$\begin{array}{ccc} 1+1 & \xrightarrow{[\![\iota_1,\iota_2]\!]} & 1+1 \\ {\scriptstyle 1_1+1_1}\Big\downarrow & & \Big\downarrow{\scriptstyle [\![\top,\bot]\!]} \\ 1+1 & \xrightarrow{[\![\top,\bot]\!]} & \Omega \end{array}$$

Step (3) Now we note that the left arrow and top arrow in that third square are in fact both equal to the identity arrow (why?) to arrive at a final pullback:

[1] OK, that was very hand-wavy! But see also, e.g., the answer by Tony Dolezal at tinyurl.com/co-pullback.

$$\begin{array}{ccc} 1+1 & \xrightarrow{1_{1+1}} & 1+1 \\ {\scriptstyle 1_{1+1}}\downarrow & \lrcorner & \downarrow{\scriptstyle [\![\top,\bot]\!]} \\ 1+1 & \xrightarrow{[\![\top,\bot]\!]} & \Omega \end{array}$$

We can then simply appeal to Theorem 92 to conclude that $[\![\top, \bot]\!]$ is monic, as required. But the devil is in the details. $\square$

46 Logic in a topos

In §22.6 I introduced the idea of the intersection of subobjects, and intimated that we could go on in a similar vein to define unions and complements, giving us an algebra of subobjects. Then in §23.3, we saw that alongside the truth-seeking arrow $\top: 1 \to \Omega$ we can define a false-seeking arrow $\bot: 1 \to \Omega$, and can also define a negation-like arrow $\neg: \Omega \to \Omega$. We can also take this idea further, and go on to introduce accompanying categorial notions of conjunction, disjunction, and a conditional.

That gives us two major themes to explore, which turn out to be closely entwined, namely *algebras of subobjects* and *logic done inside a topos.* It's convenient to tackle the logic first. So in this chapter, inter alia, I aim to go some way towards demystifying a summary slogan that you may well have come across, along the lines of 'the internal logic of a topos is intuitionistic'.

46.1 'Intuitionistic logic'?

(a) Focus on propositional logic. Then 'classical logic' will denote the familiar two-valued, truth-functional, calculus. While 'intuitionistic logic' is what you get if (i) you take a particularly natural set of inference rules governing the four basic connectives – rules which reflect the separate meaning of each – but (ii) you do not add the classical 'double negation' rule that allows us to infer A from $\neg\neg A$ (for any proposition A).

To many, that will be no news. It might be helpful though for some readers if I say a very little more to fill out those headlines.

So, to fix ideas, let's take our language for propositional logic to be built from the four connectives $\wedge, \vee, \Rightarrow, \neg$ together with the absurdity constant $\bot$.[1] These connectives are governed by the following *introduction* rules (which tell us how to deduce a formula whose main connective is as displayed):

$$\frac{A \quad B}{A \wedge B} \qquad \frac{A}{A \vee B} \quad \frac{A}{B \vee A} \qquad \begin{array}{c}[A]\\ \vdots \\ C \\ \hline A \Rightarrow C\end{array} \qquad \begin{array}{c}[A]\\ \vdots \\ \bot \\ \hline \neg A\end{array}$$

[1] We are already using the familiar arrow notation '$\to$' for, well, any arrow: so let's instead use '$\Rightarrow$' for conditionals or for specifically conditional-like arrows. (There will be no danger of confusion with our earlier use of double arrows to signal natural transformations.)

The rule for conjunction needs no explanation. Nor does the bipartite rule for disjunction. The rule for the conditional tells us that, given a proof of C from the assumption A (and perhaps other assumptions), we can drop/discharge the assumption A, and deduce $A \Rightarrow C$ from the assumptions which remain in play. Likewise the rule for negation tells us that, given a proof of absurdity from the assumption A, we can drop that assumption, and deduce $\neg A$ from the assumptions which remain.

Then there are corresponding *elimination* rules (which tell us what we can deduce from a formula whose main connective is as displayed):

$$\frac{A \wedge B}{A} \qquad \frac{A \wedge B}{B} \qquad \frac{A \vee B \quad \begin{matrix}[A] \\ \vdots \\ C\end{matrix} \quad \begin{matrix}[B] \\ \vdots \\ C\end{matrix}}{C} \qquad \frac{A \Rightarrow C \quad A}{C} \qquad \frac{A \quad \neg A}{\bot}$$

The disjunction rule here encapsulates a version of 'proof by cases': if, temporarily assuming A and B in turn, we can prove C either way, then given $A \vee B$ we can infer C. Do note, by the way, that we could alternatively introduce negation by the definition $\neg A =_{\text{def}} A \Rightarrow \bot$: defined like this, negation would still obey the same rules (why?).

Now, the introduction rule for the conditional (for example) might be said to fix the meaning of the connective by telling us what it takes to establish $A \Rightarrow C$, in this case a derivation of C from A. So if you are given $A \Rightarrow C$, that's in effect a promissory note that there is a derivation of C from A; and this promissory note combined with the assumption A then entitles you to conclude C. Which is exactly what the elimination rule says. In a sense, then, given the meaning-fixing introduction rule, the corresponding elimination rule is automatically justified. Similar points apply to the rules for the other connectives.

To these rules for the connectives, we want to add one further rule, one that in effect tells us that $\bot$ really *is* absurd – namely, for arbitrary A,

$$\frac{\bot}{A}$$

Yes: if we suppose $\bot$ is true, anything goes and chaos ensues!

(b) So we have put on the table a bunch of 'natural deduction' rules. There are introduction rules fixing the meaning of each connective; these rules come with matching elimination rules; and we have added the absurdity rule. However, if we stop here, we fall short of full classical logic: we only get what's known as intuitionistic propositional logic (I'll explain the label). In particular, as we'll see in a moment, from the given rules we *can't* derive the familiar classical principle that we can infer A from $\neg\neg A$. Likewise we can't derive the classical law of excluded middle which makes $A \vee \neg A$ a theorem for any A (otherwise, given the premiss $\neg\neg A$, you could appeal to the law $A \vee \neg A$, note that your premiss rules out the second disjunct, and conclude that A must be true after all).

Of course, we can play about with the laws of classical logic to our heart's content, for the sheer fun of it or in order to explore what depends on what. But why might you think that not adopting the classical double negation rule could, in the right circumstances, be the right stance?

Well, suppose (just suppose!) that – with respect to a particular domain of enquiry – you hold that there is no more to being a truth of that domain than being warrantedly assertible and no more to being false than being warrantedly rejectable. For example, an anti-Platonist about mathematics might claim that there is no more to being true than being constructively provable, and no more to being false than being constructively disprovable.

Now, a little reflection suggests that the introduction rules for the connectives should all remain acceptable even if you do think of truth as a matter of being warrantedly assertible and correspondingly think of valid inference as a matter of preserving warrant. And since the matching elimination rules 'come for free' and the absurdity rule simply pins down the notion of absurdity, this means that intuitionistic logic should be acceptable to our anti-Platonist. But note: we could sometimes be in a position in which we can warrantedly rule out getting to *disprove* A (so we could be in a position to warrantedly assert $\neg\neg A$), without thereby being in a position to positively *prove* A (so we couldn't warrantedly assert A). In that situation, the move from $\neg\neg A$ to A doesn't preserve warranted assertibility. (This shows that the classical double negation principle is indeed independent of the intuitionistic rules.)

(c) In short, it can be argued that intuitionistic logic is the appropriate logic e.g. for pursuing constructive mathematics where truth doesn't outrun warranted assertibility (so it is, in particular, an appropriate logic for someone who endorses a version of the radical kind of constructivism that has its roots in Brouwer's 'intuitionist' philosophy of mathematics, hence the name for the logic). The double negation rule in truth *is* an extra, that doesn't automatically come along with the basic introduction and elimination rules.

But if we do add that extra then, yes, we recover the familiar classical propositional calculus. For example, we can derive the law of excluded middle. Though I should mention a surprising complication: instead of adding the double negation rule, there are other ways of augmenting intuitionistic logic giving us an infinite number of distinct propositional logics that sit between intuitionistic and classical logic in strength!

There is a great deal more to be said, both technically and conceptually, about intuitionistic logic. But these very speedy remarks will have to suffice.[2]

46.2 Negation again

(a) Let's remind ourselves of two definitions from §23.3, but now talking of a topos rather than of a Cartesian closed category with a subobject classifier (as

[2]For a little more on intuitionistic logic, you could usefully look at Chapter 8 of Smith (2022) and the further references there. For some hints about constructive mathematics, see tinyurl.com/nlab-cons. More specifically, on intuitionism as a philosophy of mathematics see the entry in the *Stanford Encyclopedia of Philosophy*, tinyurl.com/sep-int.

always, $(\Omega, \top)$ denotes the relevant subobject classifier):

Definition 95*. In a topos, $\bot\colon 1 \to \Omega$ is the characteristic arrow of $(0, !\colon 0 \to 1)$, i.e. the unique arrow that makes this a pullback diagram:

$$\begin{array}{ccc} 0 & \longrightarrow & 1 \\ {\scriptstyle !}\big\downarrow & \lrcorner & \big\downarrow{\scriptstyle \top} \\ 1 & \xrightarrow{\ \bot\ } & \Omega \end{array}$$

△

Definition 96*. In a topos, the arrow $\neg\colon \Omega \to \Omega$ is the characteristic arrow of $(1, \bot)$, making this a pullback:

$$\begin{array}{ccc} 1 & \longrightarrow & 1 \\ {\scriptstyle \bot}\big\downarrow & \lrcorner & \big\downarrow{\scriptstyle \top} \\ \Omega & \xrightarrow{\ \neg\ } & \Omega \end{array}$$

△

And here are updates of two simple theorems we met before. First,

Theorem 111*. *In a non-degenerate topos,* $\bot \neq \top$. □

Second, $\top$ and $\bot$ are related as you'd expect:

Theorem 112*. *In a topos,* $\neg\bot = \top$ *and* $\neg\top = \bot$. □

(b) That second theorem trivially implies that $\neg\neg\top = \top$ and $\neg\neg\bot = \bot$. So let's ask: can double negations be ignored more generally? Given an arbitrary characteristic arrow $\chi\colon X \to \Omega$ in a topos, will we always have $\neg\neg\chi = \chi$ (equivalently, $\neg\neg = 1_\Omega$)?

That holds in Set where Ω is a simple two-element set $\{T, F\}$, a characteristic function $\chi\colon X \to \Omega$ sends each member of X to one of T or F, and $\neg\colon \Omega \to \Omega$ maps $T \mapsto F$, $F \mapsto T$. But for a simple example of topos where we can't drop double negations, consider what happens in M_2.

(i) Recall from Theorem 113 that this topos of sets-equipped-with-idempotent-functions has the subobject classifier whose object Ω is (Ψ, ψ), where $\Psi = \{T, \frac{1}{2}, F\}$ and ψ maps $T \mapsto T, \frac{1}{2} \mapsto T, F \mapsto F$. So how does the arrow $\neg\colon (\Psi, \psi) \to (\Psi, \psi)$ work?

(ii) It must send T to F and vice versa (why?).

(iii) But how does it act on $\frac{1}{2}$? Well, by the definition of an M_2 arrow, we require $\neg \circ \psi = \psi \circ \neg$. So noting that $\psi(\frac{1}{2}) = T$, it follows that $\psi \circ \neg(\frac{1}{2}) = \neg \circ \psi(\frac{1}{2}) = \neg T = F$.

(iv) Hence, looking at the definition of ψ, $\neg(\frac{1}{2})$ must equal F. But then $\neg\neg(\frac{1}{2}) = T \neq \frac{1}{2}$.

Hence in M_2 it *isn't* a universal law that $\neg\neg = 1_\Omega$.

The same is true of Graph, and confirming this can be your first challenge in this chapter. (We already have our first indication, then, that the laws governing 'logical arrows' like $\neg$ in a topos won't in general be fully classical.)

(c) An important reality check. Note that Theorem 112* tells us that this square commutes:

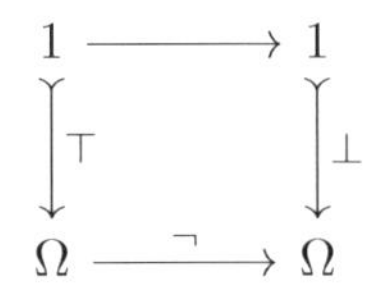

However, unlike the square in Defn. 96*, this *can't* always be a pullback square. Otherwise we could paste the two squares together and then immediately use the pullback lemma to show that we always have $\neg\neg = 1_\Omega$ after all. So, for another small challenge, give a direct proof that this last diagram doesn't depict a pullback in M_2.

46.3 Conjunction

(a) Let's continue thinking about 'logical' constructions in a topos. And to avoid boring repetition, do take all the remaining definitions and theorems in this chapter to apply to toposes.

So: consider next the arrow from 1 to the product $\Omega \times \Omega$ defined by the following product diagram:

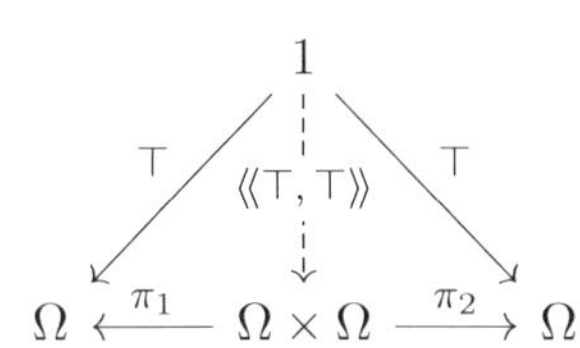

Being monic like any arrow from 1, $\langle\!\langle \top, \top \rangle\!\rangle$ gives us a subobject of $\Omega \times \Omega$, which will therefore have its own characteristic arrow. So we can introduce:

Definition 152. The 'conjunction' arrow $\wedge \colon \Omega \times \Omega \to \Omega$ is the characteristic arrow of the subobject $(1, \langle\!\langle \top, \top \rangle\!\rangle)$, making this a pullback:

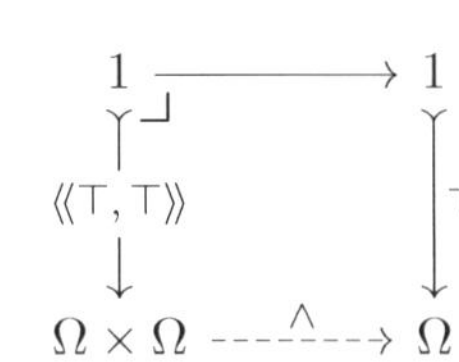

△

I claim that, as the notation suggests, $\wedge$ does behave like a conjunction. Why?

Consider the situation in Set, where Ω is a two-membered set we can think of as $\{\textit{true}, \textit{false}\}$ or, in brisker notation, $\{T, F\}$. Then $\langle\!\langle \top, \top \rangle\!\rangle$ sends the sole element of the singleton 1 to the pair of values $\langle T, T \rangle$. With that input, the diagram tells us, $\wedge$ gives the output T. And – since we are dealing with a limiting case in forming a pullback square – $\wedge$ only gives the output T for that pair of inputs.

(b) Now suppose that, in some topos, $\varphi, \psi \colon X \to \Omega$ are parallel arrows into the classifying object Ω. Then, by the usual product construction, there is an arrow $\langle\langle \varphi, \psi \rangle\rangle \colon X \to \Omega \times \Omega$. So there will be a composite arrow $\wedge \circ \langle\langle \varphi, \psi \rangle\rangle \colon X \to \Omega$. We can nicely re-notate this composite, using infix notation:

Definition 153. For arrows $\varphi, \psi \colon X \to \Omega$, $\varphi \wedge \psi =_{\text{def}} \wedge \circ \langle\langle \varphi, \psi \rangle\rangle$. △

(And similarly for the other binary 'logical arrows' $\vee$ and $\Rightarrow$ which we'll meet.)

Then, using that suggestive notation, we have in particular

Theorem 237. $\top \wedge \top = \top$, *while* $\top \wedge \bot = \bot \wedge \top = \bot \wedge \bot = \bot$.

Challenge: show at least that $\top \wedge \bot = \bot$ (proof in §46.7). But we will leave the other three cases to look after themselves.

(c) We can readily show that conjunction is commutative, i.e. for any φ, ψ, we have $\varphi \wedge \psi = \psi \wedge \varphi$.

Suppose o is the order-swapping isomorphism which maps any product object $X \times Y$ to $Y \times X$. And now consider the following diagrams:

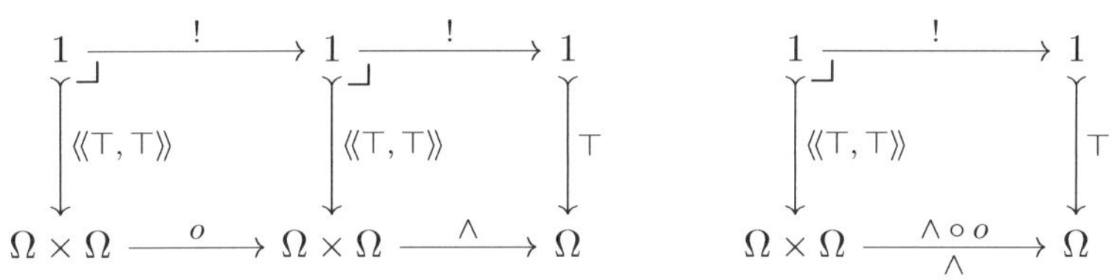

In the left-hand diagram we have pasted a square which evidently commutes (and which is easily seen to be a pullback) together with the pullback which defines $\wedge$. By the now familiar pullback lemma, this means that the diagram on the right is a pullback square with the bottom arrow $\wedge \circ o$. But by definition, that bottom arrow is $\wedge$, so $\wedge \circ o = \wedge$. (A categorial version of a familiar thought: roughly, the order of conjuncts is not logically significant.)

But we can easily show that $o \circ \langle\langle \varphi, \psi \rangle\rangle = \langle\langle \psi, \varphi \rangle\rangle$ (exercise!). Hence

$$\varphi \wedge \psi = \wedge \circ \langle\langle \varphi, \psi \rangle\rangle = \wedge \circ o \circ \langle\langle \varphi, \psi \rangle\rangle = \wedge \circ \langle\langle \psi, \varphi \rangle\rangle = \psi \wedge \varphi$$

(d) Let's have another example of how a familiar logical law can be reflected by a categorial counterpart using logical arrows.

Logically, we have $X \wedge \neg X$ is equivalent to $\bot$, for any proposition X. Categorially, I claim we have the corresponding equation $\chi \wedge \neg\chi = \bot_X$ for any characteristic arrow $\chi \colon X \to \Omega$ (where $\bot_X$ is the composite arrow $\bot \circ !_X$).

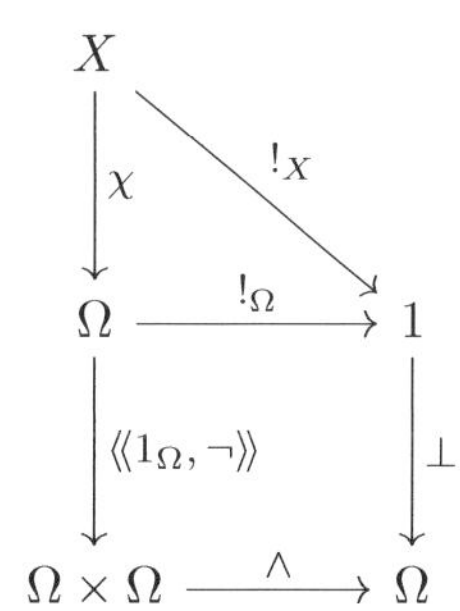

For consider the diagram. The top triangle commutes. The bottom square *ought* to commute – the hand-waving thought being that taking any element from Ω, pairing it with its negation, and taking the conjunction of that pair should always return falsity.

Assuming the bottom square commutes, and applying Theorem 50 which tells us that $\langle\!\langle 1_\Omega, \neg\rangle\!\rangle \circ \chi = \langle\!\langle \chi, \neg\chi\rangle\!\rangle$, we get the desired result $\chi \wedge \neg\chi = \bot_X$.

Which leaves you with a trickier challenge: prove that our bottom square *does* commute (hint: think how the composite $\wedge \circ \langle\!\langle 1_\Omega, \neg\rangle\!\rangle$ can appear along the bottom of a pasted-together pair of pullback squares).

46.4 Disjunction and the conditional

So far so straightforward. And for some purposes we could leave things here. But having introduced a conjunction arrow, an obvious next question is: what about disjunction and the conditional? The required definitions are rather more complicated and we need to do more work to motivate them.

(a) Let's start again from Set. We want to define a function $\vee\colon \Omega \times \Omega \to \Omega$ that sends each of the pairs $\langle T, T\rangle$, $\langle T, F\rangle$, and $\langle F, T\rangle$ to T and sends $\langle F, F\rangle$ to F. And to follow the same general pattern of definition as for $\wedge$, we want to find – for some appropriate X – a corresponding monic arrow $m\colon X \to \Omega \times \Omega$, so that this is a pullback square, call it (D):

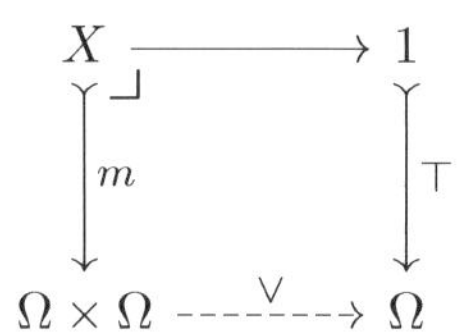

How should we proceed?

Consider the following diagrams (where $\top_\Omega$ is the composite $\Omega \xrightarrow{!} 1 \xrightarrow{\top} \Omega$ which sends every element of Ω to T):

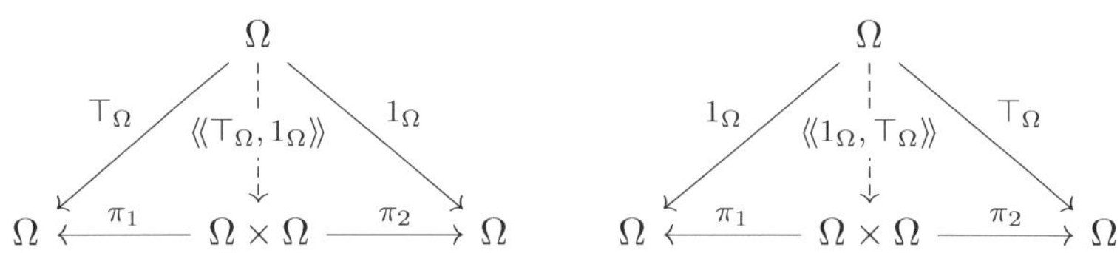

(i) (ii)

The product diagram (i) defines the arrow $\langle\!\langle \top_\Omega, 1_\Omega\rangle\!\rangle$ that sends T and F in Ω to the pairs $\langle T, T\rangle$ and $\langle T, F\rangle$ respectively. And (ii) defines $\langle\!\langle 1_\Omega, \top_\Omega\rangle\!\rangle$ that sends T and F in Ω to $\langle T, T\rangle$ and $\langle F, T\rangle$ respectively.

Now use brute force to put things together like this, using a coproduct:

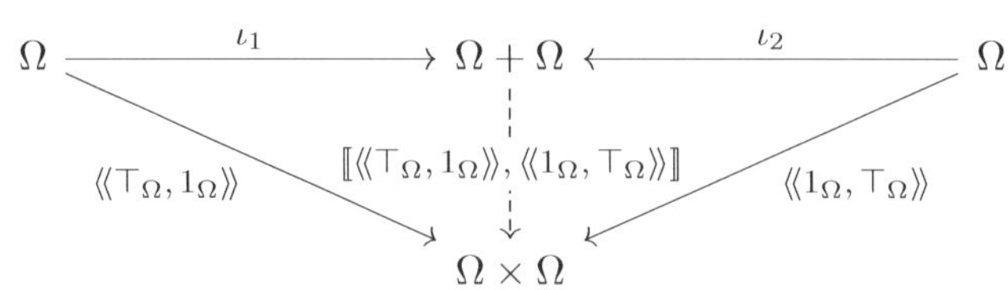

Then the mediating arrow $[\![\langle\!\langle \top_\Omega, 1_\Omega \rangle\!\rangle, \langle\!\langle 1_\Omega, \top_\Omega \rangle\!\rangle]\!]$ of this coproduct diagram will in Set be a function sending the four elements of $\Omega + \Omega$ to the desired three elements of $\Omega \times \Omega$ (three because $\langle T, T \rangle$ gets hit twice).

OK: we are almost there. However, we can't simply take X in our pullback diagram (D) to be $\Omega + \Omega$, and take the desired arrow $m\colon X \to \Omega \times \Omega$ to be our just-defined mediating arrow – because *that* arrow won't be monic (in Set, it maps four elements to three). Bother!

What to do? Try brute force again. Why not take the epi-mono factorization of $[\![\langle\!\langle \top_\Omega, 1_\Omega \rangle\!\rangle, \langle\!\langle 1_\Omega, \top_\Omega \rangle\!\rangle]\!]$ through its image, relying on Defn. 91 and Theorem 233 (which I promised would be useful)? Then we *can* extract a monic m that still hits the right three T/F pairs in $\Omega \times \Omega$. And hence, last step, $\vee$ can then be defined as the characteristic arrow for *this* monic, sending those three pairs of values to T. As we want.

Generalizing this line of thought from Set motivates the following (albeit not immediately appealing) definition for a disjunction arrow in any topos:

Definition 154. The 'disjunction' arrow $\vee\colon \Omega \times \Omega \to \Omega$ is the characteristic arrow of the image (I, m) of the arrow $[\![\langle\!\langle \top_\Omega, 1_\Omega \rangle\!\rangle, \langle\!\langle 1_\Omega, \top_\Omega \rangle\!\rangle]\!]$, making this a pullback:

$$\begin{array}{ccc} I & \longrightarrow & 1 \\ {\scriptstyle m}\downarrow & & \downarrow{\scriptstyle \top} \\ \Omega \times \Omega & \overset{\vee}{\dashrightarrow} & \Omega \end{array}$$

△

Using the infix notation $\varphi \vee \psi$ for the composite arrow $\vee \circ \langle\!\langle \varphi, \psi \rangle\!\rangle$, we have a result parallel to Theorem 237 – i.e., we can now prove

Theorem 238. $\top \vee \top = \top \vee \bot = \bot \vee \top = \top$, *while* $\bot \vee \bot = \bot$.

Challenge: to help fix ideas about $\vee$, show that $\top \vee \bot = \top$. Again, we'll leave the other cases to look after themselves.

(b) We can also introduce a conditional arrow. Here's some hand-waving motivation. In elementary logic, in a context in which we are given $R \Rightarrow S$, the propositions $R \wedge S$ and R imply each other. As we might put it, the conditional $R \Rightarrow S$ is just enough to 'equalize' $R \wedge S$ and R. Or, thinking of a conditional as operating on an ordered pair of propositions, we could also say that $\Rightarrow$ acting on $\langle R, S \rangle$ equalizes (i) $\wedge$ acting on that pair and (ii) the first projection function acting on the same pair.

Now moving to a topos framework, here (i) $\wedge$ is an arrow $\Omega \times \Omega \to \Omega$. And we are going to want to equalize this with (ii) the first projection arrow $\pi_1\colon \Omega \times \Omega \to \Omega$ that is part of the standard product package. We can of course rely on the fact that, in a topos, all equalizers exist. So, using neutral notation, we can define $(\Theta, \gg)$ as an equalizer giving us a commuting fork

$$\Theta \xrightarrow{\ \gg\ } \Omega \times \Omega \underset{\pi_1}{\overset{\wedge}{\rightrightarrows}} \Omega$$

where every other fork which shares the same prongs $\wedge$ and π_1 factors uniquely through it.

Again, we are almost there. However, if a conditional arrow in a topos is to be the same type of arrow as $\wedge$ and $\vee$, we want it to be another arrow from $\Omega \times \Omega$ to Ω. But our newly introduced $\gg$ is an arrow from Θ (whatever that is) to $\Omega \times \Omega$. Bother again!

What to do? Well, like any equalizing arrow, $\gg$ is monic. Therefore $\gg\colon \Theta \rightarrowtail \Omega \times \Omega$ is a subobject of $\Omega \times \Omega$. Hence it will have a characteristic arrow which *is* an arrow from $\Omega \times \Omega$ to Ω. This gives us the sort of arrow we wanted and so we have motivated the following definition:

Definition 155. The 'conditional' arrow $\Rightarrow\colon \Omega \times \Omega \to \Omega$ is the characteristic arrow of the subobject of $\Omega \times \Omega$ which equalizes $\wedge\colon \Omega \times \Omega \to \Omega$ and the first projection arrow $\pi_1\colon \Omega \times \Omega \to \Omega$. △

Using the infix notation $\varphi \Rightarrow \psi$ for a composite arrow $\Rightarrow \circ \langle\!\langle \varphi, \psi \rangle\!\rangle$, this time we can now show

Theorem 239. $(\top \Rightarrow \top) = (\bot \Rightarrow \top) = (\bot \Rightarrow \bot) = \top$, *while* $(\top \Rightarrow \bot) = \bot$.[3]

Challenge: to help fix ideas about $\Rightarrow$, prove $(\bot \Rightarrow \top) = \top$.

46.5 Varieties of internal logic

(a) Talk of 'the internal logic' of a topos has an established use, more sophisticated than we need here.[4] I'm going to borrow the term to use more loosely, but in a related way, to refer just to the principles governing the relationships between the four logical arrows which we have introduced. So what does the internal logic of a topos, in this limited sense, look like?

With more or less fiddly diagram-chasing, we can establish a lot of familiar-looking laws. We have already seen how the connectives interact with $\top$ and $\bot$, noted that we can show that conjunction is commutative and seen that conjoining something with its negation gives us falsity. It's a lot messier to show, for example, that $\varphi \wedge (\psi \vee \chi) = (\varphi \wedge \psi) \vee (\varphi \wedge \chi)$, or that $\neg(\varphi \vee \psi) = \neg\varphi \wedge \neg\psi$. But in principle, we can prove these equations by chasing round complicated diagrams.

However, let's not get hung up on the details here. The initial point I want to make is simply that quite a few of the familiar logical laws governing the ordinary connectives $\neg, \wedge, \vee$ and $\Rightarrow$ do carry over to apply to their arrow-theoretic counterparts: logical theorems become arrow identities.

But, as we've already seen, not all of them. Yes, there *are* elementary toposes which satisfy special conditions which ensure that their internal logic is one

[3] Don't jump to the conclusion that $\Rightarrow$ is in some sense 'truth-functional'. Compare: even in intuitionistic logic, with $\top$ (the 'verum' logical constant) defined as $\neg\bot$, and with $\Rightarrow$ the conditional, we have $\top \Rightarrow \top$, $\bot \Rightarrow \top$, and $\bot \Rightarrow \bot$ all theorems, and so all provably equivalent to $\top$; and $\top \Rightarrow \bot$ is provably equivalent to $\bot$. But the intuitionistic conditional is not truth-functional.

[4] See tinyurl.com/int-logic.

where we can eliminate double negations: Set is a paradigm example. But in other cases, like M_2 and Graph, the double negation rule can fail. In other words, if all we know is that we are in a topos, we can't conclude that the internal negation, conjunction, disjunction and conditional arrows obey analogues of *all* the familiar rules governing the usual logical connectives.

In a word, the internal 'propositional' logic of a topos need not be *classical.*

(b) More positively, though, it turns out that in any topos, the rules governing our four logical arrows ensure that identities between logical arrows will track *intuitionistic* logical equivalences.

I don't propose to further elaborate on this claim or to justify it here and now, mainly because for our purposes in these notes we'll be more interested in a closely related claim that we meet in the next chapter. But let me make four quick observations now:

(1) The fact that just laying down the standard introduction rules for the connectives (and matching elimination rules) doesn't result in classical logic makes it entirely unsurprising that simply laying down the rules for introducing the counterparts of connectives as logical arrows in a topos won't give us analogues of all the classical rules. Perhaps, from the start, we shouldn't expect to get an analogue of 'double negation elimination'.

(2) But since the very natural introduction rules (with the elimination rules which come for free) do give us intuitionistic logic, it perhaps shouldn't be a surprise either that their topos-theoretic counterparts should by default give us an intuitionistic internal logic. That too is what we should expect, if we can assume that our definitions for the categorial logical arrows are sensible ones – and we'll say a little more in their defence in the next chapter.

(3) But just as we can add additional rules to intuitionistic propositional logic to give us stronger logics, we can also add further assumptions about a topos that will make its internal logic stronger-than-intuitionistic, up to and including recovering classical logic.

(4) However, a topos whose internal logic is (only) intuitionistic can provide a suitable arena in which we can develop constructive mathematics. An example is the so-called effective topos Eff, a world in which all functions between natural numbers (and between the reals constructed from them) are recursive.[5]

46.6 Classical toposes

We have seen that the internal logic of a topos might not be classical in the sense that we don't get the identity $\neg\neg = 1_\Omega$. An obvious question to ask is therefore:

[5]The first couple of pages of van Oosten (2008) or Bernadet and Graham-Lengrand (2013) *might* give you a hint of what this involves.

what does it take, then, to ensure that this identity *does* hold? More generally, what does it take to get a 'classical' topos whose internal logic is indeed classical?

(a) We might have guessed that, to get a classical logic in a topos, it is enough for its truth-value object to be like Set's in having exactly two point elements. But we now know that this isn't going to work. M_2 is a simple example of a topos which is bivalent in *that* sense (as we noted at the very end of §23.4) but where $\neg\neg \neq 1_\Omega$ (as we showed in §46.2).

Let's try another idea, giving a different spin to the idea of being two-valued. In Set, a natural candidate to play the role of a two-valued classifying object Ω is the coproduct $1 + 1$. So what about looking at other toposes where the truth-value object is isomorphic to $1 + 1$? Will that force a classical logic?

Recall, Theorem 235 tells us that in any topos the arrow $[\![\top, \bot]\!] \colon 1+1 \to \Omega$ is monic. Intuitively, if Ω is to be isomorphic to $1+1$, this monic mediating arrow should in fact provide the required isomorphism (think of the situation in Set). And, though we won't rely on this, we do have:

Theorem 240. *In a topos, $\Omega \cong 1+1$ if and only if $[\![\top, \bot]\!] \colon 1+1 \xrightarrow{\sim} \Omega$.*

Proof sketch. For the non-trivial direction, we need a lemma: in a topos, if an arrow from the truth-value object in a subobject classifier to itself is monic, it is an isomorphism.[6]

Suppose $(\Omega, \top)$ is any subobject classifier, and that there is an isomorphism $j \colon \Omega \xrightarrow{\sim} 1+1$. Put $\top' = j \circ \top$, $\bot' = j \circ \bot$, and then $(1+1, \top')$ is also a subobject classifier by Theorem 114. So $[\![\top', \bot']\!] \colon 1+1 \to 1+1$ is monic by Theorem 235 and hence an automorphism by our lemma. Then the original $[\![\top, \bot]\!] = j^{-1} \circ [\![\top', \bot']\!]$ by the dual of Theorem 50, so is an isomorphism $1+1 \xrightarrow{\sim} \Omega$. □

Let's try, then, the following definition:

Definition 156. A *classical* topos is one where the arrow $[\![\top, \bot]\!] \colon 1+1 \to \Omega$ is an isomorphism. △

And yes, we can immediately show what we wanted:

Theorem 241. *In a classical topos, $\neg\neg = 1_\Omega$.*

Proof. The dual of Theorem 50 tells us that $e \circ [\![f, g]\!] = [\![e \circ f, e \circ g]\!]$. So in particular, in any topos $\neg\neg \circ [\![\top, \bot]\!] = [\![\neg\neg\top, \neg\neg\bot]\!] = [\![\top, \bot]\!]$.

But if $[\![\top, \bot]\!]$ is an isomorphism, post-composing each end of the equation with $[\![\top, \bot]\!]^{-1}$ immediately gives us $\neg\neg = 1_\Omega$. □

That's a nice result to carry forward: and we'll say more about classical toposes in the next chapter.

[6] Todd Trimble gives an accessible proof at tinyurl.com/todd-monic.

46.7 Challenges!

(a) You were asked first to show that in Graph it isn't always the case that $\neg\neg\chi = \chi$: in other words, $\neg\neg$ is not the identity on Ω.

Proof. Here again is a portrait of the object Ω of Graph's subobject classifier:

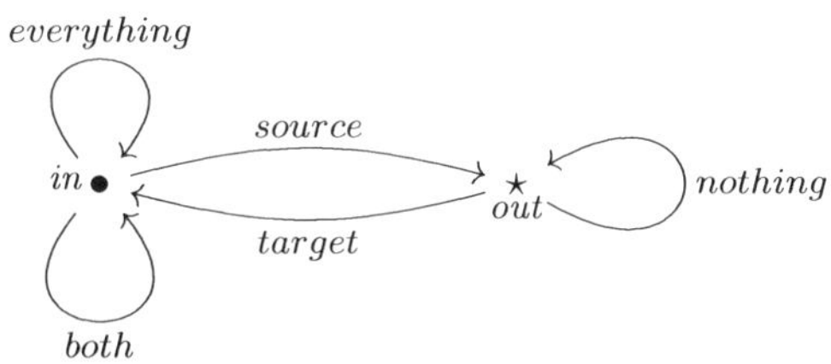

How, then, does the negation arrow $\neg\colon \Omega \to \Omega$ work in Graph?

It must send a node of Ω to the opposite one. And since loops must go where their nodes go, $\neg$ must send both loops on *in* to the single loop on *out*, while sending the loop on *out* to one of the loops on *in* (in fact, to the 'everything' loop which is the target of $\top$ – think about it!). Therefore $\neg \circ \neg\colon \Omega \to \Omega$ sends the 'both' loop to the 'everything' loop, and hence isn't the identity on Ω.

Spelling the point out a bit further:

(i) Take a graph X and one of its subobjects, i.e. some $(S, s\colon S \rightarrowtail X)$ with its characteristic arrow $\chi_s\colon X \to \Omega$. Then a given node in X will be sent to opposite nodes in Ω by χ_s and $\neg\chi_s$.

(ii) And this means that any edges in X which get sent by χ_s to one of the loops on *in* in Ω will get sent by $\neg\chi_s$ to the sole loop on *out*. While edges in X which get sent by χ_s to the loop on *out* in Ω will have to be sent by $\neg\chi_s$ to the *everything* loop on the *in* node.

(iii) So now consider an edge in X which is sent by χ_s to the *both* loop in Ω (this signifies, recall, that both the nodes of the edge X are in S, but the joining edge itself isn't). Then $\neg\chi_s$ will send that edge to the *nothing* loop in Ω. And hence, negating again, $\neg\neg\chi_s$ will send that same edge back to the *everything* loop in Ω.

(iv) Hence double-negating the characteristic arrow of the subobject (S, s) gives us the characteristic arrow of the result of adding to S any edges from X not already present as edges between existing nodes in S.

In sum: in Graph, we again *don't* always have $\neg\neg\chi = \chi$. □

(b) Next you were asked to show that this is *not* a pullback in M_2:

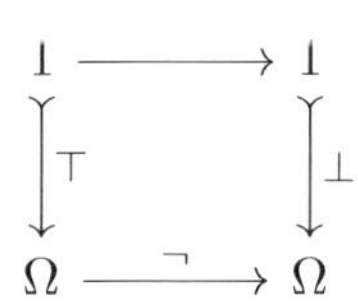

Proof. In the notation of §23.4, we are looking at the inner square of

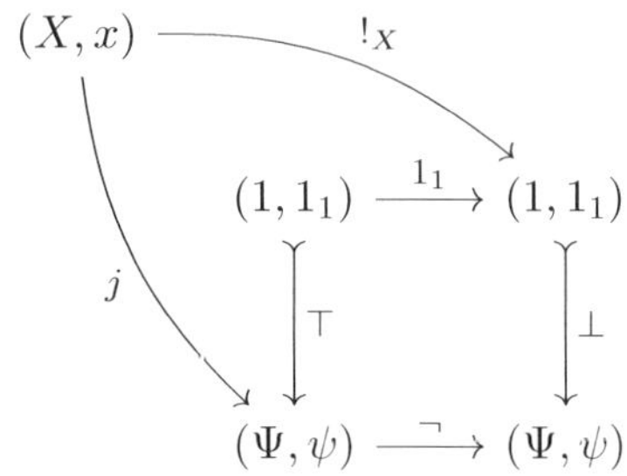

where

(i) $\Psi = \{T, \frac{1}{2}, F\}$, and $\psi\colon \Psi \to \Psi$ is the map $T \mapsto T, \frac{1}{2} \mapsto T, F \mapsto F$;

(ii) $\top\colon 1 \to \Psi$ sends the sole member of 1 to T, $\bot$ sends that same sole member to F;

(iii) $\neg\colon \Psi \to \Psi$ is the map $T \mapsto F, \frac{1}{2} \mapsto F, F \mapsto T$.

And to show that this isn't a pullback square, we want to find a wedge with vertex (X, x) as drawn, such that (i) $\neg \circ j = \bot \circ !_X$ but also such that (ii) we *can't* find a unique arrow from (X, x) to the north-west corner of the square which would make the whole diagram commute.

That's easy. Put $X = \{T, \frac{1}{2}\}$, and let $x\colon X \to X$ be ψ restricted to X, so sending both members to T. Let j be the inclusion map from X to Ψ. Trivially, $j \circ x = \psi \circ j$, so j really is an arrow from (X, x) to (Ψ, ψ).

Then we have (i) $\neg \circ j = \bot \circ !_X$, since both composites send both members of X to F. But (ii) the only possible arrow from (X, x) to $(1, 1_1)$ which makes the top triangle commute, namely $!_X$, does not make the bottom triangle commute. So we are done. □

(c) The next challenge was to prove

Theorem 237 (part). $\top \wedge \bot = \bot$, *i.e.* $\wedge \circ \langle\!\langle \top, \bot \rangle\!\rangle = \bot$.

Proof. We want to show that the composite arrow along the bottom of the following diagram equals $\bot$:

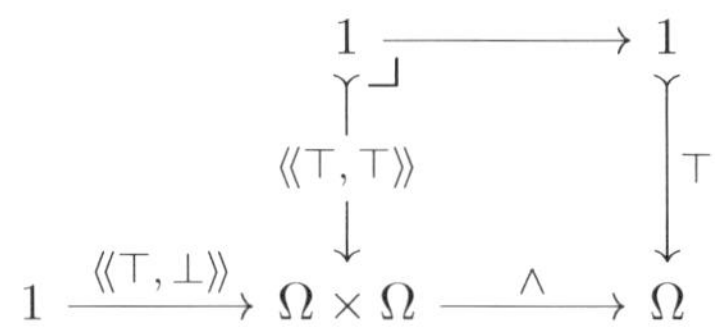

And here's the plan of action. Fill in the missing two sides of the half-formed square in the only sensible way, and then show that this results in another pullback square, so we arrive at the diagram (i) below.

But that's equivalent to the following diagram (ii) – just recall that pasting together two pullbacks gives us a pullback:

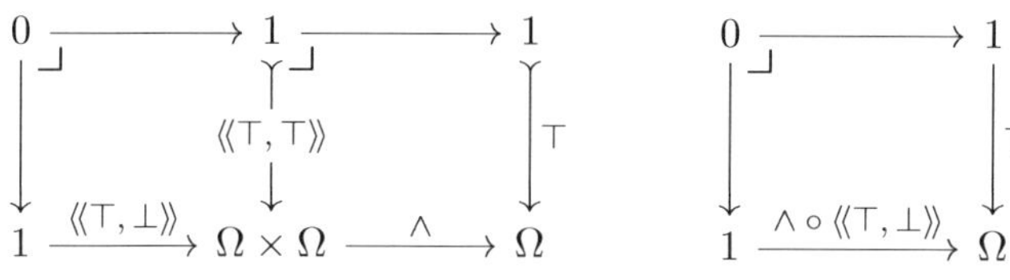

(i) (ii)

By definition, however, $\bot$ is the unique lower arrow making (ii) a pullback. So, as we wanted to show, $\wedge \circ \langle\langle \top, \bot \rangle\rangle = \bot$.

To complete the proof, then, we need to confirm that the left half of (i) is a pullback square. OK: stare hard at the following diagram:

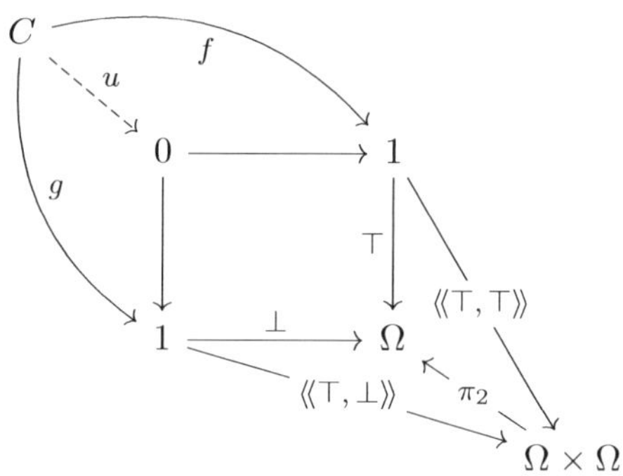

By hypothesis, the inner square commutes (by the definition of $\bot$), as do the two added lower triangles (by the definitions of $\langle\langle \top, \top \rangle\rangle$ and $\langle\langle \top, \bot \rangle\rangle$).

Now suppose that f and g are such that $\langle\langle \top, \top \rangle\rangle \circ f = \langle\langle \top, \bot \rangle\rangle \circ g$. Then, of course, $\pi_2 \circ \langle\langle \top, \top \rangle\rangle \circ f = \pi_2 \circ \langle\langle \top, \bot \rangle\rangle \circ g$, and therefore $\top \circ f = \bot \circ g$. But then, because the inner square is a pullback, there must be a unique arrow u making the diagram commute.

Which means that the left-hand square in diagram (i) *is* a pullback. □

(d) A reality check.

Like any product object, $\Omega \times \Omega$ is defined only up to (a unique) isomorphism. So, correspondingly, (1) the arrow $\langle\langle \top, \bot \rangle\rangle \colon 1 \to \Omega \times \Omega$ is only uniquely fixed once we have chosen our product object.

Similarly, of course, (2) the arrow $\wedge \colon \Omega \times \Omega \to \Omega$ is also only uniquely fixed once we have chosen our product object.

However, what we have just shown is that, having fixed on a particular product object $\Omega \times \Omega$, and having thereby uniquely determined both the arrows $\langle\langle \top, \bot \rangle\rangle$ and $\wedge$ in matching ways (i.e. fixed them with respect to the *same* product object), it will always be the case that $\wedge \circ \langle\langle \top, \bot \rangle\rangle = \bot$.

Likewise for analogous results.

(e) Moving on, you were next asked to show that the following square commutes (and evidently, if it commutes, we can use that fact to get another derivation of the previous theorem):

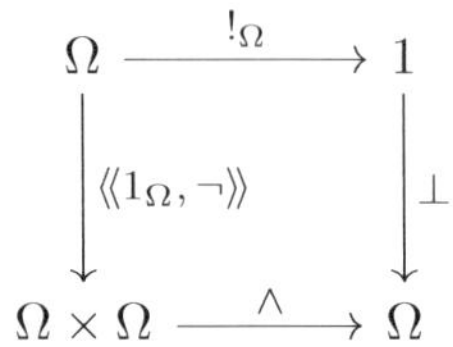

Proof. We want to think about the composite arrow

$$\Omega \xrightarrow{\langle\!\langle 1_\Omega, \neg \rangle\!\rangle} \Omega \times \Omega \xrightarrow{\;\wedge\;} \Omega$$

The only thing we know about the second component $\wedge$ is that it is defined by the pullback square on the right below, so we'll have to make use of that:

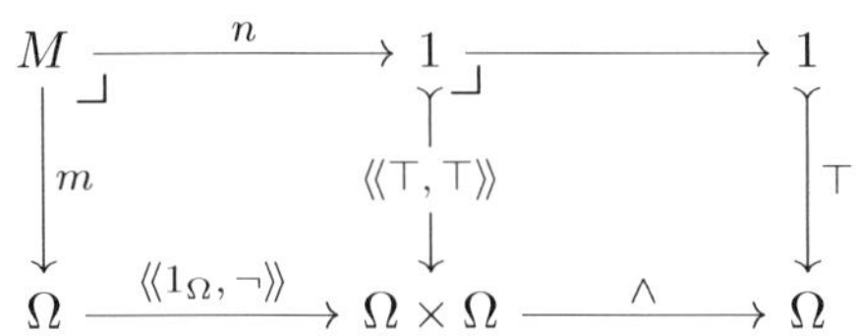

And what else can we do on the left but complete a square by pulling back $\langle\!\langle \top, \top \rangle\!\rangle$ along $\langle\!\langle 1_\Omega, \neg \rangle\!\rangle$ to get mystery arrows $m\colon M \to \Omega$ and $n\colon M \to 1$?

So what can we deduce about M, m and n? By Theorems 40 and 50 we have

$$\langle\!\langle m, \neg m \rangle\!\rangle = \langle\!\langle 1_\Omega, \neg \rangle\!\rangle \circ m = \langle\!\langle \top, \top \rangle\!\rangle \circ n = \langle\!\langle \top \circ n, \top \circ n \rangle\!\rangle$$

and hence $m = \top \circ n$ and $\neg m = \top \circ n$, and hence $\top \circ n = \bot \circ n$. Which means the bent outer 'square' of this next diagram commutes, where the inner square is our old friend, the pullback defining $\bot$:

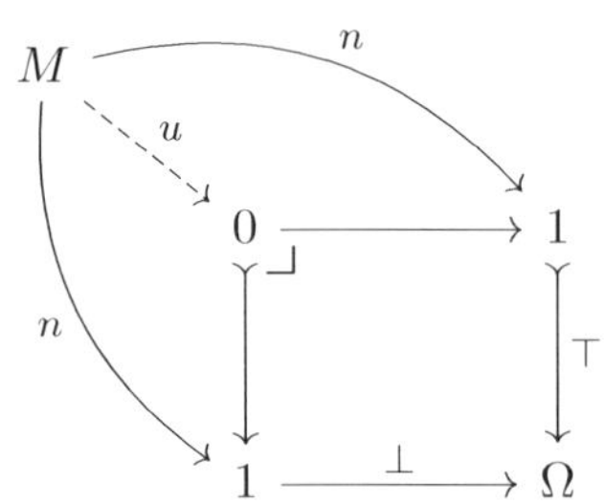

Hence, by the definition of a pullback, there must be an arrow $u\colon M \to 0$, which by Theorem 78 is an isomorphism, and M is initial. And now we are motoring!

Putting M as initial, and using the pullback lemma on the previous two-pullback diagram, we get a pullback square (A):

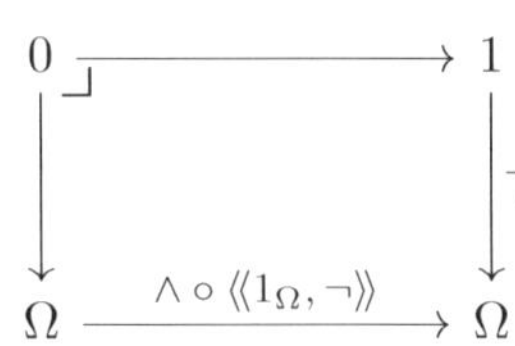

The obvious(?) next step is to find a pullback square which looks just the same except that it has $\bot \circ 1_\Omega$ as the bottom arrow. So consider:

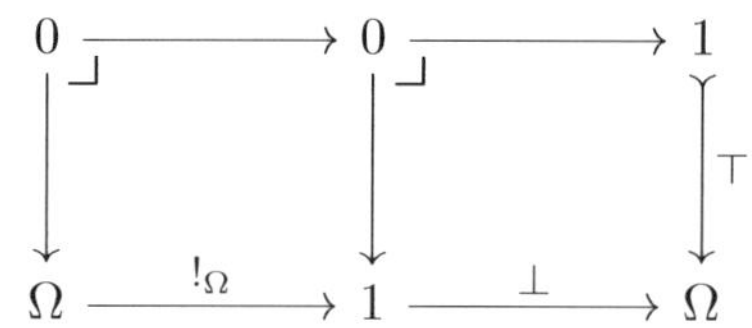

The right-hand square is the pullback defining $\bot$, and the left-hand square is also a pullback (the vertex of a wedge making a commuting square with the opposite corner with vertex 1 will have to be initial because it has an arrow to 0). So the pullback lemma gives us another pullback square (B):

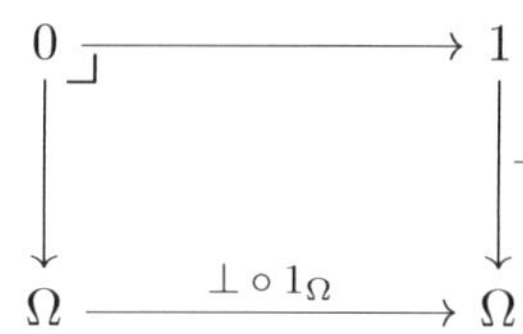

Comparing (A) and (B), we can conclude $\wedge \circ \langle\!\langle 1_\Omega, \neg \rangle\!\rangle = \bot \circ 1_\Omega$. □

(f) The next challenge is to derive

Theorem 238 (part). $\top \vee \bot = \top$, *i.e.* $\vee \circ \langle\!\langle \top, \bot \rangle\!\rangle = \top$.

Proof Stare at this diagram. The big triangle commutes because it is half of our diagram defining the arrow $[\![\langle\!\langle \top_\Omega, 1_\Omega \rangle\!\rangle, \langle\!\langle 1_\Omega, \top_\Omega \rangle\!\rangle]\!]$, whose epi-mono factorization is $m \circ e$. The square commutes by Defn. 154. The composite zig-zag from 1 to 1 must equal 1_1. So the whole diagram commutes and the composites along the two paths from 1 to Ω are equal.

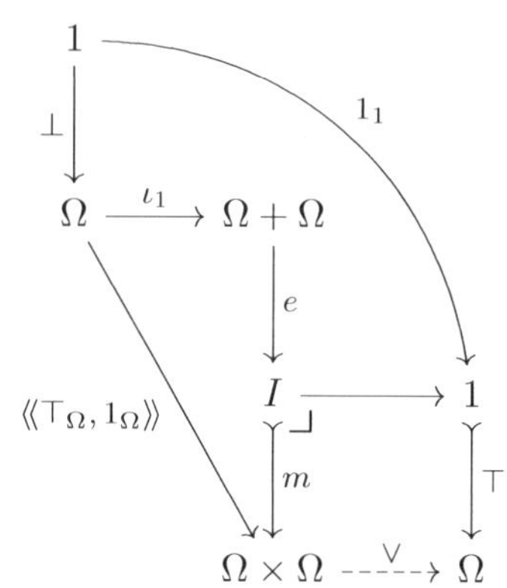

Hence $\vee \circ \langle\!\langle \top, \bot \rangle\!\rangle$, which equals $\vee \circ \langle\!\langle \top_\Omega, 1_\Omega \rangle\!\rangle \circ \bot$ by Theorem 50, must indeed equal $\top$. □

(g) The final challenge was to prove

Theorem 239 (part). $\bot \Rightarrow \top \; = \; \top$.

Proof. We want to show $\Rightarrow \circ \langle\!\langle \bot, \top \rangle\!\rangle = \top$. So contemplate the following diagram (what else?):

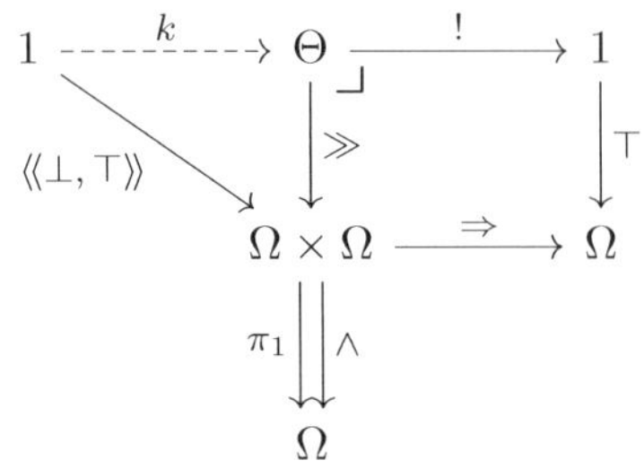

Here $(\Theta, \gg)$ is the equalizer for the parallel arrows $\wedge$ and π_1. And the top right square is the pullback defining $\Rightarrow$.

Now, we know that $\wedge \circ \langle\!\langle \bot, \top \rangle\!\rangle = \bot$ (from Theorem 237). And $\pi_1 \circ \langle\!\langle \bot, \top \rangle\!\rangle = \bot$ (from the product diagram defining $\langle\!\langle \bot, \top \rangle\!\rangle$). Hence $\langle\!\langle \bot, \top \rangle\!\rangle$ is the handle of a commuting fork with prongs $\wedge$ and π_1. And hence this fork must factor through the equalizer $(\Theta, \gg)$ for $\wedge$ and π_1 via some unique arrow k as drawn.

The top arrows compose to the identity on 1, so the two routes from 1 at the top left to Ω on the right give us, as wanted, $\Rightarrow \circ \langle\!\langle \bot, \top \rangle\!\rangle = \top$. □

The same line of proof will also apply to the cases where the diagonal arrow is $\langle\!\langle \top, \top \rangle\!\rangle$ or $\langle\!\langle \bot, \top \rangle\!\rangle$, again forming a commuting fork with $\wedge$ and π_1. But here's another challenge: how we can show $\Rightarrow \circ \langle\!\langle \top, \bot \rangle\!\rangle = \bot$?

47 Subobjects in a topos

It is an oh-so-familiar story, when told in a set-theoretic idiom. If we start from the set of natural numbers $\mathbb{N}$, and allow ourselves to freely form subsets, products, quotients, powersets, etc., then we get a framework in which we can reconstruct (almost) all 'ordinary' mathematics.

But it now seems that there should be another version of this story to be told in a categorial idiom. If we are in a topos, we similarly have subobjects, products, quotients, power objects, etc., available. So if we look at toposes that are equipped with a natural numbers object (N, z, s) – with N playing the role of the set of natural numbers – we should also be able to reconstruct much familiar mathematics in such toposes.

How is the story going to run? To prepare the ground further, we need to know more about the behaviour of subobjects in toposes. We gave a pointer at the end of Chapter 22, and now is the time to make good on the promise there to explore the 'algebra of subobjects'.

47.1 Defining the intersection and union of subobjects

(a) In §22.6 we motivated the following definition (since we are now assuming we are working with toposes, we can take it that we have pullbacks available):

Definition 93 If (R, r) and (S, s) are subobjects of X, then any $(R \cap S, r \cap s)$ is an *intersection* of them, where $R \cap S$ is the vertex of a pullback over the corner formed by r and s,

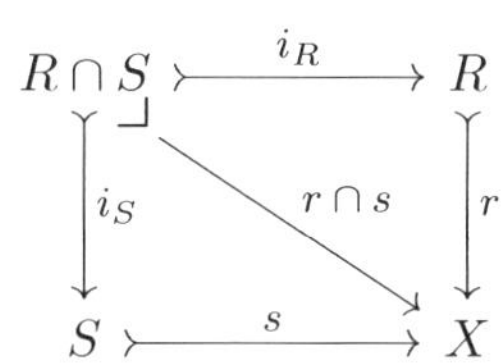

and $r \cap s \colon R \cap S \to X$ is the resulting diagonal, equalling the composite arrow round the square on either path. △

We now motivate a companion definition of unions. Recall our (pre-categorial) observation about disjoint unions and ordinary unions of sets back in §8.6. Suppose R and S are both subsets of X, let $R \sqcup S$ be the union of disjoint copies of R and S, and consider the function $j\colon R \sqcup S \to X$ which sends each element of $R \sqcup S$ to the original element of X it is a copy of. Then, we noted, j has an epi-mono factorization $m \circ e$, where the monic $m\colon R \cup S \hookrightarrow X$ is the inclusion function that sends every element of the union to itself.

Let's categorify, first in **Set**. Suppose $(R, r\colon R \rightarrowtail X)$ and $(S, s\colon S \rightarrowtail X)$ are subobjects of X. Form their coproduct (recall §11.7 on coproducts and disjoint unions) and consider the arrow $[\![r, s]\!]\colon R + S \to X$. Then this has an epi-mono factorization $m \circ e$ with a monic $m\colon R \cup S \rightarrowtail X$. (Why?)

And now we generalize:

Definition 157. If (R, r) and (S, s) are subobjects of X, then any $(R \cup S, r \cup s)$ is a *union* of them, where $r \cup s\colon R \cup S \to X$ is the monic in an epi-mono factorization of $[\![r, s]\!]$:

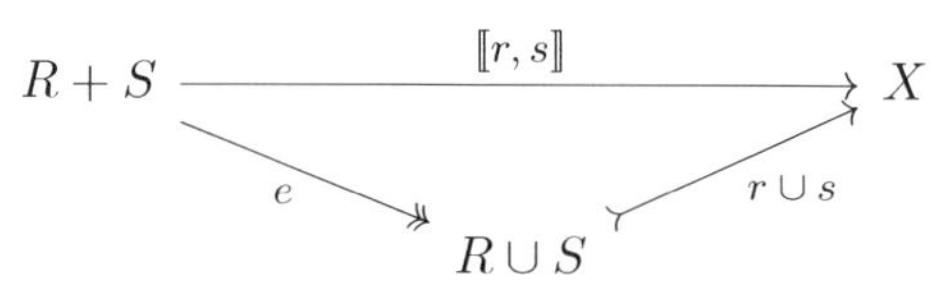

△

Theorem 233 tells us that the required factorization always exists, and Theorem 234 tells us that $R \cup S$ is unique up to isomorphism. So far so good!

(b) When we first defined intersections, we went on to prove Theorem 109, showing that our definition ensured that (as we want) intersections are greatest lower bounds. We can now prove the equally desirable companion result:

Theorem 242. *If (R, r) and (S, s) are both subobjects of some X, they have a supremum, a least upper bound, namely their union $(R \cup S, r \cup s)$.*

Proof. By definition of the coproduct, $r = [\![r, s]\!] \circ \iota_1$ (where ι_1 is the first injection into $R + S$). Hence $r = (r \cup s) \circ (e \circ \iota_1)$. Hence $r \preccurlyeq r \cup s$. Likewise $s \preccurlyeq r \cup s$. So $r \cup s$ is an upper bound.

We want next to show it is a *least* upper bound. In other words, suppose for another subobject $q\colon Q \rightarrowtail X$, both $r \preccurlyeq q$ (i.e. $r = q \circ i$, for some i) and $s \preccurlyeq q$ (i.e. $s = q \circ j$, for some j): we need to show $r \cup s \preccurlyeq q$.

By the dual of Theorem 50, $[\![r, s]\!] = [\![q \circ i, q \circ j]\!] = q \circ [\![i, j]\!]$. But $[\![i, j]\!]$ will have its own epi-mono factorization $m' \circ e'$ through some intermediate object I. Therefore, since the monics q and m' compose to a monic, we have two overall epi-mono factorizations of $[\![r, s]\!]$:

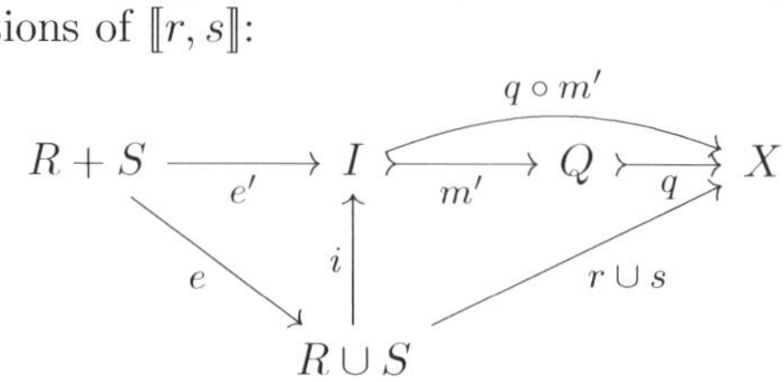

Theorem 234 then tells us that there is an isomorphism $i\colon R \cup S \to I$ making the diagram commute. So $r \cup s = q \circ (m' \circ i)$. Whence $r \cup s \preccurlyeq q$. □

(c) A portmanteau theorem now packages together some further predictable facts about the behaviour of intersections and unions:

Theorem 243. *In a topos, where q, r, s are subobjects of X:*

(1) $s \cap s \equiv s$ *and* $s \cup s \equiv s$,
$r \cap s \equiv s \cap r$ *and* $r \cup s \equiv s \cup r$.

(2) $q \cap (r \cup s) \equiv (q \cap r) \cup (q \cap s)$,
$q \cup (r \cap s) \equiv (q \cup r) \cap (q \cup s)$.

(3) $s \cap 0_X \equiv 0_X$ *and* $s \cap 1_X \equiv s$,
$s \cup 0_X \equiv s$ *and* $s \cup 1_X \equiv 1_X$.

(4) If $r \preccurlyeq s$, *then* $(q \cap r) \preccurlyeq (q \cap s)$. *Hence if* $r \equiv s$, *then* $(q \cap r) \equiv (q \cap s)$. *Likewise if* $r \equiv s$, *then* $(q \cup r) \equiv (q \cup s)$.

(5) $r \preccurlyeq s$ *if and only if* $r \equiv r \cap s$,
$r \preccurlyeq s$ *if and only if* $s \equiv r \cup s$.

(6) $\chi_r \circ q = \chi_s \circ q$ *iff* $r \cap q \equiv s \cap q$.

Only the second of (1)'s four parts might require half a moment's thought. Just recall $s \cup s$ is by definition a monic in an epi-mono factorization of $[\![s, s]\!]$. But that also factors as $[\![1, 1]\!] \circ s$, where s is monic (by the dual of Theorem 50). Whence $s \cup s \equiv s$ (by Theorem 234).

The distributive laws (2) can be derived by hacking through from first principles in tedious ways – though I think in this case we may forgive ourselves if we take the result on trust.

All the parts of (3) are easy. For example, $0_X \preccurlyeq s \cap 0_X$ by Theorem 106, while $s \cap 0_X \preccurlyeq 0_X$ by Theorem 109, hence $s \cap 0_X \equiv 0_X$. Likewise for the other parts.

I'll leave (4) for now as a challenge to prove.

Then (5) is an easy corollary of (1) and (4). For suppose $r \preccurlyeq s$. Then, for intersections, $r \equiv r \cap r \preccurlyeq r \cap s$. But $r \cap s \preccurlyeq r$. Hence $r \equiv r \cap s$. And it is trivial that if $r \equiv r \cap s$ then $r \preccurlyeq s$. Similarly for the part of (5) about unions.

Finally, think of the situation in Set. When is $\chi_r \circ q \colon Q \to \Omega$ equal to $\chi_s \circ q \colon Q \to \Omega$? When χ_r and χ_s agree on objects in $q[Q]$.

The composite $\chi_r \circ q \colon Q \to \Omega$ sends a member of Q to *true* just in case it is also in R.

Therefore that composite arrow's image is the same as the image of $\chi_{r \cap q} \colon X \to \Omega$. It follows that $\chi_r \circ q = \chi_s \circ q$ if and only if $\chi_{r \cap q} = \chi_{s \cap q}$ and hence, by Theorem 110, if and only if $r \cap q \equiv s \cap q$. Another challenge: now prove the claim (6) holds in any topos.

47.2 Alternative definitions?

(a) There is another, equally natural, approach to intersections and unions.

Suppose (R, r) and (S, s) are subobjects of some object X. These subobjects have characteristic arrows χ_r and χ_s respectively. Hence we can form a corner

$X \xrightarrow{\chi_r \wedge \chi_s} \Omega \xleftarrow{\top} 1$, where $\chi_r \wedge \chi_s$ (i.e. $\wedge \circ \langle \chi_r, \chi_s \rangle$) is the kind of composite arrow defined in Defn. 153.

As with any corner in a topos, we can form a pullback square from it. And since the arrow $\top\colon 1 \to \Omega$ is monic, its pullback along $\chi_r \wedge \chi_s$ is also a monic arrow – so we get a subobject (J, j) of X as follows:

$$\begin{array}{ccc} J & \xrightarrow{\quad !\quad} & 1 \\ {\scriptstyle j}\big\downarrow & \lrcorner & \big\downarrow{\scriptstyle \top} \\ X & \xrightarrow[\ \chi_r \wedge \chi_s\ ]{} & \Omega \end{array}$$

Think how this works in **Set**. The image of J under the injective function j will be a subset of X whose members are sent to *true* by both χ_r and χ_s. Hence $j[J]$ must be contained in the intersection of $r[R]$ and $s[S]$. And, since our square is a pullback, making $j[J]$ a limiting case, it will in fact be the whole intersection.

This motivates a natural generalization that we can apply in any topos:

Definition 158. Suppose (R, r) and (S, s) are subobjects of X with characteristic arrows χ_r and χ_s. Then any subobject $(R \cap S, r \cap s)$ which results from pulling back $\top\colon 1 \to \Omega$ along $\chi_r \wedge \chi_s$ is an *intersection* of (R, r) and (S, s):

$$\begin{array}{ccc} R \cap S & \xrightarrow{\quad !\quad} & 1 \\ {\scriptstyle r \cap s}\big\downarrow & \lrcorner & \big\downarrow{\scriptstyle \top} \\ X & \xrightarrow[\ \chi_r \wedge \chi_s\ ]{} & \Omega \end{array}$$

△

And here's the entirely predictable companion definition:

Definition 159. With $(R, r), (S, s)$ and χ_r, χ_s as before, any subobject $(R \cup S, r \cup s)$ which results from pulling back $\top\colon 1 \to \Omega$ along $\chi_r \vee \chi_s$ is a *union* of (R, r) and (S, s):

$$\begin{array}{ccc} R \cup S & \xrightarrow{\quad !\quad} & 1 \\ {\scriptstyle r \cup s}\big\downarrow & \lrcorner & \big\downarrow{\scriptstyle \top} \\ X & \xrightarrow[\ \chi_r \vee \chi_s\ ]{} & \Omega \end{array}$$

△

For consider again how things work in **Set**. The image of $(R \cup S)$ under $r \cup s$ must be contained in the union of $r[R]$ and $s[S]$. And, since our square is a pullback, making $r \cup s[R \cup S]$ a limiting case, it will actually be the whole union.

(b) Two quick comments. First, note that, according to these definitions, $\chi_{r \cap s} = \chi_r \wedge \chi_s$ and $\chi_{r \cup s} = \chi_r \vee \chi_s$.

Second, to repeat: defining an arrow as being the result of a pullback does not fix it uniquely. That's why our definitions again talk about 'an intersection', 'a union'. But as Theorem 108 tells us, different candidate intersections for a pair of subobjects will be equivalent: and likewise for unions.

(c) We now have two well-motivated definitions of intersection on the table. Fortunately we don't have to choose:

Theorem 244. *$(R \cap S, r \cap s)$ is an intersection of (R, r) and (S, s) according to Defn. 158 if and only if it is an intersection by Defn. 93 too.*

I'll leave it as a teasing challenge for now to find a proof strategy for half this: show that an intersection according to our original definition is an intersection according to the new one.

Similarly, we don't have to choose between our definitions of unions:

Theorem 245. *$(R \cup S, r \cup s)$ is a union of (R, r) and (S, s) according to Defn. 159 if and only if it is a union by Defn. 157 too.*

I am not going to spell out a tedious proof of this last theorem,[1] but here's one very quick comment. It is very natural to define the union of two subobjects by invoking a notion of disjunction. And the fact that our categorial story about disjunction as a logical arrow $\vee$ in fact leads to a definition of union which tallies with an alternative, independently attractive, account a mark in favour of our (not immediately intuitive) definition of $\vee$.

(d) We can in fact define unions of subobjects in a topos in a third, equivalent, way.

Just as §20.2 told us how to define the intersection of two subsets of a given set X in Set using a pullback, §20.5 told us how we can define the union of two subsets using a pushout. The idea, recall, was to take the wedge shaped $R \leftarrow R \cap S \rightarrow R$ defining the intersection and then form its pushout: $R \cup S$ will be the vertex of the opposite corner. We can now generalize:

Definition 160. In a topos, if (R, r) and (S, s) are subobjects of X, then any $(R \cup S, r \cup s)$ is a union of them, where $R \cup S$ is the vertex of a pushout from the wedge formed by $R \xleftarrow{i_S} R \cap S \xrightarrow{i_R} R$ as defined in Defn 93 (in §47.1),

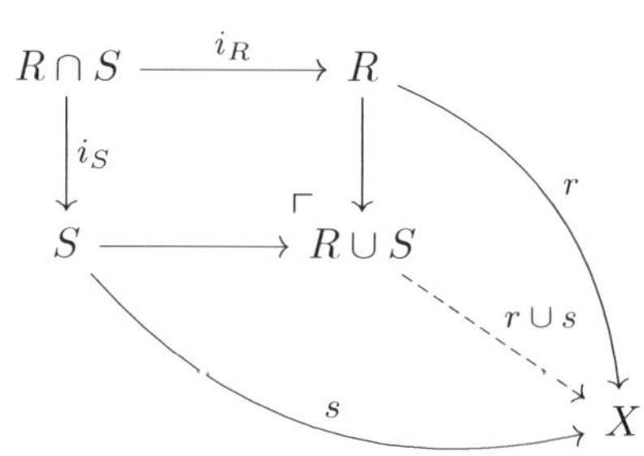

[1] See Goldblatt (1984, pp. 148-151).

and $r \cup s \colon R \cup S \to X$ is the unique arrow making the diagram commute. △

Note, by the way, that in this construction we start from a corner of the shape $R \to X \leftarrow S$; we pull back to get a wedge with the shape $R \leftarrow R \cap S \to S$; and the pushout of this wedge then gives us a corner $R \to R \cup S \leftarrow S$. So the round trip starting from the original corner and proceeding via a pullback followed by a pushout does *not* necessarily return us to where we started.

Our new definition is in good order because the outer paths from $R \cap S$ to X are equal by the definition of $R \cap S$. And hence by the definition of a pushout there must be a unique arrow from $R \cup S$ to X making the diagram commute. Further it can be shown that, in a topos, the arrow $r \cup s$ is monic. We then get an equivalence theorem:

Theorem 246. *$(R \cup S, r \cup s)$ is a union of (R, r) and (S, s) according to Defn. 160 if and only if it is a union by the original Defn. 157 too.*

The proof this time is simple enough: the basic plan is to find the obvious candidate maps in each direction between the objects $R \cup S$ defined in the two ways, and show they are inverses, giving us the required isomorphism. I'll leave it for now as a challenge to fill in the details (you can assume that $r \cup s$ in our new definition is indeed monic).

47.3 Complements: three definitions

(a) If I give you a set R and a set S, then – in standard set theory – it always makes perfectly good sense to ask what the intersection of R with S is. By contrast, our categorial story only defines what it is for $(R \cap S, r \cap s)$ to be an intersection of (R, r) and (S, s) *when those are subobjects of the same object.* Likewise for unions. So we might put it this way: the standard set-theoretic notions of intersection and union are *global*, applying to any two sets (however unrelated), while the categorial notions apply only *locally* to pairs of subobjects whose arrows have the same target. We'll return in the final chapter to think some more about the significance of this.

By comparison, the notion of the complement of a set is already local. If I give you a set S, then in standard set-theory it makes no sense to ask outright what its complement is. Only when S is presented as a subset of some set X can we ask for its complement. We then (as is entirely familiar!) say: if S is a subset of X, then its complement with respect to X is the set $\overline{S} \subseteq X$ such that $S \cap \overline{S} = \emptyset$ and $S \cup \overline{S} = X$.

This motivates a parallel categorial definition:

Definition 161. Suppose $s \colon S \rightarrowtail X$ and $s^* \colon S^* \rightarrowtail X$ are both subobjects of X. Then s^* is a *classical complement* of s iff $s \cap s^* \equiv 0_X$ and $s \cup s^* \equiv 1_X$. △

(b) The first thing to say is that, while a couple of subobjects of a given object X in a topos will always have an intersection and a union (since a topos must have the relevant pullbacks, etc.), a subobject needn't have a classical complement.

Here's a simple illustration. Think in informal terms about graphs and consider the mini-graph

$$\bullet \longrightarrow \star$$

This has only two subgraphs apart from itself and the null graph, namely the solitary node $\bullet$ and the solitary node $\star$. Consider the latter subobject. The complete original graph isn't a complement of that (as the intersection won't be null). Neither the null graph nor the solitary node $\bullet$ is a complement (as the union won't be the whole original). There are no other options. And this pre-categorial reflection about graphs carries over, as you'd expect, to the categorial story about Graph: objects of the category typically have subobjects which lack classical complements.

We will now, though, define two related notions, first what I'll call *negation complements* and then secondly *pseudo-complements*.[2] It will then be a theorem – in this section – that these two notions in fact come to the same. And there will then be another theorem – in the next section – that in the special case of a classical topos, negation-complements (or pseudo-complements) are the real deal, are true classical complements. There's work to be done.

(c) We defined intersections and unions in terms of pullbacks using the categorial versions of conjunction and disjunction. So one obvious idea is to define some notion of complement in terms of a pullback using the categorial version of negation (what more natural? after all, the informal idea of complement of S as a subset of X is the collection of X-elements which are *not* in S).

Definition 162. If $s\colon S \rightarrowtail X$ is a subobject of X with the characteristic arrow χ_s, then a *negation complement* of s is a subobject $\overline{s}\colon \overline{S} \rightarrowtail X$ which results from pulling back $\top$ along $\neg\chi_s$ (or equivalently, results from pulling back $\bot$ along χ_s):

$$\begin{array}{ccc} \overline{S} & \xrightarrow{\;!\;} & 1 \\ {\scriptstyle \overline{s}}\downarrow & & \downarrow{\scriptstyle \top} \\ X & \xrightarrow{\neg\chi_s} & \Omega \end{array} \qquad \textit{or} \qquad \begin{array}{ccc} \overline{S} & \xrightarrow{\;!\;} & 1 \\ {\scriptstyle \overline{s}}\downarrow & & \downarrow{\scriptstyle \bot} \\ X & \xrightarrow{\chi_s} & \Omega \end{array}$$

(i) (ii) △

Note that $\chi_{\overline{s}} = \neg\chi_s$. And to see that these definitions (i) and (ii) are equivalent, simply consider the following diagram (iii):

$$\begin{array}{ccccc} \overline{S} & \xrightarrow{\;!\;} & 1 & \xrightarrow{\;!\;} & 1 \\ {\scriptstyle \overline{s}}\downarrow & & \downarrow{\scriptstyle \bot} & & \downarrow{\scriptstyle \top} \\ X & \xrightarrow{\chi_s} & \Omega & \xrightarrow{\;\neg\;} & \Omega \end{array}$$

[2] Careful: some, e.g. Goldblatt, call my 'negation complements' simply 'complements'.

Suppose (i) is a pullback. That is equivalent to supposing the whole rectangle (iii) is a pullback. But the right-hand square of (iii) is the pullback defining $\neg$; so by the pullback lemma, its left-hand square (ii) is a pullback. Alternatively, if we are given (ii) as a pullback, the pullback lemma tells us that the whole rectangle (iii), so equivalently (i), is a pullback.

(d) Here's another way of characterizing complements that works for sets: the complement of S as a subset of X is the largest subset of X whose intersection with S is empty. There is a categorial analogue for this idea: we'll say that the pseudo-complement of $s\colon S \rightarrowtail X$ is a greatest subobject of X whose intersection with s is 0_X:

Definition 163. Suppose $s\colon S \rightarrowtail X$ and $s^\star\colon S \rightarrowtail X$ are subobjects of X. Then $s^\star$ is a *pseudo-complement* of s (with respect to X) iff, for any other subobject r of X, $r \preccurlyeq s^\star$ if and only if $r \cap s \equiv 0_X$. △

(e) Here's a quick trio of simple results:

Theorem 247. *In a topos,*

(1) If s_1^* and s_2^* are both classical complements of s, then $s_1^* \equiv s_2^*$; and if s_1^* is a classical complement of s and $s_1^* \equiv s_2^*$, then s_2^* is also a classical complement of s.

(2) If $\overline{s}_1$ and $\overline{s}_2$ are both negation complements of s, then $\overline{s}_1 \equiv \overline{s}_2$; and if $\overline{s}_1$ is a negation complement of s and $\overline{s}_1 \equiv \overline{s}_2$, then $\overline{s}_2$ is also a negation complement of s.

(3) If $s_1^\star$ and $s_2^\star$ are both pseudo-complements of s, then $s_1^\star \equiv s_2^\star$; and if $s_1^\star$ is a pseudo-complement of s and $s_1^\star \equiv s_2^\star$, then $s_2^\star$ is also a pseudo-complement of s.

For the first part of (1), note that if s_1^* and s_2^* are both complements of s, we can then apply assorted parts of Theorem 243 to get

$$\begin{aligned} s_1^* &\equiv 1 \cap s_1^* \equiv (s \cup s_2^*) \cap s_1^* \equiv (s \cap s_1^*) \cup (s_2^* \cap s_1^*) \equiv \\ &0_X \cup (s_2^* \cap s_1^*) \equiv (s_2^* \cap s) \cup (s_2^* \cap s_1^*) \equiv s_2^* \cap (s \cup s_1^*) \equiv s_2^* \cap 1_X \equiv s_2^*. \end{aligned}$$

The remaining claims are more or less immediate, though it might help fix ideas to check them before reading on.

(f) And now for the announced main theorem in this section:

Theorem 248. *In a topos, $\overline{s}$ is a negation-complement of X's subobject s if and only if it is also a pseudo-complement of s (with respect to X).*

For the 'only if' direction, we need to prove that for any $r\colon R \rightarrowtail X$, (i) if $r \preccurlyeq \overline{s}$ then $r \cap s \equiv 0_X$, and (ii) if $r \cap s \equiv 0_X$, then $r \preccurlyeq \overline{s}$. Which leaves (iii) the 'if' direction as an easy corollary, given Theorem 247.

Proof of (i). By Theorem 243, if $r \preccurlyeq \overline{s}$ then $r \cap s \preccurlyeq \overline{s} \cap s \equiv 0_X$. But $0_X \preccurlyeq r \cap s$. So $r \cap s \equiv 0_X$. □

Proof of (ii). Given $r \cap s \equiv 0_X$ the left-hand square of diagram (i) is a pullback (by definition of the intersection):

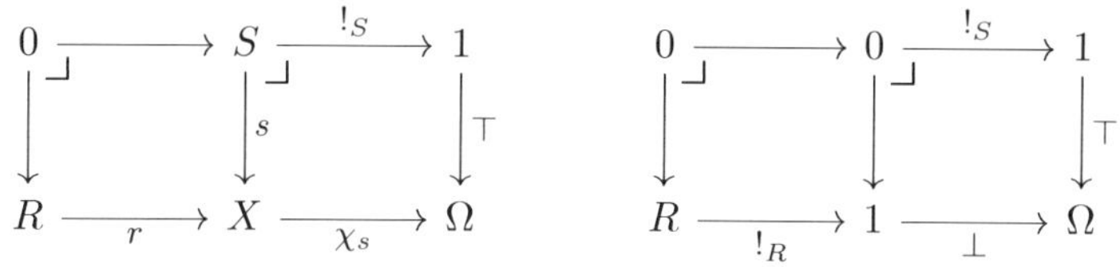

(i) (ii)

The right-hand square of (i) is also a pullback (by definition of χ_s). So the overall rectangle is a pullback by the pullback lemma. So $\chi_s \circ r$ is the characteristic arrow of the unique monic $!\colon 0 \to R$.

Again, the left-hand square of diagram (ii) is a pull-back (why?). And the right-hand square is the pullback defining $\bot$. So the overall rectangle in (ii) is again a pullback by the pullback lemma. Therefore $\bot \circ\, !_R$ is also the characteristic arrow of $!\colon 0 \to R$.

Hence $\chi_s \circ r = \bot \circ\, !_R$, and therefore

$$\chi_{\overline{s}} \circ r = \neg\chi_s \circ r = \top \circ\, !_R = \chi_r \circ r$$

with the last equation from the definition of χ_r. And now we can apply Theorem 243 to get $\overline{s} \cap r \equiv r \cap r$.

But that implies $r \equiv r \cap r \equiv \overline{s} \cap r \preccurlyeq \overline{s}$. So we are done! □

Proof of (iii). We want to show that every pseudo-complement is also a negation complement.

Suppose $s^\star$ is a pseudo-complement of X's subobject s. But s must also have a negation complement $\overline{s}$ (since the defining pullback is always available in a topos). And by (i) and (ii) we know that $\overline{s}$ is also a pseudo-complement of s. But then Theorem 247 (3) tells us that $s^\star \equiv \overline{s}$, and Theorem 247 (2) tells us that $s^\star$ is a negation complement of s. □

47.4 Classical complements

We know that subobjects in a topos needn't always have a classical complement. But negation complements (and hence pseudo-complements) are always available. So negation complements needn't be classical complements.

However, we can derive two significant theorems, the main business of this section. When classical complements *do* exist, they are negation complements. And in a classical topos, negation complements always *are* classical complements.

(a) First, then, we have:

Theorem 249. *In a topos, if s^* is a complement of X's subobject s, then s^* is also a negation complement of s.*

I'll leave the proof for now as a minor challenge (tackled in §47.7).

Now, we know that negation-complements need not be true complements, i.e. we don't always have both $s \cap \overline{s} \equiv 0_X$ and $s \cup \overline{s} \equiv 1_X$. And it is the second of those that can fail, because the first condition in fact always holds:

Theorem 250. *In a topos, if s is a subobject of X, then $s \cap \overline{s} \equiv 0_X$.*

Proof. Consider this diagram:

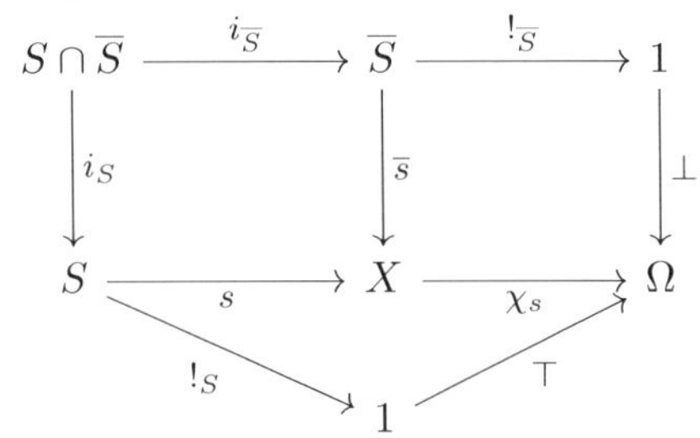

The left square is a pullback defining the intersection, the right square is a pullback defining the negation complement of s. The bottom triangle comes from the pullback defining the characteristic arrow. So that means the outer curved square in this next diagram commutes:

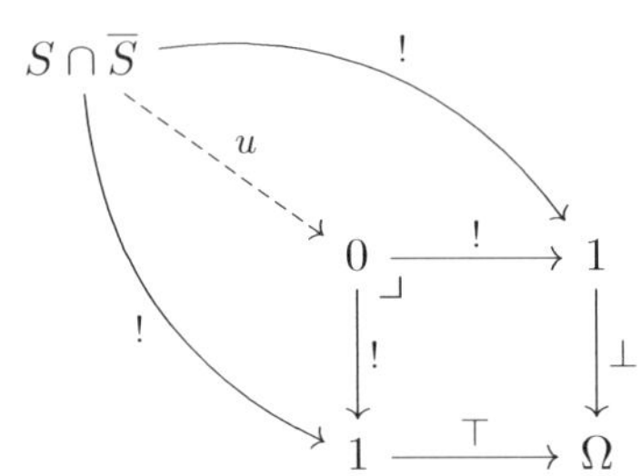

But the inner square is the pullback defining $\bot$ reflected across the diagonal. Hence there is a (unique) arrow $u\colon S \cap \overline{S} \to 0$. Theorem 78 then tells us that u is an isomorphism, making $S \cap \overline{S}$ an initial object. Whence we immediately have $s \cap \overline{s} \equiv 0_X$. □

(b) It is worth noting that we can prove the following special case of the last theorem more directly:

Theorem 251. *In any topos, $\top \cap \overline{\top} \equiv 0_\Omega$.* □

Proof. Consider these three simplest of diagrams:

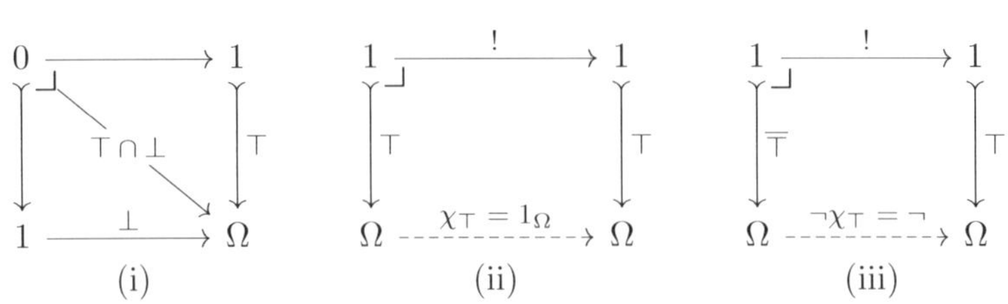

In (i), the pullback which defines $\bot$, the diagonal defines the intersection $\top \cap \bot$, showing that $\top \cap \bot = 0_\Omega$. (ii) tells us that the characteristic arrow of $\top$ as a subobject of Ω is simply 1_Ω. (iii) applies the definition of a negation complement to (ii), to define a negation complement $\overline{\top}$ as a pullback of $\top$ along $\neg$. But of course, pulling back $\top$ along $\neg$ gives us $\bot$. So we've shown that $\overline{\top} \equiv \bot$ and hence from (i) $\top \cap \overline{\top} \equiv 0_\Omega$. $\square$

(c) What about the union $\top \cup \overline{\top}$? Thinking about the case of M_2, for example, we know that we don't have $\top \cup \overline{\top} \equiv 1_\Omega$ in all toposes.

Theorem 252. $\top \cup \overline{\top} \equiv 1_\Omega$ *in a topos if and only if it is classical.*

Proof of 'if'. By Defn. 157, $\top \cup \overline{\top}$, i.e. $\top \cup \bot$, is the monic in an epi-mono factorization of $[\![\top, \bot]\!]$. But in any topos $[\![\top, \bot]\!]$ is already monic (by Theorem 235), so provides the monic in one of its own epi-mono factorizations, and so (by Theorem 234) $\top \cup \overline{\top} \equiv [\![\top, \bot]\!]$. But in a classical topos, the arrow $[\![\top, \bot]\!]$ is an isomorphism, so $\top \cup \overline{\top}$ must also be an isomorphism.

But then, trivially, $(\top \cup \overline{\top}) = 1_\Omega \circ (\top \cup \overline{\top})$ and $1_\Omega = (\top \cup \overline{\top}) \circ (\top \cup \overline{\top})^{-1}$. Hence $\top \cup \overline{\top} \preccurlyeq 1_\Omega$ and $1_\Omega \preccurlyeq \top \cup \overline{\top}$. So $\top \cup \overline{\top} \equiv 1_\Omega$. $\square$

Proof of 'only if'. As before, $\top \cup \overline{\top} \equiv [\![\top, \bot]\!]$. So if $\top \cup \overline{\top} \equiv 1_\Omega$, then $[\![\top, \bot]\!] \equiv 1_\Omega$. So, by Theorem 105, $[\![\top, \bot]\!]$ and 1_Ω factor through each other by an isomorphism, and hence – since 1_Ω is itself an isomorphism – so too is $[\![\top, \bot]\!]$. $\square$

(d) Re-using the line of argument for half the last theorem, we then get an equally easy generalization:

Theorem 253. *In a classical topos, for any subobject* $s\colon S \rightarrowtail X$, $s \cup \overline{s} \equiv 1_X$

Proof. This time, consider these three simple diagrams:

$$\begin{array}{ccc} S & \xrightarrow{\;s\;} & X \\ \big\downarrow & & \big\downarrow{\scriptstyle \chi_s} \\ 1 & \xrightarrow{\;\top\;} & \Omega \end{array} \qquad \begin{array}{ccc} \overline{S} & \xrightarrow{\;\overline{s}\;} & X \\ \big\downarrow & & \big\downarrow{\scriptstyle \chi_s} \\ 1 & \xrightarrow{\;\bot\;} & \Omega \end{array} \qquad \begin{array}{ccc} S + \overline{S} & \xrightarrow{\;[\![s, \overline{s}]\!]\;} & X \\ \big\downarrow & & \big\downarrow{\scriptstyle \chi_s} \\ 1 + 1 & \xrightarrow{\;[\![\top, \bot]\!]\;} & \Omega \end{array}$$

The first two diagrams are the pullbacks defining χ_s and $\overline{s}$, reflected about the diagonal. Then we apply the pullbacks/coproducts lemma Theorem 236 to get the third pullback diagram. But by the classical assumption, the bottom arrow is an isomorphism. And by Theorem 90 the pullback of an isomorphism is an isomorphism, so $[\![s, \overline{s}]\!]$ is an isomorphism. Recycling the line of argument from the previous 'if' proof, it follows that $s \cup \overline{s} \equiv 1_X$. $\square$

In sum, then, Theorems 250 and 253 tell us that classical toposes are classically complemented.

47.5 Relative pseudo-complements

Suppose r and s are subobjects of X. We can define their intersection $r \cap s$ (up to equivalence of arrows) by pulling back $\top$ along the conjunctive arrow $\chi_r \wedge \chi_s$. Likewise, we can define their union $r \cup s$ by pulling back $\top$ along the disjunctive arrow $\chi_r \vee \chi_s$. Our next question is: what happens if we pull back $\top$ along the conditional arrow $\chi_r \Rightarrow \chi_s$? Let's have a definition:

Definition 164. Suppose (R, r) and (S, s) are subobjects of X with characteristic arrows χ_r and χ_s. Then a *pseudo-complement of r relative to s*, notated $(R \supset S, r \supset s)$, is a subobject of X which results from pulling back $\top \colon 1 \to \Omega$ along $\chi_r \Rightarrow \chi_s$, as in

$$\begin{array}{ccc} R \supset S & \xrightarrow{\quad ! \quad} & 1 \\ {\scriptstyle r \supset s}\downarrow & \lrcorner & \downarrow{\scriptstyle \top} \\ X & \xrightarrow{\chi_r \Rightarrow \chi_s} & \Omega \end{array}$$

△

And let's note two theorems. First,

Theorem 254. *In a topos, if r, s are subobjects of X, then $(R \supset S, r \supset s)$ an equalizer of $\chi_{r \cap s}$ and χ_r.*

I'll leave this for now as a challenge to prove. But knowing that $r \supset s$ is an equalizer makes it easier to prove our second theorem:

Theorem 255. *Suppose r and s are subobjects of X in a topos. Then for any subobject x of X, $x \preccurlyeq r \supset s$ iff $x \cap r \preccurlyeq s$.*[3]

This result explains our terminology, by the way. A pseudo-complement of a subobject r is a greatest subobject whose intersection with r is 0_X. A pseudo-complement of r relative to s is a greatest subobject whose intersection with r is 'less than or equal to' s.

Proof. Note we have

$$\begin{array}{lll} x \cap r \preccurlyeq s & \text{iff } (x \cap r) \cap s \equiv x \cap r & \text{(by Theorem 243 (5))} \\ & \text{iff } (r \cap s) \cap x \equiv r \cap x & \text{(by associativity and commutativity)} \\ & \text{iff } \chi_{r \cap s} \circ x = \chi_r \circ x & \text{(by Theorem 243 (6))} \end{array}$$

[3] This should remind you of the elementary logical fact that $X \vdash R \supset S$ iff $X \wedge R \vdash S$ where the variables now stand in for propositions and $\supset$ is the material conditional.

So it is enough to prove that (1) if $x \preccurlyeq r \supset s$, then $\chi_{r \sqcap s} \circ x = \chi_r \circ x$, and (2) if $\chi_{r \sqcap s} \circ x = \chi_r \circ x$ then $x \preccurlyeq r \supset s$.

To show (1). Suppose $x \preccurlyeq r \supset s$, so there is some k such that $x = (r \supset s) \circ k$. But by the definition of the equalizer $r \supset s$, $\chi_{r \sqcap s} \circ (r \supset s) = \chi_r \circ (r \supset s)$, hence $\chi_{r \sqcap s} \circ ((r \supset s) \circ k) = \chi_r \circ ((r \supset s) \circ k)$, hence $\chi_{r \sqcap s} \circ x = \chi_r \circ x$.

To show (2). Suppose $\chi_{r \sqcap s} \circ x = \chi_r \circ x$. Then the top fork here is indeed a commuting fork:

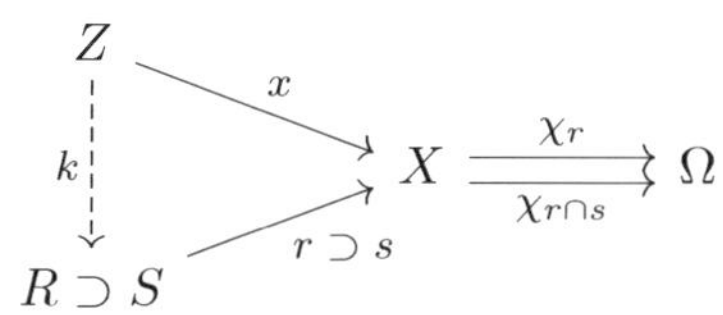

Since $r \supset s$ is an equalizer, there is a unique k making the diagram commute. Hence $x = r \supset s \circ k$, and therefore $x \preccurlyeq r \supset s$. □

47.6 Lattices of (equivalence classes of) subobjects

(a) With the notion of relative pseudo-complements now in place, we have everything we need to link up with a perhaps familiar idea from pre-categorial mathematics.

A *lattice* comprises some partially ordered objects such that every pair of objects has a greatest lower bound and a least upper bound. More officially – and because I don't want to be distracting at this point, I'll use set talk rather than plural locutions – we can say:

Definition 165. A set of objects L, equipped with a partial order $\leqslant$ and two binary operations $\sqcap$ and $\sqcup$, form a *lattice* $(L, \leqslant, \sqcap, \sqcup)$ iff, for any $x, y \in L$, $x \sqcap y$ is their infimum with respect to $\leqslant$, and similarly $x \sqcup y$ is their supremum. This lattice is

(i) *distributed* iff for any $x, y, z \in L$, $x \sqcap (y \sqcup z) = (x \sqcap y) \sqcup (x \sqcap z)$ and $x \sqcup (y \sqcap z) = (x \sqcup y) \sqcap (x \sqcup z)$;
(ii) *bounded* iff there is a $\leqslant$-minimum object 0 and a maximum 1;

And, given it is bounded, the lattice is

(iii) *complemented* when, for any $x \in L$, there is an object in L which we'll denote $\overline{x}$ such that $x \sqcap \overline{x} = 0$ and $x \sqcup \overline{x} = 1$.
(iv) *pseudo-complemented* when, for any $x \in L$, there is an $\overline{\overline{x}} \in L$ such that, for any $z \in L$, $z \leqslant \overline{\overline{x}}$ iff $x \sqcap z = 0$;
(v) *relatively pseudo-complemented* when, for any $x, y \in L$ there is a object $x \sqsupset y \in L$ such that, for any $z \subset L$, then $z \leqslant (x \sqsupset y)$ iff $z \sqcap x \leqslant y$. △

Elementary exercises show that, in any lattice, $\sqcap$ and $\sqcup$ are commutative and associative; that each of the distributivity conditions implies the other; that a bounded, distributive lattice is (relatively) pseudo-complemented if it is complemented (but not necessarily vice versa); and much more.

(b) Let's pick out two particular types of lattice that are of special interest here:

Definition 166. A bounded, distributed, relatively pseudo-complemented lattice is called a *Brouwerian* (or *Heyting*) lattice.

A bounded, distributed, complemented lattice is called a *Boolean* lattice. △

Note: a Boolean lattice is a fortiori also Brouwerian (since a complemented lattice also has all relative pseudo-complements). A paradigm example of a Brouwerian lattice that needn't be Boolean is provided by the lattice of open sets of a topological space. And a paradigm example of a Boolean lattice is the lattice of subsets of a given set.

Now, the labels we've given our two sorts of lattice flag up their intimate connection with Brouwer's/Heyting's intuitionistic logic and classical, Boolean, two-valued logic respectively. Here I simply report the headline news:[4]

Theorem 256. *Suppose – to fix ideas – we set up an intuitionistic (classical) propositional logic with the logical connectives* $\wedge$, $\vee$, $\Rightarrow$ *and the falsum* $\bot$ *built in, and negation defined by* $\neg A = A \to \bot$.

Then a formula is an intuitionistic (classical) theorem if and only if it is mapped to the top element 1 when interpreted in any Brouwerian (Boolean) lattice. In other words, if we assign objects in the chosen lattice to the propositional variables, interpret the three connectives $\wedge$, $\vee$, $\Rightarrow$ *by* $\sqcap$, $\sqcup$, $\sqsupset$ *respectively and interpret the falsum as denoting* 0, *then the formula always evaluates to 1.*

(c) What has this to do with toposes?

Evidently, the subobjects of some object X in a topos do *not* form a lattice – at least when subobjects are defined our way as individual monic arrows. That's because the relation $\preccurlyeq$ we defined on such subobjects of X is only a preorder. However, it is equally evident what we need to do to a get a lattice into play: consider instead equivalence classes of subobjects-as-monics.[5]

So let's introduce

Definition 167. Use '$[s]$' as notation for the class of monics equivalent to s. Then, for any equivalence classes of subobjects of some X, namely $[r]$ and $[s]$, we define

(1) $[r] \sqcap [s] =_{\mathrm{def}} [r \cap s]$

(2) $[r] \sqcup [s] =_{\mathrm{def}} [r \cup s]$

And further,

[4]The classic reference for such results is the wonderful, though old-school, Rasiowa and Sikorski (1963). See also e.g. Dummett (2000, §§5.1–5.3).

[5]'Ah! So sometimes we *do* want to talk about equivalence classes of subobjects.' Yes, and I didn't deny that back in §22.5. But I'm not particularly inclined to say that their appearance here is a good reason to want to go back to officially redefine subobjects from the outset as *being* equivalence classes.

(3) $\overline{[r]} =_{\text{def}} [\overline{r}]$
(4) $[r] \leqslant [s]$ is defined to hold iff $r \preccurlyeq s$. △

We had better check, however, that these definitions do work!

For (1), suppose $r \equiv r'$ so $[r] = [r']$, and similarly suppose $s \equiv s'$ so $[s] = [s']$. Then we need $[r] \sqcap [s] = [r'] \sqcap [s']$ if $\sqcap$ is to be well-defined. But part of Theorem 243 applied twice gives us that. We can similarly show that (2), (3) and (4) are in good order.

Then, putting various theorems we have to hand together with our new definitions, we can arrive at the following rather neat summary:[6]

Theorem 257. *Let $Sub(X)$ be the set of the equivalence classes of subobjects of X in some topos. Then $(Sub(X), \leqslant, \sqcap, \sqcup)$ is a bounded, distributed, (relatively) pseudo-complemented lattice. If $\top$ in the topos has a complement, such lattices will not only be (relatively) pseudo-complemented but complemented.* □

Of course, it should go without saying, for different objects X in a topos, the resulting lattices $(Sub(X), \leqslant, \sqcap, \sqcup)$ can be of wildly different sizes, upwards from the one-element lattice of subobjects of the initial object.

(d) So now we see that there are two ways that intuitionistic logic, in particular, gets into the story about toposes. First, consider the internal logic of a topos – still meaning for our limited purposes here just the set of laws governing the defined logical arrows $\wedge$, $\vee$, $\Rightarrow$, $\neg$. Then, as we explained in Chapter 46, this internal logic of a topos will include at least the intuitionistic theorems involving those arrows' logical counterparts, perhaps more. And we have now noted that any lattice of equivalence classes of the subobjects of a given object in a topos is (at least) a Brouwerian lattice and so, in the sense we indicated, it provides a model for propositional intuitionistic logic. To explore these intriguing ideas further, see Goldblatt's classic book again.

47.7 Challenges!

We'll pause here to wrap up some of the technicalities and derive the theorems whose proofs were left as challenges.

They all have elementary proofs – elementary in the sense of calling on little more than definitions and familiar basic facts about pullbacks and the like – though some derivations are a bit involved. By all means skip. But it might help to fix ideas to work through some of the proofs.

(a) The first pair of challenges was to prove

Theorem 243 (parts). In a topos, where q, r, s are subobjects of X:

(i) If $r \preccurlyeq s$, then $(q \cap r) \preccurlyeq (q \cap s)$.

[6] Let's not worry about issues of size, so that we can happily enough count $Sub(X)$ as a set.

(ii) $\chi_r \circ q = \chi_s \circ q$ iff $r \cap q \equiv s \cap q$.

Proof of (i). Stare at the following diagram! The front square (with sides $qsba$) is the pullback square defining $q \cap s$. The receding square (with sides $qrdc$) is the pullback square defining $q \cap r$. Since by assumption $r \preccurlyeq s$, we know there is an arrow $h \colon R \to S$ making the bottom triangle commute.

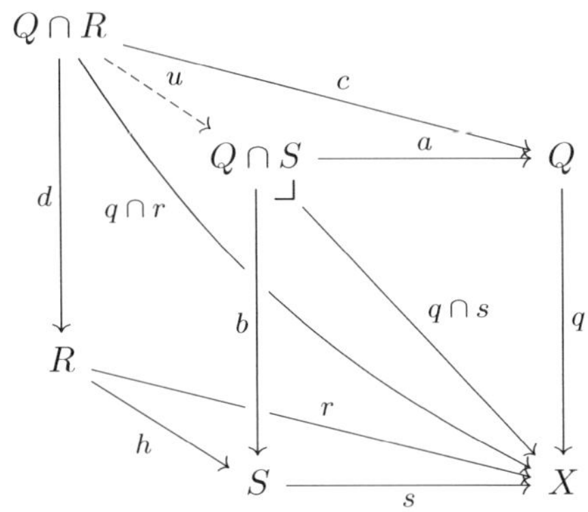

But that means the wedge $S \xleftarrow{h \circ d} Q \cap R \xrightarrow{c} Q$ forms a commuting square with the opposite corner $S \xrightarrow{s} X \xleftarrow{q} Q$. Hence, since the front square is a pullback, there will be an arrow $u \colon Q \cap R \to Q \cap S$ making the diagram commute, giving us $q \cap r = (q \cap s) \circ u$. Hence, as we wanted to show, $(q \cap r) \preccurlyeq (q \cap s)$. □

Proof of (ii). Consider the left-hand diagram, pasting together two pullbacks:

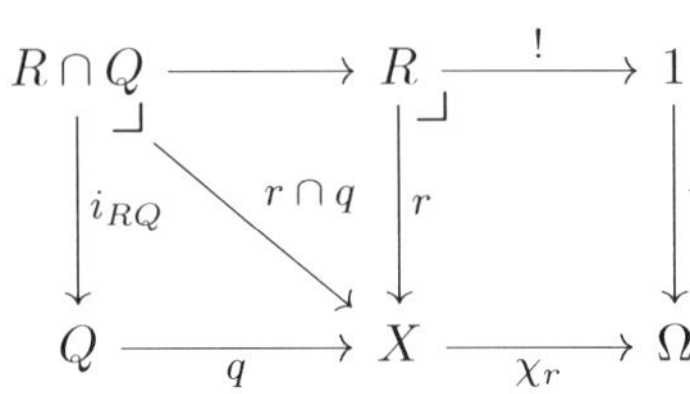

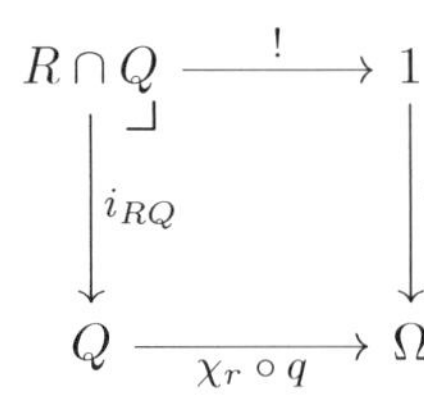

So by the pullback lemma, the right-hand diagram is a pullback, from which it follows that $\chi_{i_{RQ}} = \chi_r \circ q$. And exactly similarly, we'll get $\chi_{i_{RS}} = \chi_s \circ q$.

Therefore $\chi_r \circ q = \chi_s \circ q$ iff $\chi_{i_{RQ}} = \chi_{i_{RS}}$ iff $i_{RQ} \equiv i_{RS}$.

But $r \cap q = q \circ i_{RQ}$ and $s \cap q = q \circ i_{SQ}$. So, remembering q is monic, it is immediate that $\chi_r \circ q = \chi_s \circ q$ iff $r \cap q \equiv s \cap q$. □

(b) The next challenge is to prove

Theorem 244 (part). *If $(R \cap S, r \cap s)$ is an intersection of (R, r) and (S, s) according to the original Defn. 93 (where we pull r back along s) then it is also an intersection according to Defn. 158 (where we pull $\top$ back along $\chi_r \wedge \chi_s$).*

Proof strategy. We ultimately need to get the arrow $\wedge \circ \langle\!\langle \chi_r, \chi_s \rangle\!\rangle$ as used in Defn. 158 into play. In other words, we want to be looking at a diagram that contains the composite

$$X \xrightarrow{\langle\!\langle \chi_r, \chi_s \rangle\!\rangle} \Omega \times \Omega \xrightarrow{\ \wedge\ } \Omega.$$

So, looking at the right-hand arrow, let's complete its defining square

$$\begin{array}{ccccc} & & 1 & \xrightarrow{\ !\ } & 1 \\ & & \big\downarrow{\scriptstyle \langle\!\langle \top, \top \rangle\!\rangle} & \lrcorner & \big\downarrow{\scriptstyle \top} \\ X & \xrightarrow{\langle\!\langle \chi_r, \chi_s \rangle\!\rangle} & \Omega \times \Omega & \xrightarrow{\ \wedge\ } & \Omega \end{array}$$

The obvious strategy now is to try to use Defn. 93 to show that the left-hand square below is also a pullback:

$$\begin{array}{ccccc} R \cap S & \xrightarrow{\ !_{R\cap S}\ } & 1 & \xrightarrow{\ !\ } & 1 \\ \big\downarrow{\scriptstyle r \cap s} & \lrcorner & \big\downarrow{\scriptstyle \langle\!\langle \top, \top \rangle\!\rangle} & \lrcorner & \big\downarrow{\scriptstyle \top} \\ X & \xrightarrow{\langle\!\langle \chi_r, \chi_s \rangle\!\rangle} & \Omega \times \Omega & \xrightarrow{\ \wedge\ } & \Omega. \end{array}$$

For then, by the pullback lemma, the outer rectangle will be a pullback. But that rectangle is equivalent to the pullback square in Defn. 158, and we'll be done.

First we'll prove that that square at least commutes. So consider

$$\begin{aligned} \pi_1 \circ \langle\!\langle \chi_r, \chi_s \rangle\!\rangle \circ (r \cap s) &= \chi_r \circ (r \cap s) && \text{(by definition of } \langle\!\langle \chi_r, \chi_s \rangle\!\rangle) \\ &= \chi_r \circ r \circ i_R && \text{(by Defn. 93)} \\ &= \top \circ\ !_R \circ i_R && \text{(by definition of } \chi_r) \\ &= \top \circ !_{R\cap S} && \text{(arrows to 1 are identical)} \\ &= \pi_1 \circ \langle\!\langle \top, \top \rangle\!\rangle \circ !_{R\cap S} && \text{(by definition of } \langle\!\langle \top, \top \rangle\!\rangle) \end{aligned}$$

Exactly similarly

$$\pi_2 \circ \langle\!\langle \chi_r, \chi_s \rangle\!\rangle \circ (r \cap s) = \pi_2 \circ \langle\!\langle \top, \top \rangle\!\rangle \circ !_{R\cap S}$$

So by Theorem 45 $\langle\!\langle \chi_r, \chi_s \rangle\!\rangle \circ (r \cap s) = \langle\!\langle \top, \top \rangle\!\rangle \circ !_{R\cap S}$ and our square commutes.

Now, this isn't yet all that we want: we need that square to be a pullback. But having done the main work, let's allow ourselves to hand-wave: everything is derived from limits (pullback squares and products) so the resulting square ought to give us another limiting case too. (Exercise for enthusiasts: do better than hand-wave!) □

(c) A teasing challenge was to show that defining the union of subobjects (R, r) and (S, s) by a pushout construction is equivalent to defining it by taking an epi-mono factorization of $[\![r, s]\!]$:

Theorem 246. *$(R \cup S, r \cup s)$ is a union of (R, r) and (S, s) according to Defn. 160 if and only if it is a union by the original Defn. 157 too.*

Proof. To avoid notational tangles, use the notation (U, u) for the union introduced by the pushout construction, and use (I, m) for the union as originally defined by an epi-mono factorization via the usual image construction. Then we have the following commuting diagrams:

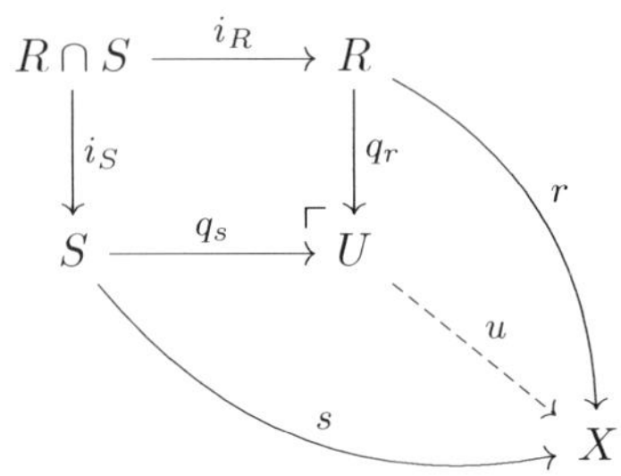

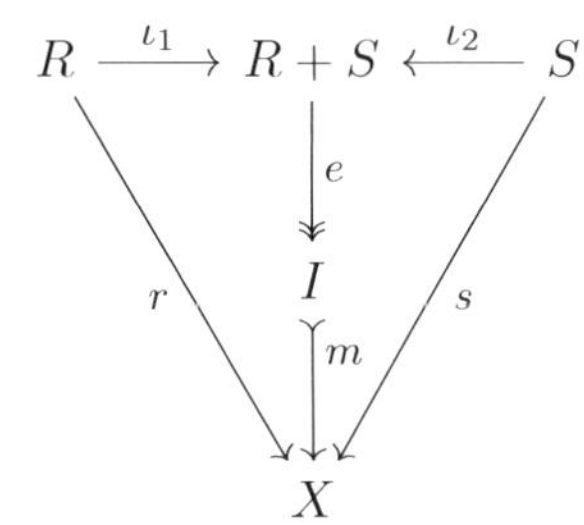

And what we want to show is that there is an isomorphism $\varphi\colon U \xrightarrow{\sim} I$ such that $u = m \circ \varphi$. Take this in stages.

(1) Combining the diagrams, the outer parts of this next diagram commute, so by the universal property of pushout, there is a unique arrow $\varphi\colon U \to I$ making the whole diagram commute:

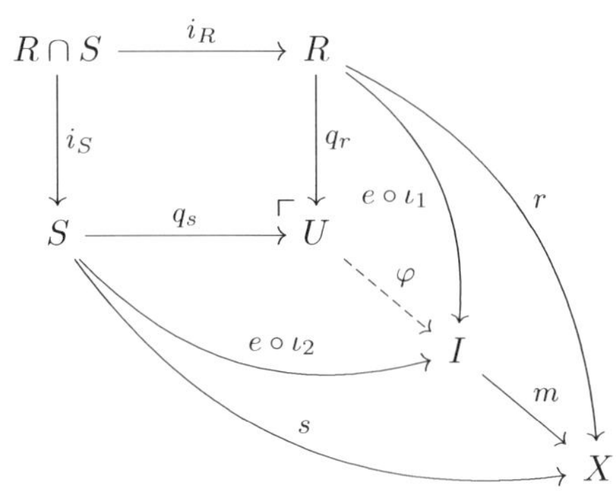

And $u = m \circ \varphi$ by the uniqueness of the pushout arrow to X.

(2) Now we construct a map $\psi\colon I \to U$. Combining those first two diagrams again we get the diagram on the left below minus the arrow from $R + S$ to U.

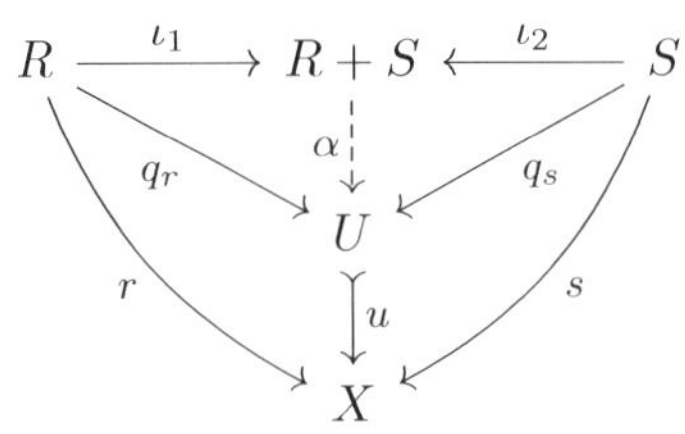

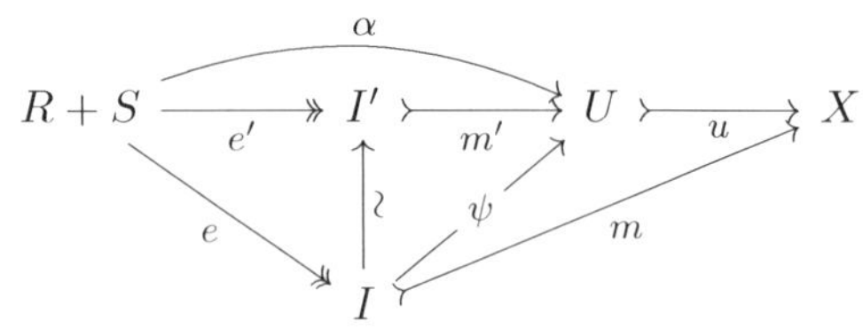

By the universal property of the coproduct, we then can add a (unique) arrow $\alpha\colon R + S \to U$ making the diagram commute. This α will itself have an epi-mono factorization $m' \circ e'$ giving us the diagram on the right. $R + S$ will now have two epi-mono factorizations (recall, we are assuming u is monic, and hence $u \circ m'$ is monic). And so there will be a (unique) isomorphism between their intermediary objects. So now a key move: define $\psi\colon I \to U$ to be the composite of that isomorphism with m' as shown.

(3) We now prove that ψ is a two-sided inverse for φ. First, $\varphi \circ \psi = 1_I$. For consider the diagram on the left:

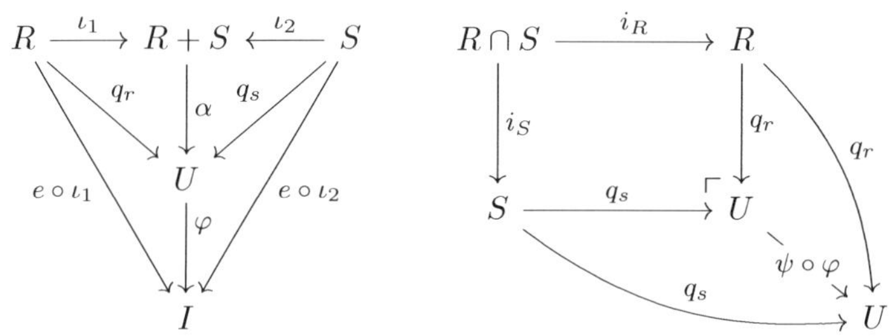

Given commuting triangles from previous diagrams, this first diagram commutes. So $\varphi \circ \alpha$ completes a coproduct diagram for the arrows $e \circ \iota_1$, $e \circ \iota_2$. But evidently, e completes the same diagram. So by the universal property of coproducts, $\varphi \circ \alpha = e$. Hence

$$\varphi \circ \psi \circ e = \varphi \circ \alpha = e = 1_I \circ e.$$

But e is an epimorphism so can be cancelled on the right. Which shows that $\varphi \circ \psi = 1_I$.

Now consider the second diagram above. This commutes too, because (again from earlier diagrams)

$$\psi \circ \varphi \circ q_r = \psi \circ e \circ \iota_1 = \alpha \circ \iota_1 = q_r$$

and similarly for q_s. But that second diagram would trivially commute with the arrow from U to U replaced with the identity arrow. Therefore by the universal property of the pushout which tells us that the arrow completing the pushout diagram is unique, $\psi \circ \varphi = 1_U$.

Hence $\varphi\colon U \to I$ has a two-sided inverse and is an isomorphism, as we wanted to prove. □

(d) You were then asked to show that

Theorem 249. *In a topos, if s^* is a complement of X's subobject s, it is also a negation complement of s.*

By assumption, $s \cap s^* \equiv 0_X$ and $s \cup s^* \equiv 1_X$. Since pseudo-complements are negation complements, it is enough to show that, for any subobject r of X, (i) if $r \preccurlyeq s^\star$ then $r \cap s \equiv 0_X$ and (ii) if $r \cap s \equiv 0_X$ then $r \preccurlyeq s^\star$. Using parts of Theorem 243, we can argue like this:

Proof of (i). If $r \preccurlyeq s^\star$ then $r \cap s \preccurlyeq s^\star \cap s \equiv 0_X$. But of course $0_X \preccurlyeq r \cap s$. So $r \cap s \equiv 0_X$. □

Proof of (ii). If $r \cap s \equiv 0_X$, then

$$(s^\star \cup r) \equiv (s^\star \cup r) \cap 1_X \equiv (s^\star \cup r) \cap (s^\star \cup s) \equiv s^\star \cup (r \cap s) \equiv s^\star \cup 0_X \equiv s^\star.$$

But $r \preccurlyeq (s^\star \cup r)$. Hence $r \preccurlyeq s^\star$. □

(e) The final challenge was to show

Theorem 254. *In a topos, if r, s are subobjects of X, then $(R \supset S, r \supset s)$ an equalizer of $\chi_{r\cap s}$ and χ_r.*

Proof. Suppose $(R \supset S, r \supset s)$ is the pullback of $\top$ along $\chi_r \Rightarrow \chi_s$. And consider this diagram:

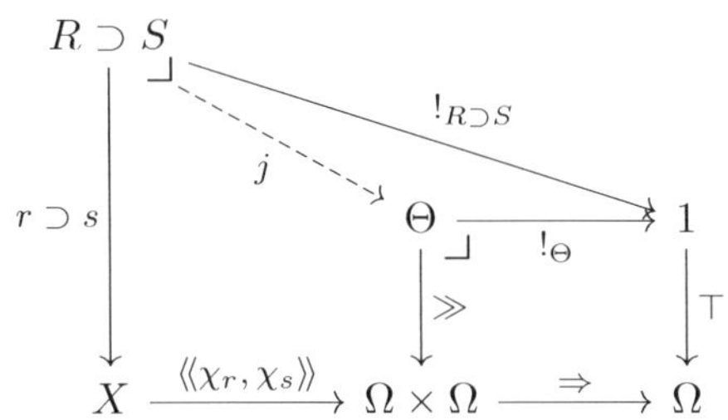

With $\gg$ the equalizer of $\wedge, \pi\colon \Omega \times \Omega$, the inner square is the pullback square defining $\Rightarrow$. The outer rectangle is pullback square giving us $(R \supset S, r \supset s)$. Since the wedge $\Omega \times \Omega \longleftarrow R \supset S \longrightarrow 1$ gives us a commuting square with opposite corner of the inner pullback square, there is a unique arrow $j\colon R \supset S \to \Theta$ making the diagram commute.

But this means (since there is a unique arrow from $R \supset S$ to the terminal 1) that – ignoring the arrow currently labelled $!_{R\supset S}$ – we have two commuting squares, with the right-hand one a pullback, and the overall rectangle is a pullback. So by the pullback lemma the left-hand square must be a pullback too.

We'll now use this fact to show that $r \supset s$ equalizes χ_r and $\chi_{r\cap s}$. Inspect the next diagram:

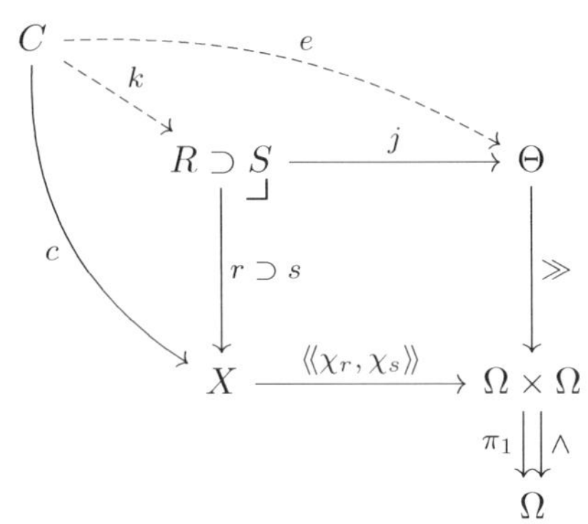

First we remark

$$\chi_r \circ r \supset s = \pi_1 \circ \langle\!\langle \chi_r, \chi_s \rangle\!\rangle \circ r \supset s = \pi_1 \circ \gg \circ j = \wedge \circ \gg \circ j = \\ \wedge \circ \langle\!\langle \chi_r, \chi_s \rangle\!\rangle \circ r \supset s = \chi_{r \cap s} \circ r \supset s$$

So composing $r \supset s$ with χ_r and $\chi_{r \cap s}$ results in equal arrows. To show that $r \supset s$ is *the* equalizer of χ_r and $\chi_{r \cap s}$ consider any other arrow $c\colon C \to X$ such that $\chi_r \circ c = \chi_{r \cap s} \circ c$.

In other words, suppose $\pi_1 \circ (\langle\!\langle \chi_r, \chi_s \rangle\!\rangle \circ c) = \wedge \circ (\langle\!\langle \chi_r, \chi_s \rangle\!\rangle \circ c)$. But that means we have a commuting fork with handle $\langle\!\langle \chi_r, \chi_s \rangle\!\rangle \circ c$ and prongs π_1 and $\wedge$, which must therefore factor through the equalizer of those prongs by a unique arrow $e\colon C \to \Theta$. And then the wedge $X \longleftarrow R \supset S \longrightarrow \Theta$ gives us a commuting square with the opposite corner of the pullback, so there must be a unique $k\colon C \to R \supset S$ making the diagram commute.

So as we wanted to show, the fork with handle c and prongs χ_r and $\chi_{r \cap s}$ factors uniquely through $r \supset s$, which is our desired equalizer. □

48 Well-pointed toposes, with choice

Some toposes provide non-classical worlds whose internal logic is intuitionistic. However, we'll be very much concentrating on toposes which give us arenas in which we can develop mainstream classical mathematics. In particular, we'll be looking at toposes which are *well-pointed* and where (optionally) a version of the *Axiom of Choice* applies.

This chapter prepares the ground for discussions in the final chapter by showing that (1) well-pointed toposes are classical, (2) they support a nice notion of 'membership' between point elements of X and subsets of X, and (3) we can add an Axiom of Choice in two equivalent ways.

48.1 Well-pointedness and its implications

(a) Back in Defn. 44, we characterized a category as well-pointed if its arrows behave like functions at least in this respect: the identity of the arrows is fixed by how they act on point elements. Now, it is not built in to the definition of a topos that its arrows need be function-like in this respect. However, if we are looking for worlds in which to (re)construct standard mathematics, it is natural to impose this condition; and toposes that *are* well-pointed do share a number of particularly nice features. Adding a non-degeneracy condition for convenience, here again is the definition we need:

Definition 168. A topos is *well-pointed* iff it is non-degenerate and whenever parallel arrows $f, g\colon X \to Y$ agree on all point elements – i.e. whenever $f \circ \vec{x} = g \circ \vec{x}$ for all $\vec{x}\colon 1 \to X$ – then $f = g$. △

(b) And here are two simple lemmas which will immediately be useful:

Theorem 258. *In a well-pointed topos, any non-initial object has at least one element.*

Proof. A non-initial object X in a topos is the target of at least two arrows, namely $0_X : 0 \to X$ and $1_X : X \to X$, which are distinct if the topos is non-degenerate, having different sources by assumption. And these are both monic arrows (by Theorems 78 and 16).

So we know that in a topos there must be a couple of related parallel arrows, their characteristic arrows $\chi_0\colon X \to \Omega$ and $\chi_1 : X \to \Omega$. Theorem 110 then tells us that these arrows must be distinct (otherwise the subobjects $(0, 0_X)$ and $(X, 1_X)$ would be equivalent, and so $0 \cong X$, contrary to hypothesis).

Hence, by well-pointedness, the parallel arrows $\chi_0, \chi_1 : X \to \Omega$ must act differently on some element of X. Implying that X must have at least one element! □

As an immediate corollary, it follows that in a well-pointed topos, an 'empty' object with no point elements is initial.

Theorem 259. *In a well-pointed topos, the only subobjects of a terminal object (up to equivalence) are* $(0, !)$ *and* $(1, !)$.

Proof. Suppose $m: X \rightarrowtail 1$ is a monic arrow targeting the terminal 1.

One possibility is that X is an initial object 0, with m the unique arrow $!: 0 \to 1$, giving us the subobject $(0, !)$.

Otherwise X is non-initial, and by the previous theorem there is a point element $\vec{x}: 1 \to X$. But then we have $m \circ \vec{x}$ is an arrow from 1 to itself, which must be the unique arrow $!: 1 \to 1$, which is the identity on 1. Hence m is a monic with a right inverse, and hence is an isomorphism by Theorem 22. So the subobject (X, m) will be equivalent to $(1, !)$. □

(c) We can now prove the following key result:

Theorem 260. *If a topos is well-pointed, its truth-value object* Ω *has just two elements,* $\top$ *and* $\bot$ *(in a word, the topos is bivalent).*

Proof. Take any truth-value-seeking arrow $v: 1 \to \Omega$. Make a corner with $\top: 1 \to \Omega$, and form its pullback:

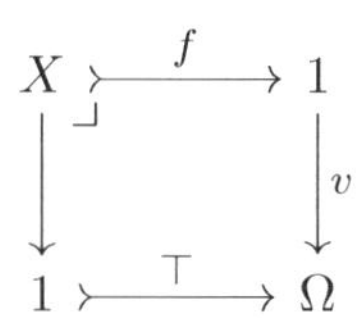

If $X \cong 0$, we simply have the pullback defining *false*, and $v = \bot$.

Otherwise $X \ncong 0$, and by the previous theorem there is an arrow $\vec{x}: 1 \to X$. But then we note that for any parallel g, h such that $g \circ f = h \circ f$ we will have $g \circ f \circ \vec{x} = h \circ f \circ \vec{x}$ and hence $g = h$ (because $f \circ \vec{x}: 1 \to 1$ has to be the identity on the terminal object). Hence f is right-cancellable, so epic.

But f is also monic, being the pullback of a monic up along v. Hence f is an isomorphism by Theorem 116, and so $X \cong 1$, which makes $v = \top$. □

(d) Here's an immediate corollary. Since $\neg\top = \bot$ and $\neg\bot = \top$ and since (as we now know) in a well-pointed topos $\top$ and $\bot$ are the only point-elements of Ω, it follows that $\neg: \Omega \to \Omega$ has no fixed points in the sense of Theorem 81. Hence we can infer a nice version of Cantor's Theorem:

Theorem 261. *In a well-pointed topos, there can be no point-surjection from an object* X *to* Ω^X. □

So, there will be an unending sequence $X, \Omega^X, \Omega^{\Omega^X}, \Omega^{\Omega^{\Omega^X}}, \ldots$, with no surjective arrow from one to the next.

(e) Now, in general, a topos can be bivalent without being complemented (and in fact vice versa). But being well-pointed keeps things tidy:

Theorem 262. *A well-pointed topos is classical, and hence complemented.*

Proof. The 'hence' part follows from the results in §47.4, so we just need to establish the first part by showing that $[\![\top, \bot]\!]$ is an isomorphism. And we do that by first showing that $[\![\top, \bot]\!]$ is epic.

Consider then the following diagram:

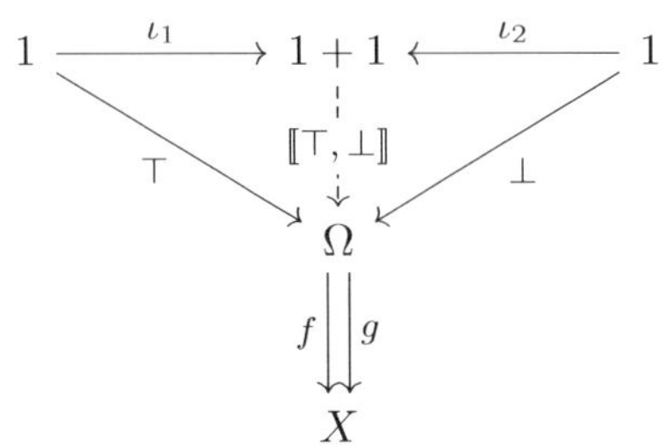

Now, suppose $f \circ [\![\top, \bot]\!] = g \circ [\![\top, \bot]\!]$.

Then $f \circ [\![\top, \bot]\!] \circ \iota_1 = g \circ [\![\top, \bot]\!] \circ \iota_1$, hence $f \circ \top = g \circ \top$. Likewise, $f \circ \bot = g \circ \bot$. Which means, given the previous theorem, that the parallel arrows f and g act the same way on *all* elements of their source, and hence (by the assumption of well-pointedness) $f = g$. Therefore, since f and g were arbitrary, $[\![\top, \bot]\!]$ is epic.

But $[\![\top, \bot]\!]$ is also monic by Theorem 235. Hence it is an isomorphism by Theorem 116. So the topos is, by definition, classical. □

48.2 Members of subobjects

So we now know that, in a well-pointed topos, the negation-complements, intersections and unions of subobjects of an object X will all behave in a parallel way to the complements, intersections and unions of subsets of given set X.

And the present section will press this subobject/subset parallel further by defining a notion of *membership*, holding between point elements of X and subobjects of X (which we have learnt to think of as arrows). We'll find some important respects in which this topos-theoretic notion of membership behaves in ways parallel to the familiar set-theoretic notion.

(a) It will be helpful first to prove a preliminary lemma, which generalizes Theorem 19 which told us that, an exceptional case apart, monics are right inverses in Set:

Theorem 263. *In a well-pointed topos, any monic $s \colon S \rightarrowtail X$ with $S \ncong 0$ is a right inverse, i.e. there is an arrow $u \colon X \to S$ such that $u \circ s = 1_S$.*

Proof. Because our topos is complemented, the upper square in the following diagram is a pushout (applying Defn. 160 to get the union of s and $\overline{s}$):

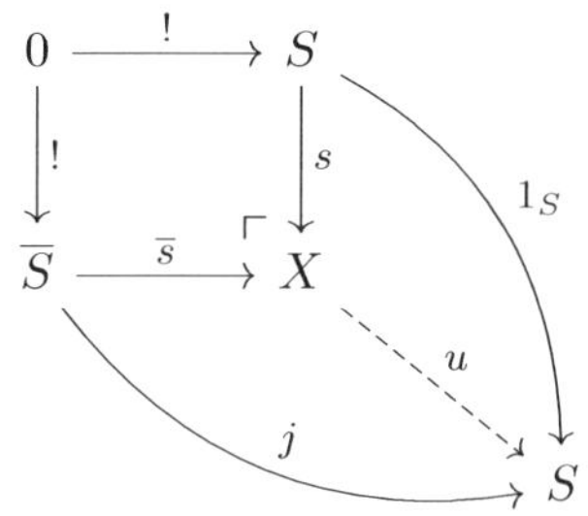

S is non-initial, so Theorem 258 tells us that there is an arrow $\vec{x}\colon 1 \to S$; and trivially, there is an arrow $!\colon \overline{S} \to 1$. So there is an arrow $j = \vec{x} \circ !\colon \overline{S} \to S$.

Hence the outer bent square with opposite vertices 0 and S exists and must commute since there is a unique arrow from the initial 0 to S. Therefore, since the inner square is a pushout, there must be an arrow u making the whole diagram commute, giving us $u \circ s = 1_S$. □

(b) So to the main definition. Suppose $\vec{x}\colon 1 \to X$ is an element of X: when shall we say that this element-as-arrow is a 'member' of a given subobject-as-arrow? This seems inviting:

Definition 169. If $\vec{x}\colon 1 \to X$ is an element of X, and $(S, s\colon S \rightarrowtail X)$ is a subobject of X, then $\vec{x}$ *is a member of* (S, s) iff there is some arrow $1 \to S$ making this diagram commute:

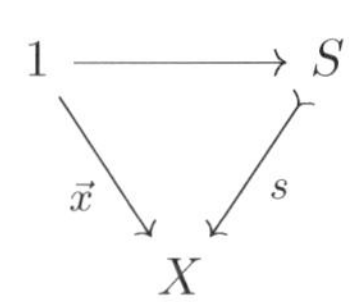

△

The arrow $1 \to S$, if it exists, is unique – for if both $f, g\colon 1 \to S$ make the triangle commute, we'd have $s \circ f = s \circ g$, and therefore $f = g$ since s is monic.

Let's immediately have two quick theorems which show that, so defined, 'membership' behaves as we might hope. The first follows immediately from previous definitions, the second has the simplest of proofs:

Theorem 264. *The members of the maximum subobject of X, $(X, 1_X)$, are all and only the elements $\vec{x}\colon 1 \to X$. While the minimum subobject $(0, 0_X)$ has no members.* □

So with only the slightest abuse of language, we can say that elements of an object X are its members.

Theorem 265. *If $\vec{x}$ is a member of (R, r) and (R, r) is included in (S, s), then $\vec{x}$ is a member of (S, s).*

Proof. Just look at the diagram:

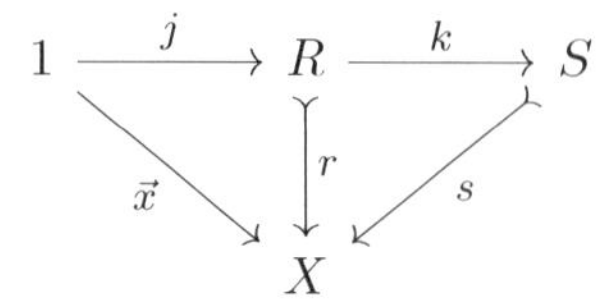

Since $\vec{x}$ is a member of (R, r) there is some j making the left triangle commute; and since $r \preccurlyeq s$ there is some k making the right triangle commute. So $k \circ j$ makes the outer triangle commute, witnessing that $\vec{x}$ is a member of (R, r). □

(c) We also have another welcome result:

Theorem 266. *Suppose (R, r) and (S, s) are subobjects of X in a well-pointed topos. Then: if every member of (R, r) is a member of (S, s) then (R, r) is included in (S, s).*

Proof. First take the case where R is initial. Then since it is initial, there is a unique arrow $j : R \to S$, giving us an arrow $s \circ j : R \to X$. But then $s \circ j = r$, since there is exactly one arrow from an initial R to X. So $r \preccurlyeq s$, and the theorem's conditional conclusion follows.

Now suppose R is not initial, and assume every member of r is a member of s. By Theorem 258 there is an arrow $k : 1 \to R$. Which makes $r \circ k$ a member of r. Therefore s has a member, and that requires there to be an arrow $1 \to S$, and hence S isn't initial.

Then since s is monic, and its source isn't initial, Theorem 263 tells us that there is a $u : X \to S$ such that $u \circ s = 1_S$. Now put $i = u \circ r$. I claim that $s \circ i = r$, so $r \preccurlyeq s$.

It is enough to show that for every $p : 1 \to S$, $s \circ i \circ p = r \circ p$. But note that $\vec{x} = r \circ p : 1 \to X$ is, trivially, a member of r, and hence by our assumption it is also a member of s. Hence there is some $j : 1 \to S$ such that $\vec{x} = s \circ j$.

Putting everything together we have

$$s \circ i \circ p = s \circ u \circ r \circ p = s \circ u \circ \vec{x} = s \circ u \circ s \circ j = s \circ j = \vec{x} = r \circ p.$$

So we are done! □

(d) Ordinarily, a *subset* of X is a *member* of its powerset $\mathcal{P}X$. What is the categorial analogue of this?

We've been here before in §24.1. To repeat, there is a natural association between subobjects $(S, s : S \rightarrowtail X)$ and their characteristic arrows $\chi_s : X \to \Omega$ (with subobjects getting the same characteristic arrow if and only if they are equivalent, as per Theorem 110). And there is a naturally arising bijection between those characteristic arrows and arrows $\chi'_s : 1 \times X \to \Omega$. And then, taking exponential transforms, there's an equally natural bijection between *those* arrows and arrows $\widetilde{\chi'_s} : 1 \to \Omega^X$ (as per Defn. 72). In short, every subobject of X corresponds in a natural way to a member of (the maximum subobject of) the power object Ω^X, with subobjects being assigned the same member if and only if they are equivalent.

(e) Here's another case where we can neatly parlay set-theoretic membership talk into categorial membership talk.

In the ordinary set-theoretic treatment, a binary relation between objects in the sets X and Y is defined as a *subset* $R \subseteq X \times Y$, and we write xRy just when $\langle x, y \rangle$ is a member of R.

Categorially, we can now say a binary *relation* R_r between X and Y is a monic $r\colon R \rightarrowtail X \times Y$. And suppose $x\colon 1 \to X$ and $y\colon 1 \to Y$ are elements of X and Y respectively, so that the mediating arrow of the product is $\langle\!\langle x, y \rangle\!\rangle\colon 1 \to X \times Y$. Then we can write $xR_r y$ when $\langle\!\langle x, y \rangle\!\rangle$ is a member of (R, r). We can then go on to define, for example, what it is for R_r to be an equivalence relation in very much the usual way. But we needn't follow up this idea here.

48.3 A reality check

Let's pause for a reality check. We have two related notions in play, the idea of being a *point element* and the idea of being a *subobject member*. They are not only different notions but typically feature in claims with a different status.

To say that the arrow $\vec{x}$ is a point element of X is (usually) to make a *typing judgement*. It doesn't (usually) report a discovery or require a proof but simply specifies the type of thing we are dealing with, namely an arrow whose source is an initial object 1 and whose target is X.[1]

By contrast, to say that $\vec{x}$ is a member of some subobject (S, s) is (usually) to report putative news about $\vec{x}$ – a proposition that will (usually) stand in need of demonstration as true or false – namely that there is a point element $\vec{j}$ of S such that $\vec{x} = s \circ \vec{j}$.

48.4 Choice

We will take up some of those ideas about subobjects and their members in the next chapter. But first, another theme.

(a) If we are working in a well-pointed topos with a natural numbers object, we will be able to implement a substantial amount of ordinary mathematics, as we also discuss in the next chapter. And we can implement *more* ordinary mathematics if we make another familiar and very basic principle available, namely some version of the Axiom of Choice. "Much ink has been spilled over this axiom", to echo the laconic remark of Lawvere and Rosebrugh in their category-theoretic text *Sets for Mathematics*: and this certainly isn't the place to spill yet more. So I'll simply assume that you know something about choice and how it can have a role in fairly elementary mathematical arguments.

[1]Sometimes, type theorists are inclined to say that such typing judgements are not true or false: but I think that is an unnecessarily unhappy way of putting things.

Now, we've already mentioned the topic much earlier: Theorem 19 tells us that the proposition

(C1) *Every epimorphism has a right inverse* (i.e. is a left inverse, or – in other jargon – 'splits')

is a version of the Axiom of Choice for Set. And (C1) can now serve as our choice principle more generally, across toposes.

(b) Interestingly, however, the original proposal for a categorial choice principle due to Lawvere was different: it was in effect the claim that

(C2) *For any arrow* $f\colon X \to Y$ *(where* $X \not\cong 0$*), there is a* $g\colon Y \to X$ *such that* $f \circ g \circ f = f$.

But there is no real divergence here:

Theorem 267. *In a well-pointed topos, the choice principles (C1) and (C2) are equivalent.*

Proof: (C1) implies (C2). In a topos, every arrow has an epi-mono factorization (by Theorem 233). So given an arrow $f\colon X \to Y$, there will be an epic $e\colon X \twoheadrightarrow Z$ and mono $m\colon Z \rightarrowtail Y$ such that $f = m \circ e$.

Given (C1), e has a right inverse, so there is an arrow $j\colon Z \to X$ such that $e \circ j = 1_Z$.

Now also assume $X \not\cong 0$. Then we also have $Z \not\cong 0$, or else by Theorem 78 the arrow e would be an isomorphism making $X \cong 0$ after all. Since $Z \not\cong 0$, we can apply Theorem 263, and the monic m is a right inverse, so there is a $k\colon Y \to Z$ such that $k \circ m = 1_Z$.

Put $g = j \circ k$. And then we have

$$f \circ g \circ f = (m \circ e) \circ (j \circ k) \circ (m \circ e) = m \circ (e \circ j) \circ (k \circ m) \circ e = m \circ e = f$$

So in sum, given (C1), then (C2) follows.[2] □

Proof: (C2) implies (C1). Suppose $X \cong 0$. Then $f\colon X \to Y$ is monic by Theorem 78; so if it is also epic then it is an isomorphism by Theorem 116 and hence has a right inverse.

So suppose $X \not\cong 0$, then by (C2) for some g we have $f \circ g \circ f = 1_Y \circ f$. Hence if f is epic, we can right-cancel and derive $f \circ g = 1_Y$, so f has a right inverse, giving us (C1). □

[2] I take this surprisingly fiddly proof from Kim (1996). See also this Mathoverflow answer where Mike Shulman offers basically the same proof: tinyurl.com/ShulmanAC.

49 ETCS

The practice of topos theory quickly spawned an associated philosophy ... whose chief tenet is the idea that, like a model of set theory, any topos may be taken as an autonomous universe of discourse or 'world' in which mathematical concepts can be interpreted and constructions performed.
John Bell, 2001

Category theory, as developed in Parts I and II of these notes, gives us conceptual tools for organizing mathematics and discovering the commonalities and relationships between structures of different kinds. That alone is reason enough to study it. But right back in §1.2(c) and again in §4.3(c), I trailed the idea that any sufficiently rich category constitutes an arena in which we can develop a good deal of mathematics; indeed it has been argued that some toposes provide a *better* such arena than conventional set theory. We now have the resources needed to begin to flesh out and assess those ideas. In particular, we will consider the kind of topos described by Lawvere's ETCS, the Elementary Theory of the Category of Sets.[1]

49.1 Classical arenas, ssets and ffunctions

(a) Suppose we are working in a well-pointed topos with a natural numbers object. Then we have an object N (with associated 'zero' and 'successor' arrows) which behaves like a set of natural numbers. Subobjects of N will behave like set-like collections of natural numbers. And we can define intersections, unions, and complements of these subobjects, which behave in a familiar classical way.

We can also form a power object of N which behaves like the powerset of the natural numbers, and this powerset will have its own subobjects behaving like various sets of sets of numbers. We can form power objects of these too, and so on. And since we are in a topos, we can in effect quotient by equivalence relations.

It seems, then, that in any well-pointed topos with a natural numbers object we should be able to emulate the kind of construction of (say) real numbers

[1] I am particularly indebted in this chapter to discussions with Rowsety Moid.

familiar from conventional set theory where we build up from sets of natural numbers. Or at least, that's the hopeful thought.

(b) Let's have a definition (though it's certainly not a standard one):

Definition 170. We will call a well-pointed topos with a natural numbers object a *classical arena*. We will call the objects of a given classical arena *ssets* and call its arrows *ffunctions*. △

Here 'classical arena' encapsulates a promise that any such topos will provide an arena in which we can develop classical mathematics (or at least, most 'ordinary' non-set-theoretic mathematics).

We know that the ingredients of a classical arena can have some of the features of sets and functions. But we don't want to beg too many questions at this stage, hence my suggestion that we pro tempore call the objects and arrows of an arena 'ssets' and 'ffunctions'. On the one hand, a story about what we can do in classical arenas will then retain a familiar enough look to help guide constructions and proofs as we go along. On the other hand, the deviant spelling is a standing reminder not to jump too readily to the assumption that ssets really are sets as conventionally conceived or that ffunctions really are functions.

(c) But can we back up that hopeful thought that a classical arena of ssets and ffunctions provides us with enough to construct e.g. a field of classical real numbers in a way that parallels the familiar sort of construction in a universe of sets that we find in a hundred conventional set-theory texts?

Yes. And we don't have to resort here to hand waving and mere promissory notes. The job has actually been carried out, in full detail but very accessibly, by Tom Leinster (2024) who has – in place of a conventional set-theory course – given and written up a course on ssets and ffunctions, developing inter alia an account of the reals. And what does he assume about ssets and ffunctions? He gives axioms that simply restate that we are in a classical arena, i.e. in a well-pointed topos with a NNO (and perhaps with choice). These axioms, though not quite in Leinster's words, are:

(1) The usual categorial axioms hold for ssets and ffunctions; so, in particular, composition of ffunctions is associative, and the identity ffunctions act as identities.

(2) There exists a terminal sset.

If 1 is terminal, then we'll say a ffunction $\vec{x}\colon 1 \to X$ is an element of X. Then,

(3) If $f, g\colon X \to Y$ are ffunctions such that $f \circ \vec{x} = g \circ \vec{x}$ for all elements $\vec{x}$ of X, then $f = g$.

(4) There exists an empty sset, i.e. a sset that has no elements.

(5) Let X and Y be ssets. Then there exists a categorial product of X and Y.

(6) Let X and Y be ssets. Then there exists a sset Y^X equipped with an evaluation ffunction ev as in Defn. 72.

(7) Let $f \colon X \to Y$ be a ffunction, and let $\vec{y}$ be an element of Y. Then there exists a pullback of $\vec{y}$ along f.[2]

(8) There exists a subobject classifier.

(9) There exists a natural number object.

And then we can add an optional choice principle,

(10) Every epic ffunction has a right inverse.

I've used category-theoretic terminology there, but that's only for brevity's sake. Leinster explicates all his axioms in more basic terms, e.g. by spelling out what counts as a category-style product or a pullback, or what counts as a natural numbers object, without ever mentioning categories. And then Leinster manages to make his proofs and constructions starting from his given axioms cleave pretty closely to the look-and-feel of the conventional set-theoretic reasoning of 'ordinary mathematics'.

To be sure, any reader of Leinster's notes who knows a little category theory – you, for example! – will recognize categorial themes being played out, and spot that his own versions of axioms (1) to (8) tell us that ssets and ffunctions form a well-pointed topos. But his mission is not to engage with category theory more widely but to concentrate on showing that, as I might put it, a classical arena does enable versions of some familiar key mathematical constructions.

(d) And so, in his different terminology and notation, Leinster retells and somewhat expands our account of the algebra of subobjects – now an algebra of subssets – in a well-pointed topos. Then he says more about relations, and about equivalence relations in particular. He discusses the now-familiar beginnings of arithmetic when a NNO is provided. He also gives, as we did, a close analogue of the usual construction of the integers using equivalence classes of pairs of naturals.

Then, going beyond anything that we have done here, Leinster also constructs an arithmetic on the integers, defines the rationals, and defines Dedekind cuts on the rationals to give us a complete ordered field implementing the reals.

Having already encountered half the story – though in a more general categorial framework – you shouldn't be too surprised to learn that the project can be continued. And in his notes, Leinster works through all the details, if you want them, in a couple of hundred enviably lucid pages, with definitional and notational choices along the way which make for a roughly comparable smoothness and elegance to the usual set-theoretic story. Important though all this is, I perhaps need not say more here since Leinster has done the work for us.[3] We get, in short, an impressively worked-out practical demonstration that a classical arena really does provide a setting in which we can begin to pursue classical mathematics.

[2]In the context of the other axioms, we get all other pullbacks too.

[3]For part of the story, you can also see Lawvere and Rosebrugh (2003).

49.2 Non-classical arenas?

A brief aside. As noted before, some toposes provide non-classical worlds, ones whose internal logic is intuitionistic. In §46.5 I mentioned in passing the effective topos Eff. Let me equally briefly mention another, particularly interesting, example: there are non-classical toposes which model so-called *smooth infinitesimal analysis*. That's a theory which adds to the axioms for a field a collection Δ of 'nilsquare' numbers, i.e. a collection of numbers r such that (i) each $r^2 = 0$, yet also such that, although (ii) not every member of Δ equals 0, (iii) no member is definitely distinct from 0.

Which sounds pretty weird. And of course, it's flatly inconsistent with the classical law $\forall x(x = 0) \lor \exists x(x \neq 0)$. Yet smooth infinitesimal analysis can be elegantly – and consistently! – developed in a non-classical framework to give us a surprisingly intuitive theory for a differential and integral calculus. And the theory in fact has topos-theoretic roots in the work of Lawvere.[4]

Fascinating though that story is, however, we certainly can't pursue it here. I'm just flagging up the point that our focus here on classical arenas necessarily gives only a partial story about the varieties of mathematics that can be done in different toposes.

49.3 ETCS

(a) Leinster's presentation defining what I've non-standardly called a classical arena has an even more conventional look-and-feel to it than I have so far described, because (of course, of course!) he doesn't talk of *ssets* and *ffunctions*, but calls the data of a classical arena simply *sets* and *functions*.

Now, there are significant differences between Leinster's sset theory and conventional set theory. For example, as I noted in §48.2, in the topos-theoretic framework, we have to distinguish between being an *element* (of an object) from being a *member* (of a subobject). Thus x is an element of the sset X iff it is an arrow $x\colon 1 \to X$; but it is a member of the sub-sset $(S, s\colon S \rightarrowtail X)$ iff there is an arrow $j\colon 1 \to S$ such that $x = s \circ j$. However, we can notationally minimize the element/member distinction by symbolizing the first by $x \in X$ and the second by e.g. $x \in_X S$. And then we can further recover ordinary set-theoretic symbolism by dropping the subscript and letting context do the work. Which is Leinster's ploy. You may or may not think that this is slightly sneaky! – but it undoubtedly makes his unfolding account look as close as possible to a regular construction of the reals in a framework of sets and functions.

(b) Leinster's assumption that he really has given a theory of sets goes back to the seminal paper 'An elementary theory of the category of sets' by Lawvere

[4] So this is a very different theory from Robinson's classical but non-standard analysis which is familiar to any logician. There's a lovely presentation in the accessible short book by Bell (2008). Note too that, in this case, adding the law of excluded middle to the intuitionistic logic of our theory doesn't give us a stronger theory but outright inconsistency. An interesting discussion of the philosophical implications of this phenomenon can be found in Shapiro (2014).

(1964).[5] The proposal there is that a suitable category of sets-for-applications is indeed provided by what I'm calling a classical arena with choice. Other category theorists can be equally emphatic that ETCS really is a theory of sets.[6] Moreover, it is claimed that ETCS is closer to what we actually need for mathematical applications than a more conventional set theory.

Two issues then arise, which we do well to keep quite separate:

(A) How similar are the ssets and ffunctions of ETCS to the sets and functions of conventional mathematics? Are they in fact so very different that it is misleading to use the same terminology for both without due qualification?

(B) Irrespective of our answer to (A), does a theory of ssets and ffunctions actually do better than a standard theory of sets and functions in supplying the needs of 'ordinary' (non-set-theoretic) mathematics?

The next section says something about (A), and some initial arguments concerning (B) are then discussed in the following section.

I will mention a few more relevant technical facts as we go along; however, to a significant extent, what follows in the final sections of these notes is no longer straight exposition but is a brief and tentative foray into disputed territory, tangling with matters of judgement and interpretation. So make of the ensuing remarks what you will.

49.4 Not really an account of sets?

(a) We have recalled the entirely novel element/member distinction needed by a theory like ETCS. Now consider the following facts:

(1) A subset of the set X, in the ordinary sense, can also be a subset of Y, where $X \neq Y$. However, an ETCS sub-sset of a sset X, i.e. some object equipped with a monic targeting X, which is to say some $(R, r\colon R \rightarrowtail X)$, cannot also be the very same item as a sub-sset of Y which is some $(S, s\colon S \rightarrowtail Y)$, unless $X = Y$. (They can't even be equivalent subobjects either unless $X = Y$, for the simple reason that equivalence is only defined as a relation between subobjects of the same object.)

[5]Note, by the way, the definite article in Lawvere's title. That seems entirely misplaced. On any sensible view, there are lots of interesting, differently structured, categories of sets. There are a multitude of different universes of sets providing models for ZFC. And then there are, for example, categories of NF sets – genuinely different, as is confirmed in a forthcoming *JSL* paper by Thomas Forster and Nathan Bowler which gives, at last, a proof for the folklore claim that NF is not synonymous with any theory of a cumulative hierarchy.

So it is notable that in Lawvere's later book with Rosebrugh, *Sets for Mathematics*, while the authors do still fall back into talking about 'the category of sets', sometimes their axiomatic theory is rather more carefully said to characterize '*a* category of abstract sets and arbitrary mappings' (2003, p. 113).

[6]So: "ETCS is a set theory. It is not a membership-based set theory like ZF. It is a function-based set theory." (McLarty 2004, p. 39). See also Mac Lane (1997, Appendix), Trimble (2011), Leinster (2012), McLarty (2017).

(2) A member of a subset of X, in the ordinary sense, can be a member of a subset of Y, where $X \neq Y$. However, a categorial member of a subobject of a sset X, i.e. an appropriate arrow $\vec{x}\colon 1 \to X$, cannot also be categorial member of a subobject of Y, i.e. an appropriate arrow $\vec{y}\colon 1 \to Y$, again unless the arrows have the same target, so $X = Y$.

(3) In ordinary set theory, any set can be both an element (of one set) and a subset (of another). But a categorial element of the sset X is an arrow of the type $\vec{x}\colon 1 \to X$ and a categorial subobject of Y, on the austere view, is a monic $s\colon S \rightarrowtail Y$. These can't be the same unless $S = 1$ and $X = Y$. (And note too that, in the special case of an arrow $\vec{x}\colon 1 \to X$, this trivially counts as a categorial member of the subobject-as-bare-monic $\vec{x}\colon 1 \to X$. But in standard set theory, nothing is a member of itself.)

(4) In an ordinary kind of set-theoretic framework, given an element x of some set X, we can sensibly ask 'and what, if any, are the elements of x?'. The answer may be 'none', if x is an urelement or is the empty set: but the question always makes sense. However it would make no sense at all to say that a categorial element of sset X, i.e. an arrow $\vec{x}\colon 1 \to X$, has elements.

(b) Lawvere goes further, not only denying that an element of a sset can itself have elements, but insisting that it has no properties at all other than distinctness from other elements of the sset. As already noted (at the end of fn. 5), he refers to the objects of ETCS as 'abstract sets', and he explains:

> [A]n abstract set may be conceived of as a bag of dots which are devoid of properties apart from mutual distinctness. Further, the bag as a whole [is] assumed to have no properties except cardinality, which amounts to just the assertion that it might or might not be isomorphic to another bag. (Lawvere 1994, p. 5)

Remember the Dedekindian idea I touched on in §2.8 according to which, in addition to more 'concrete' Klein groups, there is a purely abstract Klein group whose four elements have no properties over and above being distinct from each other and being interrelated by the group operation according to the appropriate table.[7] And now Lawvere offers the equally mysterious proposal there is a purely abstract four-element set whose elements have no properties at all over and above being distinct from each other. Does this really make sense? Let's not try to wrestle with this metaphysical conundrum. What is quite clear is that the 'bag of dots' picture certainly doesn't correspond to the notion of set that we ordinarily deploy in everyday mathematics; when we talk in Analysis 101 of a set of natural numbers (or a set of reals, etc.), we are assuredly not talking of a set of elements that are entirely devoid of all properties apart from mutual distinctness!

[7] This idea is sufficiently puzzling for Michael Dummett famously to call it 'mystical': see p. 296 of his (1991), a modern classic in the philosophy of mathematics.

(c) Leinster also adds to the list of differences between ssets and sets as ordinarily conceived:

> An important feature of the approach we take is that *we never ask whether two sets are equal*, just as a group theorist would never ask whether two groups are equal – only whether they're isomorphic. (Leinster 2024, p. 16, his emphasis.)

But many will find this injunction *very* odd. It is usually thought that the extensionality principle is definitive of the concept of set: A and B are equal as sets, are one and the same set, if and only if every member of A is a member of B and vice versa. So banning talk of A and B as being the same or different would rule out stating the extensionality principle. But as George Boolos remarks in his classic paper on justifying the axioms of set theory, while other principles might be debatable,

> [A] theory that did not affirm that the objects with which it dealt were identical if they had the same members would only by charity be called a theory of *sets* alone.[8]

Correspondingly, the extensionality principle typically appears very early on in elementary set theory texts, while the authors are still explaining what it is they are aiming to talk about.

After all, a root idea underlying conventional set theory is that, starting with some objects, we can form the set of them, with the identity of the prior objects fixing which set we get. Begin with a different selection of objects, and we get a different set. What more basic? Yet it seems that ETCS can't get a handle on this essential root idea:

> In ETCS ... there is no such thing as 'forming a set of' anything. A set in ETCS is not a collection of pre-existing things, be they functions or anything else.

It might well be said, however, that a theory in which there is no notion of 'forming a set of' is surely not a theory of sets. Again,

> Functions are the primary citizens of ETCS. ... [E]lements and membership can be interpreted in ETCS but it is strange and perhaps misleading to give them center stage by calling ETCS a set theory. ... [F]rom the perspective of someone who works with set theories all the time, it's very difficult for me to use 'set theory' to describe a theory that does not rest on an extensional membership relation and does not even have one as a formal part.

It is difficult to resist this summary conclusion:

[8] Boolos (1971, p. 28 of the reprint). The quoted observation is echoed by other philosophers of mathematics such as Potter (2004, p. 33) and Incurvati (2020, p. 11).

> The basic objects of ZFC and ETCS are both called 'sets', but they behave so differently that it can be confusing to use the same name for both.[9]

Consequently it has become common to mark the differences between standard set theories and Lawvere's theory by using 'material set theory' to refer to theories based on a global membership relation, and 'structural set theory' to refer to theories like ETCS.

We might still quibble. But having done the important thing here, having stressed some of the key differences, I suspect that there is little profit to be had in arguing any further about question (A), about whether the objects of a structural set theory are sufficiently set-like to still be appropriately called sets, albeit now in a qualified way. So let's move on.

49.5 Capturing what we need?

Fortunately, we don't need to come to a verdict on (A) in order to go on to consider question (B): does something like ETCS do better than a standard set theory in meeting the needs of 'ordinary' (non-set-theoretic) mathematics?

(a) We can perhaps summarize points (1) and (2) in the last section like this. A categorial subobject is, as it were, *local* to the unique object it is a subobject of. And likewise a categorial member of a subobject is *local* to the unique object that the relevant subobject is a subobject of. Now, is this stern locality of the categorial notions as contrasted with their less restrictive set-theoretic counterparts a troublesome bug – or is it a feature that we should, on further reflection, positively embrace?[10]

Well, when we look at how set-talk is deployed in ordinary mathematics, we do find that we are in fact usually working in some limited ambient universe comprising, e.g., the natural numbers, or the reals, or the points of a given space, etc. Then the sets that will concern us will, in the first instance, all be subsets of that background ambient universe, and the members of these subsets will all be elements of that same ambient universe. (Then we build up powersets, etc., but still from the same local ingredients.)

Colin McLarty makes this point:

> Throughout mathematics it is crucial to know which elements $x \in A$ of a set A are members of which subsets $S \subseteq A$. We say this relation

[9]The first of these last three quotes is from Mike Shulman (who is warmly disposed towards category theory). His remark is a brief contribution to a long and fascinating discussion at the n-Category Café on an earlier short paper by Tom Leinster (2012) advocating ETCS as a way of 'Rethinking Set Theory': see tinyurl.com/cafe-rethink. The second quote is from François G. Dorais in the same discussion thread; and on functions being the 'primary citizens' compare the quotation from Dana Scott at the end of §4.2. The third quote is from Shulman (2013).

[10]It would be rather nice to call ETCS a 'local set theory' – except that this term has come to denote a close cousin, given an explicitly type-theoretic twist and a non-classical logic, as in Bell (1988).

> is *local* to elements and subsets of the ambient set A. For example, arithmeticians need to know which natural numbers $n \in \mathbb{N}$ are in the subset of primes $Pr \subset \mathbb{N}$. On the other hand, nearly no one ever asks whether the imaginary unit $i \in \mathbb{C}$ is also a member of the unit sphere $S^2 \subset \mathbb{R}^3$, because they do not lie inside of any one natural ambient. (McLarty 2017, p. 11)

So perhaps a theory which only defines what is for an element of X to be a member of this or that subset of X, and only defines what it is for one subset of X to be included in another subset of X, is good enough for at least many mathematical purposes. And it might be said that Leinster's detailed development of his user-friendly version of ETCS is, so to speak, proof of concept.

(b) And what about points (3) and (4), highlighting that the categorial story doesn't accommodate downward membership chains? Well again, it might be argued that in non-set-theoretic mathematics we very often take it that we are dealing, in the first place, with a set of elements – say, the points of a topological space – for which the question 'and what are the elements of *those* elements?' just doesn't arise.

This time, here's Leinster in his earlier 'Rethinking Set Theory' pressing the point:

> [I]n the framework of ZFC, the elements of a set are always sets too. Thus, given a set X, it always makes sense in ZFC to ask what the elements of X's elements are. Now, a typical set in ordinary mathematics is $\mathbb{R}$. But accost a mathematician at random and ask them 'what are the elements of π?', and they will probably assume they misheard you, or ask you what you're talking about, or else tell you that your question makes no sense. If forced to answer, they might reply that real numbers have no elements. But this too is in conflict with ZFC's usage of 'set': if all elements of $\mathbb{R}$ are sets, and they all have no elements, then they are all the empty set, from which it follows that all real numbers are equal. (Leinster 2012, p. 1)

Again, our categorial story where questions about 'elements of elements' don't arise might be said – *is* said by fans of ETCS – to be in pleasing accord with at least *some* ordinary non-set-theoretic ways of thinking.

(c) On reflection, however, McLarty's and Leinster's observations do not weigh particularly heavily against conventional set theory.

(1) First, to pick up a simple but pivotal point already noted in §4.1, fn. 5, the sort of relatively naive set theory for ordinary mathematical applications which is outlined in the introductory chapters of a hundred textbooks on topology, analysis, algebra, etc., typically allows elements that aren't assumed to be sets, i.e. allows urelements or individuals. We can then start with a universe of natural numbers as urelements, or can start with the reals as given, or with geometric points, or whatever: then we use the

apparatus of our set-theory-for-applications to build up sets of those chosen urelements (and then sets of *them*, etc.).[11]

It is certainly *not* ordinarily assumed that the elements of resulting sets like $\mathbb{R}$ are always themselves more sets. Nor will the question normally arise whether an element of (say) some set of interest of those doing complex analysis is also an element of (say) some set of interest to topologists (since different local realms of individuals will be involved).

(2) So, to emphasize, if we want to regiment and round out that sort of common-or-garden set-theory-for-applications, we'll *not* immediately get Leinster's "framework of ZFC", a theory of pure sets. Rather we'll land on a theory which allows urelements, but which will be neutral about the existence and character of any such urelements; and arguably this theory will lack some of the strength of ZFC too – so perhaps it will be a version of ZU, Zermelo set theory with urelements, augmented with a choice principle.[12]

[11] Evidence? I'll mention three modern classics. First, Sutherland's *Introduction to Metric and Topological Spaces* (2009) – the first text I took off my shelves – seems quite typical of many books which dive in, expecting their readers to be already acquainted with some 'naive' set theory. Early on, the reals are assumed to be given as a complete ordered field, so that we then can talk about the set $\mathbb{R}$ (and can apply familiar set-theoretic operations to it). There is no assumption at all that the reals, the elements of $\mathbb{R}$, are themselves sets. Later we meet the idea of a topological space, consisting of a non-empty set X equipped with a topology. The elements of X are taken as structureless points, again with no assumption at all that they are sets. It isn't sets 'all the way down': on the contrary, the implied low-level set-theory for applications evidently allows sets to have non-sets, individuals, as urelements.

Second, for a paradigm text that is more explicit about its set-theoretic assumptions, consider *Topology* by Munkres (2000) (already mentioned in §4.1, fn. 5). McLarty (2017) rightly notes the comparative modesty of the informal set theoretic principles laid out in the long opening chapter. Oddly, though, McLarty writes that "Munkres hardly denies that all objects are sets". Yet Munkres does deny just that – his informal set theory explicitly allows urelements which aren't sets: "The objects belonging to a set may be of any sort. One can consider the set of all even integers, and the set of all blue-eyed people in Nebraska, and the set of all decks of playing cards in the world."

For a third example, consider Tao's *Analysis I* (2016). The author aims to be more careful than usual in laying out his set-theoretic assumptions. So after an initial chapter on the natural numbers, he gives us a substantial chapter on sets. But he is explicit that the set theory in question is in itself agnostic about whether or not are objects which aren't sets, neither ruling in nor ruling out urelements. Which given that Tao is emphatic that natural numbers aren't sets but wants, of course, to talk about sets of numbers, is how things have to be.

[12] ZU is the base-line set theory presented by the indispensable Potter (2004). See also the excellent presentation by Moschovakis (2006, pp. 23–29).

It's a nice question, however, just how strong a set theory *does* suffice for non-set-theoretic mathematics. Mac Lane famously argues in his book *Mathematics, Form and Function* that something even weaker than Zermelo with choice will do: "For most Mathematics, the appropriate axioms for set theory seem to be ZBQC: The Zermelo axioms with [unlimited] comprehension replaced by bounded comprehension and with choice added" (1986, p. 373). Mathias almost as famously has counter-argued that there are statements of analysis that count as belonging to ordinary enough mathematics but which require stronger set-theoretic assumptions for their only known proofs (2000; 2001). We can't adjudicate the boundaries of 'ordinary mathematics' here; and for our purposes, we won't need to.

Interestingly, Tao does describe the set theory in his (2016) as Zermelo-Fraenkel set theory with Choice. But first, he seems to significantly overshoot what is needed for his purposes; and

(3) The key point to note, then, is that to get from the usual sort of set-theory-for-applications to standard ZFC – going from perhaps some version of Zermelo set theory with urelements to a much richer theory of pure sets – evidently involves two key steps, which are quite independent of each other and which are quite differently motivated – namely (i) adding some axioms and (ii) disallowing urelements.

Apropos of (i), it is natural to conceive of sets as coming in levels (a base level, perhaps empty, of urelements; sets of them; then sets of what we've got so far; then sets of everything we've now got; keep on going . . .). Label the levels with ordinals, and we can show that Zermelo's axioms only give us levels up to $\omega + \omega$ (though that is arguably more than enough for ordinary mathematical applications). But, once we are interested in sets for their own sake, why not go higher? An inviting *maximizing* principle is: for every new ordinal, there is a new level in the hierarchy. Fraenkel's Axiom of Replacement (the additional axiom that takes us from Z to ZF) gives us this and more.

Apropos of (ii), we find that (nearly) all the fascinating and challenging mathematics that we get in enrichments of ZU is independent of the presence or absence of urelements. So, again once we are focusing on sets for their own sake, why not avoid unnecessary (albeit minor) complications? It is inviting to *simplify* by adopting a principle of purity and now disallow urelements.[13]

But note, these two moves (i) and (ii) arguably do take us away from the kind of base-line set theory apt for ordinary mathematical applications.

In sum, yes, we can acknowledge McLarty's and Leinster's points that we ordinary work in some locally ambient universe, and allow sets to have elements-lacking-elements. And we can happily grant that a theory of pure sets like ZFC is not what we need to play the role of a set theory for applications in ordinary mathematics. But so far we have no reason to suppose that the role will be better played by a novel type of structural set theory like ETCS as opposed to, say, some version of the conventional set theory ZU.

49.6 Foundations?

(a) We can think of a set-theory-for-applications as introducing a *superstructure* of sets enabling us to talk about pluralities of any chosen individuals (and pluralities of pluralities, etc.). But a theory of pure sets like ZFC is claimed, changing

second, the resulting strong theory is in fact not the usual pure ZFC but the theory ZFCU we get when urelements are still allowed.

[13]For a classic statement, see Shoenfield (1967, p. 238) who restricts his attention to a theory of pure sets because this is "sufficient to illustrate all the problems which arise in the general case" where we allow urelements. See also, for just one other example of a common line, Lévy (1979, p. 4): urelements "are not essential to what we shall do and, therefore, will not be considered".

the architectural metaphor, to provide a *foundation* for mathematics. Different roles and not to be confused.[14]

How, though, should we unpack that talk of ZFC as a foundation? There's a particularly illuminating paper on 'Set-theoretic foundations' by Penelope Maddy (2017) in which she describes a number of interconnected but different foundational roles a set theory can be thought to play.[15] Let me highlight four which – I agree with Maddy – are of positive value.

(1) *Elucidation* In embedding informal mathematics into set theory we don't just get set-theoretic surrogates for informal notions, we get improvements, set-theoretic replacements of imprecise notions with precise ones, which enable more rigorous proofs. (Consider, for just one example, the development of the notion of a function.)

(2) *Shared Standard* Moreover, the possibility of regimenting an informal proof as a formal derivation in set theory serves as a needed shared standard of what actually should count as a proof.

(3) *Risk Assessment* How do we test the consistency of proposed axioms for particular theories? Maddy quotes Vladimir Voevodsky on the role of set theory in proving the consistency of his 'univalent foundations' programme: "Set theory will remain the most important benchmark of consistency. ... each new addition to the ... language will require formal 'certification' by showing, through formally constructed interpretation, that it is at least as consistent as ZFC."

(4) *Generous Arena* But it isn't enough that we know that our separate mathematical theories are internally consistent. We crucially want to be able to transfer results from one area of mathematics to apply in another. Maddy here quotes John Burgess, "Interconnectedness implies that it will no longer be sufficient to put each individual branch of mathematics separately on a rigorous basis ... To guarantee that rigor is not compromised in the process of transferring material from one branch of mathematics to another, it is essential that the starting points of the branches being connected should ... be compatible. ... The only obvious way to ensure compatibility of the starting points ... is ultimately to derive all branches from a common, unified starting point." Set theory provides the sort of generous arena in which those branches can be brought to consistently co-exist (a point I already made in §4.1).

There is of course much more to be said here about these themes – perhaps I'd highlight (4) as a particular driver towards considering a theory of pure sets like

[14] I borrow 'superstructure' vs 'foundation' from my one-time colleagues Oliver and Smiley (2016, §14.6), though I am more warmly disposed than they are to the idea of ZFC playing a foundational role in some good senses.

[15] An earlier paper by Marquis (1995) also explores the idea of set-theoretic foundations and discusses their relation to category theory.

ZFC – and Maddy's own exploration is extremely helpful. But hopefully just the raw headlines which I have given will already strike a chord with you.

(b) Now here's the category theorist Todd Trimble, in the course of casting aspersions on ZFC to set us up for the idea that ETCS is much to be preferred. He starts

> When you get right down to it, the idea that everything in mathematics (like say the number e) is a 'set' is just plain bizarre, and actually very far removed from the way mathematicians normally think. And yet this is how we are encouraged to think, if we are asked to take ZFC seriously as a foundation. (Trimble 2011)

But that goes wrong from the outset. For ZFC to play the sort of foundational roles Maddy elucidates, it is *not* at all necessary to suppose that everything 'really' is a set. It is, for just one example, enough to be able to implement the reals (including e, of course) in its universe of pure sets by constructing some complete ordered field. And if we can find set-theoretic surrogates for widgets and wombats living happily together in the generous arena provided by a model of ZFC, then we will be en route to seeing how to marry up widget-theory and wombat-theory without worries about incompatibility. Of course, it will then make sense to ask questions about a widget-surrogate and a wombat-surrogate – like 'do they have a non-empty intersection?' – that it would be absurd to ask about the original widget and wombat. However, that's not a bug (or, as Trimble puts it, 'extraneous dreck and driftwood') but an inevitable feature of finding surrogates living together in a single generous arena.

Trimble does back off a little:

> One might argue that all expressions and theorems of normal mathematics are interpretable or realizable in the single theory ZFC, and that's really all we ever asked for – the details of the actual implementation (like, 'what is an ordered pair?') being generally of little genuine interest to mathematicians .... But this would seem to demote ZFC foundations, for most mathematicians, to a security blanket – nice to know it's there, maybe, but otherwise fairly irrelevant to their concerns. ... But if there really is such a disconnect between how a mathematician thinks of her materials at a fundamental level and how it specifically gets coded up ... in ZFC, ... we might re-examine just how appropriate ZFC is as 'foundations' of our subject.

Is giving Voevodsky a consistency proof a mere matter of offering a security blanket? Hardly. Nor can we similarly just dismiss the other foundational roles that Maddy identifies for a set theory.

Still, we can quite happily agree that much of the time, perhaps all the time, our working mathematician won't have foundational concerns. But she'll probably want a modicum of set theory for applications. Fine. She can add to her account of widgets a superstructure of sets built from them as urelements; and she can add to her account of wombats a superstructure of sets built from *them* as urelements, using the same topic-neutral apparatus of sets. No 'coding up' is called for, and this won't generate such dreck and driftwood as issues about intersections of widgets and wombats. (Though our mathematician is also nicely primed if her interests *do* turn more foundational, and she wants to implement widgets-and-sets-of-widgets together with wombats-and-sets-of-wombats in a single arena: just find suitable pure sets to play the role of widgets and pure sets to play the role of wombats.)

There is nothing here that need yet pull us away from a conventional theory of material sets towards the radical re-thinking of ETCS.

(c) Where do these preliminary skirmishes leave us? Cautious, perhaps, about being too quick to criticize conventional set theory. So far, while allowing for the legitimate points in the quotes from McLarty, Leinster and Trimble, it seems that we can still say that ZU (more or less) serves very well in the role of set-theory-for-ordinary-applications, while ZFC (or some extension thereof) serves very well in the role of giving us a generous arena when we have more 'foundational' concerns. However, that *of course* doesn't rule out ETCS also playing one or other or both of these roles just as well, if not better in some respects that we haven't yet considered; nor does it rule out ETCS having quite other attractions. So what more is there to say?

49.7 Foundations of another kind?

There is an often-quoted remark by Lawvere from a paper written a few years after he introduced ETCS:

> Foundations will mean here the study of what is universal in mathematics. Thus Foundations in this sense cannot be identified with any "starting-point" or "justification" for mathematics, though partial results in these directions may be among its fruits. But among the other fruits of Foundations so defined would presumably be guidelines for passing from one branch of mathematics to another and for gauging to some extent which directions of research are likely to be relevant. (Lawvere 1969a, p. 281)

Maddy (2017, §2) discusses some similar remarks from Mac Lane and McLarty – indeed the latter's 2013 paper is actually titled 'Foundations as truths which organize mathematics'. And of course, with the benefit of half a century of hindsight, we can now readily agree that category-theoretic ideas provide a way of organizing (much) mathematics and understanding commonalities and relationships between branches in a way that can guide our further investigations. We perhaps needn't argue over whether it is wise to co-opt 'foundations' as a label for this role.

However, why should it follow from the fact the category theory excels at playing an organizational role that categorial set theory is preferable to conventional set theory for the roles that it usually plays? Neither Maddy nor I is persuaded (but read her paper and some of the references therein).

49.8 Questions, questions, ...

(a) There are also technical issues that I should mention here. For a start, on further investigation, a classical arena with choice not only gives us a framework in which we can implement much ordinary non-set-theoretic mathematics, but one in which we can also by clever constructions implement a standard material set theory equivalent to Mac Lane's preferred weak version of Zermelo set theory. And if we add another axiom to ETCS to recover the power of the conventional Axiom of Replacement, we actually end up with a theory which is equivalent to (in the sense of mutually intertranslatable with) ZFC.[16]

The category theorists know this perfectly well, of course! But given this equivalence, it becomes a delicate matter to say just what the change of perspective will buy us, if we regard ETCS as opposed to ZFC as the better way of thinking about set-like collections.

Now, in simple cases, we do often think of readily intertranslatable theories as just giving us two different perspectives on the same structures. For a trite example, consider theories of Boolean algebras and of Boolean rings. On the other hand, do we regard the intertranslatability of Peano Arithmetic with a theory of strings of characters and with a theory of hereditarily finite sets as showing we have here three theories about the same things?[17] Perhaps we should resist the thought that intertranslatability is in general enough to make two theories 'come to the same'.[18] But then just what *should* we say about the intertranslatability of (augmented) ETCS and ZFC, which start off as seemingly significantly different theories (even if, by my lights, some category theorists downplay the contrasts)? I know of no good story to tell.

(b) Some have wondered whether ETCS piggy-backs on standard set theory in some way – for discussion, see e.g. the eminently lucid Linnebo and Pettigrew (2011). Leinster, however, urges that the category theorist should simply be bold here. After all, the conventional set theorist cheerfully says 'there are some objects (called "sets"), and a relation on them (called "membership"), that together satisfy so-and-so axioms' (citing Zermelo's axioms and usually some more). Such a theorist doesn't suppose that there must be another lower level of whatnots, as it were propping up this universe of sets: the story is supposed to bottom out

[16]For more on replacement, see episodes 12 and 12.5 in particular of the very illuminating series of blog posts by Leinster (2021, with additional contributions by Michael Shulman). Mac Lane and Moerdijk (1992, §VI.10) will also give you a sense of how the equivalence between augmented ETCS and ZFC is proved, even if you don't at this stage work through all the details of the described construction.

[17]On strings, see the classic Quine (1946) and also Corcoran et al. (1974). On hereditarily finite sets, see Kaye and Wong (2007).

[18]A suggestion also canvassed by Julia Kameryn Williams in a blog post about ETCS and ZFC: tinyurl.com/jkw-etcs.

with these sets and the membership relation between them. Well, why shouldn't the enthusiast for a categorial approach say, analogously, 'there are some objects (call them "ssets" if you must) and there are some morphisms between them (call them "ffunctions"), that together satisfy such-and-such axioms'? – where our theorist cites the axioms for a classical arena with choice. And, she continues, this time the story bottoms out here, with the objects and arrows of this category. What's not to like?

But of course, no set theorist will simply say 'there are some sets and a membership relation, satisfying the first-order ZFC axioms' and stop there, supposing that she has specified a unique universe. ZFC has a multitude of non-equivalent models – indeed, a set theorist can have great fun running up models to your taste e.g. to make the cardinality of the continuum pretty much anything you want. And then a major debate arises among the set theorists, between those who hold that, all the same, there is a canonical universe of (pure) sets of which first-order ZFC (a non-categorical theory in the logical sense) can only give a partial account, and those who hold that we have to countenance a whole multiverse of models of ZFC, none of which has a special standing. Likewise, we can't really suppose that the first-order theory that says 'there are some ssets and ffunctions making a classical arena' will suffice to pin down a unique base case of a classical arena (even taking that just to mean uniqueness up to some sensible notion of equivalence of categories). But then how should we handle the thought that there can be a whole teeming multiverse of different such classical arenas, different models of (augmented) ETCS? Which is the category Set? Or does 'Set' somehow ambiguously denote any of them? Perhaps at least some category theorists will be happier with a multiverse than a traditional set theorist. But again we lack – or at least, *I* lack – a good story to tell here.

(c) And at this point I will have to leave such issues about the interpretation and significance of categorial 'structural set theory' hanging tantalizingly in the air. There are troublesome questions which also need to be explored about the relation between such a set theory *in* category theory and set theory *for* category theory (to borrow the title of Shulman's impressive 2008 paper about the issues of 'size' and more which arise in tackling more advanced category theory). Looking a bit further afield, there are other questions too about how a categorial treatment of collections where elements are *typed* (an X-element can't be the same thing as a Y-element unless $X = Y$) relates to varieties of type theory.

But such – hardly introductory! – matters must be for another day.

And now, where next?

Reading what I have just written, I now believe
I stopped precipitously, so that my story seems to have been
slightly distorted, ...
Louise Glück, 'Afterword'

This is already a long book, and I must stop somewhere. So, if you've got this far, what to read next?

For a brisker text, covering much of the same ground as Parts I and II but in a quite different order, Tom Leinster's *Basic Category Theory* (2014) is first rate. A notch up in difficulty, and making many more connections across mathematics, Emily Riehl's *Category Theory in Context* (2017) is also excellent. On the logical topics of Part III, starting at a similar level but then exploring beyond, it is still hard to beat Robert Goldblatt's admirably accessible *Topoi* (1984).

Those three books have the added attraction of being freely downloadable (for links, see the bibliography). Other downloadable options include the old but particularly clear book by Michael Barr and Charles Wells, *Category Theory for Computing Science* (1995). More briefly and more recently, Paolo Perrone's *Notes on Category Theory* (2021) could be very useful for consolidating your understanding of topics in Parts I and II. The first part of Birgit Richter's *From Categories to Homotopy Theory* (2020), which goes rather further, can also be recommended.

I should highlight three more books. If approached without any prior acquaintance with its topic, Steve Awodey's *Category Theory* (2010) makes for a bumpier ride than the author intended. But it should now be entirely accessible and helpful. If you want to tackle something significantly more challenging, in a much terser idiom, there's Colin McLarty, *Elementary Categories, Elementary Toposes* (1992) which will in particular tell you more about the topics of Part III. And then, of course, there is the classic by Saunders Mac Lane, *Categories for the Working Mathematician* (1997).

What else? You will find more links to freely downloadable books and notes at various levels at www.logicmatters.net/categories. Which should be enough to launch you under way!

Bibliography

Adámek, J., Herrlich, H., and Strecker, G., 2009. *Abstract and Concrete Categories: The Joy of Cats.* Mineola, NY: Dover Publications. URL http://www.tac.mta.ca/tac/reprints/articles/17/tr17.pdf. Originally published 1990.

Agore, A., 2023. *A First Course in Category Theory.* Cham, Switzerland: Springer.

Aluffi, P., 2009. *Algebra: Chapter 0.* Providence, RI: American Mathematical Society.

Arbib, M. A. and Manes, E. G., 1975. *Arrows, Structures, and Functors: The Categorical Imperative.* New York: Academic Press.

Awodey, S., 2004. An answer to Hellman's question: "Does category theory provide a framework for mathematical structuralism?". *Philosophia Mathematica*, 12: 54–64.

Awodey, S., 2010. *Category Theory.* Oxford: Oxford University Press, 2nd edn.

Barr, M. and Wells, C., 1985. *Toposes, triples, and theories.* New York: Springer-Verlag. URL http://www.tac.mta.ca/tac/reprints/articles/12/tr12.pdf.

Barr, M. and Wells, C., 1995. *Category Theory for Computing Science.* New York: Prentice Hall, 2nd edn. URL http://www.tac.mta.ca/tac/reprints/articles/22/tr22.pdf.

Beardon, A. F., 2005. *Algebra and Geometry.* Cambridge: Cambridge University Press.

Bell, J. L., 1988. *Toposes and local set theories: an introduction.* Oxford: Clarendon Press.

Bell, J. L., 2008. *A Primer of Infinitesimal Analysis.* Cambridge: Cambridge University Press, 2 edn.

Benacerraf, P., 1965. What numbers could not be. *Philosophical Review*, 74: 47–73.

Bernadet, A. and Graham-Lengrand, S., 2013. A simple presentation of the effective topos. URL http://arxiv.org/abs/1307.3832.

Bollobás, B., 1998. *Modern Graph Theory.* New York: Springer.

Boolos, G., 1971. The iterative conception of set. *Journal of Philosophy*, 68: 215–232. Reprinted in his *Logic, Logic and Logic* (Harvard UP, 1998).

Booth, D. and Ziegler, R. (eds.), 1996. *Finsler Set Theory: Platonism and Circularity.* Basel: Birkhäuser Verlag.

Borceux, F., 1994. *Handbook of Categorical Algebra 1, Basic Category Theory.* Cambridge: Cambridge University Press.

Burgess, J., 2015. *Rigor and Structure.* Oxford: Oxford University Press.

Church, A., 1956. *Introduction to Mathematical Logic.* Princeton, NJ: Princeton University Press.

Corcoran, J., Frank, W., and Maloney, M., 1974. String theory. *Journal of Symbolic Logic*, 39: 625–636.

Crole, R. L., 1993. *Categories for Types.* Cambridge: Cambridge University Press.

Dummett, M., 1991. *Frege: Philosophy of Mathematics.* London: Duckworth.

Dummett, M., 2000. *Elements of Intuitionism.* Oxford: Clarendon Press, 2nd edn.

Dummit, D. S. and Foote, R. M., 2004. *Abstract Algebra*. Hoboken, NJ: John Wiley, 3rd edn.

Eilenberg, S. and Mac Lane, S., 1942. Natural isomorphisms in group theory. *Proceedings of the National Academy of Sciences of the United States of America*, 28: 537–543.

Eilenberg, S. and Mac Lane, S., 1945. General theory of natural equivalences. *Transactions of the American Mathematical Society*, 58: 231–294.

Enderton, H. B., 1977. *Elements of Set Theory*. New York: Academic Press.

Finsler, P., 1926. Über die Grundlegung der Mengenlehre, I. *Mathematische Zeitschrift*, 25: 683–713. Reprinted and translated in Booth and Ziegler 1996: 103–132.

Fong, B. and Spivak, D. I., 2019. *An Invitation to Applied Category Theory: Seven Sketches in Compositionality*. Cambridge: Cambridge University Press. URL `http://arxiv.org/abs/1803.05316`.

Forster, T., 1995. *Set Theory with a Universal Set*. Oxford: Clarendon Press, 2nd edn.

Forster, T., Lewicki, A., and Vidrine, A., 2019. Category theory with stratified set theory. URL `https://arxiv.org/pdf/1911.04704.pdf`.

Freyd, P., 1965. The theories of functors and models. In J. W. Addison, L. Henkin, and A. Tarski (eds.), *The Theory of Models*, pp. 107–120. North-Holland Publishing Co.

Freyd, P., 1970. Homotopy is not concete. URL `http://www.math.upenn.edu/~pjf/homotopy.pdf`.

Goedecke, J., 2013. Category theory. URL `http://www.julia-goedecke.de/pdf/CategoryTheoryNotes.pdf`.

Goldblatt, R., 1984. *Topoi: The Categorial Analysis of Logic*. Amsterdam: North-Holland, revised edn. URL `https://tinyurl.com/GoldblattTopoi`.

Grandis, M., 2018. *Category Theory and Applications: A Textbook for Beginners*. New Jersey: World Scientific.

Hellman, G., 2003. Does category theory provide a framework for mathematical structuralism? *Philosophia Mathematica*, 11: 129–157.

Hellman, G., 2005. Structuralism. In S. Shapiro (ed.), *The Oxford Handbook of Philosophy of Mathematics and Logic*, pp. 536–562. Oxford University Press.

Hilbert, D., 1900. Mathematical problems. *Göttinger Nachrichten*, pp. 253–297. URL `https://www.gutenberg.org/ebooks/71655`.

Hinze, R. and Marsden, D., 2023. *Introducing String Diagrams*. Cambridge: Cambridge University Press.

Hungerford, T. W., 1974. *Algebra*. New York: Springer.

Imada, K., 2019. The equivalence of typed lambda calculi and cartesian closed categories. URL `https://tinyurl.com/imada-lambda`.

Incurvati, L., 2020. *Conceptions of Set and the Foundations of Mathematics*. Cambridge: Cambridge University Press.

Johnstone, P. T., 1997. *Topos Theory*. New York: Academic Press.

Johnstone, P. T., 2002. *Sketches of an Elephant: A Topos Theory Compendium, Vol. 1*. Oxford: Clarendon Press.

Kaye, R. and Wong, T. L., 2007. On interpretations of arithmetic and set theory. *Notre Dame Journal of Formal Logic*, 48: 497–510.

Kim, I. S., 1996. On the axiom of choice in a well-pointed topos. *J. Korea Soc. of Math. Edu.: Pure and Applied Mathematics*, 3: 131–139.

Krömer, R., 2007. *Tool and Object: A History and Philosophy of Category Theory*. Basel: Birkhäuser.

Kunen, K., 1980. *Set Theory: An Introduction to Independence Proofs.* Amsterdam: Elsevier.

Kunen, K., 2012. *The Foundations of Mathematics.* London: College Publications.

Lambek, J. and Scott, P. J., 1986. *Introduction to Higher Order Categorical Logic.* Cambridge: Cambridge University Press.

Lawvere, F. W., 1964. An elementary theory of the category of sets. *Proceedings of the National Academy of Sciences*, 52: 1506–1511.

Lawvere, F. W., 1969a. Adjointness in foundations. *Dialectica*, 23: 281–296.

Lawvere, F. W., 1969b. Diagonal arguments and Cartesian closed categories. In *Lecture Notes in Mathematics*, vol. 92, pp. 134–145. Springer-Verlag. URL `http://www.tac.mta.ca/tac/reprints/articles/15/tr15.pdf`.

Lawvere, F. W., 1994. Cohesive toposes and Cantor's 'lauter Einsen'. *Philosophia Mathematica*, 2: 5–15.

Lawvere, F. W. and Rosebrugh, R., 2003. *Sets for Mathematics.* Cambridge: Cambridge University Press.

Lawvere, F. W. and Schanuel, S. H., 2009. *Conceptual Mathematics: A first introduction to categories.* Cambridge: Cambridge University Press, 2nd edn.

Leinster, T., 2000. The Yoneda Lemma: what's it all about? URL `https://www.maths.ed.ac.uk/~tl/categories`.

Leinster, T., 2012. Rethinking set theory. URL `https://arxiv.org/abs/1212.6543`.

Leinster, T., 2014. *Basic Category Theory.* Cambridge: Cambridge University Press. URL `https://arxiv.org/abs/1612.09375`.

Leinster, T., 2021. Large sets. URL `https://tinyurl.com/large-sets`.

Leinster, T., 2024. Axiomatic set theory. URL `https://tinyurl.com/lein-set`.

Lévy, A., 1979. *Basic set theory.* Mineola, N.Y.: Dover Publications (reprint 2002). URL `http://www.loc.gov/catdir/description/dover031/2002022292.html`.

Linnebo, Ø., 2022. Plural Quantification. In *The Stanford Encyclopedia of Philosophy.* Spring 2022 edn.

Linnebo, Ø. and Pettigrew, R., 2011. Category theory as an autonomous foundation. *Philosophia Mathematica*, 19.

Mac Lane, S., 1986. *Mathematics, form and function.* New York: Springer. MR:816347. Zbl:0675.00001.

Mac Lane, S., 1997. *Categories for the Working Mathematician.* New York: Springer, 2nd edn.

Mac Lane, S. and Moerdijk, I., 1992. *Sheaves in Geometry and Logic: A First Introduction to Topos Theory.* New York, Heidelberg, Berlin: Springer.

Maddy, P., 2017. Set-theoretic foundations. *Contemporary Mathematics*, 690: 289–322.

Marquis, J.-P., 1995. Category theory and the foundations of mathematics: Philosophical excavations. *Synthese*, 103: 421–447.

Marquis, J.-P., 2006. What is category theory? In G. Sica (ed.), *What is Category Theory?*, pp. 221–256. Polimetrica.

Marquis, J.-P., 2008. *From a Geometrical Point of View: A Study of the History and Philosophy of Category Theory.* New York: Springer.

Mathias, A. R. D., 2000. Strong statements of analysis. *Bulletin of the London Mathematical Society*, 32: 513–526.

Mathias, A. R. D., 2001. The strength of Mac Lane set theory. *Annals of Pure and Applied Logic*, 110: 107–234.

May, J. P., 1999. Stable algebraic topology, 1945–1966. In I. James (ed.), *History of Topology*, pp. 665–724. Amsterdam: North-Holland.

Mazur, B., 2008. When is one thing equal to some other thing? In B. Gold and R. Simons (eds.), *Proof and Other Dilemmas: Mathematics and Philosophy*. Mathematical Association of America. URL `http://tinyurl.com/mazur-equal`.

McKay, T. J., 2006. *Plural Predication*. Oxford: Clarendon Press.

McLarty, C., 1992. *Elementary Categories, Elementary Toposes*. Oxford: Oxford University Press.

McLarty, C., 2004. Exploring categorical structuralism. *Philosophia Mathematica*, 12: 37–53.

McLarty, C., 2013. Foundations as truths which organize mathematics. *Review of Symbolic Logic*, 6: 76–86.

McLarty, C., 2017. The roles of set theories in mathematics. In E. Landry (ed.), *Categories for the Working Mathematician*, pp. 1–17. Oxford: Oxford University Press.

Mendelson, E., 1964. *Introduction to Mathematical Logic*. Princeton, NJ: van Nostrand.

Moschovakis, Y. N., 2006. *Notes on Set Theory*. New York: Springer, 2nd edn.

Munkres, J. R., 2000. *Topology*. Prentice Hall, 2nd edn.

Oliver, A. and Smiley, T., 2006. What are sets and what are they for? *Philosophical Perspectives*, 20: 123–155.

Oliver, A. and Smiley, T., 2016. *Plural Logic*. Oxford: Oxford University Press, 2nd edn.

Paré, R., 1974. Colimits in topoi. *Bulletin of the American Mathematical Society*, 80: 556–561.

Pareigis, B., 1970. *Categories and Functors*. New York: Academic Press.

Perrone, P., 2021. *Notes on Category Theory*. URL `https://arxiv.org/abs/1912.10642`.

Perrone, P., 2023. *Starting Category Theory*. World Scientific.

Pierce, B. C., 1991. *Basic Category Theory for Computer Scientists*. Cambridge, MA: MIT Press.

Potter, M., 2004. *Set Theory and its Philosophy*. Oxford: Oxford University Press.

Quine, W. V. O., 1946. Concatenation as a basis for arithmetic. *Journal of Symbolic Logic*, 11: 105–14.

Quine, W. V. O., 1963. *Set Theory and Its Logic*. Cambridge, MA: The Belknap Press of Harvard University Press.

Rasiowa, H. and Sikorski, R., 1963. *The Mathematics of Metamathematics*. Warsaw: Państwowe Wydawnictwo Naukowe.

Richter, B., 2020. *From Categories to Homotopy Theory*. Cambridge: Cambridge University Press. URL `https://tinyurl.com/cat-to-hom`.

Riehl, E., 2017. *Category Theory in Context*. Mineola, NY: Dover Publications. URL `https://emilyriehl.github.io/files/context.pdf`.

Roman, S., 2017. *An Introduction to the Language of Category Theory*. Cham, Switzerland: Birkhäuser Verlag.

Russell, B. A. W., 1903. *The Principles of Mathematics*. Cambridge: Cambridge University Press.

Schubert, H., 1972. *Categories*. New York, Heidelberg, Berlin: Springer.

Scott, D. S., 1980. Relating theories of the λ-calculus. In J. Hindley and J. Seldin (eds.), *To H. B. Curry, Essays on Combinatory Logic, Lambda Calculus and Formalism*, pp. 403–450. London: Academic Press.

Sellars, W., 1963. Philosophy and the scientific image of man. In *Science, Perception and Reality*. London: Routledge & Kegan Paul.

Shapiro, S., 1997. *Philosophy of mathematics : structure and ontology.* Oxford: Oxford University Press, 1st ed. edn.

Shapiro, S., 2014. *Varieties of Logic.* Oxford: Oxford University Press.

Shoenfield, J. R., 1967. *Mathematical logic.* Reading, Mass.: Addison-Wesley Pub. Co.

Shulman, M., 2008. Set theory for category theory. *arXiv.* URL http://arxiv.org/abs/0810.1279.

Shulman, M., 2013. From set theory to type theory. *The n-Category Café.* URL https://tinyurl.com/shul-sett.

Simmons, H., 2011. *An Introduction to Category Theory.* Cambridge: Cambridge University Press.

Simpson, S. G., 2009. *Subsystems of Second Order Arithmetic.* Cambridge: Cambridge University Press, 2nd edn.

Smith, P., 2022. *Beginning Mathematical Logic: A Study Guide.* Cambridge: Logic Matters.

Spivak, D. I., 2014. *Category Theory for the Sciences.* Cambridge, MA: MIT Press.

Streicher, T., 2004. *Introduction to Category Theory and Categorical Logic.* URL https://tinyurl.com/StrCCL.

Sutherland, W. A., 2009. *Introduction to Metric and Topological Spaces.* Oxford: Oxford University Press, 2nd edn.

Tao, T., 2016. *Analysis I.* Singapore: Springer, 3rd edn.

Taylor, P., 1999. *Practical Foundations of Mathematics.* Cambridge: Cambridge University Press.

Trimble, T., 2011. ETCS I. URL https://tinyurl.com/trimble-etcs.

van Oosten, J., 2008. *Realizability: An Introduction to its Categorical Side.* Amsterdam: Elsevier.

Yanofsky, N. S., 2003. A universal approach to self-referential paradoxes, incompleteness and fixed points. *Bulletin of Symbolic Logic*, 9: 362–386. URL https://arxiv.org/pdf/math/0305282.

Yanofsky, N. S., 2024. *Monoidal Categories: Unifying Concepts in Mathematics, Physics, and Computing.* Cambridge, MA: MIT Press.

Index

Some special notation

Categories

Categorial definitions

Printed in Dunstable, United Kingdom

64504632R10292